LIDC Contributions on Antitrust Law, Intellectual Property and Unfair Competition

Series Editor

International League of Competition Law (LIDC), Competition Law (LIDC), Lausanne, Switzerland

In each volume of the LIDC Contributions on Antitrust Law, Intellectual Property and Unfair Competition, two topics are discussed, one focusing on competition law, the other on intellectual property and/or unfair competition.

Practitioners and academics from different industrial countries contribute national reports on each topic, followed by an analysis and a comparison of the respective national situations as well as an international report. The national reports include essential and hard-to-find case law analysis. The common trends and principles identified in the international reports serve as a basis for more general recommendations, some of which have been presented to national and international authorities.

The series is produced by the International League of Competition Law (LIDC), an association that brings together national associations for antitrust law, intellectual property and unfair competition law. Created in 1930 by a German lawyer who subsequently became Professor of Law at the University of Saint Louis in the United States, the LIDC has a long standing reputation for scientific exploration, debate and work.

Pranvera Këllezi • Pierre Kobel •
Bruce Kilpatrick
Editors

Sustainability Objectives in Competition and Intellectual Property Law

Editors
Pranvera Këllezi
Këllezi Legal
Geneva, Switzerland

Pierre Kobel
Athena Law
Geneva, Switzerland

Bruce Kilpatrick
Linklaters LLP
London, UK

ISSN 2199-742X   ISSN 2199-7438   (electronic)
LIDC Contributions on Antitrust Law, Intellectual Property and Unfair Competition
ISBN 978-3-031-44868-3   ISBN 978-3-031-44869-0   (eBook)
https://doi.org/10.1007/978-3-031-44869-0

This work was supported by International League of Competition Law (LIDC).

© The Editor(s) (if applicable) and The Author(s) 2024. This book is an open access publication.

Open Access This book is licensed under the terms of the Creative Commons Attribution 4.0 International License (http://creativecommons.org/licenses/by/4.0/), which permits use, sharing, adaptation, distribution and reproduction in any medium or format, as long as you give appropriate credit to the original author(s) and the source, provide a link to the Creative Commons license and indicate if changes were made.

The images or other third party material in this book are included in the book's Creative Commons license, unless indicated otherwise in a credit line to the material. If material is not included in the book's Creative Commons license and your intended use is not permitted by statutory regulation or exceeds the permitted use, you will need to obtain permission directly from the copyright holder.The use of general descriptive names, registered names, trademarks, service marks, etc. in this publication does not imply, even in the absence of a specific statement, that such names are exempt from the relevant protective laws and regulations and therefore free for general use.

The publisher, the authors, and the editors are safe to assume that the advice and information in this book are believed to be true and accurate at the date of publication. Neither the publisher nor the authors or the editors give a warranty, expressed or implied, with respect to the material contained herein or for any errors or omissions that may have been made. The publisher remains neutral with regard to jurisdictional claims in published maps and institutional affiliations.

This Springer imprint is published by the registered company Springer Nature Switzerland AG
The registered company address is: Gewerbestrasse 11, 6330 Cham, Switzerland

Paper in this product is recyclable.

Preface

Businesses are increasingly embracing sustainability objectives, driven by the international community. This volume of LIDC contributions focuses on the question of whether and to what extent competition and intellectual property laws incorporate sustainability objectives. Competition and intellectual property law are certainly not the only tools for addressing sustainability issues, but they can play a role in moving towards a more sustainable society.

Sustainability has gained prominence in competition law in all jurisdictions covered in this volume. The reports focus on classic questions such as whether sustainability agreements restrict competition and, if so, to what extent they can be exempted on efficiency grounds. The contributions raise many questions, in particular the treatment of non-market efficiencies. The soft law and case law produced by competition authorities are examined, while presenting the leadership role of some competition authorities in the field, from advocacy to policy papers and sustainability guidelines. The authors call for more individual guidance to provide more transparency and clarity to industry, advisors and society at large on sustainability issues, with guidelines or even block exemptions dealing with sustainability providing even greater legal certainty. The International Report on Sustainability and Competition Law, prepared by Julian Nowag, associate professor at Lund University and associate at the Oxford Centre for Competition Law and Policy, summarises the findings on the core provisions of competition law, the prohibition of coordinated practices, the abuse of a dominant position and merger regulation.

On intellectual property, the contributions examine various important issues around sustainability, such as the need for intellectual property rights to remain technology-neutral, ways to promote the use of sustainable technologies and incentives for licensing, ways to promote the dissemination of sustainable technologies, including compulsory licensing, cross-licensing, open source or FRAND licensing, or replacing the destruction of counterfeit goods by recycling. The reports also discuss greenwashing and how it can be addressed through revisions to trademarks and related rights. An International Report on Sustainability and Intellectual Property, prepared by Chris de Mauny, Partner, Bird & Bird LLP, London, examines trends and draws conclusions based on contributions from eight

European countries and Brazil. The report could not be included in this volume, but is available for free download from lidc.org or by contacting the author.

This volume provides a well of information as to new trends, new legislation, regulations, decisions, and enforcement policies regarding these fascinating questions. It takes a huge amount of work and commitment from the national reporters, and particularly the international reporters, to produce their studies and we would like to thank them for their enormous contribution, which is greatly appreciated. The purpose of this book is to share their learnings with a wider audience across the members of the LIDC and its constituent national organisations and with academics, practitioners, and students across the world. We hope you enjoy reading it as much as we enjoyed listening to the debate and discussions which took place in Milan.

The editors would like to thank all the authors for their contributions and their patient collaboration during the editing of this book. They would like to express their sincere gratitude to the Members of the Bureau, of the Council and of the Scientific Committee for their kind support and encouragement during the preparation of this book.

Geneva, Switzerland   Pranvera Këllezi
Geneva, Switzerland   Pierre Kobel
London, UK   Bruce Kilpatrick

Contents

Part I Sustainability Objectives in Competition Law

1 Sustainability and Competition Law: An International Report 3
Julian Nowag

2 Sustainability and Competition Law in Austria 27
Adrian Kubat and Adnan Tokić

3 Sustainability and Competition Law in Belgium 51
Jan Blockx

4 Sustainability and Competition Law in Brazil 55
José Mauro Decoussau Machado

5 Sustainability and Competition Law in Czech Republic 61
Radka MacGregor Pelikánová

6 Sustainability and Competition Law: A French Perspective 73
Michaël Cousin, Alexandre Marescaux, Lauren Mechri,
and Guillaume Melot

7 Sustainability and Competition Law in Germany 83
Eckart Bueren and Jennifer Crowder

8 Sustainability and Competition Law in Hungary 127
András M. Horváth

9 Sustainability and Competition Law in Italy 163
Elisa Teti

10 Sustainability and Competition Law in Malta 179
Clement Mifsud-Bonnici

11 Sustainability and Competition Law in Switzerland 187
Johana Cau and Alexandra Telychko

12 Sustainability and Competition Law in the United Kingdom 203
Simon Holmes, Nicole Kar, and Lucinda Cunningham

Part II Sustainability Objectives in Intellectual Property Law

13 Sustainability and Intellectual Property in Austria 247
Georg Kresbach

14 Sustainability and Intellectual Property in Brazil 263
Lucas Bernardo Antoniazzi

15 Sustainability and Intellectual Property in Germany 271
Thomas Hoeren, Tabea Ansorge, and Oliver Lampe

16 Sustainability and Intellectual Property in Hungary 307
Bálint Halász, Ádám Liber, Dániel Arányi, Fruzsina Nagy,
and Olivér Németh

17 Sustainability and Intellectual Property in Italy 323
Marina Cristofori

18 Sustainability and Intellectual Property in Malta 335
Philip Mifsud and Sasha Muscat

19 Sustainability and Intellectual Property in Sweden 347
Martin Zeitlin

20 Sustainability and Intellectual Property in Switzerland 369
Eugenia Huguenin-Elie

21 Sustainability and Intellectual Property in United Kingdom 387
Ben Hitchens, Caitlin Heard, and Joel Vertes

Editors and Contributors

About the Editors

Pranvera Këllezi is an attorney at law in Geneva and a member of the Swiss Competition Commission, Switzerland. She represents companies and public organisations in business law, antitrust and competition, data protection, as well as in public economic law. She is member of the LIDC board responsible for publications.

Pierre Kobel practiced in all fields of law in different law firms in Geneva, Lausanne, and in the USA, as well as with the Federal administration in Bern and as a deputy judge in Geneva. He is frequently invited as a guest lecturer and is a recognised practitioner in the field of antitrust, intellectual property, and dispute resolution.

Bruce Kilpatrick is a partner at Linklaters LLP, based in London. He advises a range of clients on competition law, utility regulation, merger control, and EU state aid matters. He has particular expertise in the energy, transport, retail, and financial services sectors.

List of Contributors

Tabea Ansorge Westfälische Wilhelms-Universität, Münster, Germany

Lucas Bernardo Antoniazzi Di Blasi, Parente & Associados, Rio de Janeiro, Brazil

Dániel Arányi Bird & Bird, Budapest, Hungary

Jan Blockx University of Antwerp, Antwerpen, Belgium

Eckart Bueren Georg-August-University Göttingen, Göttingen, Germany

Johana Cau Lenz & Staehelin, Geneva, Switzerland

Michaël Cousin Addleshaw Goddard, Paris, France

Marina Cristofori Studio Legale Jacobacci & Associati, Milan, Italy

Jennifer Crowder Georg-August-University Göttingen, Göttingen, Germany

Lucinda Cunningham University of Oxford and Competition Appeal Tribunal, Oxford, UK

Linklaters, London, UK

Matrix Chambers, London, UK

Bálint Halász Bird & Bird, Budapest, Hungary

Caitlin Heard CMS Cameron McKenna Nabarro Olswang LLP, London, UK

Ben Hitchens CMS Cameron McKenna Nabarro Olswang LLP, London, UK

Thomas Hoeren Westfälische Wilhelms-Universität, Münster, Germany

Simon Holmes University of Oxford and Competition Appeal Tribunal, Oxford, UK

Linklaters, London, UK

Matrix Chambers, London, UK

András M. Horváth Hegymegi-Barakonyi and Fehérváry Baker & McKenzie, Budapest, Hungary

Eugenia Huguenin-Elie Lenz & Staehelin, Geneva, Switzerland

Nicole Kar University of Oxford and Competition Appeal Tribunal, Oxford, UK

Linklaters, London, UK

Matrix Chambers, London, UK

Georg Kresbach Wolf Theiss Rechtsanwälte GmbH & Co KG, Vienna, Austria

Wolf Theiss Rechtsanwälte GmbH & CoKG, Wolf Theiss Rechtsanwälte GmbH & CoKG, Vienna, Austria

Adrian Kubat University of Vienna, Wien, Austria

Oliver Lampe Westfälische Wilhelms-Universität, Münster, Germany

Ádám Liber PROVARIS Attorneys at Law, Budapest, Hungary

Radka MacGregor Pelikánová Metropolitan University Prague, Strašnice, Czech Republic

José Mauro Decoussau Machado Pinheiro Neto Advogados, São Paulo, Brazil

Alexandre Marescaux Fidal, Paris, France

Lauren Mechri Fieldfisher, Paris, France

Guillaume Melot CMS Francis Lefebvre, Paris, France

Philip Mifsud Ganado Advocates, Valletta, Malta

Clement Mifsud-Bonnici Ganado Advocates, Valletta, Malta

Sasha Muscat Ganado Advocates, Valletta, Malta

Fruzsina Nagy Bird & Bird, Budapest, Hungary

Olivér Németh PROVARIS Attorneys at Law, Budapest, Hungary

Julian Nowag Lund University, Lund, Sweden

Centre for Competition Law and Policy Oxford University, Oxford, UK

Alexandra Telychko Faculty of Law, University of Geneva, Geneva, Switzerland

Elisa Teti Rucellai&Raffaelli, Milano, Italy

Adnan Tokić University of Vienna, Wien, Austria

Joel Vertes CMS Cameron McKenna Nabarro Olswang LLP, London, UK

Martin Zeitlin Advokatfirman MarLaw, Stockholm, Sweden

List of Abbreviations

AUD	Australian dollar
AI	Artificial intelligence
Berne Convention	The Berne Convention for the Protection of Literary and Artistic Works of 9 September 1886, as amended
Brussels Convention	Brussels Convention Relating to the Distribution of Programme-Carrying Signals Transmitted by Satellite, 1974
BGBl.	Bundesgesetzblatt (Germany)
BGH	Bundesgerichtshof (Germany)
BGN	Bulgarian lev
B2B	Business to business
B2C	Business to consumers
bn	billion
BRL	Brazilian Real (reais)
c./ca.	Circa
cf.	Compare
CFI	Court of First Instance of the ECJ (before 1 December 2009)
CFREU	The Charter of Fundamental Rights of the European Union, OJ 2010 C 83, p. 389
CHF	Swiss franc
CJEU	Court of Justice of the European Union (after 1 December 2009)
CMLR	Common Market Law Review
CR_n	Concentration Ratio measuring the percentage market share held by n largest undertakings
DCT(s)	Digital Comparison Tool(s)
De minimis Notice	Commission Notice on agreements of minor importance which do not appreciably restrict competition under Article 101(1) of the Treaty on the Functioning of the European Union (de minimis), OJ 2014 C 291, p. 1

xiii

Directive 97/7	Directive 97/7 of the European Parliament and of the Council of 20 May 1997 on the protection of consumers in respect of distance contracts, OJ 1997 L 144, p. 19 (no longer in force)
Directive 96/09 or Database Directive	Directive 96/9 of the European Parliament and of the Council of 11 March 1996 on the legal protection of databases, OJ 1996 L 77, p. 20
Directive 2000/31 or Directive on electronic commerce	Directive 2000/31 of the European Parliament and of the Council of 8 June 2000 on certain legal aspects of information society services, in particular electronic commerce, in the Internal Market, OJ 2000 L 178, p. 1
Directive 2001/29 or InfoSoc Directive	Directive 2001/29 of the European Parliament and of the Council of 22 May 2001 on the harmonisation of certain aspects of copyright and related rights in the information society, OJ 2001 L 167, p. 10
Directive 2001/83	Directive 2001/83 of the European Parliament and of the Council of 6 November 2001 on the Community code relating to medicinal products for human use, OJ 2001 L 311, p. 67
Directive 2004/48	Directive 2004/48 of the European Parliament and of the Council of 29 April 2004 on the enforcement of intellectual property rights, OJ 2004 L 157, p. 45
Directive 2005/29 or Unfair Commercial Practices Directive	Directive 2005/29 of the European Parliament and of the Council of 11 May 2005 concerning unfair business-to-consumer commercial practices in the internal market and amending Council Directive 84/450, Directives 97/7, 98/27 and 2002/65 of the European Parliament and of the Council and Regulation 2006/2004 of the European Parliament and of the Council, OJ 2005 L 149, p. 22
Directive 2006/115 or Rental and Lending Rights Directive	Directive 2006/115 of the European Parliament and of the Council of 12 December 2006 on rental right and lending right and on certain rights related to copyright in the field of intellectual property, OJ 2006 L 376, p. 28
Directive 2009/24 or Software Directive	Directive 2009/24 of the European Parliament and of the Council of 23 April 2009 on the legal protection of computer programs, OJ 2009 L 111, p. 16
Directive 2011/83	Directive 2011/83/EU of the European Parliament and of the Council of 25 October 2011 on consumer rights, amending Council Directive 93/13/EEC and Directive 1999/44/EC of the European Parliament and of the Council and repealing Council Directive 85/577/EEC and Directive 97/7/EC of the European Parliament and of the Council OJ 2011 L 304, p. 64

List of Abbreviations

Directive 2012/28 or Orphan Works Directive	Directive 2012/28 of the European Parliament and of the Council of 25 October 2012 on certain permitted uses of orphan works, OJ 2012 L 299, p. 5
Directive 2014/104	Directive 2014/104 of the European Parliament and of the Council of 26 November 2014 on certain rules governing actions for damages under national law for infringements of the competition law provisions of the Member States and of the European Union, OJ 2014 L 349, p. 1
Directive 2019/1 or ECN+ Directive	Directive 2019/1 of the European Parliament and of the Council to empower the competition authorities of the Member States to be more effective enforcers and to ensure the proper functioning of the internal market (ECN+ Directive), OJ 2019 L 11, p. 3
DKK	Danish krone
DM	Deutsche mark
e.g.	for example
EC	European Community
ECHR	Council of Europe, European Convention for Human Rights of 4 November 1950
ECJ	European Court of Justice (before 1 December 2009)
ECR	European Court Reports
ECtHR	European Court of Human Rights
EDPB	European Data Protection Board
EU	European Union
ff	and following
GBP	Pound sterling (UK)
GATT 1947	General Agreement on Tariffs and Trade 1947
GATT 1994	General Agreement on Tariffs and Trade 1994
GC	General Court of the CJEU (after 1 December 2009)
GDP	Gross Domestic Product
GVH	Hungarian Competition Authority (Gazdasági Versenyhivatal)
Guidelines on the effect on trade concept	Commission Notice - Guidelines on the effect on trade concept contained in Articles 81 and 82 of the Treaty, OJ 2004 C 101, p. 81
Guidelines on Vertical Restraints	Guidelines on vertical restraints, OJ 2022 C 248, p. 1
Guidance on the implementation of the UCP Directive	Commission staff working document of 25 May 2016. Guidance on the implementation/application of Directive 2005/29 on unfair commercial practices accompanying the document communication from the commission to the European parliament, the Council, the European economic and social Committee and the Committee of the regions a comprehensive approach to stimulating cross-border e-commerce for Europe's citizens and businesses (swd/2016/0163 final)

ha	hectare
Hague Agreement	Hague Agreement Concerning the International Registration of Industrial Designs, 1925
HRK	Croatian Kuna (*hrvatska kuna*)
HMT	Hypothetical monopolist test
HUF	Hungarian Forint (*Magyar forint*)
i.e.	id est (that is)
Id./Idem	the same as previously mentioned
IP	Intellectual property
kg	kilogram
Lisbon Agreement	Lisbon Agreement for the Protection of Appellations of Origin and their International Registration of October 31, 1958, as revised at Stockholm on July 14, 1967, and as amended on September 28, 1979
Locarno (Classification) Agreement	Locarno Agreement Establishing an International Classification for Industrial Designs, 1979
m	million
m^2	*square metre*
min	*minutes*
Madrid Agreement	*Madrid Agreement Concerning the International Registration of Marks, 1979*
MFN	*Most Favoured Nation*
NAAT-rule	The non-appreciable affectation of trade rule
Nice (Classification) Agreement	Nice Agreement Concerning the International Classification of Goods and Services for the Purposes of the Registration of Marks, 1979
Notice on the relevant market	Commission Notice on the definition of relevant market for the purposes of Community competition law, OJ 1997 C 372, p. 5
OECD	Organisation for Economic Co-operation and Development
OJ	Official Journal
p./pp.	page(s)
para/paras	paragraph(s)
Phonograms Convention	Convention for the Protection of Producers of Phonograms Against Unauthorized Duplication of Their Phonograms, 1971
PTL	Patent Law Treaty, 2000
Paris Convention	Paris Convention for the protection of industrial property, 1883
pt/pts	point(s)
R&D	Research and development
Regulation 2659/2000	Commission Regulation 2659/2000 of 29 November 2000 on the application of Article 81(3) of the Treaty to categories of research and development agreements, OJ 2000 L 304, p. 7 (no longer in force)

Regulation 1400/2002	Commission Regulation 1400/2002 of 31 July 2002 on the application of Article 81(3) of the Treaty to categories of vertical agreements and concerted practices in the motor vehicle sector, OJ 2002 L 203, p. 30 (no longer in force)
Regulation 1/2003	Council Regulation 1/2003 of 16 December 2002 on the implementation of the rules on competition laid down in Articles 81 and 82 of the Treaty, OJ 2003 L 1, p. 1
Regulation 2006/2004	Regulation 2006/2004 of the European Parliament and of the Council of 27 October 2004 on cooperation between national authorities responsible for the enforcement of consumer protection laws (the Regulation on consumer protection cooperation), OJ 2004 L 364, p. 1 (no longer in force)
Regulation 139/2004	Council Regulation 139/2004 of 20 January 2004 on the control of concentrations between undertakings, OJ 2004 L 24, p. 1
Regulation 110/2008	Regulation 110/2008 of the European Parliament and the Council of 15 January 2008 on the definition, description, presentation, labelling and protection of geographical indications of spirit drinks, OJ 2008 L 39, p. 16
Regulation 330/2010	Commission Regulation 330/2010 of 20 April 2010 on the application of Article 101(3) of the Treaty on the Functioning of the European Union to categories of vertical agreements and concerted practices, OJ 2010 L 102, p. 1 (no longer in force)
Regulation 1151/2012	Regulation 1151/2012 on quality schemes for agricultural products and foodstuffs, OJ 2012 L 343, p. 1
Regulation 608/2013	Regulation 608/2013 of the European Parliament and of the Council of 12 June 2013 concerning customs enforcement of intellectual property rights and repealing Council Regulation 1383/2003, OJ 2013 L 181, p. 15
Regulation 1308/2013	Regulation 1308/2013 of the European Parliament and of the Council of 17 December 2013 establishing a common organisation of the markets in agricultural products and repealing Council Regulations 922/72, 234/79, 1037/2001 and 1234/2007, OJ 2013 L 347, p. 671
Regulation 251/2014	Regulation 251/2014 of the European Parliament and of the Council of 26 February 2014 on the definition, description, presentation, labelling and the protection of geographical indications of aromatized wine products, OJ 2014 L 84, p. 14

Regulation 316/2014	Commission Regulation 316/2014 of 21 March 2014 on the application of Article 101(3) of the Treaty on the Functioning of the European Union to categories of technology transfer agreements, OJ 2014 L 93, p. 17
Regulation 2016/679 or GDPR or General Data Protection Regulation	Regulation 2016/679 of the European Parliament and of the Council of 27 April 2016 on the protection of natural persons with regard to the processing of personal data and on the free movement of such data, and repealing Directive 95/46/EC (General Data Protection Regulation), OJ 2016 L 119, p. 1
Regulation 2019/1150 or P2B Regulation	Regulation 2019/1150 of the European Parliament and of the Council of 20 June 2019 on promoting fairness and transparency for business users of online intermediation services, OJ 2019 L 186, p. 57
Regulation 2022/720	Commission Regulation 2022/720 of 10 May 2022 on the application of Article 101(3) of the Treaty on the Functioning of the European Union to categories of vertical agreements and concerted practices, OJ 2022 L 134, p. 4
Rome Convention	International Convention for the Protection of Performers, Producers of Phonograms and Broadcasting Organizations, 1961
RON	Romanian leu
SEK	Swedish Krona
SIEC	Significant Impediment of Effective Competition
Singapore Treaty	Singapore Treaty on the Law of Trademarks, 2006
SMEs	Small and medium size enterprises
SMP	Significant market power
SSNIP	Small but Significant and Non-transitory Increase in Price
TEC	Treaty Establishing the European Community
TFEU	Treaty on the Functioning of the European Union
TLT	Trademark Law Treaty, 1994
TRIPs	Agreement on Trade-Related Aspects of Intellectual Property Rights, Annex 1C of the Marrakesh Agreement Establishing the World Trade Organization, signed in Marrakesh, Morocco on 15 April 1994
UAH	Ukrainian hryvnia
UK	United Kingdom
US/USA	United States of America
v	versus
Vienna (Classification) Agreement	Vienna Agreement Establishing an International Classification of the Figurative Elements of Marks, 1985

WCT/WIPO Copyright Treaty	World Intellectual Property Organization Copyright Treaty, 1996
WPPT	WIPO Performances and Phonograms Treaty, 1996

Part I

Sustainability Objectives in Competition Law

Sustainability and Competition Law: An International Report

1

Julian Nowag

1.1 Introduction

While recent years have been dominated by Covid-19 and now the war in Ukraine, we are witnessing numerous ongoing tragedies. The last few decades have seen a dramatic increase in freak weather events caused by climate change. We have seen dramatic biodiversity loss of the global wildlife population at an average of 69% since 1970 on top of general environmental degradation. These problems are a part of a broader sustainability challenge that humankind faces. As such, sustainability is defined as 'development that meets the needs of the present without compromising the ability of future generations to meet their own needs' by the UN in 1987.[1] This definition acknowledges that issues of environmental protection, economic development, and social progress are interconnected by shifting the focus from environmental problems to the causes of these problems. While a trade-off might occasionally be needed, it does not require balancing growth against environmental sustainability. Economic growth is still possible while preserving the environment; moreover, that growth may be important for ecological protection and social equity.[2]

Sustainability has been on the agenda of international organisations and countries for years, yet the world of competition law has only recently started talking about the subject. As such, we have also seen the concept of sustainability increasingly

---

[1] *See* G.H. Brundtland, Report of the World Commission on Environment and Development: Our Common Future , United Nations 1987, available at https://sustainabledevelopment.un.org/content/documents/5987our-common-future.pdf. Accessed 29 November 2022.

[2] K. E. Portney, Sustainability, MIT Press 2015 p. 6.

J. Nowag (✉)
Lund University, Lund, Sweden

Centre for Competition Law and Policy Oxford University, Oxford, UK
e-mail: julian.nowag@jur.lu.se

© The Author(s) 2024
P. Këllezi et al. (eds.), *Sustainability Objectives in Competition and Intellectual Property Law*, LIDC Contributions on Antitrust Law, Intellectual Property and Unfair Competition, https://doi.org/10.1007/978-3-031-44869-0_1

embraced by the business community,[3] and we can observe a drive for more sustainable business activity.[4] This development might have to do with the 2030 Agenda and goal number 12.6 in particular. This goal highlights the need for companies to adopt more sustainable business practices.

While it is clear that addressing sustainability concerns is not the primary concern of competition and the task of competition agencies, questions still arise. What role can competition law play in supporting the transition to a more sustainable operation of markets? And it might not be surprising that competition authorities around the world are anticipating more frequent questions around sustainability and competition.[5] Some agencies, such as the Dutch,[6] Greek,[7] German,[8] UK,[9] and EU[10] authorities, have already become active by adopting policy papers etc. Similarly,

---

[3] Business Roundtable (19 August 2019), Business Roundtable Redefines the Purpose of a Corporation to Promote 'An Economy That Serves All Americans', available at https://www.businessroundtable.org/business-roundtable-redefines-the-purpose-of-a-corporation-to-promote-an-economy-that-serves-all-americans. Accessed 29 November 2022.

[4] A. Winston, Sustainable Business Went Mainstream in 2021, Harvard Business Review 2021, available at https://hbr.org/2021/12/sustainable-business-went-mainstream-in-2021. Accessed 29 November 2022; S. Bonini and S. Görner, The business of sustainability, McKinsey & Company 2011, available at https://www.mckinsey.com/business-functions/sustainability/our-insights/the-business-of-sustainability-mckinsey-global-survey-results. Accessed 29 November 2022.

[5] Hungarian Competition Authority, Sustainable Development and Competition Law Survey Report, Special Project for the 2021 ICN Annual Conference, International Competition Network Conference 2021, available at https://www.gvh.hu/pfile/file?path=/en/gvh/Conference/icn-2021-annual-conference/sustainability_survey_REPORT_2ndEd_2021_09_30_final_PUBLIC.pdf1&inline=true. Accessed 29 November 2022.

[6] Authority for Consumers & Markets, Second draft version: Guidelines on Sustainability Agreements – Opportunities within competition law, Autoreit Consument & Markt 2021, available at https://www.acm.nl/en/publications/second-draft-version-guidelines-sustainability-agreements-opportunities-within-competition-law. Accessed 29 November 2022.

[7] Hellenic Competition Commission, Competition Law & Sustainability, Hellenic Competition Commission 2022, available at https://www.epant.gr/en/enimerosi/competition-law-sustainability.html. Accessed 29 November 2022.

[8] Bundeskartellamt (Federal Cartel Office), Offene Märkte und nachhaltiges Wirtschaften – Gemeinwohlziele als Herausforderung für die Kartellrechtspraxis, Bundeskartellamt 2020, available at https://www.bundeskartellamt.de/SharedDocs/Publikation/DE/Diskussions_Hintergrundpapier/AK_Kartellrecht_2020_Hintergrundpapier.pdf?__blob=publicationFile&v=2. Accessed 29 November 2022.

[9] Competition and Markets Authority, Sustainability agreements: CMA issues information for businesses, Press Release (29 November 2022), available at https://www.gov.uk/government/news/sustainability-agreements-cma-issues-information-for-businesses. Accessed 29 November 2022.

[10] European Commission, Directorate-General for Competition, Competition policy brief, 2021-01, September 2021, European Commission (2021), available at https://data.europa.eu/doi/10.2763/962262. Accessed 29 November 2022.

1 Sustainability and Competition Law: An International Report

the OECD's Competition Division has discussed the issue,[11] followed up with discussions about environmental protections[12] and the measurement of environmental benefits for antitrust purposes.[13]

Against this background, the International League of Competition Law (LIDC) had decided to explore the topic of sustainability and competition law at the LIDC Congress on 20–22 October 2022 in Milano. Given that this is a new and emerging topic, it was expected that many jurisdictions would not have pre-existing cases or considerable experience with the subject. Hence, the national reporters were asked to not only provide cases but also suggest how their national system would and could address the interaction between competition law and sustainability.

The report is based on the submissions of 11 countries, namely Austria, Belgium, Brazil, Czech Republic, France, Germany, Hungary, Italy, Malta, Switzerland, and the United Kingdom. It aims at summarising the findings regarding core competition law provisions, the prohibition of coordinated practices, the abuse of a dominant position, and merger regulation. It leaves aside closely related areas, such as false advertising or unfair competition, yet mentions them in a section on other related subjects.

The report first sets out the general legal framework on the interaction between competition law and sustainability in the different countries. Here, the focus is on legal obligations regarding sustainability that might affect the application of competition law. This section is aimed at addressing questions such as: what can be said about the role of sustainability in competition law according to the different national laws? Should sustainability play a role? Can it play a role? And would that be purely up for the competition authority to decide? The first part of the report, therefore, sets out national rules that are important to the interaction between sustainability and competition law.

The second part of the report addresses the core question by looking at specific interactions between competition law and sustainability, exploring cases in the areas of co-operation, cartels, abuse/monopolisation, and mergers. It is subdivided into two sections. The first part explores cases where sustainability played a role in the enforcement of competition law in form of a 'shield',[14] in other words, whether

---

[11] OECD, Sustainability and Competition, OECD Competition Committee Discussion Paper 2020, available at https://www.oecd.org/daf/competition/sustainability-and-competition-2020.pdf. Accessed 29 November 2022.

[12] OECD, Environmental Considerations in Competition Enforcement, OECD Competition Committee Discussion Paper 2021, available at https://www.oecd.org/daf/competition/environmental-considerations-in-competition-enforcement-2021.pdf. Accessed 29 November 2022.

[13] N. Watson, Measuring environmental benefits in competition cases, OECD Roundtable on Environmental Considerations in Competition Enforcement 2021, available at https://one.oecd.org/document/DAF/COMP(2021)14/en/pdf. Accessed 29 November 2022.

[14] On the distinction between sword and shield see S. Holmes, Climate Change, sustainability, and competition law, Journal Antitrust Enforcement 2020(8), p. 355, J. Nowag, Competition Law's Sustainability Gap? Tools For an Examination and a Brief Overview, Nordic Journal of European Law 2022(1) pp. 150-152, J. Nowag, Environmental Integration in Competition and Free-Movement Laws, Oxford University Press 2017, pp. 1-12.

companies that take action to foster sustainability can rely on that fact before competition authorities. This can take different forms: for example, the agency might find that competition is not restricted, that on balance the restriction can be outweighed, or that the case is not taken up based on priorities. The second part addresses the role of competition law as a 'sword.' This pertains to cases where protecting competition was in turn expected to be beneficial to sustainability, such as protecting competition in industries crucial to sustainability.

The third part addresses a concern that is often heard in the debate, namely 'greenwashing'. The report presents a broad definition of greenwashing and summarises the reported instances.

The fourth part deals with the policy side. It explores how agencies, beyond specific cases, have become active in the area, for example by setting out priorities, guidelines, working papers, and individual guidance and strengthening capacity. Similarly, it discusses any existing future plans of agencies and legislatures.

The final part briefly highlights some of the contributions that the reports have mentioned in other closely related areas of law, such as false advertising or unfair competition.

1.2 General Framework

In this section, we look at the different legal frameworks of the participating jurisdictions and how the relevant general framework of their national legal order regarding sustainability interacts with competition law. While the majority of countries do not have specific clauses that address sustainability—the revised version of the Austrian competition law and the Hungarian being the exception—many countries have provisions in their constitution or in other areas of the law that require an engagement with sustainability. The different reports also make clear that the jurisdictions are at different positions as to the level of importance attached to climate change and sustainability more generally. It also becomes clear that occasionally different definitions of sustainability are used by the competition authorities, although the UN sustainable development goals play a significant role.

Some country reports, like the Italian one, show how sustainability is being linked to markets, competition, and even industrial policy. The *Italian* report highlights the link to the concept of ecological transition, which in turn stems from the EU Recovery and Resilience Facility of 12 February 2021 (Regulation 2021/241) and the implementing of April 2021. Ecological transition for Italy is then linked to increasing 'the competitiveness of the production system of goods and services, stimulating new entrepreneurial activities, and encouraging the creation of stable employment'.[15]

---

[15] *See* National Recovery and Resilience Plan (hereafter NRRP), p. 14. The NRRP, approved on 13 July 2021 by the European Council, is the document that the Italian government prepared to illustrate to the European Commission how Italy intends to invest the funds allocated at the

1 Sustainability and Competition Law: An International Report

In terms of sustainability and competition law, the Italian authority highlights the complementary role of competition law to regulation and taxation in fostering an environmentally sustainable growth model. It highlights how competitive pressure fosters allocative efficiency also in terms of natural resources, the importance of innovation to limit CO2 emissions, and the use of renewable energy.[16] The authority sees its task in balancing the dynamic and competitive markets with promoting investments by companies in terms of environmental sustainability. Further, they claim to be ready to develop the application of competition law in co-operation with the EU Commission and other agencies to expand the range of available instruments in support of sustainable and competitive development.[17]

The *Czech* report highlights their alignment with EU law and presents a 'long journey ahead' with slow moves and possibly even reluctance towards the inclusion of sustainability in its policies and law.[18] For example, the Czech Republic adopted a first strategy for sustainable development only in 2004.[19] Its updated version sets out key principles and goals in various sectors.[20] As the report highlights, an overview of the legislative acts seems to show that the inclusion of sustainability is not a proactive legislative choice but rather a result of compliance with EU law. The report also highlights that while not mentioning sustainability in competition law, the consideration seems possible as far as EU law does so.

While no specific legislative acts address the interaction between competition law and sustainability, the *Belgian* competition authority sees the current legislative framework as allowing sustainability considerations, including 'out-of-market' considerations, to be taken into account.[21] It would do so by considering the total and the consumer welfare benefit and the general economic interest.[22] On the

---

European level under the Next Generation EU programme. The NRRP is divided into six missions subdivided into 16 functional components to achieve the economic and social objectives that are part of the Government's strategy.

[16] IAA Annual Report of 31 March 2021, pp. 15 ff.

[17] IAA Annual Report of 31 March 2021, pp. 15 ff.

[18] R. MacGregor Pelikánová, Corporate Social Responsibility Information in Annual Reports in the EU – Czech Case Study, Sustainability 2019 (11), p. 237, available at https://doi.org/10.3390/su11010237.

[19] Český Statistický Úřad (Czech Statistic Office), Resolution Nr. 1242 - Udržitelný rozvoj v ČR (Sustainable development in the Czech Republic), available at https://www.czso.cz/csu/czso/13-1134-07-2006-1_1___uvod. Accessed 4 June 2022.

[20] Ministerstvo Pro Místní Rozvoj (Regional Development Ministry), Základní dokumenty (Fundamental documents), available at https://www.mmr.cz/cs/ministerstvo/regionalni-rozvoj/informace,-aktuality,-seminare,-pracovni-skupiny/psur/uvodni-informace-o-udrzitelnem-rozvoji/zakladni-dokumenty. Accessed 4 June 2022.

[21] *See* Government of Belgium, Note on the Environmental Considerations in Competition Enforcement, OECD Competition Committee meeting 1-3 December 2021 (DAF/COMP/WD(2021)47), available at https://one.oecd.org/document/DAF/COMP/WD(2021)47/en/pdf, p. 2, 5. Accessed 29 November 2022.

[22] *See* Government of Belgium, Note on the Environmental Considerations in Competition Enforcement, OECD Competition Committee meeting 1-3 December 2021 (DAF/COMP/WD(2021)47),

judicial level, the Court of First Instance of Brussels found that the various Belgian governments have been negligent in their climate policy,[23] yet it stopped short of imposing specific obligations.

Brazil also does not have any specific provision in its competition law that addresses sustainability. Brazilian competition law instead seems more focused on 'the freedom of initiative, free competition, social function of property, consumer protection and repression of the abuse of economic power', as the Brazilian report highlights.[24] Yet the report also highlights that competition law can be interpreted to foster sustainability as competition law seeks to 'promote human rights, efficiency and consumer welfare',[25] particularly in view of the Federal Constitution.[26]

The Brazilian economic order, as foreseen by the Constitution, is 'founded on the valorisation of human work and free initiative (or free enterprise), [and] aims, through social justice (Article 170 of the Constitution), to contribute to the achievement of human dignity'.[27] Moreover, Article 225, which is a part of the economic rights section of the Constitution, sets out a right for everyone to an ecologically balanced environment. It thereby imposes a duty upon the government and the community. This duty is to defend and preserve the environment for present and future generations, also by means of antitrust, as the report highlights. The report also highlights how the absence of a position by the CADE—the Brazilian Competition Authority—might create legal uncertainty that might deter companies from taking action. Yet CADE seems 'attentive to foreign initiatives' such as the Dutch proposal, and it is likely that CADE will take a position on these matters in the future.

As the *French* report highlights, sustainability (with a particular view to environmental sustainability) is not specifically mentioned in the French competition law but plays a role in enforcement. The report highlights how environmental quality and performance are now an established parameter on which undertakings compete.[28] More specifically, the report sets out that Article L. 420-4, which is similar to Article 101(3) TFEU, is seen by the French competition authority as a basis for providing exemptions for sustainability initiatives by companies where its conditions are met, as the authority considers environment protection as a form of economic progress.[29] In the field of mergers, the competition authority takes a similar view, as economic progress is part of the assessment of a merger's benefits according to Article

---

available at https://one.oecd.org/document/DAF/COMP/WD(2021)47/en/pdf, p. 2-3. Accessed 29 November 2022.

[23] Judgment of the Brussels Court of First Instance of 17 June 2021, *Klimaatzaak*, available at https://prismic-io.s3.amazonaws.com/affaireclimat/18f9910f-cd55-4c3b-bc9b-9e0e393681a8_1 67-4-2021.pdf. Accessed 29 November 2022.

[24] *See* Brazilian report section 1.2.

[25] *See* Brazilian report section 1.2.

[26] *See* articles 3, items II and III, 5 and 170, item IV, 225

[27] *See* Brazilian report section 1.3.

[28] *See* B. Coeuré, Closing speech, New Frontiers of Antitrust, 21 June 2022.

[29] FCA, Opinion n°99-A-22, 4 December 1999, p. 10.

1 Sustainability and Competition Law: An International Report

L. 430-6. Thus, the agency has qualified the ability of the parties to abandon technologies that damage the environment as a benefit and have used it, *inter alia*, to approve a merger.[30]

Overall, the report notes that while economic progress is a broad concept and encompasses environmental benefits, the French agency focuses on the principle of proportionality, which it implements in 'a cautious manner'.[31] The assessment by the agency examines whether 'the environmental goal exists'. And if it does, the agency examines whether it is not already pursued by the applicable regulation.[32] It also makes sure that the environmental benefits may not be achieved by means that are less restrictive to competition.[33] Moreover, these benefits must be passed on to the consumer. The president of the agency announced that the agencies will be improving the assessment of the passing on to consumers of such environmental benefits.[34]

For the future, the report highlights two possible routes for French competition law. First, it could adopt a specific aggravating factor in its fining practice where negative impacts on the environment exist. Second, French competition law may provide a specifically lenient treatment for measures that provide environmental benefits beyond what is already happening. The French report highlights that this would be difficult, given the mandate of the competition authority to protect competition, but that a better translation of environmental benefits and harms into the framework of competition law can certainly be a path forward.

While there are no specific articles in the *German* competition act related to sustainability at present,[35] the German competition authority relies on the UN sustainable development goals, including their environmental and social dimensions.[36] But most importantly, German competition law would need to be interpreted in line with EU law. The German report highlights that the German Constitution requires the State to protect the environment (Articles 1, 2(1), and 20a of the Basic Law). Article 20a has recently been the focus of the debate and while predominantly addressing that the legislature also binds public authorities when they interpret the relevant statutes.[37] This obligation to take into account the effects on the

---

[30] FCA, Opinion n°94-A-18, 17 May 1994, Metaleurop/Heubach & Lindgens.

[31] *See* French report section 1.2.

[32] FCA, Opinion n°95-A-08 or 9 May 1995, CEAC/Exeide.

[33] FCA, Opinion n°94-A-31 of 6 December 1994, Elimination of used oils. See also *See* French report section 2.2.

[34] B. Coeuré, Closing speech, New Frontiers of Antitrust, 21 June 2022.

[35] *See* however below under Section 5 where a possible change in the law is addressed.

[36] Bundeskartellamt, Offene Märkte und nachhaltiges Wirtschaften – Gemeinwohlziele als Herausforderung für die Kartellrechtspraxis, October 2020, p. 5; Bundeskartellamt, Jahresbericht 2020/21, p. 46; Bundeskartellamt, Offene Märkte und nachhaltiges Wirtschaften – Gemeinwohlziele als Herausforderung für die Kartellrechtspraxis, October 2020, p. 5.

[37] *See* German report section 2.2.1. The report points to S. Huster and S. Rux, in: Epping and Hillgruber (eds), BeckOK Grundgesetz, 51st ed, C.H. Beck 15.5.2022, GG Art. 20a paras 27, 32-33; H. Schulze-Fielitz, in: Dreier (ed), Grundgesetz Kommentar, Vol. 2, 3rd ed, Mohr

climate when making a decision is also spelled out in Section 13(1)1 of the Climate Protection Act. Yet this obligation might create challenges when implementing competition law where the original objective of the law is different.

The *UK* report highlights the importance that the UK government and their competition authority have attached to sustainability and climate change in particular. The report also emphasises the influence that EU competition law has even though it is now legally possible for the UK to diverge from EU competition law. Against this background, the UK competition authority has become increasingly active in this area. It cites sustainability as a 'strategic priority' in its latest annual reports with a particular focus on the transition to a low carbon economy. The agency thus aims to prioritise cases that could impede the net zero goals of the UK. Moreover, the agency has issued a guidance note for sustainability issues in antitrust, mergers, and market regimes with a focus on environmental sustainability. In this sense, the agency has not made any reference to the broader UN sustainable development goals.

The report from *Malta* highlights that its Constitution has, since 2018, provided for an obligation to 'protect and conserve the environment and its resources for the benefit of the present and future generations and shall take measures to address any form of environmental degradation in Malta, including that of air, water and land, and any sort of pollution problem and to promote, nurture and support the right of action in favour of the environment'.[38–39] This is a positive obligation imposed upon the government, but it does not seem to be judicially enforceable. Yet it remains 'fundamental to the governance of the country and it shall be the aim of the State to apply these principles in making laws'.[40]

In line with the Brundtland report, sustainability is defined under the Sustainable Development Act as 'development that meets the needs of the present without compromising the ability of future generations to meet their own needs'. Like many competition laws, the Maltese competition law does not make reference to sustainability. Yet the report highlights 'that sustainability can be part of the assessment to be made by the Office for Competition within the Malta Competition and Consumer Affairs Authority [. . .], and by Civil Court (Commerce Section) and other courts and tribunals where applicable (hereinafter the 'Malta courts'), when assessing claims of breaches of competition law and proposed concentrations'.[41]

Siebeck 2015, GG Art. 20a paras 67, 74-75; R. Scholz, in: Dürig, Herzog and Scholz (eds), Grundgesetz-Kommentar, 96th supplement November 2021, C.H. Beck 2021, GG Art. 20a paras 46, 56; H. Jarass, in: Jarass and Pieroth (eds), Grundgesetz für die Bundesrepublik Deutschland, 17th ed, C.H. Beck 2022, GG Art. 20a paras 18, 21; K.-P. Sommermann, in: v. Münch and Kunig (eds), Grundgesetz-Kommentar, Vol. 1, 7th ed, C.H. Beck 2021, GG Art. 20a, paras 41, 46-48.

[38] Article 9(2) Constitution of Malta.

[39] *See* Maltese report section 1.

[40] Article 21 Constitution of Malta.

[41] *See* Maltese report section 2.1. Malta operates a prosecutorial model, with the competition authority bringing cases to the court.

In *Switzerland*, sustainable development, as understood by the Brundtland Commission, is a principle that binds the Federal State and the Cantons via Article 2 of the Federal Constitution. This article makes sustainable development a national goal to be achieved. Moreover, Article 72 demands that the various parts of the State strive for '[. . .] a balanced relationship between nature and its ability to renew itself, on the one hand, and the demands placed on it by the human race, on the other'.

Sustainable development is pursued by national sustainability plans, and some Cantons have also adopted their own law to govern action at that level of the Swiss state. While competition law does not specifically address sustainability, the national report highlights that Articles 2 and 73 of the Constitution should guide the national competition agencies. In competition law, Article 5(2)(a) of the Cartel Act might gain particular relevance as it mentions the more rational use of resources as a reason for economic efficiency that can justify restrictions of competition. The report also highlights that 'the term "resources" includes (i) entrepreneurial resources, such as money, (ii) public resources, (iii) natural resources and (iv) knowledge.[42]

Austria is the only jurisdiction that specifically addresses sustainability in its competition law. Thus, the Austrian reports highlights Austria's recent changes to their competition law in 2021, which was aimed to address an environmentally sustainable and carbon-neutral economy. The change included a provision that guides the authorities in the definition of consumer benefit. The provision now states:

> Consumers are also allowed a fair share of the resulting benefit, if the benefit derived from improving the production or distribution of goods or promoting technical or economic progress contributes significantly to an environmentally sustainable or carbon-neutral economy.

This definition is particularly relevant for the exemption for agreements deemed to restrict competition. The report, however, notes that notwithstanding this legislative clarification, uncertainty remains which stems from questions about the exact scope and the relationship of Austrian competition law to EU competition law.

The *Hungarian* report notes that, just as in Austria, a specific environmental protection can be found as a ground for exempting agreements, the equivalent to Article 101(3) TFEU. In particular, Article 17 sets out:

> An agreement is exempted from the prohibition pursuant to Article 11 [the prohibition of anti-competitive agreements] provided that
>
> i. it contributes to a more reasonable organisation of production or distribution, the promotion of technical or economic progress, or the improvement of competitiveness or of the protection of the environment.
> ii. it allows trading parties not participating in the agreement a fair share of the resulting benefit;

---

[42] *See* Swiss report section 2.3.

iii. the concomitant restriction or exclusion of competition does not exceed the extent necessary to attain

iv. economically justified common goals; and

v. it does not enable the exclusion of competition in respect of a substantial proportion of the goods concerned.

Moreover, sustainability is also constitutionally enshrined. Under the heading 'Foundations', Article Q(1) sets out 'sustainable development of humanity' as an aim of the state. Moreover, Article XXI provides for a right to a healthy environment. These provisions also have bearing on the application of competition law by the courts. As the Hungarian report points out, Article 28 of the Fundamental Law explains that '[i]n the course of the application of law, the courts shall in principle interpret the laws in accordance with their objective and with the Fundamental Law. [...].'

1.3 Specific Interaction and Cases

In this section, we look at the specific interaction between competition law provisions on co-operation, cartels, abuse, monopolisation, mergers, and sustainability covering general interaction and specific cases. The section is subdivided into two parts. The first part covers the shield situation, where companies take action aimed at fostering sustainability and the extent to which they are able to rely on sustainability before competition authorities. This can either take the form of a finding that competition law was not infringed in the first place, for example, because the relevant entities or activities were not subject to competition law or because competition was not restricted or because the sustainability benefits have the potential to outweigh the competitive harm. The second section addresses the sword situation. These are situations where the agency or courts became active using the competition rules protecting competition, and this competition in turn was expected to be beneficial to sustainability, for example by protecting competition in industries crucial to sustainability. In other words, the section addresses the question of whether and to what extent it is possible to use competition law to address unsustainable business practices, such as abuses of monopoly power. Three reports of the reports submitted by the 11 jurisdictions state that there have been no cases concerning sustainability and competition law. The *Czech* report highlights that no cases were found involving competition law and sustainability. Furthermore, even cases in other areas of law where sustainability is of concern have been rare. *Brazil* also has not reported any cases involving sustainability, while the agency is said to be taking an active interest in the development in other jurisdictions. Similarly, the *Belgian* report highlights that the Belgian authority has not dealt with sustainability concerns, although it would in general be willing to address these in its competition analysis.

1.4 Shield

Several of the national reports mention cases or explain how sustainability might function as a shield in their jurisdictions.

In *France*, an exception similar to Article 101(3) TFEU is contained in Article L. 420-4. But this exception applies not only to agreements but also to abuses. As early as 1973, the agency found that the prevention of environmental pollution was part of the definition of 'economic progress', on which the article relies. The first case where environmental protection serves as a ground for exemption under this provision appears in a 1988 case relating to salt and the related environmental protection concerns. Another case can be identified in 1999 regarding used batteries. However, as the report points out, more recent cases in this regard are lacking.

In the merger context, in a case from 1994, the agency accepted that the abandonment of the most polluting production techniques would be technical progress that could justify the authorisation of a merger. A year later, in another case, the agency, however, clarified that the sole compliance with existing environmental protection laws would not be considered as such economic and technical progress.

The *German* report highlights that exceptions existed in German competition law that might have captured sustainability matters, such as an exemption from the taking back or disposal of goods or services. These were abolished by the legislature. Yet the legislature highlighted that these changes were meant to bring German competition law in line with EU law rather than to affect substantial change. The legislative documents also highlight that a change is not to be expected as European competition law and policy would already be required to take environmental protection into account.

In Germany, there has been a debate about the extent to which Section 2(1) of the competition act (which sets out the prohibition on anti-competitive agreements) can be interpreted more narrowly to allow sustainability initiatives. Where such an interpretation is based on climate law, a full climate impact assessment would be required.

The report notes that the German competition authority's position seems to have developed over time. Until 1995, the agency used an approach that allowed for the balancing of competition restrictions with public interests, such as public health and environmental protection. This approach has been described as a full balancing within the substantive assessment, sometimes as an exercise in enforcement discretion in terms of priorities. After sustained critique, the competition authority moved away from the substantive balancing approach and instead adopted an approach based on its enforcement discretion and priorities. In other words, anti-competitive action can be 'tolerated' even if there is a restriction of competition. This exercise of discretion can be used either for avoiding the allocation of enforcement recourses for minor restrictions or in cases where the strict enforcement of competition law would create serious clashes between protecting competition and other objectives. A prime example mentioned in the report is the competition authority's activity in the waste management areas. The German legislature enacted a new code to address the recycling of plastic waste. It had been warned by the agency that this might create serious problems with competition law. Yet the legislature still went ahead without changing the competition law framework.

The report highlights a major downside of this approach—the lack of legal certainty. The competition agency can change its mind at any time, and private enforcement stays a relevant risk.

Three recent cases are highlighted by the report: the decision regarding Fairtrade, the Living Wages initiative, and the Initiative Tierwohl case. In these cases, the agency decided that the restrictions of competition—in the case of Initiative Tierwohl, after further changes suggested by the agency—did not restrict competition to an extent that would warrant the competition agency to bring a case.

Yet the report highlights how sustainability might be considered as part of the substantive assessment of Section 2 of the German competition act, which is the national equivalent of Article 101(3) TFEU, and how this section should be applied in conformity with EU law. The report further elaborates that the current view in Germany is that the benefit would therefore have to occur within the same group of consumers that are affected by the restrictions. Notwithstanding, there is a lively debate. The developments in the Netherlands and the more generous approach in the Commission's draft guidelines have been received with great interest.

While there are no reported cases of sustainability defences regarding unilateral conduct, the report notes regarding mergers that sustainability efficiencies could theoretically be taken into account even though no cases exist yet. This lack of cases is, however, not surprising, given that, so far, no efficiency argument has ever been successful before the German competition authority. Yet a different tool can come into play in this regard: the approval by the relevant minister on public interest grounds, even where the German competition authority has prohibited the merger. The ministry mentions environmental concerns as one of the interest grounds that can be relied upon.[43] In the case of Miba/Zollern, such an argument had been successful, while in the E.ON/Ruhrgas merger, the relevant minister did not take up the arguments of the parties in this regard but approved the merger based on other reasons. In the Miba/Zollern case, the main argument was one of benefits in terms of 'know-how and innovation potential for energy transition and sustainability', which would outweigh restrictions of competition that would only be minor as a level of competition remained.

The *UK* report highlights that sustainability concerns have only played a limited role in the actual enforcement work or private enforcement. Beyond a sector enquiry into vehicle charging points, no cases have been reported. Thus, the report provided ideas on how sustainability might function as a shield under UK competition law with a particular focus on agreements that do not restrict competition within the meaning of the Chapter I prohibition on anti-competitive agreements and exemptions for such agreements satisfying the conditions of Section 9, the equivalent of Article 101(3) TFEU. The report also highlights the possibility that certain

---

[43] BMWi, Verf. v. 19.08.2019, I B 2 -20302/14-02, para 164.

sustainability agreements, such as climate-change-related ones, could be excluded altogether from the scope of Chapter I, as witnessed for certain agreements during the Covid crisis.

With regard to the exception under Section 9, the report highlights the debate around the UK Supreme Court's judgment in Sainsbury's Supermarkets Ltd v Visa Europe. The debate concerns the question of whether full compensation of the affected customers is a requirement. The report highlights that the competition authority announced that it would apply the 'fair share' criterion with a degree of 'flexibility' and rely on Section 60A to depart from EU precedent as applied by the Supreme Court, if needed. The report also highlights that the competition authority has not positioned itself clearly with regard to collective benefits and seems in this regard to be behind the EU Commission, which set out its approach in the draft guidelines on horizontal co-operation.

While no cases of using sustainability as a defence of abuse under Chapter II have been reported, the agency highlighted this possibility and suggested a case-by-case assessment.

In the area of mergers, the report notes that the UK competition agency has recognised that environmental benefits can constitute merger efficiencies in its recently updated merger guidelines. These can be assessed as rivalry-enhancing efficiencies, such as research and development (R&D), or as relevant customer benefits, as long as they are merger specific. While rivalry-enhancing efficiencies are typically related to the market, which is affected by the merger, the relevant customer benefits are not and can appear in other markets. As the agency mentions, relevant customer benefits may include lower energy costs or a lower carbon footprint of the firm's products. The agency has more recently assessed such claims in two merger cases (Cargotec/Konecranes and Pennon Group plc/Bristol Water) but found that the parties had not provided sufficiently detailed evidence.

The *Maltese* report highlights that sustainability might function as a shield in cases involving Article 5 of the Competition Act, particularly under Article 5(3), which is the equivalent of Article 101(3) TFEU. Similarly, it might function as an objective justification under the abuse prohibition and efficiency gain in merger review. While there are no reported cases and no guidance from the agency, the guidance by the EU Commission will be of particular relevance once adopted. The Competition Act imposes a duty on the competition authority and the courts that these 'shall have recourse [...] to relevant decisions and statements of the European Commission including interpretative notices on the relevant provisions of the TFEU and secondary legislation relative to competition [...]'.

The *Austrian* report notes that not all sustainability agreements are within the scope of the prohibition on anti-competitive agreements, such as certain environmental standardisation agreements. Moreover, it underlines the possibility of considering sustainability under Section 2(1) of the Cartel Act, the equivalent of Article 101(3) TFEU. Previously, it was considered that out-of-market efficiencies could not readily be considered under this provision. The amendments to the Cartel Act are meant to provide legal certainty. They clarify that out-of-market environmental sustainability benefits are assumed to be passed on to consumers. In other words, as the legislative material notes, environmental sustainability and contributions to net zero always benefit society at least in the long run and thereby compensate the affected consumers as part of society. This also distinguishes environmental sustainability from other elements of sustainability. In these cases, Austrian law would still require the direct compensation of the consumers affected. Yet, in any case, the exemption for environmental sustainability cannot apply to hardcore restrictions like price fixing.

The new guidelines of the authority set out specific steps for the exemption criteria to be fulfilled. There must be technological or economic progress, and it must contribute to environmental sustainability (e.g. carbon neutrality). This contribution must be significant while not imposing indispensable restrictions or eliminating competition in respect of a substantial part of the product.

There is a debate as to the extent that this exemption is broader and will be applied in practice. The exemption covers only cases without a cross-border element to which Article 101 TFEU would apply.

The *Swiss* contribution reports that the use of practices such as more sustainable packaging material can be seen as efficiency gain that is positive for the consumers within the meaning of Article 5(2)(a) of the Cartel Act. This is the case where they are sufficiently linked to agreement. But even where there are doubts as to the permissibility under that article, Article 8 allows for the Federal Council to provide an exemption on public interest grounds. The report also describes several cases where sustainability played a role in Article 5(2)(a), with cases as early as 1997 concerning the disposal of electronic waste. It also highlights a project where the agency found that the envisioned CO_2 savings were not sufficient to justify the restriction of competition.

The report notes how this practice seems to contrast sharply with merger decisions. There, the competition authority seems to not take account of such benefits. In particular, the report underscores the 'Swiss H2 Generation' case, where the agency mentioned the importance of hydrogen for the global energy transition but then decided the case purely on grounds of market shares and the potential for creating a dominant position.

The *Hungarian* report highlights that under the specific exemption for environmental protection, numerous agreements could be exempted, such as those that 'jointly develop a production technology that reduces energy consumption; [the sharing] of infrastructure with a view to reducing the environmental footprint of a production process; [the joint] purchase [of] products having a limited environmental footprint as an input for their production; an agreement [. . .] to purchase from suppliers observing certain sustainability principles'. Yet, as the report notes, hardcore restrictions cannot benefit from this clause. The report also highlights that the competition authority, although sympathetic in principle, has not yet fully accepted such a defence where the parties ex post tried to mount environmental protection as a defence.[44] Yet, as Hungarian competition law needs to be compliant with EU law,

---

[44] Although occasionally the fine has been lowered.

1 Sustainability and Competition Law: An International Report

where a cross-border situation exists, the report expects further developments once the EU horizontal guidelines and its sustainability chapter are adopted.

The picture is similar in abuse cases. Environmental protection was occasionally argued as a defence, but these cases were mostly not decided so that it was not clear whether and to what extent these arguments would have convinced the agency. The report comments that the agency would be open to assess such arguments but that the benefits would have to be clearly shown.

For mergers, the report envisions the possibility that sustainability benefits could be argued as part of the efficiency defence and that the agency would examine them seriously or that such an agreement could be exempted by the government under the public interest exception of Section 24/A of the Competition Act. Yet such cases have not been observed yet.

1.5   Sword

It might be well known that the environmental but more generally the sustainability aspects of a product are part of the quality or performance of a good or service. Thus, they can be a parameter on which undertakings compete. In this sense, competition on these aspects may be hindered, and it might fall under the prohibitions of anti-competitive conduct or be prohibited as an anti-competitive merger. In these cases, we speak of competition law used as a sword.

Of those countries that submitted reports on the substantive interaction between competition law and sustainability, three countries did not report cases or could not address the matter in more detail. This is the case regarding the *Austrian* report. The *Maltese* report also mentions that it is conceivable that competition law might function as a sword in the fight against unsustainable practices, particularly with regard to the abuse prohibition. Similarly, the *Swiss* report did not record any cases where competition law was used to attack unsustainable practices. Moreover, the Swiss report expressed doubt as to whether the competition authority would take account of such negative effects on sustainability, given its mandate to protect competition.

In *France*, the agency has considered sustainability in several cases where competition law was used as a sword in the context of their provision addressing anti-competitive agreements, unilateral conduct, and mergers. In the context of anti-competitive agreements, the agency has fined companies that have prevented competition based on the sustainability element. This happened in a case involving manufacturers of PVC and linoleum flooring that agreed not to compete and not to provide customers with information on the environmental performance of their products. A currently ongoing case concerns an agreement of manufacturers of food containers not to advertise and market BPA-free products before a French law banning the use of BPA in food packaging would come into effect.

With regard to dominance, the report mentions a case regarding Nespresso machines. The agency took steps to ensure that not only Nespresso-branded capsules could be used in these machines. And while this case did not specifically mention sustainability, it paved the way for things such as biodegradable capsules. Similarly, one could imagine that coffee produced in other, more sustainable ways might now be available, such as coffee produced with higher levels of social sustainability.

In merger control, the report highlights the indirect usage of sustainability as a sword. For example, the report highlights how the agency in one case defined sustainable products as a distinct product market. This allowed the agency to identify specific problems in certain regions that could then be addressed in the form of commitments.

In terms of the sword, the *German* report highlights that in several non-sustainability-related cases, the courts have taken a more extensive approach to ensure compliance with the Constitution and fundamental rights in particular. In these cases, the competition provision was interpreted more expansively. This interpretation meant that behaviour was prohibited where it would usually be questionable as to whether it is covered by the prohibitions.

In this line, the report highlights the discussion regarding the German prohibitions on abuse and whether these might be used to address unsustainable practices. The report highlights that this might be possible where the conditions of the Facebook case are fulfilled. In other words, there must be a sufficiently close connection between the violation of a certain law, competition, and harm to consumers.

The report also highlights earlier cases where intervention by the agency had positive impacts on sustainability whether intended or not, particularly highlighting cases from the waste management sector involving Duales System Deutschland (DSD).

In terms of the sword, the *UK* report highlights that the agency has started to investigate arrangements regarding the recycling of old vehicles and explained its actions with its commitment to prioritise environmental sustainability cases. This seems to be mainly the case with regard to anti-competitive agreements as the report also highlights that the abuse provision has not been substantially featured in the sustainability debate in the UK. The report, however, makes suggestions on how the UK competition authority could address unsustainable and exploitative actions by dominant companies where the unstainable practice has a reasonable nexus to competition.

The sword question has also been relevant in the context of mergers. The competition authority has highlighted in its new merger guidelines that competition between parties can be sustainability related (e.g. energy efficiency, sustainability innovation) and that such competition would need to be protected. This can take the form of remedies or prohibition decisions. In the *Cargotec Corporation / Konecranes Plc* case, the agency in fact found that the companies were competing on environmental-performance-related matters. But there was still sufficient evidence that the transaction would not result in a substantive lessening of competition. Importantly, the UK competition authority has committed to conducting at least a market inquiry in markets relevant to the net zero target within the next fiscal year.

In *Hungary*, the report notes that the idea of competition as a sword seems possible within certain limitations. And while no cases have been observed with

1 Sustainability and Competition Law: An International Report   19

regard to agreements, abuse, or mergers, the further development of EU law in this regard will be influential. Moreover, in the areas of digital markets, the competition authority has shown that it is willing to examine 'non-price effects, namely, the merger's effect on quality, variety and innovation'.[45]

1.6   Greenwashing

Greenwashing is a concern that is often raised in the context of the debate about sustainability and competition law. Thus, the country rapporteurs were specifically asked about cases of illegal anti-competitive conduct that occurred in the context of sustainability initiatives. These involved cases where companies gave the false impression of pursuing a sustainability initiative when they really served as a cover for anti-competitive behaviour and cases of genuine sustainability initiatives that served as a springboard for other anti-competitive behaviour. While these cases can clearly be distinguished on a theoretical level, a certain overlap can also be found, in particular where genuine sustainability initiatives 'spill over' into anti-competitive practices such as price fixing. Hence, these cases are grouped together here.

Regarding the issue of green washing, many reports highlight the possibility of addressing such matters under unfair competition, consumer protection law, or rules of false advertising. In the more specific competition law sense, greenwashing refers to situations where a sustainability benefit was claimed but the actual agreement pursues alternative, anti-competitive objectives; in other words, these are disingenuous sustainability initiatives.

Even with this broad definition of greenwashing and specific questions about it, not many jurisdictions reported cases, including those jurisdictions where sustainability is a rather established parameter that might exempt anti-competitive agreements. Reports of such situations have been received from Brazil and Germany. In France, the competition agency does seem to look into such cases.

The *Brazilian* report highlights one case involving an association in the sand industry. The case concerned the unloading and storage at a sand terminal and also involved environmental concerns. The association served as a springboard to share sensitive commercial information. The remedies imposed in that case addressed competition and environmental concerns.

The *German* report mentions two cases of potential greenwashing where companies made claims in proceedings that were subsequently refuted by the agency. For example, in the 2007 DSD case, the report cites the head of the competition authority, who said: 'We will not allow anti-competitive cartel agreements to be made under the guise of environmental protection.'[46] This

---

[45] *See* Hungarian report Section 4.2.1.2.

[46] Bundeskartellamt, Bundeskartellamt prohibits purchasing cartel in the container glass industry, Press Release (1 June 2007). Available at https://www.bundeskartellamt.de/SharedDocs/Meldung/

statement needs to be seen as a response to the companies' claim in the proceedings that the monopolistic structure that they created for recycling purposes was needed in order to achieve a functioning recycling system. In a more recent example, the agency provided individual guidance. In Agrardialog Milch,[47] the aim was to establish a unified surcharge for producers of raw milk as the cost of production often exceeds the price of milk. The competition authority rejected the initiative as it did not set out any criteria for milk production that fosters sustainability but instead set up a price agreement at the expense of consumers.

While the *French* report does not reveal any specific case, it mentions that the French competition agency is looking actively into possible issues of greenwashing to ensure that anti-competitive activities are not taking place under the guise of sustainability initiatives.

1.7 Agency and Legislative Activities in the Areas of Competition and Sustainability

Having set out the general framework for sustainability and competition and having looked at how the relevant competition laws treat specific cases, the rapporteurs were also asked about activities of agencies, such as priorities, guidelines, working papers, individual guidance, or capacity building. Moreover, they were asked whether their legislatures have become or will become active in the area.

Overall, we can see that several agencies have provided individual guidance or adopted policy documents. However, many seem to be waiting for the EU Commission's guidance on horizontal co-operation agreements which will again contain a section on sustainability agreements, after a lively debate of the draft guidelines.

While the *Czech* submission does not report any cases, the Office for the Protection of Competition addresses these issues in documents on its webpage. For example, it has a policy document called 'Application of rules for public support in the environmental field'[48] that addresses environmental elements of sustainability with a focus on efficiency and energy efficiency. The document describes different options for exemptions with a specific focus on EU law, such as de minimis, individual notification to the European Commission, regulation of block exemptions, and the adjustment of payment for services in the general economic

---

EN/Pressemitteilungen/2007/01_06_2007_Beh%C3%A4lterglasindustrie.html.   Accessed 29 November 2022.

[47] Bundeskartellamt, Fallbericht v. 08.03.2022, B2-87/21 32. Available at https://www.bundeskartellamt.de/SharedDocs/Entscheidung/DE/Fallberichte/Kartellverbot/2022/B2-87-21.pdf?__blob=publicationFile&v=2. Accessed 29 November 2022.

[48] ÚOHS (Office for the Protection of Competition), Application of Rules for Public Support in the Environmental Field, available at https://www.uohs.cz/cs/verejna-podpora/vybrane-oblasti-verejne-podpory/aplikace-pravidel-verejne-podpory-v-oblasti-zivotniho-prostredi.html. Accessed 29 November 2022.

1 Sustainability and Competition Law: An International Report

interest. In 2022, the agency also organised an international conference where sustainability was one of the key areas of discussion.[49] The report expects further impetus in the sustainability area from the EU Council presidency, which the Czech Republic will hold from July until 2022.

The *Belgian* competition authority has made the green and circular economy a strategic priority for competition policy in 2022 in order to stimulate innovation and technical developments in this area while expressing a willingness to look more closely at issues where the competition rules might hinder sustainability initiatives.[50] Thus, the agency offers to provide individual guidance to sustainability initiatives.[51] It is also supportive of the initiatives of the Dutch and Greek competition authorities in the field of sustainability.[52] Yet the Belgian competition authority also highlights the importance of guidance at the EU level.[53]

In *Brazil*, the agency has so far not adopted any policy measures, but the report notes that the agency is observing the guideline drafts in Europe and is discussing the adoption of a policy. The report also notes hesitancy among Brazilian businesses to engage in sustainability initiatives due to fear of prosecution under competition law.[54]

---

[49] ÚOHS (Office for the Protection of Competition), The Office for the Protection of Competition is preparing the first year of the May Conference on Public Procurement to be held on 18 and 19 May 2022. The conference will take place at the Brno headquarters of the Office, Press Release (20 April 2022), https://www.uohs.cz/en/information-centre/press-releases/public-procurement/3321-may-2022-public-procurement-conference.html. Accessed 4 June 2022.

[50] *See* Government of Belgium, Note on the Environmental Considerations in Competition Enforcement, OECD Competition Committee meeting 1-3 December 2021 (DAF/COMP/WD(2021)47), available at https://one.oecd.org/document/DAF/COMP/WD(2021)47/en/pdf, p. 6. Accessed 15 July 2022; See BCA priority note 2022, p. 3. Available at https://www.abc-bma.be/sites/default/files/content/download/files/2022_politique_priorites_ABC.pdf. Accessed 29 November 2022.

[51] *See* Government of Belgium, Note on the Environmental Considerations in Competition Enforcement, OECD Competition Committee meeting 1-3 December 2021 (DAF/COMP/WD(2021)47), available at https://one.oecd.org/document/DAF/COMP/WD(2021)47/en/pdf, p. 6. Accessed 29 November 2022.; See BCA priority note 2022, p. 3. Available at https://www.abc-bma.be/sites/default/files/content/download/files/2022_politique_priorites_ABC.pdf. Accessed 29 November 2022.

[52] *See* Government of Belgium, Note on the Environmental Considerations in Competition Enforcement, OECD Competition Committee meeting 1-3 December 2021 (DAF/COMP/WD(2021)47), available at https://one.oecd.org/document/DAF/COMP/WD(2021)47/en/pdf, p. 6. Accessed 29 November 2022.

[53] *See* Government of Belgium, Note on the Environmental Considerations in Competition Enforcement, OECD Competition Committee meeting 1-3 December 2021 (DAF/COMP/WD(2021)47), available at https://one.oecd.org/document/DAF/COMP/WD(2021)47/en/pdf, p. 6. Accessed 29 November 2022.

[54] L. P. Rocha Silva and P. S. Amaral Mello, Cooperation agreements among competitors to promote sustainable measures: which should be CADE's role?, 2020, available at https://www.migalhas.com.br/depeso/331669/acordos-de-cooperacao-entre-concorrentes-para-promocao-de-medidas-de-desenvolvimento-sustentavel%2D%2Dqual-deve-ser-o-papel-do-cade. Accessed 12 June 2022.

The *French* report does not note any specific guidance issues. Yet in May 2020, a group of independent regulators, including the competition agency, issued a document on the regulatory challenges of the Paris Agreements and the climate emergency.[55] The report notes that '[t]his document, which is not an instrument of soft law but rather a roadmap, expresses the regulator's commitment to help undertakings adapt to the changes caused by the global warming'.[56] And while the document contains the clarification that 'the mandate given to the [French Competition Authority] is to protect competition, not to protect the environment', the president of the Authority has made it clear that the agency is working on the matter to ensure that the passing on of environmental benefits to consumers is assessed better.[57]

The *German* report notes that the agency has given individual guidance in sustainability-related cases, such as the Fairtrade label, Living Wages initiative for the Ecuadorian banana sector, and Initiative Tierwohl. And while it published a background paper[58] and addressed the issue in its annual report[59] and a press release,[60] it has not published guidelines. Similarly, the independent expert body, the Monopoly Commission, has addressed the issue in its annual report.[61] The agency has expressed resistance to the issue through more than individual guidance by means of guidelines.[62] It argued that there is not enough case law yet, that the effects are case specific and decisions need to be made on a case-by-case basis, that such guidance creates a risk of false negatives, and that it would not be the primary task of the agency to protect sustainability but competition and such decision should be made by the legislature. In fact, the German Federal Ministry for Economic Affairs and Climate Action, which is in charge of competition policy, has forwarded

---

[55] Accords de Paris et urgence climatique : enjeux de régulation, Accessed 29 November 2022.

[56] *See* French report section 1.3.

[57] B. Coeuré, Closing speech, New Frontiers of Antitrust, 21 June 2022.

[58] Hintergrundpapier - Arbeitskreis Kartellrecht, 1 Oct 2020 available at https://www.bundeskartellamt.de/SharedDocs/Publikation/DE/Diskussions_Hintergrundpapier/AK_Kartellrecht_2020_Hintergrundpapier.pdf?__blob=publicationFile&v=2. Accessed 29 November 2022.

[59] Bundeskartellamt, Tätigkeitsbericht 2019/2020 , available at https://www.bundeskartellamt.de/SharedDocs/Publikation/DE/Taetigkeitsberichte/Bundeskartellamt%20-%20T%C3%A4tigkeitsbericht%202019_2020.html. Accessed 29 November 2022.

[60] Bundeskartellamt, Achieving sustainability in a competitive environment – Bundeskartellamt concludes examination of sector initiatives, Press Release (18 Jan 2022), available at https://www.bundeskartellamt.de/SharedDocs/Meldung/EN/Pressemitteilungen/2022/18_01_2022_Nachhaltigkeit.html;jsessionid=1E2A8166777D61CFB1985BCA7F515B5E.2_cid381?nn=3591568Press/. Accessed 29 November 2022.

[61] Monopolkommission, Wettbewerb 2022 - XXIV. Hauptgutachten, available at https://www.monopolkommission.de/images/HG24/HGXXIV_Gesamt.pdf. Accessed 29 November 2022.

[62] See Bundeskartellamt, Tätigkeitsbericht 2019/2020, available at https://www.bundeskartellamt.de/SharedDocs/Publikation/DE/Taetigkeitsberichte/Bundeskartellamt%20-%20T%C3%A4tigkeitsbericht%202019_2020.html. Accessed 29 November 2022.

1 Sustainability and Competition Law: An International Report

10 points in its policy road map for competition policy until 2025.[63] These include steps to ensure a more sustainable competition policy with a legal framework that provides clarity for sustainability co-operation while addressing the risk of greenwashing. It will have to be seen what this will entail, but a change in the law might be possible.

In the *UK,* the agency has consulted, issued guidance on sustainability matters, and then established a sustainability task force. The report notes that this includes the (i) January 2021 information sheet on sustainability agreements, (ii) a market inquiry regarding electric vehicle charging points (including a subsequent investigation with commitments); (iii) provided advice to the UK Government on how the current competition and consumer law frameworks facilitate or hinder sustainability and Net Zero objectives; (iv) established a Sustainability Task Force to spearhead further engagement with these issues; and (v) under its consumer powers, has published a 'Green Claims Code' to help businesses and consumers avoid 'greenwashing' claims. Moreover, the competition agency committed to increase the number of market inquiries in the area.

The *Maltese* report notes that the agency has not issued any guidance or policy notes on the subject but should do so in the future. Moreover, it suggests that advocacy might be a particularly fruitful role as the advice given by the agency has been given a strong weight by the applicable legal framework. The report also highlights that in any competition matter, Malta's Sustainable Development Vision for 2050 will be a crucial point of reference. This 2018 policy document was published based on the Sustainable Development Act by the Ministry for the Environment, Energy, and Enterprise.

The *Italian* report does not mention any specific guidelines or policy documents. But for the authority, advocacy has been high on the agenda, and it has used its powers to propose measures where competitive pressure can help ensure more sustainable development with proposals in such areas as infrastructure for electrical vehicles, waste management, and energy market organisation.[64]

The *Swiss* agency has not issued any specific guidance or policy documents, but the report highlights that the agency has set up a 'Core Group Sustainability' that monitors the developments, both abroad and in Switzerland.

The *Hungarian* report notes the engagement of the agency with the topic as part of the European Competition Network (ECN) and in the International Competition Network (ICN), including drafting a report,[65] and they were involved in organising events. And while no substantive cases have been decided by the agency, the report

---

[63] Bundesministerium für Wirtschaft und Klimaschutz, Wettbewerbspolitische Agenda des BMWK bis 2025, 21.02.2022.

[64] *See* report S4143 of March 2021, Section V, Competition at the Service of Environmental Sustainability, S4143, p 65ff.

[65] *See* Hungarian Competition Authority, Special project for the 2021 ICN Annual Conference: Sustainable development and competition law, 30 September 2021, available at https://www.gvh.hu/en/gvh/Conference/icn-2021-annual-conference/special-project-for-the-2021-icn-annual-confer ence-sustainable-development-and-competition-law. Accessed 29 November 2022.

highlights some interesting procedural elements with regard to sustainability. Sustainability is considered in fining and in making commitments. The fining guidelines[66] and the commitment notice[67] take account of sustainability, and actions of companies to foster sustainability are described as contributing to consumer welfare. The report also mentions the possibility that the competition agency might bring a damages action in the public interests, where a large group of consumers is affected. Thus, it seems possible that the agency could bring a claim for the environmental damages caused by an anti-competitive action.

1.8 Other Competition-Law-Related Subjects and Competition

Countless legal tools might be named that address sustainability and its interaction with the market and market participants, which in turn may affect competition. While not the focus of this report, the national reports have highlighted other areas of law that are related to the core competition provisions, such as co-operation, cartels, abuse, monopolisation, and mergers. These areas of law might be relevant to the pursuit of sustainability and might directly affect competition between companies. The *Austrian*, *UK*, *Hungarian*, and *Maltese* reports noted environmental advertising, which is dealt with by consumer protection or via unfair competition law. The *Swiss* report highlighted the importance of public procurement rules and sustainable finance.

1.9 Conclusion

Upon reviewing the different national reports, it becomes clear once again that no one seems to argue that sustainability is a primary goal of competition law. However, in a vast majority of jurisdictions, competition law is required to account for sustainability impacts.[68] These obligations can be contained directly in competition laws—as in the examples from Austria and Hungary—and in specific laws—such as the climate law in Germany—or in the constitutions—as in the vast majority of the jurisdictions covered by this report.

The debate about sustainability and competition has been ongoing for some time. And to address the argument of legislative choices that is often heard in the debate, the legislature has acted in Austria and introduced a sustainability exemption. Germany might follow this example in the future. However, the Austrian experience also highlights how national changes to accommodate sustainability concerns also

---

[66] *See* Hungarian Fining Guidelines (1 January 2021) para 33(iv).

[67] *See* Commitment Notice of the Hungarian Competition Authority, 1 January 2021, para 19(h).

[68] Similar to the obligation of competition law to respect other fundamental rights enshrined in the different legal orders.

create challenges, particularly where national competition law is embedded in the European legal order.

Overall, the topic of sustainability and competition law seems to gain importance in all the jurisdictions covered. There are a number of cases reported across the jurisdictions, with some having experiences going back more than 20 years. And while it might be speculated that the total number of cases related to sustainability matters might be even higher than the number of cases covering data protection and competition, many jurisdictions still lack cases. However, all competition authorities in the jurisdictions covered seem to consider this an emerging issue that needs attention. The reports highlight that even when there are no direct experiences and activities reported by the agencies, it is expected that the agencies will become more active in the future.

The main point of difference and debate is the treatment of out-of-market efficiencies, in other words, situations where the beneficiaries in terms of sustainability do not overlap with the consumer affected by the restriction of competition.

The new EU guidelines on horizontal co-operation and their chapter on sustainability will be highly influential in all of the covered jurisdictions. The reports indicate that it will have this influence whether or not the jurisdictions are a part of the EU, whether the jurisdictions have already addressed sustainability and competition law, and whether or not they adopt a wait-and-see attitude.

The importance of these guidelines also becomes clear as the reports highlight the potential to deter measures that are beneficial from a sustainability point of view, thereby creating false positives. More legal certainty on the use of sustainability, whether as a shield or as a sword, is certainly called for. Once more certainty in the substantive field exists, the debate might also shift to more procedural questions, such as different forms of remedies, including public interest damage actions.

Based on the national reports and this international report, the LIDC adopted the following conclusions and recommendations:

- Competition law is certainly not the only, and may not be the best, tool to address sustainability issues, but it can play a role in the move towards a more sustainable society.
- The debate is beginning and/or increasing momentum in a growing number of jurisdictions. Some jurisdictions are more advanced than others in their thinking.
- Certain competition authorities are taking a lead, from advocacy on sustainability issues to publishing policy papers and guidelines. Certain authorities are proactively leading the debate and not waiting until the cases come to them.
- While there is still not an abundance of cases, the number of cases in the different jurisdiction is increasing.
- There are still numerous questions that remain unanswered, in particular—but not limited to—questions relating to the treatment of out-of-market efficiencies.
- Intervention by the legislator could address problems of (democratic) legitimacy and the balancing/weighing exercise to be undertaken by competition authorities.

- Companies should be encouraged to take up the opportunity offered by certain competition authorities to be provided with individual guidance.
- In order to provide greater transparency and clarity to industry, advisors, and society as a whole, it would be helpful if individual guidance were at least published as press release. The publication of more detailed/concrete documents other than press releases, such as individual guidance (comfort/business advice letters) would be preferrable from the perspective of transparency, legal certainty, and value for advising on matters of sustainability.
- Guidelines addressing sustainability, or even block exemptions, could provide even greater legal certainty, although this may be difficult, at least until there are sufficient examples/cases to draw upon.
- International sharing of experience between competition authorities/agencies around the world is highly recommended as they have experiences in different fields. Sustainability is a global issue, and this requires a global debate.
- The establishment of (global) best practices, possibly with a focus on standards—an area where a lot of experience exists—is strongly encouraged.

In terms of next steps and the role of the LIDC, all the factors outlined above merit further debate. Potential legislative and soft law solutions could be explored in the future. In the meantime, the dialogue on the interface of competition law and policy and sustainability should continue (between authorities, practitioners, industry, and interest groups). As an international organisation, the LIDC is well placed to lead and facilitate this discussion and will look for ways to further the international dialogue. Accordingly, the Scientific Committee shall add this to our work programme for 2023 and will explore opportunities for follow-up webinars and/or a working group on sustainability and competition.

Open Access This chapter is licensed under the terms of the Creative Commons Attribution 4.0 International License (http://creativecommons.org/licenses/by/4.0/), which permits use, sharing, adaptation, distribution and reproduction in any medium or format, as long as you give appropriate credit to the original author(s) and the source, provide a link to the Creative Commons license and indicate if changes were made.

The images or other third party material in this chapter are included in the chapter's Creative Commons license, unless indicated otherwise in a credit line to the material. If material is not included in the chapter's Creative Commons license and your intended use is not permitted by statutory regulation or exceeds the permitted use, you will need to obtain permission directly from the copyright holder.

Sustainability and Competition Law in Austria

2

Adrian Kubat and Adnan Tokić

2.1 Introduction

Sustainability and environmental protection are the pressing questions of our time. Austrian competition law keeps pace with these issues and addresses them under the light of both antitrust and unfair competition law.[1] Suitably, the amendment of the Austrian Cartel Act in 2021 took account of environmentally sustainable and carbon-neutral economies within its exemption provision. This led to a remarkable novelty in antitrust law. In particular, the legislator extended the view of consumer welfare as the Austrian exemption provision assesses the fair share for consumers in a future-oriented manner. In contrast to EU law, out-of-market efficiencies now fulfil the pass-on criterion under Austrian antitrust law.

2.2 Sustainability and Antitrust Law

"Consumers are also allowed a fair share of the resulting benefit, if the benefit derived from improving the production or distribution of goods or promoting technical or economic progress contributes significantly to an *environmentally sustainable or carbon-neutral* economy."[2]

---

[1] Adrian Kubat has written the section regarding unfair competition law, whereas the section on antitrust law has been written by Adnan Tokić.

[2] Section 2 para. 1 KartG (Austrian Cartel Act), amended by the Kartell- und Wettbewerbsrechts-Änderungsgesetz 2021, BGBl (Austrian Federal Law Gazette) I Nr 176/2021, which entered into force on 10 September 2021, Austrian equivalent to Article 101 para. 3 TFEU (emphasis by the author).

A. Kubat · A. Tokić (✉)
University of Vienna, Wien, Austria
e-mail: adrian.kubat@univie.ac.at; adnan.tokic@univie.ac.at

© The Author(s) 2024
P. Këllezi et al. (eds.), *Sustainability Objectives in Competition and Intellectual Property Law*, LIDC Contributions on Antitrust Law, Intellectual Property and Unfair Competition, https://doi.org/10.1007/978-3-031-44869-0_2

What appears to be the legal consideration of a "more sustainable economic approach", has found its way into the text of the Austrian Cartel Act since an amendment of the exemption provision in 2021.[3] This is the very first time for a Member State's antitrust law to explicitly address the issue of sustainability. The Austrian antitrust law thus plays a pioneering role in exempting agreements between undertakings that are committed to sustainability or climate neutrality. Given this unique selling point, this paper focuses on the amended exemption provision.

Most recently, the discussion on the assessment of sustainability agreements under antitrust law was animated by the European Green Deal actions.[4] Commissioner *Vestager* also picked up on these developments several times, hinting at a certain openness to the topic.[5] In addition, both national competition authorities and business representatives called for regulations to provide clarity on the compatibility of such arrangements with competition law.[6] The paper provides information on how sustainability agreements can be permitted in the light of the amendment of the Austrian Cartel Act. Furthermore, it relates the Austrian regulation to EU antitrust law as the latter is decisive for an appropriate assessment of the topic of interest. This is due to the primacy of application of EU law as well as its model effect for national antitrust laws. Beyond that, the Austrian Supreme Cartel Court has stated that the prohibition of anti-competitive agreements (Section 1 Cartel Act) and the corresponding exemption provision (Section 2 Cartel Act) are to be interpreted in the light of Article 101 TFEU.[7]

---

[3] See A. Ablasser-Neuhuber, Nationale Rechtsentwicklung, ÖBl 2021(4), p. 160; W. Barfuß, Das KaWeRÄG 2021 – Zum Geleit, ecolex 2021(10), pp. 879–880; T. Barth and S. Natlacen, Kartell- und Wettbewerbsrechts-Änderungsgesetz 2021, GesRZ 2021(4), pp. 198–199; A.-H. Bischke and S. Brack, Neuere Entwicklungen im Kartellrecht, NZG 2022(9), pp. 393–396; R. Fucik, Neues im Kartell- und Wettbewerbsrecht – Regierungsvorlage für ein KaWeRÄG, ÖJZ 2021(13), p. 593; J. P. Gruber, Das KaWeRÄG 2021 – Erster Teil: Kartellgesetz, ÖZK 2021(4), pp. 123–134; I. Hartung and A. Reidlinger, KaWeRÄG 2021: Die Anpassung von KartG und WettbG an die ECN+-Richtlinie, ecolex 2021(10), pp. 880–883; E. Hlina and H. Wollmann, KaWeRÄG 2021: Neuerungen in der Fusionskontrolle, ecolex 2021(10), pp. 887–889; A. Koprivnikar, Aktuelle Neuerungen im österreichischen Kartell- und Wettbewerbsrecht, WRP 2022(1), pp. 23–27; S. Köller-Thier, V. Strasser and G. Bauer, Neue Wege im Kartellrecht, ÖBl 2021(6), pp. 244–254; V. Robertson, Sustainability: A World-First Green Exemption in Austrian Competition Law, JECLAP 2022, available at https://doi.org/10.1093/jeclap/lpab092. Accessed 4 November 2022; V. Robertson, Kartellverbot und Nachhaltigkeit – zum neuen § 2 Abs 1 KartG (2021) available at https://doi.org/10.2139/ssrn.3957550. Accessed 4 November 2022; V. Robertson, Kartellrecht wird grün und digital, Die Presse 18 September 2021; C. Schumacher, Ein Sommer wie damals? ÖBl 2021(4), p. 145; V. Strasser, Kartellrecht und Green Deal – der österreichische Weg, WuW 2022(2), pp. 68–71; J. Zöchling, KaWeRÄG 2021 – Digitales Update für die Missbrauchskontrolle, ecolex 2021(10), pp. 885–887; M. Zöhrer and A. S. Reumann, KaWeRÄG 2021 – Ökologisierung des Kartellrechts? ecolex 2021(10), pp. 884–885.

[4] EC, The European Green Deal, COM(2019) 640 final.

[5] See D. Wouters, Which Sustainability Agreements Are Not Caught by Article 101 (1) TFEU, JECLAP 2021(3), pp. 257–270 (257).

[6] See Commission Staff Working Document, SWD(2021) 103 final, p. 19.

[7] OGH (Austrian Supreme Court) 26 June 2006, 16 Ok 51/05, *Asphaltmischanlage II*, ÖBl-LS 2007/27.

The effect of the climate crisis is increasingly felt in today's everyday life. Some therefore go so far as to speak of a "moral imperative" to counteract this global development with all possible means, antitrust law included.[8] However, we must acknowledge that an overly broad view of antitrust law is restricted.[9] After all, cost externalisation, information asymmetries or consumer cognitive biases often cause substantial market failures.[10] Sustainability agreements offer an opportunity to overcome this problem. To fall under the definition of sustainable agreements, agreements between undertakings must not exclusively aim at profit maximisation but above all make a contribution to sustainable development.[11] Such agreements are based on a modern and broad concept of sustainability, which, in addition to environmental protection, also includes, e.g., animal welfare or the pursuit of better social living conditions.[12] By their very nature, sustainability agreements that pursue such interests are based on welfare and solidarity considerations and make these their primary objective.[13]

In principle, Article 101 para. 1 TFEU prohibits agreements between undertakings, decisions by associations of undertakings and concerted practices that may affect trade between Member States and that have as their object or effect the prevention, restriction or distortion of competition within the internal market. Section 1 Cartel Act contains a corresponding prohibition. However, if such corporate conduct fulfils the conditions laid down in Article 101 para. 3 TFEU (respectively Section 2 para. 1 Cartel Act), it can be exempted from the ban on cartels. In addition, a specific exception can be found in Article 210a of the Regulation establishing a common organisation of the markets in agricultural products.[14] According to this, practices restricting competition that relate to agricultural products are permitted if they aim to apply a sustainability standard higher than that mandated by Union or national law.

The advantage of fast and unbureaucratic implementation of sustainability goals, which corporate agreements undoubtedly have over legislative measures, is offset by

---

[8] See S. Holmes, Climate change, sustainability, and competition law, JoAE 2020(8), pp. 354–405 (356).

[9] See R. Inderst and S. Thomas, Prospective Welfare Analysis – Extending Willingness-To-Pay Assessment to Embrace Sustainability, JOCLEC 2021, available at https://doi-org.uaccess.univie.ac.at/10.1093/joclec/nhab021. Accessed 4 November 2022; R. Inderst and S. Thomas, Reflective Willingness To Pay: Preferences for Sustainable Consumption in a Consumer Welfare Analysis, JOCLEC 2021(4), pp. 848–876 (871).

[10] See for the cost of market transaction R. H. Coase, The Problem of Social Cost, The Journal of Law & Economics 1960(III), p.1).

[11] A consistent definition has not yet emerged, see D. Wouters, Which Sustainability Agreements Are Not Caught by Article 101 (1) TFEU, JECLAP 2021(3), pp. 257–270 (257f).

[12] See UN Sustainable Development Goals, Resolution adopted by the General Assembly, 25 September 2015, A/RES/70/1, p. 14.

[13] A. Gerbrandy, Solving a Sustainability-Deficit in European Competition Law, World Competition 2017(4), pp. 539–562 (558).

[14] Regulation 1308/2013 of the European Parliament and of the Council of 17 December 2013 establishing a common organisation of the markets in agricultural products, OJ 2013 L 347, 671; the provision was implemented by Regulation 2021/2117 of the European Parliament and of the Council of 2 December 2021, OJ 2021 L 435, 298.

the concern that competition law could be misused for greenwashing purposes.[15] The fear is that companies will green-light agreements that in fact restrict competition to evade antitrust law. In favor of the exemption, it is frequently argued that first-mover disadvantages can be overcome by synergy effects.[16] Yet this argument is not convincing as the sustainability awareness of today's consumers is already advanced. Nowadays, consumers reward innovative corporate concepts that promote a sustainable economy. This is why it is rather correct to assume a first-mover advantage than a first-mover disadvantage in the given context.[17]

These considerations already show that the antitrust framework for the exemption of sustainability agreements is not clear and is therefore characterised by substantial uncertainties. In addition, undertakings must evaluate on a self-assessment basis whether their agreements fall under the legal exception of Article 101 para. 3 TFEU (Section 2 para. 1 Cartel Act). The lack of case law or specific guidelines contributes to further legal uncertainty. Against this background, undertakings are risk averse in promoting sustainability through agreements, which promotes inefficiency with respect to the legal exception of cartel prohibition.[18] In the following, the conditions that allow for an exemption of sustainability agreements are analysed in light of the amended Austrian antitrust law with reference to the relevant EU provisions on the basis of the four renowned criteria for exempting corporate agreements.

2.3 Exemption Under Section 2 Para. 1 Cartel Act (Article 101 Para. 3 TFEU)

Section 2 para. 1 Cartel Act as well as Article 101 para. 3 TFEU offer a possibility for promoting sustainability through agreements between undertakings. To be exempt from the prohibition of restrictive agreements, an agreement must meet four cumulative criteria: there must be (i) a contribution to improving the production or distribution of goods or to promoting technical or economic progress while (ii) allowing consumers a fair share of the resulting benefit, without (iii) imposing restrictions that are not indispensable to the attainment of these objectives on the undertakings concerned and (iv) affording such undertakings the possibility of eliminating competition in respect of a substantial part of the products in question.

---

[15] M. Gassler, Sustainability, the Green Deal and Art 101 TFEU: Where We Are and Where We Could Go, JECLAP 2021(6), pp. 430–442 (433); A. Gerbrandy, Solving a Sustainability-Deficit in European Competition Law, World Competition 2017(4), pp. 539–562 (543).

[16] S. Holmes, Climate change, sustainability, and competition law, JoAE 2020(8), pp. 354–405 (367).

[17] E. Loozen, Strict Competition Enforcement and Welfare: A Constitutional Perspective Based on Article 101 TFEU and Sustainability, CMLR 2019(56), pp. 1265–1302 (1284).

[18] G. Monti, Four Options for a Greener Competition Law, JECLAP 2020(3–4), pp. 124–132.

2 Sustainability and Competition Law in Austria

The exemption test is therefore based on an economic approach.[19] The economic aim is to allow a restriction of competition only if cheaper or better products are offered to consumers, compensating the latter for the adverse effects of the restrictions of competition.[20] In other words, an agreement must always increase consumer welfare and must not be at the expense of consumers to be efficient with respect to cost allocation and thus exemptible. The amendment of the Cartel Act led to a remarkable change regarding the exemption of cartels in Austria. In particular, the view of consumer welfare is significantly broadened as out-of-market efficiencies now take precedence over the pass-on criterion. In general, the legislative materials classify sustainability as a cross-cutting issue but do not see competition law as the primary key to solving the climate crisis. This is why the provision's central intention is to provide legal certainty. Sustainability and environmental protection aspects are thus firmly anchored in the renewed Cartel Act.[21]

2.4 Efficiencies

The first criterion requires an improvement in the production or distribution of goods or the promotion of technical or economic progress. Thus, efficiencies must necessarily be at hand in order to permit sustainability agreements. The accurate narrow view of the provision permits only the consideration of improvements in economic efficiency.[22] Non-economic benefits alone cannot justify an exemption.[23] These normative premises apply to both EU and Austrian antitrust laws.

Despite the amendment, the Austrian exemption provision continues to be based on the economic system predetermined by Article 101 para. 3 TFEU. In particular, it does not establish a new, independent sectoral exemption.[24] According to the examples given by the legislative materials, environmental sustainability is understood to include climate protection (use of renewable energy, reduction of greenhouse gas emissions), sustainable use and the protection of water resources

---

[19] With regard to EU law R. Whish and D. Bailey, Competition Law, 10th ed, Oxford University Press 2021, p. 165; with regard to European and Austrian law A. Tokić, Freistellungsfähigkeit von Nachhaltigkeitsvereinbarungen nach europäischem und novelliertem österreichischen Kartellrecht, wbl 2022(6), pp. 301–313.

[20] EC, Guidelines on the application of Article 81(3) of the Treaty (Article 101(3) AEUV), 2004/C 101/08, OJ 2008 C 101, p. 102 at recital 34.

[21] ErläutRV (Legislative materials to the government bill) 951 BlgNR (SupplementsNo.) 27. GP (Legislative period) 8.

[22] EC, Guidelines on the application of Article 81(3) of the Treaty (Article 101(3) AEUV), at recital 57 and Guidelines on Vertical Restraints, 2010/C 130/01, OJ 2010 C 130, p. 27 at recital 122.

[23] Controversial, see R. Inderst and S. Thomas, Reflective Willingness To Pay: Preferences for Sustainable Consumption in a Consumer Welfare Analysis, JOCLEC 2021(4), pp. 848–876 (873); in detail A. Tokić, Freistellungsfähigkeit von Nachhaltigkeitsvereinbarungen nach europäischem und novelliertem österreichischen Kartellrecht, wbl 2022(6), pp. 301–313.

[24] See S. Köller-Thier, V. Strasser and G. Bauer, Neue Wege im Kartellrecht, ÖBl 2021(6), pp. 244–254 (250).

(protection of the environment from the adverse effects of wastewater discharge), transition to a circular economy (promotion of the reparability and recyclability of products, increased use of secondary raw materials) and protection and restoration of biodiversity and ecosystems (sustainable forest management). The concept of climate neutrality is in itself covered by this definition but was, however, specifically included in the wording due to the particular importance of this aspect.[25]

The legislative materials state that agreements for an ecologically sustainable economy do not necessarily fall under the prohibition of cartels, e.g. if standardisation agreements are open and non-exclusive and their participation remains voluntary.[26] However, if an agreement falls within the scope of Section 1 Cartel Act, the first exemption criterion it has to meet is the existence of economic efficiencies.[27] According to the legislator, such efficiencies are given if, as a result of the agreement, the production method or the distribution is CO_2 saving or if ecologically more sustainable or less CO_2-emitting products or services are created. Examples include the use of exhaust gas or wastewater filters in production (improving the production of goods), joint distribution to reduce transport costs (improving the distribution of goods) or the production of cars that emit less CO_2 (promoting technical progress). On the surface, this refers to resource-saving practices, which also constitute objective economic advantages under EU law in the sense of the first exemption requirement.[28] If, on the other hand, economic progress is being promoted by the agreement, this is usually reflected in cost savings or a qualitative improvement in the case of sustainability-related innovations.

As set out by the legislative materials, the first condition is fulfilled by the contribution to sustainability only if it is associated with efficiencies as described. However, the new regulation denies exemptions to hardcore cartels such as price fixing or market division, even in the case of positive sustainability effects. Following the case law on Article 101 para. 3 TFEU, under Austrian law, there must also be appreciable, objective advantages to sustainability so as to compensate for the disadvantages that such an agreement entails for competition. The Austrian legislator considers positive effects on sustainability, the environment or climate for the public to be in proportion to the disadvantages of the competitive agreement on the relevant market. Thus, such effects correspond to the notion of advantages set out by EU case law.[29]

Under these considerations, the established EU assessment schema with its four exemption criteria is maintained under Austrian law. That applies in particular with regard to the efficiency requirement.[30] This is due to the fact that non-competition

---

[25] ErläutRV 951 BlgNR 27. GP 9.

[26] Sustainability agreements are similarly assessed under Union law, see EC, Guidelines on the applicability of Article 101 TFEU to horizontal co-operation agreements, 2011/C 11/01, OJ 2011 C 11, p. 59 at recital 257ff.

[27] ErläutRV 951 BlgNR 27. GP 8.

[28] F. Schuhmacher, Article 101 TFEU, In: E. Grabitz, M. Hilf and M. Nettesheim (eds), Das Recht der Europäischen Union: EUV/AEUV, 75th ed, C. H. Beck 2022, at recital 267ff.

[29] ErläutRV 951 BlgNR 27. GP 9 f.

[30] A. Koprivnikar, Aktuelle Neuerungen im österreichischen Kartell- und Wettbewerbsrecht, WRP 2022(1), pp. 23–27 (26).

interests meet the first criterion under Section 2 para. 1 Cartel Act insofar as they are linked to objective economic advantages.[31] The examples cited by the Austrian legislator underline this fact as all such examples promote sustainability. In addition, there is always an improvement in the production or distribution of goods or in technical or economic progress. However, no accurate statement can be extracted from the legislative materials that mere sustainability benefits would replace economic efficiencies at the level of the first criterion. The legislator's intention remains unclear in this respect. This leads to the conclusion that Section 2 para. 1 Cartel Act continues to require economic efficiencies as a first prerequisite. Thus, the legislator has not changed the assessment system in this context. A contribution to an ecologically sustainable or climate-neutral economy still increases efficiency only if it leads to economic benefits.

2.5 Fair Share for Consumers

Pursuant to Section 2 para. 1 Cartel Act, consumers are now allowed a fair share of the resulting benefit of an agreement if the benefit derived from improving the production or distribution of goods or promoting technical or economic progress contributes significantly to an *environmentally sustainable or carbon-neutral* economy. The Austrian legislator takes a broad view of the second criterion, according to which consumers must be allowed a fair share of the resulting benefit. The legislative materials state that the effects of an ecologically sustainable or climate-neutral economy always benefit the public as a whole, even if this may only be the case with a time lag for future generations. In the view of the legislator, this is justified because the consumer group concerned in each case represents a part of the public and therefore benefits from such an environmentally sustainable or carbon-neutral economy.[32]

This is an essential difference to EU law. Pursuant to Article 101 para. 3 TFEU, the benefits of an agreement must compensate for the disadvantages of the consumer groups specifically affected.[33] In contrast, welfare increases are now recognised under Section 2 para. 1 Cartel Act, even if the advantages of an agreement do not directly benefit the same group of affected consumers. The same stands for benefits that only accrue in the future. Out-of-market efficiencies that are exclusively associated with welfare increases for future generations are thus taken into account when examining the fair share for consumers in Austria. At the same time, the need

---

[31] Reaching a different conclusion V. Robertson, Kartellverbot und Nachhaltigkeit – zum neuen § 2 Abs 1 KartG (2021) available at https://doi.org/10.2139/ssrn.3957550. Accessed 4 November 2022; V. Strasser, Kartellrecht und Green Deal – der österreichische Weg, WuW 2022(2), pp. 68–71 (70).

[32] ErläutRV 951 BlgNR 27. GP 8 f.

[33] See for instance E. Van den Brink and J. Ellison, Article 101(3) TFEU: the Roadmap for Sustainable Cooperation, In: S. Holmes, D. Middelschulte and M. Snoep (eds), Competition Law, Climate Change & Environmental Sustainability, Concurrences 2021, pp. 39–54 (43).

for rapid action on climate and environmental protection is met. In this way, the Cartel Act allows for a broader view of consumer compensation compared to primary EU law.

Sustainability aspects beyond ecologically sustainable or climate-neutral benefits are not explicitly covered by the wording of Section 2 para. 1 Cartel Act. However, the conclusion that, e.g., better social working conditions or increased animal welfare cannot be taken into account at all would not be accurate.[34] In conjunction with a corresponding economic advantage at the level of the first criterion described above, they can be included in the exemption test and, in qualitative economic terms, also increase consumer welfare. However, the disadvantages of the consumers specifically affected must in that case still be compensated for. This does not detract from the consideration of such sustainability advantages as long as they lead to a better product for the affected consumer group, even if the product becomes more expensive.

If the benefits of a sustainability agreement occur at a later point in time, they must be discounted in the course of their assessment. This requires the use of an appropriate discount rate for the purpose of quantification.[35] The legislator's position is noteworthy in this respect, especially since the quantification of sustainable efficiencies is seen as one of the greatest challenges in the given context. The legislative materials assume that, at least for the time being, not every contribution to environmental sustainability can be represented in exact figures. This is also not necessary in all cases to be able to carry out an appropriate balancing of interests between the restriction of competition and ecological sustainability. The assessment cannot rely on purely quantitative aspects. If the effects of the restriction of competition are expected to be only slightly detrimental whereas the contribution to an environmentally sustainable or carbon-neutral economy is evidently positive, it is - according to the legislator - possible to dispense with quantifying the environmental benefit.[36] The wording of Section 2 para. 1 Cartel Act allows for such an understanding as the contribution to an environmentally sustainable or carbon-neutral economy must be "substantial" to provide a fair share for consumers. Within the framework of a required objective assessment *ex ante*, a predominant probability of

---

[34] Similarly V. Robertson, Sustainability: A World-First Green Exemption in Austrian Competition Law, JECLAP 2022, available at https://doi.org/10.1093/jeclap/lpab092. Accessed 4 November 2022; V. Strasser, Kartellrecht und Green Deal – der österreichische Weg, WuW 2022(2), pp. 68–71 (69).

[35] J. Nowag, OECD Competition Committee Discussion Paper, 2020, p. 24; R. Inderst and S. Thomas, Prospective Welfare Analysis – Extending Willingness-To-Pay Assessment to Embrace Sustainability, JOCLEC 2021, available at https://doi.org/10.1093/joclec/nhab02 (19). Accessed 4 November 2022.

[36] ErläutRV 951 BlgNR 27. GP 10; see already before Autoriteit Consument & Markt, Second Draft Version: Guidelines on Sustainability Agreements (2021) at recital 54ff, available at https://www.acm.nl/sites/default/files/documents/second-draft-version-guidelines-on-sustainability-agreements-oppurtunities-within-competition-law.pdf; Accessed 4 November 2022 similarly D. Wouters, Which Sustainability Agreements Are Not Caught by Article 101 (1) TFEU, JECLAP 2021(3), pp. 257–270 (268).

providing a fair share to the consumers is necessary. The benefits to be passed on and the time of their occurrence must also be specified.[37]

2.6 Indispensability of the Restrictions

In order to be exempted, the restriction of competition must be indispensable, i.e. it must not be (economically) feasible to achieve the agreement's sustainability goals through less restrictive manners. The legislative materials refer in this context to an agreement regarding compressed and less waste-generating packaging. Such packaging would in the view of the legislator not be able to prevail over conventional packaging on the market without the agreement.[38] This example underlines the fact that in the sustainability context, new products are regularly brought to the market. They may not be able to withstand competitive pressure if a single company manufactures, distributes or introduces them to the market. Yet this inability is a case not of first-mover disadvantage but of classic market failure. Due to information asymmetries, consumers lack relevant knowledge regarding product packaging and might thus wrongly infer a smaller content from a smaller package.[39] If the companies were to provide – economically efficient – sufficient information about the environmentally friendly packaging technology and made it clear that this has no effect on content quantity, the supposed competitive disadvantage would turn into a first-mover advantage.

Of course, one might think of cases in which sustainability agreements provide advantages over individual entrepreneurial initiatives. In principle, it should be noted in the proportionality test that such agreements are usually only indispensable for achieving objectives in a start-up-like phase and may become inadmissible over time. This is particularly the case with new legislative measures that raise the general level of protection in environmental and sustainability issues.[40] In these cases, the benefits of an agreement can only exceed the indispensability threshold if they go beyond the minimum level of protection standards mandated by law.[41] In this respect, the assessment of indispensability represents a strict greenwashing test in the sustainability context.

---

[37] F. Schuhmacher, Article 101 TFEU, In: E. Grabitz, M. Hilf and M. Nettesheim (eds), Das Recht der Europäischen Union: EUV/AEUV, 75th ed, C. H. Beck 2022, at recital 322.

[38] ErläutRV 951 BlgNR 27. GP 10.

[39] O. Schley and M. Symann, Art. 101 Abs. 3 AEUV goes green, WuW 2022(1), pp. 2–7 (3).

[40] O. Schley and M. Symann, Art. 101 Abs. 3 AEUV goes green, WuW 2022(1), pp. 2–7 (6); D. Wouters, Which Sustainability Agreements Are Not Caught by Article 101 (1) TFEU, JECLAP 2021(3), pp. 257–270 (262).

[41] Austrian National Competition Authority, Stellungnahme der Bundeswettbewerbsbehörde zum Entwurf des Kartell- und Wettbewerbsrechts-Änderungsgesetzes (2021) p. 39, available at https://www.bwb.gv.at/fileadmin/user_upload/BWB__XXVII_SNME_97529_1_Volltext.pdf. Accessed 4 November 2022; similarly Article 210a Regulation 1308/2013.

2.7 No Elimination of Competition

The contribution of antitrust law to the development of a sustainable economy becomes particularly effective when as many companies as possible participate in private-sector measures. However, the admissibility of an agreement must not allow companies to eliminate competition for a substantial part of the market. The legislative materials use an agreement between transport companies to switch to more ecologically sustainable fuel in the future as an example for this criterion.[42] While price-fixing agreements with regard to tickets would not be justified, the obligation to use a certain fuel only, but not to agree on ticket prices or to adhere to exclusive routes, constitutes no hardcore restriction. Competition cannot be eliminated by such agreements, so that the market structure is preserved and sufficient room remains for potential competition. Against this background, an individual exemption is legitimate if the above-mentioned criteria are fulfilled.

2.8 Evaluation of the Amended Austrian Cartel Act

The Austrian National Competition Authority (NCA) has published Draft Guidelines on the Application of Section 2 para. 1 Cartel Act on Sustainability Agreements. These "Sustainability Guidelines" provide a specific assessment schema for the exemption of sustainable agreements between undertakings. It determines five criteria, which an agreement has to fulfil in order to fall under the sustainability exception. The agreement therefore must (i) lead to efficiencies that (ii) contribute to an environmentally sustainable or carbon-neutral economy, while (iii) such contribution is significant and the agreement does not (iv) impose on the undertakings concerned restrictions that are not indispensable to the attainment of these objectives as well as it does not (v) afford such undertakings the possibility of eliminating competition in respect of a substantial part of the products in question.[43]

Despite its Sustainability Guidelines, the Austrian NCA is quite critical of the amendment of Section 2 para. 1 Cartel Act. In its statement on the (at the time) proposed competition law, it points to free competition as a guarantor of innovation and emphasises that empirical studies have regularly found that sustainability cartels result, in fact, in low investments in sustainability. Moreover, antitrust law could serve the undertakings as a tool to evade the implementation of more comprehensive environmental protection measures. Therefore, the NCA doubts the usefulness of a legal exception to the ban on cartels to achieve sustainability goals. Instead, it argues

---

[42] ErläutRV 951 BlgNR 27. GP 10.

[43] Austrian National Competition Authority, Leitlinien zur Anwendung von § 2 Abs 1 KartG auf Nachhaltigkeitskooperationen (Nachhaltigkeits-LL) (2022) p. 26, available at https://www.bwb.gv.at/fileadmin/user_upload/Nachhaltigkeits-LL_fuer_oeff_Konsultation_01.06.2022.pdf. Accessed 4 November 2022.

2 Sustainability and Competition Law in Austria

for the prioritised prosecution of environmentally harmful agreements.[44] Such agreements were already examined under EU law.[45]

This critical view is offset by the urgent need for effective and targeted action in environmental and sustainability matters. Moreover, the amended Cartel Act continues to require economic efficiencies as the first criterion for an exemption. As a result, Section 2 para. 1 Cartel Act rightly retains the economic system of resource allocation predetermined by Article 101 para. 3 TFEU, including its four exemption criteria. The Austrian regulation primarily differs from EU law with respect to the assessment of the fair share for consumers. Pursuant to the amended provision, a contribution to an ecologically sustainable or climate-neutral economy no longer has to occur on the specific market where competition is restricted. This legislative decision aims to meet the needs of the current generation without compromising the ability of future generations.[46] This forward-looking understanding of the pass-on condition is utterly welcome.

According to the opinion expressed here, the renewed provision does not open a gateway for recognising non-competition interests without linking them to economic advantages. While sustainability agreements by competitors were already permissible under the former legal situation if these conditions were strictly met, the amendment now contains a clear commitment by the legislator to the compatibility of sustainability and antitrust law. Apart from the expanded view of consumer welfare, there are ultimately no significant legal changes. Rather, the amendment plays an important role in clarifying the law and strengthening legal certainty. Due to the primacy of EU competition law, some do not expect a more frequent application of Austrian antitrust law in the future. The broadly inter-State clause fits this view.[47] However, this view ignores the fact that environmentally sustainable initiatives often occur on a regional level and do not necessarily affect trade between Member States.[48] Thus, an assessment of sustainability agreements under national law is not unlikely and welcome.

Ultimately, sustainability agreements regularly run the risk of violating competition law due to the narrow view of the legal exemption criteria. Therefore, they can only make a limited contribution to sustainable economic development and to the

---

[44] Austrian National Competition Authority, Stellungnahme der Bundeswettbewerbsbehörde zum Entwurf des Kartell- und Wettbewerbsrechts-Änderungsgesetzes (2021) p. 40.

[45] EC 8 July 2021, COMP/AT.40178, *Car Emission* (delay in the introduction of exhaust gas purification systems); EC 8 February 2017, COMP/AT.40018, *Car battery recycling* (agreement not to recycle car batteries); EC 13 April 2011, COMP/AT.39579, *Consumer detergents* (price-fixing agreements in an environmental initiative concerning washing powder); CJEU 8 November 1983 joined cases 96-102, 104, 105, 108 und 110/82, *IAZ*, ECLI:EU:C:1983:310 (market entry barriers to prevent parallel imports instead of labeling for environmentally friendly washing machines).

[46] ErläutRV 951 BlgNR 27. GP 9.

[47] J. P. Gruber, Das KaWeRÄG 2021 – Erster Teil: Kartellgesetz, ÖZK 2021(4), pp. 123–134 (124).

[48] V. Strasser, Kartellrecht und Green Deal – der österreichische Weg, WuW 2022(2), pp. 68–71 (70).

solution to the climate crisis because of the risk aversion of undertakings in this context. Legislative measures are, on the other hand, much better suited to achieving these goals and are of utmost importance, given the urgency of the issue. This is also confirmed by microeconomic studies,[49] according to which there are indeed cases in which entrepreneurial cooperation initially leads to positive results. However, as shown above, these constellations rarely occur, which is why legislative measures on sustainability issues are more efficient. One of the main economic reasons is that companies basically have no incentive to invest in sustainability unless the law induces them to do so.

As a result, the amendment of the Cartel Act makes it clear that – regarded only from a competition law point of view – no change in law or treaty is necessary for antitrust law to contribute to sustainable economic development and climate change mitigation. Nevertheless, a clarification of the current law along the lines of the Austrian model is the best way to overcome undertakings' risk aversion towards sustainable agreements. On the EU level, this refers to a renewal of the Commission's block exemption regulations and guidelines.[50] In this context, the new version of the Cartel Act can serve as an impetus for a modernisation of the Union-wide application practice of antitrust law.

2.9 Environmental Advertising and Unfair Competition Law

The fact that climate and environmental protection is of utmost importance for the majority of the population is no longer a secret and has become a more important goal than ever since the increasing climate change. Nonetheless, it is equally evident that climate and environmental protection does not stop at the advertising industry, and consumers are therefore confronted with a variety of environment-related advertisements in their everyday shopping, whether it is the plastic bottle that supposedly consists of 50% recycled plastic from the sea,[51] the "all-natural" jam[52] or the organic bread[53] from the local bakery. Such advertising slogans have a certain attraction and can ultimately lead to making a decision to purchase depending on the supposed positive aspects of the product for the environment. Where these environmental claims are actually true, no particular problems arise as it is in the nature of competition to emphasise positive aspects of one's own product and thus achieve corresponding competitive advantages. The same applies to (truthful) references to

---

[49] See J. Nowag, OECD Competition Committee Discussion Paper, 2020, p. 18.

[50] As this paper was submitted in 2022, the Commission's Block Exemption Regulations as well as the Guidelines on the Applicability of Article 101 TFEU to horizontal co-operation agreements amended in 2023 were not considered in the paper.

[51] OGH 4 Ob 114/18g, *Ocean Bottle II*, ecolex 2018/449 (M. Horak).

[52] CJEU 4 April 2000, case C-465/98, *Darbo*, ECLI:EU:C:2000:184.

[53] OGH 4 Ob 94/09s, *Österreichs bester Biobäcker*, ÖBL-LS 2010/7.

2 Sustainability and Competition Law in Austria

the environmental commitment of a company, which could subsequently improve its reputation compared to competitors.

However, environmental advertising is certainly a problem when the information does not correspond to reality or is deliberately formulated in a very vague and ambiguous way to imply supposed environmental compatibility.[54] In times of ecological transition and increasing public awareness of the need for climate protection, this so-called greenwashing is booming and is therefore reason enough to be featured in this paper. The strongest weapon against this form of advertising in Austria has always been the Unfair Competition Act (UWG),[55] which will be examined in more detail below based on a selection of relevant cases.

2.10 Applicability of the UWG to Environmental Advertising

In order to discuss any particularities with regard to environmental advertising, it must first be clarified under which conditions the UWG applies at all. The general requirement for the application of the Unfair Competition Act is acting in the course of trade. According to the case law of the Austrian Supreme Court (OGH),[56] this is to be understood as any independent activity aimed at making a profit, insofar as it goes beyond purely private or official activities. Thus, a broad understanding is appropriate, which in particular not only is based on the intention to make a profit but also includes charitable activities. Furthermore, it is not only those who promote their own competition who are acting in the course of trade but also those who engage in conduct that is objectively suited to promote or hinder competition by others.[57] Thus, for example, a municipality may also be acting in the course of trade if it wishes to promote the use of solar energy and, for this purpose, recommends a certain provider of solar energy systems in an announcement, which, of course, favours the provider's advancement.[58] Furthermore, according to case law, a person who is paid by an entrepreneur for advertising activities or who, even if free of

---

[54] In detail A. Anderl and A. Appl, In: Wiebe and Kodek (eds), UWG, 2nd ed, Manz 2016, Sec 2 paras 250f; A. Anderl and A. Ciarnau, Green & Blue Washing – Die Grenzen des Marketings, In: Zahradanik and Richter-Schöller (eds), Handbuch Nachhaltigkeitsrecht, Manz 2021, pp. 73-100; E. Artmann, Wettbewerbsrecht und Umweltschutz, Manz 1997; M. Hofer and I. Amschl, Greenwashing und UWG: Ein Überblick, NR 2021(4) pp. 421-427; G. Kucsko, Über irreführende Umweltengel, ecolex 1990(2) pp. 93-64; R. Mahfoozpour and G. Staber, Spannungsfeld "Greenwashing": Eine Rechtsprechungsübersicht, NR 2022(1), pp. 62-69; F. Rüffler, Umweltwerbung und Wettbewerbsrecht I, ÖBl 1995(6) pp. 243-259; F. Rüffler, Umweltwerbung und Wettbewerbsrecht II, ÖBl 1996(1), pp. 3-12.

[55] Bundesgesetz über den unlauteren Wettbewerb 1984 – UWG, BGBl 448/1984.

[56] Exemplarily OGH 4 Ob 2007/96t, *Cliniclowns*, ÖBl 1996, 191; countless other decisions at L. Wiltschek and M. Horak, UWG, 8th ed, Manz 2016; Sec 1 UWG paras 131f.

[57] R. Heidinger, In: Wiebe and Kodek (eds), UWG, 2nd ed, Manz 2016, Sec 1 paras 99f.

[58] BGH I ZR 54/11, *Solarinitiative*, GRUR 2013, 301.

charge, exclusively emphasises the advantages of certain products in a euphoric manner typical of advertising is acting in the course of trade.[59] According to the German case law, which is comparable with regard to the legal situation, it is sufficient to set a hyperlink in order to be within the advertising sphere of the respective entrepreneur, which is particularly relevant for all forms of native advertising.[60]

Nevertheless, the objectives pursued by the act are irrelevant as the subjective element of acting for the purpose of competition was eliminated in the course of the 2007 amendment to the UWG, and therefore the intention to promote competition associated with it may no longer be taken into account.[61] Instead, the UWG follows a functional understanding, which demands an interpretation exclusively on the ground of the protective purposes[62] of the UWG (protection of competitors, consumer protection and the interest of the general public in undistorted competition).[63] Non-competitive interests are therefore excluded, which for the environmental advertising relevant here means that the effect on environmental interests alone is not sufficient for the applicability of the UWG but that there must also be a corresponding market connection of the action.[64] The same applies in principle to the case of an animal protection association that points out animal suffering associated with battery hen farming and calls on people to buy free-range eggs instead.[65] Here in particular, however, the fundamental rights dimension must also be taken into account as there is an obvious conflict with freedom of expression (Article 10 ECHR), which also protects exaggerated and controversial statements. [66] The OGH therefore excluded the call in question from the scope of application of the UWG.[67] Although the appeal may have an indirect effect on competition, the result of the OGH is nevertheless convincing as the freedom of expression imposes a systematic barrier that does not conflict with a functional interpretation.[68] Conversely, the woven fur manufacturer who refers to animal suffering caused by real

---

[59] BGH I ZR 90/20, *Influencer I*, GRUR 2021, 1400; I ZR 126/20, *Influencer III*, GRUR-RS 2021, 26632.

[60] BGH I ZR 90/20, *Influencer I*, GRUR 2021, 1400; I ZR 126/20, *Influencer III*, GRUR-RS 2021, 26632; see A. Kubat, Drei neue Nachrichten aus Karlsruhe zum Influencer-Marketing, ÖBl 2022(3) pp. 100-108 (102f).

[61] R. Heidinger, In: Wiebe and Kodek (eds), UWG, 2nd ed, Manz 2016, Sec 1 para 140.

[62] R. Heidinger, In: Wiebe and Kodek (eds), UWG, 2nd ed, Manz 2016, Sec 1 para 5.

[63] H.-G. Koppensteiner, Wettbewerbsrecht II, 2nd ed, Orac 1987, pp. 232f; F. Schuhmacher, Überlegungen zum Handeln im geschäftlichen Verkehr, wbl 2016(11), pp. 601-612 (606); F. Rüffler, Umweltwerbung und Wettbewerbsrecht I, ÖBl 1995(6), pp. 243-259 (p. 243).

[64] F. Rüffler, Umweltwerbung und Wettbewerbsrecht I, ÖBl 1995(6), pp. 243-259 (p. 244).

[65] OGH 4 Ob 99/92, *Tierschutzverein*, wbl 1993, 195.

[66] In detail W. Berka, Der Schutz der freien Meinungsäußerung im Verfassungsrecht und Zivilrecht, ZfRV 1990(1), pp. 35-60.

[67] OGH 4 Ob 99/92, *Tierschutzverein*, wbl 1993, 195.

[68] See F. Schuhmacher, Überlegungen zum Handeln im geschäftlichen Verkehr, wbl 2016(11), pp. 601-612 (604).

2 Sustainability and Competition Law in Austria

fur at a trade fair is very much subject to the UWG[69] because he uses the statement as an advertising medium for the sale of his own products, and commercial communication enjoys less protection with regard to freedom of expression.[70] Especially in the typical forms of environmental advertising, i.e. advertising one's own reputation or product, this is usually unproblematic, which is why the applicability of the UWG can regularly be affirmed.[71]

2.11 Prohibition of Misleading Statements (Section 2 UWG)

The key provision for assessing the legitimacy of environment-related advertising within the UWG is Section 2, which regulates misleading business practices. According to this, it is unfair to use business practices that either contain incorrect information or are otherwise suitable to deceive market participants about certain characteristics of a product in such a way that they are induced to make a business decision that they would not have made otherwise.[72] In addition, the annex to the UWG, the so-called black list, contains a number of business practices that are unfair in any case, even without the addition of further circumstances. In certain cases, for instance when considerable pressure is exerted on consumers,[73] it would also be conceivable to apply Section 1a UWG to environment-related advertising or to consider an unfair advantage by breach of law (Section 1 UWG) if a business practice violates legal norms, like those for the protection of the environment, since this typically results in a distortion of competition.[74] However, the vast majority of the cases dealt with in case law are treated from the point of view of the prohibition of misleading statements, which is why the further explanations will also focus on this. Following the categories established by *Rüffler*,[75] a distinction will especially be made between product-related and company-related environmental advertising.

---

[69] OGH 4 Ob 94/91, *Webpelz I*, JBl 1993, 330 (W. Berka); 4 Ob 2118/95, *Webpelz II*, RdW 1996, 409.

[70] W. Berka, Der Schutz der freien Meinungsäußerung im Verfassungsrecht und Zivilrecht, ZfRV 1990(1), pp. 35-60 (p. 51).

[71] F. Rüffler, Umweltwerbung und Wettbewerbsrecht I, ÖBl 1995(6), pp. 243-259 (p. 247).

[72] In detail A. Anderl and A. Appl, In: Wiebe and Kodek (eds), UWG, 2nd ed, Manz 2016, Sec 2.

[73] See A. Anderl and A. Ciarnau, Green & Blue Washing – Die Grenzen des Marketings, In: Zahradanik and Richter-Schöller (eds), Handbuch Nachhaltigkeitsrecht, Manz 2021, pp. 73-100 (89f); M. Hofer and I. Amschl, Greenwashing und UWG: Ein Überblick, NR 2021(4), pp. 421-427 (425).

[74] F. Rüffler, Umweltwerbung und Wettbewerbsrecht I, ÖBl 1995(6), pp. 243-259 (p. 246); M. Hofer and I. Amschl, Greenwashing und UWG: Ein Überblick, NR 2021(4), pp. 421-427 (425).

[75] F. Rüffler, Umweltwerbung und Wettbewerbsrecht I, ÖBl 1995(6), pp. 243-259 (p. 245).

2.12 Product-Related Advertising

A typical form of environmental advertising is product-related advertising, i.e. highlighting the supposedly environmentally friendly aspects of a product.[76] Practically, one encounters a whole range of buzzwords, all of which serve to draw the consumer's interest in the product in question. Products are often only very generally described as "organic", "green" or "naturally pure", while sometimes reference is made to the fact that the ingredients or the production process is particularly environmentally friendly, for example if recycled materials are used or the product is even climate neutral. All these buzzwords have two things in common. Firstly, they distinguish the respective product from those that cannot provide such positive characteristics, which can regularly be a decisive factor for the purchase, and secondly, according to case law, a very strict standard must be applied to the vast majority of the catchwords in order to use them lawfully for advertising purposes as they are particularly likely to misleading.[77]

In this respect, consumers's expectations are of central importance as they ultimately provide information as to whether an action is misleading or not in an individual case. According to the case law of the OGH, this is defined by the averagely informed and averagely understanding consumer of the target group, applying a degree of attention adapted to the specific circumstances.[78] On the one hand, this can lead to the fact that certain knowledge can be assumed, for example if an advertisement is addressed to a specialised audience, but on the other hand, it can also lead to the fact that very precise information about the actual facts must be provided in order not to mislead the average consumer addressed. From the advertiser's point of view, this is aggravated by the fact that in the case of an ambiguous interpretation of the advertisement, he must accept the interpretation that is least favourable to him.[79] Finally, it must also be considered that according to Section 39 of the UWG, it is possible to mislead not only with words but with images as well, for example an environmental symbol or a graphic representation that gives the impression that the product is particularly valuable with regard to environmental protection.[80]

The OGH initially takes a strict view of advertising with buzzwords such as *organic* or *natural*.[81] The average consumer confronted with these catchwords

---

[76] F. Rüffler, Umweltwerbung und Wettbewerbsrecht I, ÖBl 1995(6), pp. 243-259 (pp. 246f).

[77] For an overview see in particular R. Mahfoozpour and G. Staber, Spannungsfeld "Greenwashing": Eine Rechtsprechungsübersicht, NR 2022(1), pp. 62-69; F. Rüffler, Umweltwerbung und Wettbewerbsrecht I, ÖBl 1995(6), pp. 243-259 (246f).

[78] A. Anderl and A. Appl, In: Wiebe and Kodek (eds), UWG, 2nd ed, Manz 2016, Sec 2 paras 66f.

[79] A. Anderl and A. Appl, In: Wiebe and Kodek (eds), UWG, 2nd ed, Manz 2016, Sec 2 para 156.

[80] C. Handig, In: Wiebe and Kodek (eds), UWG, 2nd ed, Manz 2016, Sec 38 paras 2f; F. Rüffler, Umweltwerbung und Wettbewerbsrecht I, ÖBl 1995(6), pp. 243-259 (p. 247).

[81] A. Anderl and A. Appl, In: Wiebe and Kodek (eds), UWG, 2nd ed, Manz 2016, Sec 2 para 252; E. Artmann, Wettbewerbsrecht und Umweltschutz, Manz 1997, pp. 138f; M. Görg, UWG, LexisNexis 2020, Sec 2 paras 454f; R. Mahfoozpour and G. Staber, Spannungsfeld "Greenwashing": Eine Rechtsprechungsübersicht, NR 2022(1), pp. 62-69 (pp. 64f).

regularly expects products advertised in this way to be completely free of chemicals, i.e. in particular not to contain any chemically treated additives such as modified starch or synthetic vitamins.[82] Especially the catchword *organic* is ambiguous and ultimately to be interpreted in the sense of absolute environmental compatibility.[83] This means that the product, with regard to its production, its composition as well as its disposal, would have exclusively positive characteristics with regard to the environment. It may be argued that the average consumer might well distinguish between the product itself and its manufacture or disposal and that it is therefore necessary to differentiate according to the respective phases or, even further, that environmental compatibility in general should only be understood in relative terms, meaning in comparison to other products.[84] Especially given the increasing awareness of consumers for environmental protection issues, the strict standard in the sense of absolute environmental compatibility as applied by the OGH for the catchword "organic" is nevertheless convincing. It is therefore all the more important for the advertiser to use such catchwords either only if they actually apply to the product in full or to make it sufficiently clear that only a certain part of the product is meant. Conversely, however, it would indeed be going too far to affirm that a product is misleading even if the pollutants contained in the product result from ubiquitous environmental pollution. In this sense, the ECJ also ruled that it was not misleading to advertise a jam as naturally pure if it actually contained pollutants from general environmental influences.[85] The average consumer would expect the product to be as naturally pure as possible, which exculpates ubiquitous environmental influences from the outset. This is also convincing as otherwise one would have to assume a thoroughly uninformed consumer as it is obvious that even strawberries grown organically and picked by hand are exposed to general environmental influences on a daily basis, even if these are not pleasant.[86]

---

[82] OGH 4 Ob 316/86 ÖBl 1986, 104; 4 Ob 200/05y, *Naturrein*, wbl 2006/62 (W. Schuhmacher); 4 Ob 44/13v, *Natürliches Keimlingsmehl*, ÖBl-LS 2013/52 (R. Schultes/M. Kasper).

[83] OGH 4 Ob 23/94, *Biowelt*, ecolex 1994, 480 (G. Kucsko); 4 Ob 90/94, *Bioziegel*, ecolex 1995, M. Görg, UWG, LexisNexis 2020 Sec 2 para 454; F. Rüffler, Umweltwerbung und Wettbewerbsrecht I, ÖBl 1995(6), pp. 243-259 (p. 252); R. Mahfoozpour and G. Staber, Spannungsfeld "Greenwashing": Eine Rechtsprechungsübersicht, NR 2022(1), pp. 62-69 (p. 64); H. Köhler in Köhler, Bornkamm and Feddersen (eds), UWG, 40th ed, C. H. Beck 2022, Sec 5 para 2.183; S. Weidert in Harte-Bavendamm and Henning-Bodewig (eds), UWG, 5th ed, C. H. Beck 2021, Sec 5 para 576.

[84] E. Artmann, Wettbewerbsrecht und Umweltschutz, Manz 1997, p. 142; G. Kucsko, Über irreführende Umweltengel, ecolex 1990(2), pp. 93-64 (p. 93); O. Sosnitza in Ohly and Sosnitza (eds), Gesetz gegen den unlauteren Wettbewerb, 7th ed, C. H. Beck 2016, Sec 4a para 117; for a relative approach, at least with regard to the catchword "environmentally friendly", also M. Görg, UWG, LexisNexis 2020, Sec 2 para 457.

[85] CJEU 4 April 2000, case C-465/98, *Darbo*, ECLI:EU:C:2000:184.

[86] Also F. Rüffler, Umweltwerbung und Wettbewerbsrecht I, ÖBl 1995(6), pp. 243-259 (p. 252); M. Görg, UWG, LexisNexis 2020, Sec 2 para 455.

Just as strictly, the OGH recently judged the advertising of a drinking bottle with the claim that it was made of 50% recycled plastic waste from the sea.[87] Although the defendant was able to prove that the drinking bottle was indeed made of more than 50% recycled plastic waste, this waste was collected not from the sea but from beaches. Admittedly, this may seem very strict at first glance and also raises the question of whether the element of relevance necessary for misleading is actually to be affirmed.[88] The latter asks whether the statement is also capable of influencing the consumer to make a decision that he would not have made otherwise.[89] In this respect, it can be argued that the average consumer focuses not so much on the origin of the plastic but rather on the accuracy of the claim that the drinking bottle actually consists of a certain percentage of recycled plastic.[90] However, if one takes into account that plastic from the sea is much more difficult to collect than plastic from beaches and that the recycling process of sea plastic is more complicated, then the reasoning of the OGH is ultimately convincing.[91] Furthermore, the OLG Stuttgart has to be agreed with, which states with regard to the same case that the average consumer would also consider the direct threat to the animal world and the danger of plastic entering the food cycle in the case of genuine sea plastic.[92] Conversely, however, the OGH does not apply an absolute understanding of environmental compatibility, unlike in the case of the catchword "organic". Therefore, as far as a product is actually recycled but energy is consumed in the process of recycling, the advertiser does not have to point this out separately. The average consumer would be aware of the fact that recycling requires the use of energy, and it would also make advertising with recycling processes almost impossible if one had to explain all conceivable effects on the environment.[93]

Lately, advertisements with the catchword *climate neutral* are also gaining in popularity. Rarely, however, are these products produced in a climate-neutral way, but they typically involve compensation payments for the CO_2 emissions caused during production. It is doubtful whether the average consumer, when a product is advertised as climate neutral, actually thinks of compensation payments and not rather of the production of the product without greenhouse gas emissions. While

---

[87] OGH 4 Ob 114/18g, *Ocean Bottle II*, ecolex 2018/449 (M. Horak).

[88] R. Mahfoozpour and G. Staber, Spannungsfeld "Greenwashing": Eine Rechtsprechungsübersicht, NR 2022(1), pp. 62-69 (p. 65).

[89] A. Anderl and A. Appl, In: Wiebe and Kodek (eds), UWG, 2nd ed, Manz 2016, Sec 2 para 47; E. Artmann, Wettbewerbsrecht und Umweltschutz, Manz 1997, pp. 179f; F. Rüffler, Umweltwerbung und Wettbewerbsrecht I, ÖBl 1995(6), pp. 243-259 (pp. 258f).

[90] R. Mahfoozpour and G. Staber, Spannungsfeld "Greenwashing": Eine Rechtsprechungsübersicht, NR 2022(1), pp. 62-69 (p. 65).

[91] Referring to this M. Horak, Irreführende Werbung mit Recycling-Meeresplastik, ecolex 2018(11), pp. 1009-1011 (p. 1010).

[92] OLG Stuttgart 2 U 48/18, *Ocean Bottle*, GRUR-RR 2019, 274.

[93] OGH 4 Ob 38/93, *Kästle-Öko-System*, ecolex 1993, 611; E. Artmann, Wettbewerbsrecht und Umweltschutz, Manz 1997, pp. 152f; R. Mahfoozpour and G. Staber, Spannungsfeld "Greenwashing": Eine Rechtsprechungsübersicht, NR 2022(1), pp. 62-69 (p. 66).

2 Sustainability and Competition Law in Austria

compensation payments may already be commonplace for flights, this is not the case for most other products. Therefore, the case law of the OGH is strict in the case of advertisement of a stamp as climate neutral, stating that without an explanatory reference to the compensation payment, it cannot be assumed that the average consumer is familiar with this concept in all cases.[94]

2.13 Advertising with Environmental Certificates

Since a misleading statement can be realised not only by a literal statement but, according to Section 39 UWG, also by a graphic illustration, the question also arises as to how to deal with advertising with environmental certificates.[95] First of all, reference has to be made to Nr 2 and Nr 4 of the Annex to the UWG, which contain a mandatory restriction on the use of quality labels or similar without authorisation (Nr 2)[96] or the claim of authorisation by a public or private body, although the requirements for such authorisation do not apply in reality (Nr 4).[97] In addition, the general prohibition of misleading statements under Section 2 UWG is also applicable. With regard to advertising with environmental certificates, a distinction must be made between advertising with public eco-labels (in Austria, especially the "Hundertwasserzeichen") and private eco-labels.[98]

Public eco-labels usually give the viewer the impression of objectivity to a greater extent as they are awarded by public authorities. Again, therefore, the reference to such an environmental certificate is only permissible if the conditions necessary for acquiring the certificate are fulfilled and continue to be fulfilled at the time of use.[99]

---

[94] OGH 4 Ob 202/12b, *klimaneutral*, ÖBl 2013/41; see most recently also LG Konstanz 7 O 6/21 KfH, *Klimaneutrales Heizöl*, WRP 2022, 118; LG Mönchengladbach, *Klimaneutrale Marmelade*, WRP 2022, 781; dissenting opinion OLG Koblenz, 9 U 163/11, *CO_2-neutral*, WRP 2011, 1499; see A. Anderl and A. Appl, In: Wiebe and Kodek (eds), UWG, 2nd ed, Manz 2016, Sec 2 para 253; M. Görg, UWG, LexisNexis 2020, Sec 2 para 456.

[95] E. Artmann, Wettbewerbsrecht und Umweltschutz, Manz 1997, pp. 154f; F. Rüffler, Umweltwerbung und Wettbewerbsrecht II, ÖBl 1996(1), pp. 3-12 (pp. 3f); R. Mahfoozpour and G. Staber, Spannungsfeld "Greenwashing": Eine Rechtsprechungsübersicht, NR 2022(1), pp. 62-69 (pp. 67f).

[96] A. Anderl and A. Appl, In: Wiebe and Kodek (eds), UWG, 2nd ed, Manz 2016, Annex to Sec 2 paras 25f.

[97] A. Anderl and A. Appl, In: Wiebe and Kodek (eds), UWG, 2nd ed, Manz 2016, Annex to Sec 2 paras 34f.

[98] F. Rüffler, Umweltwerbung und Wettbewerbsrecht II, ÖBl 1996(1), pp. 3-12 (p. 3); H. Köhler in Köhler, Bornkamm and Feddersen (eds), UWG, 40th ed, C. H. Beck 2022, Sec 5 para 2.190.

[99] A. Anderl and A. Appl, In: Wiebe and Kodek (eds), UWG, 2nd ed, Manz 2016, Annex to Sec 2 para 25; A. Anderl and A. Ciarnau, Green & Blue Washing – Die Grenzen des Marketings, In: Zahradanik and Richter-Schöller (eds), Handbuch Nachhaltigkeitsrecht, Manz 2021, pp. 73-100 (p. 76); E. Artmann, Wettbewerbsrecht und Umweltschutz, Manz 1997, p. 157; M. Hofer and I. Amschl, Greenwashing und UWG: Ein Überblick, NR 2021(4), pp. 421-427 (p. 424); M. Görg, UWG, LexisNexis 2020, Sec 2 para 458; R. Mahfoozpour and G. Staber, Spannungsfeld "Greenwashing": Eine Rechtsprechungsübersicht, NR 2022(1), pp. 62-69 (pp. 67); O. Sosnitza in Ohly and Sosnitza (eds), Gesetz gegen den unlauteren Wettbewerb, 7th ed, C. H. Beck 2016, Sec 4a para 119; S. Weidert in Harte-Bavendamm and Henning-Bodewig (eds), UWG, 5th ed, C. H. Beck 2021, Sec 5 para 576.

According to the German case law, a general reference to a tested *environmental compatibility* without further explanation is also inadmissible, since it is neither apparent to the consumer who carried out the test nor according to which criteria it was carried out.[100] Another issue of debate is whether a person who has received an eco-label in an admissible manner must also provide additional information in his advertising about the concrete environmental advantages that result from it in order not to run the risk of misleading the average consumer who associates the respective eco-label with even higher requirements than those actually required.[101] However, the better arguments are in favour of denying such an obligation.[102] In contrast to advertising with buzzwords such as *organic*, the average consumer will either already be familiar with certain eco-labels or be able to obtain further information about them without much effort. Nevertheless, where the advertiser is otherwise misleading through the use of an eco-label, for example by giving the impression of a unique position that does not actually apply, a corresponding explanation will have to be demanded.[103]

Regarding private eco-labels, the question must be assessed in a different manner. The authority typically associated with public eco-labels, and also the publicity of guidelines for awarding them, is not given to the same extent.[104] It is therefore all the more important to clarify which specific environmental benefits are associated with the product or the eco-label. Furthermore, if the private eco-label is designed in such a way that it resembles a public eco-label or that it gives the impression that it has been awarded by an independent third party rather than by the advertiser itself, this circumstance can also be misleading, with Section 2 UWG and Section 9 UWG providing a suitable legal basis for a claim.[105]

---

[100] OLG Koblenz, 9 U 163/11, *CO$_2$-neutral*, WRP 2011, 1499.

[101] E. Artmann, Wettbewerbsrecht und Umweltschutz, Manz 1997, pp. 158f; F. Rüffler, Umweltwerbung und Wettbewerbsrecht II, ÖBl 1996(1), pp. 3-12 (pp. 4f).

[102] F. Rüffler, Umweltwerbung und Wettbewerbsrecht II, ÖBl 1996(1), pp. 3-12 (p. 4); dissenting opinion E. Artmann, Wettbewerbsrecht und Umweltschutz, Manz 1997, p. 160.

[103] E. Artmann, Wettbewerbsrecht und Umweltschutz, Manz 1997 p. 162; O. Sosnitza in Ohly and Sosnitza (eds), Gesetz gegen den unlauteren Wettbewerb, 7th ed, C. H. Beck 2016, Sec 4a para 119.

[104] F. Rüffler, Umweltwerbung und Wettbewerbsrecht II, ÖBl 1996(1), pp. 3-12 (p. 5).

[105] E. Artmann, Wettbewerbsrecht und Umweltschutz, Manz 1997, p. 168; F. Rüffler, Umweltwerbung und Wettbewerbsrecht II, ÖBl 1996(1), pp. 3-12 (p. 5); R. Mahfoozpour and G. Staber, Spannungsfeld "Greenwashing": Eine Rechtsprechungsübersicht, NR 2022(1), pp. 62-69 (p. 68); see OLG München 6 U 1973/21 WRP 2022, 494.

2.14 Corporate-Related Advertising

Other than the product-related environmental advertising discussed above, company-related environmental advertising promotes the environmentally positive characteristics of a company separately from its products in order to promote a corresponding reputation.[106] There are plenty of potential ways of doing this, whether it be an internal waste avoidance system, a switch to renewable energy sources for heating the company's buildings, or the appointment of environmental managers to raise awareness among employees.[107] Typically, however, one encounters company-related environmental advertising wherever companies either sponsor environmental organisations or make this sponsorship dependent on the purchase of a product. If, for example, a company advertises that for every product purchased a certain percentage of the sales price will be donated to a well-known environmental protection organisation or that the planting of a tree will be financed, then this marketing strategy falls into the scope of company-related environmental advertising.

By this company-related environmental advertising, the social awareness of the consumer is regularly addressed, on the one hand, in order to link the image of the company with certain positive characteristics and, on the other hand, to influence the decision to buy in favour of one's own company by means of the purposely emotional advertising. It is therefore hardly surprising that the case law assessed the first company-related environmental advertising efforts as immoral emotional advertising according to Section 1 UWG.[108] Accordingly, while mere image advertising was regularly considered permissible, linking this image advertising with the purchase of a product was typically judged to be an immoral business practice. This case law is now outdated, especially since the consumer model has changed and the standard of immorality has been replaced by that of unfairness. It is therefore reasonable to consider permissible not only image advertising but also the linking of emotional environmental advertising with the purchase of a product.[109] What remains, however, are the same restrictions that have already been set for product-related environmental advertising. Company-related environmental advertising must therefore also be judged in particular by the standard of the prohibition of misleading statements in Section 2 UWG and is unfair if it does not correspond to the actual facts. Especially, advertising with very general buzzwords such as *environmentally*

---

[106] F. Rüffler, Umweltwerbung und Wettbewerbsrecht II, ÖBl 1996(1), pp. 3-12 (p. 8).

[107] See more examples at F. Rüffler, Umweltwerbung und Wettbewerbsrecht II, ÖBl 1996(1), pp. 3-12 (p. 8); A. Anderl and A. Ciarnau, Green & Blue Washing – Die Grenzen des Marketings, In: Zahradanik and Richter-Schöller (eds), Handbuch Nachhaltigkeitsrecht, Manz 2021, pp. 73-100 (pp. 78f).

[108] BGH I ZR 239/93, *Ölverschmutzte Ente*, GRUR 1995, 598; E. Artmann, Wettbewerbsrecht und Umweltschutz, Manz 1997, pp. 215f; F. Rüffler, Umweltwerbung und Wettbewerbsrecht II, ÖBl 1996(1), pp. 3-12 (p. 9).

[109] F. Rüffler, Umweltwerbung und Wettbewerbsrecht II, ÖBl 1996(1), pp. 3-12 (p. 10); H. Köhler in Köhler, Bornkamm and Feddersen (eds), UWG, 40th ed, C. H. Beck 2022, Sec 5 para 2.180.

friendly will therefore have to be judged just as strictly as advertising a product with the buzzword *organic* since the latter is to be understood just as ambiguously.[110]

On the other hand, it is equally outdated to assume unfairness simply because a free additional service, namely the promised environmental commitment of the company, is added to the purchase of a main service.[111] Although Section 9a UWG, which used to be relevant for these cases, contained a prohibition of such additional benefits, it has been repealed in the course of the fully harmonising effect of the UCP Directive.[112] Insofar as company-related environmental advertising therefore corresponds to the real facts and is not otherwise likely to mislead, there is no room for any restrictions based on appealing to feelings.

2.15 Advertising with Self-Evident Facts

While incorrect information is regularly misleading and thus unfair, it may seem surprising at first glance that correct information can also be misleading in individual cases. If an entrepreneur advertises his products or his image with information that is self-evident anyway but gives the consumer the impression that it is a special characteristic, this also constitutes a misleading business practice.[113] Such is the case if legally prescribed general conditions are emphasised as a special advantage.[114] If, for example, an entrepreneur advertises a 2-year warranty period for his products and emphasises this warranty period in an eye-catching manner, this represents not a special benefit of his product but only the warranty period regulated in Section 933 para 1 ABGB,[115] which must not be shortened in consumer transactions. The same applies if it is advertised that a company complies with certain ecological standards, although compliance with these standards is a legal requirement for the operation of the company anyway, or if products are advertised as vegan, whereas they are always vegan already due to natural conditions.[116] In all these cases, there is a risk of generating the impression in the consumer's perception that the product or the company has special characteristics that comparable other suppliers cannot offer. However, two aspects must be taken into account. Firstly,

---

[110]F. Rüffler, Umweltwerbung und Wettbewerbsrecht II, ÖBl 1996(1), pp. 3-12 (p. 10); see also A. Anderl and A. Ciarnau, Green & Blue Washing – Die Grenzen des Marketings, In: Zahradanik and Richter-Schöller (eds), Handbuch Nachhaltigkeitsrecht, Manz 2021, pp. 73-100 (p. 84); M. Hofer and I. Amschl, Greenwashing und UWG: Ein Überblick, NR 2021(4), pp. 421-427 (p. 425).

[111]E. Artmann, Wettbewerbsrecht und Umweltschutz, Manz 1997, pp. 231f; F. Rüffler, Umweltwerbung und Wettbewerbsrecht II, ÖBl 1996(1), pp. 3-12 (pp. 10f).

[112]A. Anderl and A. Appl, In: Wiebe and Kodek (eds), UWG, 2nd ed, Manz 2016, Sec 2 para 375.

[113]A. Anderl and A. Appl, In: Wiebe and Kodek (eds), UWG, 2nd ed, Manz 2016, Sec 2 paras 201f.

[114]A. Anderl and A. Appl, In: Wiebe and Kodek (eds), UWG, 2nd ed, Manz 2016, Sec 2 paras 202, 205.

[115]"Allgemeines Bürgerliches Gesetzbuch" (Austrian Civil Code).

[116]F. Rüffler, Umweltwerbung und Wettbewerbsrecht I, ÖBl 1995(6), pp. 243-259 (pp. 256f).

unlawful advertising with self-evident facts must be distinguished from permissible exaggerated advertising. An advertisement stating that a bicycle does not require fossil fuels in order to move forward or that an insect protection net is insecticide-free is therefore not problematic because in both cases, the average consumer recognises that it is a self-evident fact, and the advertiser is merely using it as a humorous marketing tool.[117] Secondly, the presentation in the individual case is decisive. Of course, an advertiser is free to refer to attributes that are self-evident, but the line has to be drawn where this reference is emphasised in such a way that the impression is given that it is a special characteristic of the product.[118] Furthermore, the recent Commission draft[119] regarding empowering consumers for the green transition through better protection against unfair practices and better information also refers to this. Among other provisions, this draft includes an additional per se prohibition of commercial practices that present requirements that apply by law to all products in the relevant product category as a special attribute of the trader's offer.[120]

2.16 Conclusion

Overall, with respect to antitrust law, the Austrian legislator took sustainability issues into account by amending the Austrian Cartel Act in 2021. The Austrian legislator made clear that antitrust law and sustainability are compatible and incorporated the notion of environmentally sustainable or carbon-neutral economy in the wording of the new exemption provision. This change primarily aimed at providing more legal certainty and thus reducing risk aversion of undertakings to participate in sustainability agreements. However, the new regulation does not deviate from the economic approach predetermined by Article 101 para. 3 TFEU. This applies especially to the four exemption criteria, except for the fair share for consumers. In contrast to EU competition law, out-of-market efficiencies can now be considered to fulfil the latter criterion by increasing consumer welfare under Austrian antitrust law, which is the main focus of interest for antitrust lawyers within the scope of the Austrian amendment.

Regarding unfair competition law, in general, a strict standard can be identified which Austrian case law applies to environmental advertising. Both product-related advertising and company-related environmental advertising are regularly subject to the requirements of the UWG as a business practice. Caution is advised in particular

---

[117]F. Rüffler, Umweltwerbung und Wettbewerbsrecht I, 1995(6), pp. 243-259 (p. 257).

[118]A. Anderl and A. Appl, In: Wiebe and Kodek (eds), UWG, 2nd ed, Manz 2016, Sec 2 paras 203.

[119]COM(2022) 143 final: Proposal for a Directive of the European Parliament and of the Council amending Directives 2005/29/EC and 2011/93/EU as regards empowering consumers for the green transition through better protection against unfair practices and better information.

[120]See C. Alexander, Green Deal: Verbraucherschutz und ökologischer Wandel, WRP 2022(6), pp. 657-665 (p. 662); also C. Schumacher, Ein Sommer wie damals? ÖBl 2021(4), p. 145.

when using very general catchwords such as "organic" or "natural". The OGH in this respect considers the average consumer to have an absolute understanding of environmental compatibility, meaning that the product does not in any way cause negative effects on the environment. The use of such terms is therefore not advisable if they do not in fact fully correspond to reality. The same applies to the use of public or private eco-labels as these regularly evoke certain expectations in the average consumer. In light of the EU's efforts to harmonise environmental advertising, it remains to be seen how the legal situation will develop in the future and whether the already very strict case law will have to be adapted.

Open Access This chapter is licensed under the terms of the Creative Commons Attribution 4.0 International License (http://creativecommons.org/licenses/by/4.0/), which permits use, sharing, adaptation, distribution and reproduction in any medium or format, as long as you give appropriate credit to the original author(s) and the source, provide a link to the Creative Commons license and indicate if changes were made.

The images or other third party material in this chapter are included in the chapter's Creative Commons license, unless indicated otherwise in a credit line to the material. If material is not included in the chapter's Creative Commons license and your intended use is not permitted by statutory regulation or exceeds the permitted use, you will need to obtain permission directly from the copyright holder.

Sustainability and Competition Law in Belgium

3

Jan Blockx

3.1 Introduction

This report describes the relationship between sustainability and competition law under Belgian law, including Belgian competition law, as enforced by the Belgian Competition Authority (BCA) and the Belgian courts. As a member state of the European Union, European competition law also applies in Belgium, but questions of European competition law are not covered in this national report.

The scope of the notion of sustainability is often not clearly defined in debates about the interplay between sustainability and competition law, ranging from environmental sustainability over sustainable development and social justice to animal welfare. The review of cases for this report has focused on cases that relate to environmental sustainability, i.e., where contribution or harm to one of the environmental objectives mentioned in Article 9 of the Taxonomy Regulation[1] is at stake: (i) climate change mitigation, (ii) climate change adaptation, (iii) the sustainable use and protection of water and marine resources, (iv) transition to a circular economy, (v) pollution prevention and control, (vi) the protection and restoration of biodiversity and ecosystems.

---

[1] Regulation 2020/852 of the European Parliament and of the Council of 18 June 2020 on the establishment of a framework to facilitate sustainable investment, and amending Regulation 2019/2088, OJ 2020 L 198, p. 13.

J. Blockx (✉)
University of Antwerp, Antwerpen, Belgium
e-mail: jan.blockx@uantwerpen.be

© The Author(s) 2024
P. Këllezi et al. (eds.), *Sustainability Objectives in Competition and Intellectual Property Law*, LIDC Contributions on Antitrust Law, Intellectual Property and Unfair Competition, https://doi.org/10.1007/978-3-031-44869-0_3

3.2 Legal Framework

In Belgium, there are no specific statutory provisions that concern the role of sustainability in competition law, nor are any legislative initiatives planned at the time of writing this report. The BCA nevertheless considers that the current legislative framework allows it to take sustainability considerations into account, provided that sustainability is relevant to the functioning of the markets concerned by a certain practice or is relevant for those who may be affected by the practice (including out-of-market).[2] According to the BCA, it can do so by taking into account total as well as consumer welfare in its analysis and by considering the general economic interest.[3]

With increased attention to questions of sustainability in the media, politics, and society as a whole, it can be expected that the interaction between sustainability and competition law will come more to the fore in the near future, including through pressure from the courts. While the Belgian courts have not taken such far-reaching decisions as those in, for example, the Netherlands,[4] the Court of First Instance of Brussels has also found that the various Belgian governments have been negligent in their climate policy.[5] However, the Court refrained from imposing specific obligations on the Belgian government to remedy this situation, which has led the nongovernmental organizations (NGOs) to bring an appeal against the judgment.

3.3 Enforcement Record

The BCA has explicitly stated that it has not dealt with cases in which sustainability played a role either to condemn certain practices or to defend them, and this regardless of the subfield of its competence (restrictive practices, abuse of dominance, or mergers).[6] To my knowledge, this topic has also not been raised in private litigation.

---

[2] See the Environmental Considerations in Competition Enforcement – Note by Belgium for the OECD Competition Committee meeting on 1-3 December 2021 (DAF/COMP/WD(2021)47), available at https://one.oecd.org/document/DAF/COMP/WD(2021)47/en/pdf (hereafter "Belgian note for OECD roundtable"), p. 2 and 5. Accessed 21 November 2022.

[3] Belgian note for OECD roundtable, p. 2–3.

[4] See in particular the judgment of the Dutch Supreme Court of 20 December 2019, ECLI:NL: HR:2019:2006, *Urgenda*.

[5] Judgment of the Brussels Court of First Instance of 17 June 2021, *Klimaatzaak*, available at https://prismic-io.s3.amazonaws.com/affaireclimat/18f9910f-cd55-4c3b-bc9b-9e0e393681a8_167-4-2021.pdf. Accessed 21 November 2022.

[6] Belgian note for OECD roundtable, throughout.

3.4 Advocacy and Prioritization

The BCA has expressed its support for the initiatives taken by the Dutch and Greek competition authorities to better incorporate sustainability in the application of European competition law.[7] At the same time, the BCA has insisted that there is a need for EU guidance in this respect, presumably to avoid divergent national approaches in the EU.[8]

The BCA has also indicated that the application of competition policy in the green and circular economy in Belgium is one of its strategic priorities in 2022. It pointed out that a successful competition policy is an important factor in stimulating innovation and technological developments, which can contribute to the greening of the economy. At the same time, it indicated that it is willing to develop its position on the compatibility between the competition rules and sustainability measures.[9]

The BCA has stated that it is particularly keen to provide (informal) guidance on sustainability initiatives.[10]

3.5 Conclusion

So far, sustainability concerns have not influenced the enforcement of competition law by the BCA or the national courts. However, with increased attention to this topic, this may change in the future. The BCA has also stated that it will pay increased attention to the relationship between competition law and sustainability and is open to providing guidance on sustainability initiatives.

Open Access This chapter is licensed under the terms of the Creative Commons Attribution 4.0 International License (http://creativecommons.org/licenses/by/4.0/), which permits use, sharing, adaptation, distribution and reproduction in any medium or format, as long as you give appropriate credit to the original author(s) and the source, provide a link to the Creative Commons license and indicate if changes were made.

The images or other third party material in this chapter are included in the chapter's Creative Commons license, unless indicated otherwise in a credit line to the material. If material is not included in the chapter's Creative Commons license and your intended use is not permitted by statutory regulation or exceeds the permitted use, you will need to obtain permission directly from the copyright holder.

---

[7] Belgian note for OECD roundtable, p. 6.

[8] Belgian note for OECD roundtable, p. 6.

[9] BCA, Note de priorités de l'Autorité belge de la Concurrence pour 2022 (Priority note 2022), available at https://www.abc-bma.be/sites/default/files/content/download/files/2022_politique_priorites_ABC.pdf (hereafter "BCA priority note 2022"), p. 3. Accessed 21 November 2022.

[10] Belgian note for OECD roundtable, p. 6 and BCA priority note 2022, p. 3.

Sustainability and Competition Law in Brazil

4

José Mauro Decoussau Machado

4.1 General Framework

4.1.1 Definition of Sustainability

The concept of sustainability admits different definitions. One of the most known is the one given by the UN as "meeting the needs of the present without compromising the ability of future generations to meet their own needs."[1] In the following answers, sustainability is understood as preventing the depletion of natural or physical resources so that they will remain available for society for the long term. In business and policy contexts, the concept of sustainability refers to the idea of planned and responsible development regarding the use of natural resources in economic activities to avoid or reduce environmental risks and allow the continuity of human life.

The rapporteur would like to thank his colleagues from the Study Committee in Competition from the Brazilian Intellectual Property Association – Gabriela Monteiro, Jessica Ribeiro, Lucas Antoniazzi, and Pedro Mota – for the inestimable aid with research and responding to questions.

[1] United Nations Brundtland Commission (1987) Academic Impact, Sustainability, available at https://www.un.org/en/academic-impact/sustainability. Accessed 19 November 2022.

J. M. D. Machado (✉)
Pinheiro Neto Advogados, São Paulo, Brazil
e-mail: jmachado@pn.com.br

© The Author(s) 2024
P. Këllezi et al. (eds.), *Sustainability Objectives in Competition and Intellectual Property Law*, LIDC Contributions on Antitrust Law, Intellectual Property and Unfair Competition, https://doi.org/10.1007/978-3-031-44869-0_4

4.1.2 The Role of Sustainability in Competition Law

Although the promotion of sustainable development is the biggest challenge today and is present from state regulation to the agenda of international organizations and private companies, the Brazilian Antitrust law does not bring any specific provision related to sustainability nor its role in competition.

In fact, it has a more traditional approach, focused on promoting the freedom of initiative, free competition, the social function of property, consumer protection, and the repression of the abuse of economic power.

However, as the application of competition law seeks to promote human rights, efficiency, and consumer welfare, it is possible to consider sustainability as a parameter of quality (not related to price) and innovation in the control of structures and conducts. Furthermore, given that the protection of human rights and the promotion of free competition are provided for in the Federal Constitution (Articles 3, items II and III, 5 and 170, item IV), it is possible to interpret that antitrust can offer instruments for the promotion of sustainable development.

However, reconciling antitrust and sustainability values, without risking companies being punished by antitrust authorities, can be a challenge. Many environmental, social, and governance (ESG) actions would only be achieved by companies if they—usually in the position of competitors—joined efforts to this end in interdependent relationships.

The absence of a position from the competition authority, therefore, can make these companies give up on implementing sustainability projects due to the fear of penalties. This is because eventual cooperation could create incentives for information exchange and/or the adoption of uniform commercial behavior, which could, at least in theory, facilitate collusion.

Thus, it is understood that the competition authority can apply sustainability values, but this will depend on specific analysis, on a case-by-case basis. For example, a cooperation agreement between companies with a view to adopting measures favorable to the preservation of the environment, such as the development of new ecological technology or a common mechanism for reducing the negative environmental impacts of a given economic activity, can be considered as generating environmental benefits. The same structure, however, may be considered a cartel or may lead to the exclusion of competitors that will not be able to follow the actions of these other companies.

In conclusion, it is important that the antitrust community engages in a discussion about the effective role that competition law plays in creating a more egalitarian and sustainable society. This is because for companies to know the limits of their cooperation in favor of sustainability actions, it is necessary that antitrust authorities collaborate with guidelines so that companies' efforts toward a more sustainable activity are rewarded—and not punished.

4.1.3 Specific Laws and Rules That Address the Intersection Between Sustainability and Competition Law

There is no specific legislation in Brazil that uses competition law as a tool to achieve sustainable values. However, it is possible to identify an intersection between sustainability and competition law from the interpretation of Articles 170 and 225 of the Federal Constitution.

The economic order, founded on the valorization of human work and free initiative (or free enterprise), aims, through social justice (Article 170 of the Constitution), to contribute to the achievement of human dignity. For this reason, the provision that deals with economic rights includes the defense of the environment in the list of principles. Article 225, caput, of the Constitution provides for the right of everyone to an ecologically balanced environment and imposes on the government and the community the duty to defend and preserve it for present and future generations. It is, therefore, an interest of the collectivity, the holder of the interests protected by the constitutional economic order and by antitrust.

In Brazil, the Competition Authority (Administrative Council of Economic Defense (CADE)) is attentive to foreign initiatives (such as the Dutch authority's "Proposal for a Guide on Sustainability Agreements") but has not yet expressed itself on the subject. Given the importance that the topic of sustainability has gained in the agenda of Brazilian companies, it is believed that CADE will position itself in the future.

4.2 Specific Interaction and Cases

4.2.1 Brazilian Cases Where Sustainability Played a Role in the Enforcement of Competition Law

Based on publicly available information, we did not find any case in which sustainability played any role, not as a sword nor as a shield, as mentioned in the questions. In fact, so far, in Brazil, discussions related to the interaction between sustainability and antitrust have basically been developed in a few merger cases (concentration acts). However, no concrete position has been established by the Brazilian antitrust authority yet.[2]

---

[2] For instance, in Concentration Act No. 08700.007101/2018-63 (Vale S.A. and Ferrous Resources Limited), it was stated by Reporting Commissioner Mauricio Oscar Bandeira Maia that although the case involved several regulatory, social, and environmental matters, among others, CADE should restrict its analysis to competitive aspects within the limits of its remit. One could not expect that the antitrust authority addresses all these other aspects because it does not have a remit for this and there are other bodies with specific attributions to carry out much deeper analyses about each one of these matters.

4.2.2 Brazilian Cases of Illegal Anticompetitive Conduct or Green Washing Conduct in the Context of Sustainability Initiatives

Based on publicly available information, there have been no major cases of investigation of illegal anticompetitive conduct that occurred in the context of sustainability initiatives. Thus, there have been no cases of greenwashing conduct decided by the Brazilian antitrust authority so far.

4.2.2.1 Brazilian Cases of Genuine Sustainability Initiatives Served as a Springboard for Other Anticompetitive Behavior

Between 2010 and 2015, CADE investigated anticompetitive practices of cartels and their influence on uniform behavior in the sand market of the state of Paraná (BR). The cartel formation by the competing players was facilitated by a Term for the Adjustment of Conduct ("TAC" for its acronym in Portuguese)—a kind of settlement executed between companies and the Public Prosecutors' Office—which was executed due to environmental concerns (Administrative Proceeding No. 08012.004430/2002-43).[3]

TAC included the obligations to establish an association of sand extractive industries (the Northwest Sand Extractive Industries Association or "APA" for its acronym in Portuguese) and to implement an unloading and storage terminal to be used jointly by the companies and the association. The main purpose of such a settlement was environmental preservation and the recovery of areas located on the banks of the Paraná River.

The administrative proceeding was initiated after a complaint from the Union of Retail Commerce of Hardware, Paints, Wood, Electric, Hydraulic and Construction Materials of Maringá and Region ("Simatec" for its acronym in Portuguese), which was forwarded to CADE by the Public Prosecutors' Office of the State of Paraná ("MPE/PR" for its acronym in Portuguese).

The meetings held in the context of the association were used by the companies to exchange commercially sensitive information among the competitors. Also, the association was allegedly used to impair the growth and intensification of competition in the corresponding market.

The investigated parties (including the companies,[4] the association, and its former president) executed Cease and Desist Agreements ("TCC" for its acronym in Portuguese) with CADE, and the administrative proceeding was suspended in March 2015.

---

[3]CADE, Administrative Proceeding No. 08012.004430/2002-43. Representantes: Simatec e Ministério Público do Estado do Paraná. Representada: Porto de Areia Cristo Rei Ltda. e outros. Relator Conselheiro Márcio de Oliveira Júnior.

[4]Baleal Indústria e Comércio de Areia Ltda., Porto de Areia do Lago Ltda., Porto de Areia Cristo Rei Ltda., Indústria e Comércio de Areia e Pedra Vera Cruz Ltda. – ME, Daniel de Oliveira Reis & Cia. Ltda., Vilmar Pasquali & Cia. Ltda., J.M. Lada & Cia. Ltda., Manoel Cruz Malassise Neto, Mineração Nova Londrina Ltda.

The TCCs' obligations included measures to address both the environmental and antitrust concerns discussed in the case, which had to be reconciled and aligned. In this regard, CADE's Commissioner, Gilvandro Araújo, expressly stated that "The parties presented a solution that mitigates both the environmental concern of the TAC of the Public Prosecutors' Office and the competition concern of CADE."

4.3 Agencies and Legislature

4.3.1 The Role of Agencies

CADE has not yet launched any specific guideline or working paper related to horizontal cooperation agreements for sustainable practices and antitrust. However, specialists[5] stated that CADE is attentive to guidelines issued by other jurisdictions, namely, the Dutch and German authorities, and to discussing their policies to further issue a position on the matter.

Therefore, currently, there is no guidance (either collective or individual) provided by CADE to inform the market of the boundaries between common action toward sustainable practices and collusion, the latter punishable under Article 36 of Brazilian Competition Law (Federal Law n. 12.159/11). That is why scholars and attorneys have been alerting that companies are avoiding debating, establishing, or taking common sustainable measures due to a fear that CADE might question these practices and even interpret them as violations under Brazilian law.[6]

4.3.2 The Role of Legislature and Specific Committees

Neither the Brazilian legislature nor specific committees have yet discussed the legal framework for carrying common sustainable actions by companies and their treatment under competition law.

---

[5]D. O. Andreoli and P. Puglies (2020) Competition and sustainability: reconcilable matters, available at https://www.demarest.com.br/wp-content/uploads/2020/10/Concorr%C3%AAncia-e-sustentabilidade-d%C3%A1-para-conciliar.pdf. Accessed 19 November 2022.

[6]Migalhas (2020) Cooperation agreements among competitors to promote sustainable measures: which should be Cade's role?, available at: https://www.migalhas.com.br/depeso/331669/acordos-de-cooperacao-entre-concorrentes-para-promocao-de-medidas-de-desenvolvimento-sustentavel%2D%2Dqual-deve-ser-o-papel-do-cade. Accessed 19 November 2022.

What has been discussed and proposed within the Brazilian Parliament are bills and amendments concerning green certifications that could grant access to credit[7] and preferences at public bids[8] to companies that comply with certain sustainable practices. Nevertheless, these policies promote individual actions taken by companies and do not establish the rules to foster collaboration among private actors without infringing competition law.

Open Access This chapter is licensed under the terms of the Creative Commons Attribution 4.0 International License (http://creativecommons.org/licenses/by/4.0/), which permits use, sharing, adaptation, distribution and reproduction in any medium or format, as long as you give appropriate credit to the original author(s) and the source, provide a link to the Creative Commons license and indicate if changes were made.

The images or other third party material in this chapter are included in the chapter's Creative Commons license, unless indicated otherwise in a credit line to the material. If material is not included in the chapter's Creative Commons license and your intended use is not permitted by statutory regulation or exceeds the permitted use, you will need to obtain permission directly from the copyright holder.

---

[7] On that matter, Bill n° 735/22 proposed by Deputy Carlos Henrique Gaguim establishes a green stamp for capital market and investment companies that follow sustainable practices and grant them special conditions, such as access to federal credit and financing programs under public funds and banks. The Bill will yet be debated at the Congress' Environment and Sustainable Development Committee. More information available at https://www.camara.leg.br/propostas-legislativas/2318833. Accessed 19 November 2022.

[8] Bill n° 358/2020 proposed by Senator Styvenson Valentim alters the Brazilian Public Bid Federal Law (Law n. 8.666/93) in order to establish the green stamp as a tiebreaker in Public Bids.

Sustainability and Competition Law in Czech Republic

5

Radka MacGregor Pelikánová

5.1 Introduction

The Czech Republic is a member of the United Nations (UN), the World Trade Organization (WTO), the Organisation for Economic Co-operation and Development (OECD) as well as the European Union (EU). The Czech law shares the continental law tradition and reflects the international and regional commitments and obligations of the Czech Republic, including the call for sustainable development.[1] Namely, the concept of sustainability projected on the responsibility of all stakeholders, including businesses—see corporate social responsibility (CSR)—is recognized in the Czech Republic, and its application in various spheres is expected.[2] Indeed, sustainability rests on three pillars—economic, environmental and social—and all three need to be met while their objectives should be advanced

---

The Author is grateful for the support, with respect to this publication outcome, which was provided via the Metropolitan University Prague research project no. 87-02 "International Business, Financial Management and Tourism" (2022) based on a grant from the Institutional Fund for the Long-term Strategic Development of Research Organizations.

[1] CVIK, E. D. Cvik & R. MacGregor Pelikánová, The Significance of CSR during COVID-19 Pandemic in the Luxury Fashion Industry – A Front-Line Case Study, European Journal of Business Science and Technology, 2021, 7(1), pp. 109-129. DOI: https://doi.org/10.11118/ejobsat.2021.005.

[2] R. MacGregor Pelikánová and M. Hála. CSR Unconscious Consumption by Generation Z in the COVID-19 Era – Responsible Heretics Not Paying CSR Bonus? Journal of Risk and Financial Management, 2021, 14(8), pp. 390. DOI: https://doi.org/10.3390/jrfm14080390. Accessed 1 November 2022.

R. MacGregor Pelikánová (✉)
Metropolitan University Prague, Strašnice, Czech Republic

© The Author(s) 2024
P. Këllezi et al. (eds.), *Sustainability Objectives in Competition and Intellectual Property Law*, LIDC Contributions on Antitrust Law, Intellectual Property and Unfair Competition, https://doi.org/10.1007/978-3-031-44869-0_5

simultaneously, if not synergistically, i.e. there should not be trade-offs. At the same time, it must be emphasized that the Czech Republic is not a pro-sustainability force *par excellence* and rather follows than leads these trends in the EU. These follow-ups are sometimes more and sometimes less enthusiastic, generally on time, but occasionally a national resistance can be identified.

A different picture appears when the focus goes exclusively to the competition setting, i.e. the Czech national law is fully harmonized in this arena; the Czech competition authority, i.e. the Office for the Protection of Competition, works closely with the European Commission, its DG COMP and the European Competition Network (ECN); and Czech case law does not deviate from EU case law.

In order to address the burning issues of the impact, if not the interplay, of sustainability and competition law in the Czech Republic, it must be reviewed the policy and law setting regarding competition, sustainability and their potential overlap (Sect. 5.2); the relevant cases (Sect. 5.3); and the critical institutions and agencies and trends embraced by them (Sect. 5.4).

5.2 Policy and Law Setting: The Potential Impact of Sustainability on the Protection of Competition

The law valid and applicable in the Czech Republic has sources from three legal systems – international, the EU and national. Namely, since its accession to the EU in 2004, EU law has applied on the territory of the Czech Republic, and especially regarding competition, the Czech Republic has not demonstrated resistance or contradicting trends. Basically, the only difference between national Czech competition law and EU competition law is in the intensity, i.e. effectiveness and efficiency, of their applicability. Indeed, for many years, despite the same policies and law provisions, the Czech Office for the Protection of Competition has had a rather low score of prosecuted cases ending with final sanctions.[3] In sum, regarding the public law aspects of competition protection (anti-monopoly, anti-cartel, merger, state aid), Czech Act Nr. 143/2001 Coll., on the protection of competition ("Czech Competition Protection Act"), is aligned and harmonized, if not unified, with EU law, especially Article 101 et foll. of the Treaty on the Functioning of the European Union (TFEU), and there have not been objections regarding its wording. Similarly, regarding the private law aspects of competition protection (unfair competition), Czech Act Nr. 89/2012 Coll., Civil Code ("Czech Civil Code"), is compliant with EU law; see, e.g., Directive 2005/29/EC concerning unfair business-to-consumer commercial practices in the internal market, aka the Unfair Commercial Practices Directive (UCPD).[4]

---

[3]R. MacGregor Pelikánová, Divergence of antitrust enforcements – where, and where not, to collude, Antitrust – Revue of Competition Law, 2014, 2, i-viii.

[4]R. MacGregor Pelikánová, Harmonization of the protection against misleading commercial practices: ongoing divergences in Central European countries, Oeconomia Copernicana, 2019, 10(2), pp. 239–252. DOI: https://doi.org/10.24136/oc.2019.012.

A hot topic of discussion, perhaps criticism, was merely about the manner of application of competition protection in individual cases by the Czech Office for the Protection of Competition and the courts. Namely, the Czech Office for the Protection of Competition has seen, fairly often, its cases quashed by Czech courts, in particular by the Regional Court in Brno and the Supreme Administrative Court, occasionally even by the Constitutional Court, i.e. Czech courts have had to cancel many of decisions of the Czech Office for the Protection of Competition and this is not typical. For example the CJ EU seldom cancel/change the decisions of the European Commission. To put this differently, the Czech Office for the Protection of Competition has issued a lot of controversial decisions and these decisions were cancelled by Czech Courts. Such a massive reversal of cases we do not see typically, especially not in the case of the review of CJ EU of decisions by the European Commission.[5] However, it needs to be pointed out that improvement has progressively occurred, and currently, the difference in the effectiveness and efficiency of the application of competition law has diminished, and the success rate of the Czech Office for the Protection of Competition has increased.

Well, regarding sustainability, the Czech Republic still has a long journey ahead, and this concerning both national policies and national law. Many legal norms are based on ethical principles and other concerns pointing out that the development of business activities must not be justified by market power along with the exclusive command of maximizing immediate profits and gratification at any cost.[6] At the same time, the Czech Republic has embraced a rather reluctant attitude and only slowly moves towards the inclusion of sustainability in its policies and law.[7] The input and inspiration to do so have a clear UN and EU origin, and it must be emphasized that the Czech Republic and especially Czech businesses are becoming, over time, more compliant and committed in this respect, i.e. there is not a direct resistance or violation.[8] Other stakeholders follow at varying time frames.[9]

The above-mentioned UN and EU goes back to the Report of the World Commission on Environment and Development Report: Our Common Future prepared

---

[5] R. MacGregor Pelikánová, R., The unbearable lightness of imposing e-commerce in a vertical agreement setting, Antitrust – Revue of Competition Law, 2015, 3, pp. 68-76.

[6] R. MacGregor Pelikánová, R. K. MacGregor. & M. Černek, New trends in codes of ethics: Czech business ethics preferences by the dawn of COVID-19. Oeconomia Copernicana, 2021, 12(4), pp. 973–1009. DOI: https://doi.org/10.24136/oc.2021.032.

[7] R. MacGregor Pelikánová, Corporate Social Responsibility Information in Annual Reports in the EU – Czech Case Study, Sustainability, 2019, 11, 237. DOI: https://doi.org/10.3390/su11010237.

[8] R. MacGregor Pelikánová, T. Němečková & R. K. MacGREGOR, CSR Statements in International and Czech Luxury Fashion Industry at the onset and during the COVID-19 pandemic – Slowing Down the Fast Fashion Business? Sustainability, 2021, 13(7): 3715. DOI: https://doi.org/ 10.3390/su13073715.

[9] R. K. MacGregor, W. Sroka & R. MacGregor Pelikánová, The CSR Perception of Front-line Employees of Luxury Fashion Businesses: Fun or Free for Sustainability? Organizacija, 2020, 53(3), pp. 198-211. DOI: https://doi.org/10.2478/orga-2020-0013.

by the Brundtland Commission and published as the UN Annex to document A/42/427 in 1987 ("Brundtland Report 1987"). It needs to be emphasized that the Brundtland report indicates that "Sustainable development is development that meets the needs of the present without compromising the ability of future generations to meet their own needs". Consequently, it is proposed that the concept of sustainability is analogous to the concept of *usufructus*, i.e. the right to use another´s property without changing its substance, which is extended beyond the economic realm to cover the social and environmental aspects of human activities. The Brundtland Report 1987 led to UN General Assembly resolution A/RES/60/1 adopted via World Summit 2005, along with UN Agenda 21, which were paralleled by the EU strategy Agenda 2000, aka Lisbon Agenda 2000, which was agreed upon by the European Council in order to strengthen employment, economic reform and social cohesion as part of a knowledge-based economy in the EU.[10] The following year, in 2001, the European Commission prepared its first strategy regarding sustainable development as COM/2001/0264 Communication from the Commission A Sustainable Europe for a Better World: A European Union Strategy for Sustainable Development. This proposal was discussed and approved by the European Council during the summit in Göteborg (Gothenburg) and was accepted by the European Council. In 2002, this strategy was expanded due to UN endeavours; see the summit in Johannesburg, and, even more importantly, in 2006, it was renewed, i.e. the Review of EU Sustainable Development Strategy—Renewed Strategy was adopted by the European Council ("EU Renewed Sustainability Strategy"). The determination of smart, sustainable and inclusive growth was cemented in 2010 by the European Commission with its 10-year-long strategy, aka Europe 2020, and by the UN Resolution made during a historic UN Summit in September 2015 and entitled "Transforming our world: the 2030 Agenda for Sustainable development" ("UN Agenda 2030"), which brought with it 17 Sustainable Development Goals (SDGs) and 169 associated targets and was adopted by world leaders. Importantly, SDG 8 deals with decent work and economic growth; SDG 9 means to build resilient infrastructures, promote inclusive and sustainable industrialization and foster innovation, including the increase of information system (IS)/information technology (IT) and affordable access to the Internet; and SDG 12 wants to ensure sustainable consumption and production patterns.[11] This was clearly appreciated by the European Commission of Jean-Claude Juncker; see endeavours and updates related to, e.g., Directive 2013/34/EU on the annual financial statements, consolidated financial statements and related reports of certain types of undertakings or Regulation 2019/2088 on sustainability related disclosures in the financial services sector (SFDR). However, this is even more advanced and heralded by the

---

[10] R. MacGregor Pelikánová & R. K. MacGregor, The EU puzzling CSR regime and the confused perception by ambassadors of luxury fashion businesses: A case study from Pařížská. Central European Business Review, 2020, 9(3), pp. 74-108. DOI: https://doi.org/10.18267/j.cebr.240.

[11] R. MacGregor Pelikánová & R. K. MacGregor, The EU puzzling CSR regime and the confused perception by ambassadors of luxury fashion businesses: A case study from Pařížská. Central European Business Review, 2020, 9(3), pp. 74-108. DOI: https://doi.org/10.18267/j.cebr.240.

5 Sustainability and Competition Law in Czech Republic

strong drive for a "more green Europe" by the current European Commission of Ursula von der Leyen; see, e.g., Regulation 2020/852 on the establishment of a framework to facilitate sustainable investment, and amending Regulation aka Taxonomy Regulation, which is designed to support the transformation of the EU economy to meet its European Green Deal objectives.

As mentioned above, regarding the Czech Republic, we can observe a delay and a kind of weaker trend. In 2004, the First Strategy for Sustainable Development was prepared and approved by the Czech Government via Resolution Nr. 1242.[12] This First Strategy for Sustainable Development was set as a consensual framework for sectorial policies until 2014, while endorsing certain goals targeting 2030, and is still valid.[13] In 2006, the EU moved for a renewal; see the EU Renewed Sustainability Strategy, and the Czech Republic followed by doing basically the same the next year. Namely, in 2007, the Czech Republic renewed the Strategy for sustainable development, i.e. Resolution Nr. 1434 was prepared and approved by the Czech Government ("Czech Renewed Sustainability Strategy").[14] In the year of the adoption of Europe 2020, in 2010, the Czech Government approved, through its Resolution Nr. 37, a Strategic framework for sustainable development of the Czech Republic ("Czech Strategic Framework 2010"), which reflects the mentioned renewed EU Strategy for sustainable development from 2006. The Czech Strategic Framework sets the vision of sustainable development in a general manner, along with key principles and goals, to be applied in more detail and in more specific manners in various sectors.[15] It rests on the conventional pillars—economic, environmental and social—which should be mutually well balanced and assure the prosperity of Czech society.[16] Interestingly, the Czech Strategic Framework is structured towards five priority axis: PO1 company, individual, health; PO2 economics and innovations; PO3 territory development; PO4 ecosystems and

---

[12] Český statistický úřad (Czech Statistic Office). Udržitelný rozvoj v ČR (Sustainable development in the Czech Republic). , available at https://www.czso.cz/csu/czso/13-1134-07-2006-1 1___uvod. Accessed 1 November 2022.

[13] Ministerstvo pro místní rozvoj (Regional development Ministry). Základní dokumenty (Fundamental documets). available at https://www.mmr.cz/cs/ministerstvo/regionalni-rozvoj/informace,-aktuality,-seminare,-pracovni-skupiny/psur/uvodni-informace-o-udrzitelnem-rozvoji/zakladni-dokumenty.Accessed 1 November 2022.

[14] Český statistický úřad (Czech Statistic Office). Udržitelný rozvoj v ČR (Sustainable development in the Czech Republic). , available at https://www.czso.cz/csu/czso/13-1134-07-2006-1_1___uvod. Accessed 1 November 2022.

[15] Ministerstvo pro místní rozvoj (Regional development Ministry). Základní dokumenty (Fundamental documets),available at https://www.mmr.cz/cs/ministerstvo/regionalni-rozvoj/informace,-aktuality,-seminare,-pracovni-skupiny/psur/uvodni-informace-o-udrzitelnem-rozvoji/zakladni-dokumenty.Accessed 1 November 2022.

[16] Ministerstvo pro místní rozvoj (Regional development Ministry). Základní dokumenty (Fundamental documets), available at https://www.mmr.cz/cs/ministerstvo/regionalni-rozvoj/informace,-aktuality,-seminare,-pracovni-skupiny/psur/uvodni-informace-o-udrzitelnem-rozvoji/zakladni-dokumenty. Accessed 1 November 2022.

eco-diversity; and PO5 stable and safe society.[17] In 2017, the Czech Government approved a new strategic framework—the Strategic Framework for the Czech Republic 2030 ("Czech Strategic Framework 2030"), which sets the principles of sustainable development. Sustainable development is a complex and dynamic system: diversity should be supported, human rights are to be observed, participation and transparency need to be advanced, learning needs to be boosted and global ideas with local implementations should be recognized.[18] The Czech Strategic Framework 2030 reflects and directly refers to 17 SDGs and is presented via six chapters: 1) People and society, 2) Business model, 3) Resilient eco-systems, 4) Communities and regions, 5) Global development, and 6) Good government. Regarding the overlap of business competition and sustainability, the most relevant is Chap. 2. The business model includes 2.1) business institutions (the simplification of business conduct, support for small and medium-sized enterprises (SMEs) and start-ups), 2.2) innovations, 2.3) source management, 2.4) infrastructure and 2.5) public finances. In 2018, the Czech Government approved the Implementation Plan for Czech Strategic Framework 2030.[19] Nevertheless, even the provisions of this advanced document remain rather abstract and not suitable for direct incorporation into legislative endeavors and acts.

This leads to a burning question of whether Czech law, in particular statutes, which are valid and applicable in 2022, includes a clear reference to sustainability and in particular to the impact of sustainability on competition law. A simple legislative research project using the keyword "sustainability" indicates the following statutes and their provisions:

- Act Nr. 134/2006 Coll., on public procurement – § 129 "sustainability" can be a criterion for quality selection.
- Act Nr. 256/2004 Coll., on doing business on the capital market – § 12 management of risks – "sustainability" of business models.
- Act Nr. 372/2011 Coll., on providing health services – § 73 "sustainability" of financing health care.
- Act Nr. 29/21/1992 Coll., on banks – footnote references to the Taxonomy Regulation.
- Act Nr. 240/2013 Coll., on investment companies and funds – § 220 reference to the Taxonomy Regulation.
- Act Nr. 353/2003 Coll., on consumption tax – § 3 definition and tax exemptions in relation to "sustainable" oil.

---

[17]Ministerstvo pro místní rozvoj (Regional development Ministry). Základní dokumenty (Fundamental documets), available at https://www.mmr.cz/cs/ministerstvo/regionalni-rozvoj/informace,-aktuality,-seminare,-pracovni-skupiny/psur/uvodni-informace-o-udrzitelnem-rozvoji/zakladni-dokumenty Accessed 1 November 2022.

[18]CR2030. Dokumenty ke stažení (Documents to be downloaded), available at https://www.cr2030.cz/strategie/dokumenty-ke-stazeni/. Accessed 1 November 2022.

[19]CR2030. Dokumenty ke stažení (Documents to be downloaded), available at https://www.cr2030.cz/strategie/dokumenty-ke-stazeni/. Accessed 1 November 2022.

- Act Nr. 201/2012 Coll., on the protection of air – § 16 et foll. criteria for sustainable biofuel.
- Act Nr.406/2000 Coll., on the management of energy – § 4 territorial energetic concept and the competitiveness and sustainable use of energy.
- Act Nr. 541/2022 Coll., on waste management – § 3 hierarchy of principles – feasibility and "sustainability".
- Act Nr. 277/2009 Coll., on insurance – § 8 reference to SFDR.
- Act Nr. 165/2012 Coll., on supported sources of energy – § 4 support of biofuel.
- Act Nr. 427/2011 Coll., on additional retirement insurance – footnote references to SFDR and the Taxonomy Regulation.
- Act Nr. 326/2004 Coll., on plant care – § 37 on plant variety and "sustainable" agricultural production.
- Act Nr. 248/2000 Coll., on the support of regional development – § 17e monitoring system and the life cycle of the set period of "sustainability".
- Act Nr. 115/2001 Coll., on support for the sport – § 6b dotation (financial support) and the requirement of "sustainable" sport activity for 10 years.
- Act Nr. 23/2017 Coll., on rules about budgetary discipline – § 9 budget strategies and "sustainable" public finances.
- Act Nr. 340/2006 Coll., on the activities of institutions of employment retirement insurance – footnote reference to SFDR.

A cursory overview reveals that although sustainability is included in a myriad of Czech statutes, this inclusion is not a demonstration of proactive legislation towards sustainability in general or the competition sphere in particular. Namely, it appears that the Czech Republic merely respects its commitment to its EU membership and, when needed, goes ahead with the transposition of EU law (Directives) or the recognition of directly applicable instruments of EU law (footnote references to Regulations). Arguably, this impact is most noticeable in the sphere of biofuel and financial services.

Nevertheless, in order to double-check the absence of any Czech legislative consideration or any mention of sustainability as a factor or criteria for the assessment of competition cases, both sources of Czech competition law need to be reviewed (public law on the protection of competition via the Czech Competition Protection Act and private law on the protection against unfair competition via the Czech Civil Code). The Czech Competition Protection Act is a rather short statute with only 30 provisions that have a rather technical character and closely follow EU law, especially Article 101 et foll. TFEU, without any hint regarding sustainability. At the same time, it is plausible that the consideration and assessment of monopoly, cartel and similar cases could include the criterion of sustainability since for that, an explicit legislative mandate is not necessary. The Czech Civil Code includes unfair competition regulation in its § 2976–§ 2990, and again, sustainability is not directly or indirectly mentioned, although it is imaginable that misleading advertisement (§ 2977) or misleading labelling (§ 2978) or threat to health and the environment (§ 2987 Czech Civil Code) or other types of behaviour based on the general clause (§ 2976) can violate sustainability; see, e.g., "greenwashing".

Hence, regarding sustainability, it can be summarized that Czech law is in compliance with EU law but is definitely not proactive in this field. Czech legislation does not include any definition of sustainability, and the role of sustainability in the competition law context is not specifically addressed, but still, it appears feasible. Although there are no specific rules regarding the overlap of sustainability and competition, there is great potential that this could happen based on case law. Therefore, the examination of and engagement with the judgments of the Czech Supreme Court, Czech Supreme Administrative Court and Czech Constitutional Court are indispensable.

5.3 Case Law on the Protection of Competition and the (Lack) of Sustainability Impact

Higher Czech courts have only seldom referred, and in an auxiliary manner, to the concept of sustainability. As a matter of fact, there are very few judgments mentioning sustainability as part of *ratio decidendi*, i.e. if sustainability is mentioned, it is done so rather as a part of *obiter dictum*. These rare decisions, genuinely decided based on, or in, a direct relation to sustainability, include the judgment of the Regional Court in Hradec Králové in 31 Ca 82/2000 from 31 October 2000, where it was stated that there is a public interest in the optimal development of the life of society and the co-operation of all subjects and activities in the territory, with the goal of achieving permanent sustainable development. This decision was about expropriation and land planning, and the key point was that no building activity is to be considered as mere and exclusive self-realization of the builder because building and developing land clearly have a public dimension. Well, undoubtedly, the concept of sustainability was involved but not exactly as understood by the UN and the EU, i.e. as a system of three balanced and not mutually excluding pillars – economic, environmental and social.

There are no cases manifestly founded upon the concept of sustainability. However, at the same time, the three-pillar structure underlines the importance of the interaction and balance of economic, environmental and social concerns and priorities and implies a strong potential that sustainability will be, if not directly then at least indirectly, included or implied in the sphere of competition law case law. Considering the growing importance of sustainability and CSR and their progressively growing inclusion in EU law, it is relevant to review cases from the last 5 years, ideally the very last year, that were decided by top Czech courts and are about either the public law branch of competition law or the private law branch dealing with unfair competition.

Regarding the public law branch of competition law cases, i.e. cases decided based on the Czech Competition Protection Act, the majority of recent cases decided by the highest Czech courts are about public procurement. From them, the following five cases had a strong potential to be centred around the concept of sustainability. First, the Czech Supreme Administrative Court decided in *4 As 214/2020* on 18 March 2021 (*Esox building*) concerning the rejection of an appeal against a

5 Sustainability and Competition Law in Czech Republic

decision condemning bid rigging, i.e. the co-ordination of participation and the readjustment of offers in public procurement. The sued parties tried in vain to bring forth many arguments and mitigation circumstances but none concerning sustainability. The public procurement was for the reconstruction of a sports complex, and undoubtedly, sustainability and CSR concerns could be easily advanced. However, nobody from the large group of involved parties engaged in these concerns. Second, the Czech Supreme Administrative Court decided in *7 As 245/2019* on 13 January 2020 (*Seznam.cz*) regarding free access to information in the context of the application of the Czech Competition Protection Act. Arguably, one of factors that helped the claimant's case to prosper was its engagement in various constitutional values and the balancing of those values. So unlike 4 As 214/2020 *Esox building*, in 7 As 245/2019 *Seznam.cz*, there is at least a symbolic indirect reference to sustainability. Third, the Czech Supreme Administrative Court decided in *2 As 257/2018* from 20 March 2019 (*AV Media*) about the bid rigging and "fishing expedition" of the Czech Office for the Protection of Competition and rejected all objections by refering to the concept "a document within sight". It needs to be pointed out that this case was decided while taking into consideration the case law of the CJEU, such as *T-289/11, T-290/11 & T-521/11 Deutsche Bahn AG v. Commission* and *T-325/16 České dráhy v. Commission*, and the case law of the European Court of Human Rights (ECtHR), such as *Nr. 97/11 Delta pekárny a. s. v. Czech Republic* from 2 October 2014. Fourth, the Czech Supreme Administrative Court decided in *1 As 80/2018* on 30 January 2019 (*Daich, Hochtief*) against a broad and excessive "fishing expedition". This case is to a certain extent similar to *2 As 257/2018 AV Media* and rests upon a similar case law of the CJEU and ECtHR case law, i.e. *Nr. 97/11 Delta pekárny a. s. v. Czech Republic* and *T-621/16 České dráhy v. Commission* (not *T-325/16 České dráhy v. Commission*). In this case, i.e. *1 As 80/2018 Daich, Hochtief*, the Czech Supreme Administrative Court found that the Czech Office for the Protection of Competition went too far with its "fishing expeditions", i.e. the Czech Supreme Administrative Court rejected the superficial approach and gave a clear indication that, at least in bid rigging, the multistakeholder approach, with a consideration of all aspects, is to be employed. Fifth, the Czech Constitutional Court decided in *IV. ÚS 2350/21* from 30 September 2021 (*Baby Direkt*) that there is not *per se* a right on the return of documents, which were not properly collected during a search. Manifestly, despite having potential, these cases either have not included sustainability concerns or have done it in an indirect and rather weak and implied manner. The future will show whether they can still become a precursor of a strong case law showing how sustainability can mitigate the prohibition given or implied by the Czech Competition Protection Act.

Regarding the private law branch of competition law, i.e. protection against unfair competition as stated in the Czech Civil, the majority of the recent cases decided by the highest Czech courts are about labelling, trademarks and domain names. The following five cases had a strong potential to be centered on the concept of sustainability. Firstly, the Czech Supreme Court decided in *23 Cdo 3500/2019* on 29 July 2020 (*Hemostop*) concerning the misleading labelling of alimentary health products for health that a deeper study and analysis of the *bonos mores* (good manners) of the competition is needed while referring to a number of Regulations and Directives and the case law of the CJEU, such as *C-544/10 Deutsches Weintor eG v. Land Rheinland-Pfalz*. Secondly, the Czech Supreme Court decided in *23 Cdo 1538/2019* on 30 June 2020 (*Betonepox*) about the distinctiveness and association potential in the context of trademarks and referred to the classical CJEU case law, such as *C-251/95, SABEL BV v. Puma AG, C-39/97 Canon Kabushiki Kaisha v. Metro-Goldwyn-Mayer Inc., C-342/97 Lloyd Schuhfabrik Meyer & Co. GmbH v. Klijsen Handel BV*. Indeed, *23 Cdo 1538/2019 Betonepox* brought a myriad of concerns and arguments, including constitutional aspects, but they did not involve sustainability, and this despite the involved building products, and ultimately, the court did not find a breach of *bonos mores*. Thirdly, the Czech Supreme Court decided in *23 Cdo 4554/2017* from 25 April 2018 (*NOVASTIM*) about fertilizers and their labelling while heavily referring to the EU law, such as *Regulation 1107/2009*, and clearly supporting the direct application and priority of EU Regulations. In this context and in relation to *bonos mores*, a number of social factors were considered, such as the social and health impacts on vulnerable groups. Fourthly, the Czech Supreme Court decided in *23 Cdo 4931/2017* of 16 April 2018 (*Tecomat*) that the use of the trademark of another person in advertisements, in the context of paid search engine optimization (SEO), is an infringement of the trademark and a violation of the law. Basically, the entire argumentation of this case was based on the case law of the CJEU: *C-236/08 to C-238/08 Google France SARL & Google Inc. v. Louis Vuitton Malletier SA* and *C-323/09 Interflora v. Marks & Spencer*. Fifthly, the Czech Constitutional Court decided in *II. ÚS 1641/19* from 9 January 2020 (*SEND Turist*) about the lack of the qualification excess and declined to reverse the finding of lower courts about the lack of the breach in the reputation. Manifestly, despite having potential, these cases either have not included sustainability concerns or have done so in an indirect and rather weak and implied manner. The consideration of the *bonos mores* appears a very suitable venue to bring the concept of sustainability in the arena of unfair competition decision-making. Similarly, to the decisions about the public law branch of the competition law, the highest Czech courts do heavily rely on the case law of the CJEU. Therefore, it might be assumed that once sustainability becomes an integral part of the decisions of the CJEU, it will progressively come from above down in the Czech case law. Probably, it will first appear as one of the factors for balancing and assessing, and later on will become both a shield and sword in the competition cases. This trend might be supported by the Czech Office for the Protection of Competition, i.e. this Czech Authority can be proactive and endorse the EU law and EU policies as well as Czech policies, see e.g. the Czech Strategic Framework 2030.

5.4 Agencies and Legislature for Sustainability

The agency *par excellence* to address competition and sustainability is the Czech Office for the Protection of Competition. Its www pages reveal a set of policies and guidelines in this direction. For example, the policy "Application of rules for public support in the environmental field" deals in great detail with the environmental pillar of sustainability, especially energy effectiveness and efficiency.[20] This policy includes a description of available exemptions while referring to EU law (de minimis, individual notification to the European Commission, regulation of block exemptions, adjustment payment for services in the general economic interest). Examples of other policies involving sustainability are policies on the use of EU structural funds,[21] the modernization of public procurement rules[22] and research and development.[23]

In addition, the Czech Office for the Protection of Competition attempts to raise awareness; for example, in May 2022, it organized a public procurement conference with foreign guests, including representatives of the European Commission and supervisory bodies from Slovakia, Austria and Croatia, as well as academia. During this conference, sustainability was discussed and became one of the key points.[24]

The Czech Republic had the presidency of the Council of the EU from July to December 2022, and this may further accelerate the EU's impact and the transposition of sustainability into the Czech context, especially in the applicable law. Regarding the Czech Parliament, there are a number of committees that are, or at least should be, actively involved and concerned when it comes to sustainability, such as the *Committee for European matters, Committee for social policies, Committee for public administration and regional development and Committee for environment protection.* Occasionally, members of Parliament issue statements dealing with sustainability, but typically, these endeavors are not systematic and appear more as *ad hoc* declarations.

---

[20] ÚOHS. Available at https://www.uohs.cz/cs/verejna-podpora/vybrane-oblasti-verejne-podpory/aplikace-pravidel-verejne-podpory-v-oblasti-zivotniho-prostredi.html. Accessed 1 November 2022.

[21] ÚOHS, available at https://www.uohs.cz/cs/o-uradu/cerpani-ze-strukturalnich-fondu-eu.html. Accessed 1 November 2022.

[22] ÚOHS, available at https://www.uohs.cz/cs/verejna-podpora/modernizace-pravidel-verejne-podpory.html. Accessed 1 November 2022.

[23] ÚOHS, available at https://www.uohs.cz/cs/verejna-podpora/vybrane-oblasti-verejne-podpory/aplikace-pravidel-verejne-podpory-v-oblasti-vyzkumu-vyvoje-a-inovaci.html. Accessed 1 November 2022.

[24] ÚOHS, available at https://www.uohs.cz/en/information-centre/press-releases/public-procurement/3321-may-2022-public-procurement-conference.html. Accessed 1 November 2022.

5.5 Conclusion

The Czech Republic is part of the EU, and Czech law shares the continental law tradition. Czech competition law, both its public law branch protecting against abuse of monopoly and cartels and its private law branch protecting against unfair competition, is in compliance with EU law and meets the standards of the twenty-first century. The UN's and EU's drive for sustainability is recognized in the Czech Republic and has been projected in a myriad of Czech policies and strategic documents since 2004. However, so far, the move towards direct legislation and explicit inclusion in enforceable policies has not occurred, and even Czech case law is rather cautious. As a matter of fact, the top Czech courts – the Czech Supreme Court, the Czech Supreme Administrative Court and especially the Czech Constitutional Court – have, several times, already been invited, or at least allowed, to bring up sustainability in their judicial considerations and to decide cases while considering the concept of sustainability and its practical consequences. Great opportunities in this respect are various balancing test cases addressing public law aspects of competition and *bonos mores* linked to private law aspects of competition (unfair competition). The Czech Constitution and its Charter of fundamental rights and duties, perhaps even the EU, even allow the push for sustainability, along with the highest values. Nevertheless, so far, judges are rather reserved. Considering the impact of current crises, which typically magnify pre-existing trends, and the priorities of the current European Commission, the official entry of the concept of sustainability into the enforceable Czech law should be just a matter of time, perhaps months but definitely not long years.

Open Access This chapter is licensed under the terms of the Creative Commons Attribution 4.0 International License (http://creativecommons.org/licenses/by/4.0/), which permits use, sharing, adaptation, distribution and reproduction in any medium or format, as long as you give appropriate credit to the original author(s) and the source, provide a link to the Creative Commons license and indicate if changes were made.

The images or other third party material in this chapter are included in the chapter's Creative Commons license, unless indicated otherwise in a credit line to the material. If material is not included in the chapter's Creative Commons license and your intended use is not permitted by statutory regulation or exceeds the permitted use, you will need to obtain permission directly from the copyright holder.

Sustainability and Competition Law: A French Perspective

6

Michaël Cousin, Alexandre Marescaux, Lauren Mechri, and Guillaume Melot

6.1 Sustainability in the French Competition Law Framework

6.1.1 A Concept Akin to Environmental Protection

The word "sustainability" is absent from the legislative and regulatory provisions composing the French competition law. Therefore, it is not possible to say what this concept exactly means legally speaking.

Sustainability has a broader meaning than environmental protection. It refers to an objective of development that meets the needs of the present without compromising the ability of future generations to meet their own needs.[1] Environmental protection is obviously at the core of sustainability. However, it is one aspect, among others, of sustainability. In this regard, Article 11 TFEU refers to environmental protection as a means to promote sustainable development. This conveys the idea that the latter is one means, among others, to achieve sustainability.

---

[1] OECD Competition Committee Discussion Paper, Sustainability and Competition, 2020, p. 11 ff.

M. Cousin (✉)
Addleshaw Goddard, Paris, France
e-mail: Michael.Cousin@aglaw.com

A. Marescaux
Fidal, Paris, France
e-mail: alexandre.marescaux@fidal.com

L. Mechri
Fieldfisher, Paris, France
e-mail: lauren.mechri@fieldfisher.com

G. Melot
CMS Francis Lefebvre, Paris, France
e-mail: guillaume.melot@cms-fl.com

© The Author(s) 2024
P. Këllezi et al. (eds.), *Sustainability Objectives in Competition and Intellectual Property Law*, LIDC Contributions on Antitrust Law, Intellectual Property and Unfair Competition, https://doi.org/10.1007/978-3-031-44869-0_6

That being said, when reflecting on the interactions between competition law and sustainability, it would seem relevant to focus on the environmental dimension of that concept. Indeed, the interaction between competition law and environmental protection is on the agenda of a number of competition authorities throughout the EU, let alone the EU Commission's reflexion on how competition policy may support the EU's Green Deal,[2] which purpose is to make Europe the first climate-neutral continent by 2050.

6.1.2 Sustainability in French Competition Law

While the concepts of sustainability and environmental protection are not enshrined in the French competition law rules, they play a role in competition law enforcement in France.

Firstly, it is now well established that the environmental quality or performance of a good or service is a parameter on which undertakings may compete.[3] This means that any conduct that may hinder competition on that aspect of competition is liable to fall under the prohibition.

Secondly, Article L. 420-4 of the French Code of Commerce (**FCC**) contains, similarly to the third paragraph of Article 101 of TFEU, exemption criteria applicable to agreements or concerted practices that restrict competition. "Economic progress" is one of these criteria. Yet according to the French Competition Authority (**FCA**), environmental protection is a form of economic progress, which may be the basis for an exemption, provided that the other exemption criteria are met.[4]

In the field of merger control, economic progress is also a factor liable to outweigh the harm to competition caused by a concentration. It is enshrined in Article L. 430-6 of the FCC. The FCA has, for instance, relied, *inter alia*, on the fact that a merger helped the parties to abandon technologies damaging to the environment in order to give its green light to a merger.[5]

Economic progress is a broad concept, and one could anticipate that it gives great leeway to the FCA to include environmental considerations in its competitive assessment. However, it is governed by a principle of proportionality, which the FCA implements in a cautious manner.

The FCA will typically assess whether an environmental goal exists and, if so, whether it is not already pursued by the applicable regulation.[6] It also makes sure that the environmental benefits may not be achieved by means that are less restrictive

---

[2]European Commission, Competition Policy supporting the Green Deal – Call for contributions, 13 October 2020.

[3]See B. Coeuré, Closing speech, New Frontiers of Antitrust, 21 June 2022.

[4]FCA, Opinion n°99-A-22 of 14 December 1999, p. 10.

[5]FCA, Opinion n°94-A-18 of 17 May 1994, Metaleurop/Heubach & Lindgens.

[6]FCA, Opinion n°95-A-08 or 9 May 1995, CEAC/Exeide.

of competition.[7] With regard to anti-competitive agreements or concerted practices, it must also be demonstrated that a fair share of the benefits invoked are passed on to consumers and that the agreement or concerted practice does not eliminate competition. According to the president of the FCA, the FCA is prepared to work at better assessing the passing on of environmental benefits to consumers.[8]

6.1.3 The Future of Sustainability in French Competition Law Enforcement

The FCA, like any other organisation, is keen to play a role in the fight against global warming. Given the necessity to use any means possible to achieve that goal, it is tempting to claim that the FCA could and should give more weight to environmental protection in the way it enforces competition law rules. This, however, deserves a more detailed assessment.

In the current legal context, giving more weight to environmental protection may be achieved essentially in two different ways.

First, the FCA may decide to be more severe with anti-competitive practices or concentrations that are harmful to the environment. With regard to anti-competitive practices, this would mean, in practice, that the FCA would apply a specific aggravating factor to practices having a negative impact on the environment or affecting a competitive differentiator linked to environmental protection (e.g. the environmental performances of a given product). As will be described below in this report, such an approach may already be found in the FCA's decisional practice. It is not difficult to implement in practice, given the margin of appreciation that the FCA enjoys as regards the setting of the fine in antitrust matters.

With regard to concentrations, giving more weight to environmental protection would mean that the FCA would be prepared to prohibit concentrations or submit them to remedies on the basis that they threaten to harm the environment, leaving aside other issues strictly related to market functioning. This would be much more difficult to implement in practice. The actual impact of concentrations on the environment is still unknown, and the tools do not really exist to measure it. This means that in the event that the FCA would intend to block concentrations on the basis that they harm the environment, it would struggle to actually measure, if not identify, that harm.

Second, and by contrast, the FCA may decide to be more lenient with concentrations or anti-competitive practices bringing about environmental efficiencies. In this paragraph, we assume that these efficiencies would not be of the type that benefits the consumers because if they benefit the consumers, there is no reason that the FCA does not already factor them in its competitive assessment.

---

[7] FCA, Opinion n°94-A-31 of 6 December 1994, Elimination of used oils.

[8] B. Coeuré, speech quoted in note 4 above.

Being more lenient with "eco-friendly" conducts or concentrations would probably imply that restrictions of competition not acceptable under the FCA's current approach would be accepted because they would be justified by considerations pertaining to environmental protection. Such a policy would face two hurdles. First, it would oblige the FCA to identify and measure the environmental efficiencies at hand. Yet, as already said, the relevant conceptual tools are still missing. Second, this would obviously lead the FCA to depart from the proportionality principle that governs its enforcement powers. The balance made between restrictions of competition and efficiencies would indeed have to factor in a new parameter, namely the benefits for the environment.

The FCA does not seem ready for this. Its doctrine is summarised in a document published in May 2020 and authored by an informal group of eight independent regulation authorities (including the FCA) regarding the regulatory challenges posed by the Paris Agreements and the climate emergency.[9] This document, which is not an instrument of soft law but rather a roadmap, expresses the regulator's commitment to help undertakings adapt to the changes caused by global warming. The document describes the challenges faced by the economic sectors to which a regulation applies. On that basis, the regulators point out the necessity to articulate the enforcement of their powers with the objectives of the Paris Agreement (i.e. a reduction of greenhouse gas emissions to achieve a climate-neutral world by mid-century). Yet this document makes it clear that *the mandate given to the FCA is to protect competition, not to protect the environment*. This is another way of saying that the FCA may not use powers that it does not have, like including considerations that are not strictly related to market functioning in its competition law assessments.

This is difficult to object to. Indeed, the issue of whether the FCA's powers should be extended to include an environmental dimension boils down to questioning the role of a competition authority. The issue is whether such an authority should go beyond ensuring the optimal functioning of markets. This is a matter of public policy rather than a competition law issue.

That being said, it would seem possible, without giving more weight to environmental protection in the FCA's decisional practice, to improve or facilitate the inclusion of environmental considerations in it.

One may for instance consider the development of an analytical framework helping and better translate environmental efficiencies (or positive externalities) which are not immediately market-related, into consumer benefit or harm liable to be factored in a competitive assessment.

---

[9] Autorité de la concurrence, AMF, Arcep, ART, CNIL, CRE, CSA, HADOPI, Accords de Paris et urgence climatique : enjeux de régulation, May 2020. Available at https://www.arcep.fr/fileadmin/user_upload/publications/cooperation-AAI/publication_AAI-API_Accord_de_Paris_052020.pdf. Accessed 27 October 2022.

6.2 Sustainability in the French Competition Law Decisional Practice and Case Law

While French competition law leaves little or no room for environmental protection in the enforcement of competition rules, the decisional practice of the FCA and case law give some room for sustainability in competition law enforcement.

The consideration of sustainability in decision-making practice and case law can be approached from several angles.

First, it can be considered that competition law is at the service of sustainability by sanctioning behaviour that is either not very or not at all virtuous from an environmental point of view. We can then speak of "sustainability as a sword" because the protection of the environment becomes, through competition rules, a weapon against less virtuous companies.

Second, it may be considered that in cases where there is a conflict between competition rules and the protection of sustainability, *i.e.* where anti-competitive behaviour or a merger that restricts competition is favourable to sustainability, exemptions from competition law should be granted to promote environmental protection. In this case, we can speak of "sustainable development as a shield" since the protection of the environment becomes a means of avoiding sanctions from competition authorities.

Third, we can also consider cases where illegal anti-competitive conduct occurred in the context of sustainability initiatives. In this case, we can speak of "sustainability as a trap" in that companies have engaged in anti-competitive behaviour by seeking to protect the environment.

The developments that follow will aim to determine to what extent French decision-making practice and competition case law used sustainability "as a sword" (Sect. 6.2.1) or "as a shield" (Sect. 6.2.2) or where sustainability has become a trap for undertakings willing to improve the sustainability of their operation through multilateral or unilateral initiatives that have proven to be anti-competitive (Sect. 6.2.3).

6.2.1 Sustainability "as a Sword"

6.2.1.1 Sustainability "as a Sword" in Antitrust

As said in Sect. 6.1.2. above, it is now well established that the environmental quality or performance of a good or service is a parameter on which undertakings may compete so that any conduct liable to hinder competition on that aspect of competition is liable to fall under the prohibition of anti-competitive conduct. In this case, we can consider that sustainability acts "as a sword". We can also consider that sustainability acts "as a sword" when the damage to it is taken into account as an element of gravity or aggravation in the calculation of the sanction.

The FCA has already had the opportunity, on several occasions, to sue and condemn anti-competitive agreements found to be harmful to the environment.

First, as regards anti-competitive agreements, in 2017, the FCA imposed fines of a total of €302 million on three leading manufacturers of PVC and linoleum flooring in France and a trade association.[10] The collusion classically involved many aspects of the companies' commercial policy, including prices, and exchanges of sensitive commercial information within the trade association.

Moreover, the agreement consisted in the signature of a non-competition agreement regarding communication on the environmental performances of their products: this agreement aimed to eliminate all "competitive marketing initiatives on the environmental characteristics" and "to avoid any sterile polemic relating to this or that product and to adopt a coherent marketing approach" in order to avoid any "dangerous green marketing" (free translation). For instance, a flooring distributor was refused permission to communicate with its customers on the emission rate of volatile organic chemicals per product.

These elements were taken into account at the stage of the calculation of the sanction as an element of gravity of the practice:

> This practice prevented the implementation of competition based on environmental performance, while the value of products on this point constitutes, in the sector concerned, one of the essential criteria for the choice of buyers. This agreement may also have discouraged companies from improving the technical performance of their products and from investing in innovative processes aimed at improving their environmental performance, in particular with regard to the emission of volatile organic chemicals, which are considered likely to have an impact on human health. Thus the practice in question is, by its very nature, particularly serious (free translation).[11]

Furthermore, the FCA's investigation services recently (late 2021) issued a Statement of Objections to companies that are accused of having agreed not to disclose the presence of bisphenol A (BPA) in certain food contact materials.[12] The companies involved are manufacturers of cans and other packaging that may contain BPA, food manufacturers, retailers and trade associations. These companies would have agreed not to market BPA-free products and/or advertise these new products before the entry into force of the French law banning BPA in coatings inside food packaging due to alleged adverse effects on human health. The cartel would thus have aimed at not gaining a competitive advantage from the BPA ban.

Second, as regards dominance, the FCA also prosecutes behaviours that are detrimental to sustainability. Thus, in 2014, the FCA accepted Nespresso's commitments to unlock the market for capsules compatible with its coffee machines.[13] Nespresso was accused of having tried to drive competing capsule

---

[10] FCA, decision 17-D-20 of 18 October 2017.

[11] FCA, decision 17-D-20 of 18 October 2017, point 456.

[12] Bisphenol A in food containers: the general rapporteur indicates having stated objections to 101 companies and 14 professional organisations, available at https://www.autoritedelaconcurrence.fr/en/press-release/bisphenol-food-containers-general-rapporteur-indicates-having-stated-objections-101. Accessed 27 October 2022.

[13] FCA, decision 14-D-09 of 4 September 2014.

6 Sustainability and Competition Law: A French Perspective

manufacturers out of the market through successive modifications to Nespresso machines. That had the effect of making capsules from competing manufacturers incompatible with the new models. Nespresso was also accused of disseminating information encouraging consumers to use only Nespresso-branded capsules and of developing communication in the press and in its stores encouraging consumers to use only Nespresso-branded capsules. While the FCA's decision does not explicitly mention environmental issues related to the practice, it paved the way for the development by competitors of more sustainable capsules, such as biodegradable ones.

6.2.1.2 Sustainability "as a Sword" in Merger Control

With regard to merger control, sustainability may act "as a sword" where the FCA would prohibit mergers or submit them to remedies on the basis that they threaten to harm the environment, leaving aside issues strictly related to market functioning.

To our knowledge, there is no case in which the FCA has taken into account the impact of a merger on sustainability in order to prohibit it or make it subject to commitments. Neither does it appear that the reduction of competition on innovation in favour of more sustainable products has been explicitly put forward.

However, sustainability can also be seen as a sword in merger control when green products are considered a separate market from the same product category but "non-green". Indeed, in doing so, the FCA indirectly supports sustainability by ensuring that a sufficient level of competition remains in these green markets and by guaranteeing continued innovation in them. Conversely, mixing green and "non-green" products may result in mitigating the strengthening of a dominant position in green products, which would then not be treated in the assessment of the merger.

It should be pointed out that the market definition is not actually used to fit sustainability objectives. That effect is only incidental because the delimitation of the markets within the meaning of competition law turns out to protect environmental interests. This paradigm is also to be found in the area of anti-competitive practices: also in this case, the environment is often protected not in itself but because it constitutes a parameter of competition.

On this aspect, it can be said that sustainability became a sword in the decision-making practice of the FCA, which released decisions where green products were taken into account.

For instance, in 2021, the FCA conducted an analysis with regard to the acquisition of 100 Bio c'Bon stores (a French organic food distribution network) by Carrefour (a very important French food and non-food distribution network).[14] In order to assess the impact on the competition environment, the FCA assessed the substitutability of demand between organic and conventional products, recognising the existence of a specific market for organic products:

---

[14]FCA, decision 21-DCC-161 of 10 September 2021.

> At the Authority's request, and in cooperation with the Authority, the notifying party has carried out a survey of consumers of organic products. The analysis of the responses to the survey shows, inter alia, that in the event of a 10% price increase for all organic products, a very large majority of consumers would continue to consume organic products. The survey therefore shows that the substitutability, in the eyes of consumers, between organic and conventional products is limited (free translation).[15]

The delimitation of this market for the distribution of organic food products has made it possible to identify risks of harm to competition in these markets in ten catchment areas due to very high market shares. To resolve the competition problems identified, Carrefour undertook to sell eight shops to competitors.

This decision is an example of how sustainability can become a sword by acknowledging the existence of specific markets for green products: by acknowledging the existence of an organic product market, the FCA has been able to protect competition in this specific market and, therefore, indirectly the protection of the environment and people's health.

6.2.2 Sustainability "as a Shield"

As discussed in Sect. 6.1.2 above, French law provides for provisions similar to those in EU law allowing for the exemption of certain anti-competitive practices that meet several cumulative conditions aimed at establishing a positive economic outcome of anti-competitive practices.

Indeed, like Article 101(3) TFEU, Article L. 420-1 of the FCC provides that the provisions of Article L. 420-1 of the FCC, which prohibits anti-competitive agreements, do not apply to practices for which it can be demonstrated that they have the effect of ensuring economic progress and that they allow users a fair share of the resulting benefit without giving the undertakings concerned the possibility of eliminating competition in respect of a substantial part of the products in question.

The exemption provided for in Article L. 420-4 of the FCC differs from the one contained in Article 101(3) TFEU since Article 101(3) TFEU applies only to anti-competitive agreements, whereas Article L. 420-4 of the FCC also applies to abuses of dominant positions, which are sanctioned under French law by Article L. 420-2 of the FCC.

Although, in theory, the protection of the environment is not expressly included among the criteria that may justify an anti-competitive practice under French law, the FCA nevertheless recognised very early that, with regard to the exemption of anti-competitive agreements, the protection of the environment constituted a form of "economic progress" referred to in the above-mentioned articles.

First, in an opinion of 28 March 1973 on the competition in the markets for the collection and regeneration of used oils,[16] the FCA considered that the prevention of

---

[15] Ibid, point 24.

[16] FCA, advice of 28 March 1973.

6 Sustainability and Competition Law: A French Perspective

environmental pollution may be a part of the notion of "economic progress"; however, it refused in this case to exempt anti-competitive practices as it considered that these practices meet objectives unrelated to this concern. Furthermore, in an opinion of 6 December 1994 on a draft decree regulating the elimination of used oils,[17] the FCA stated that the objective to prevent pollution related to used oils could not lead to the organisation of a sector involving a risk of serious harm to competition.

Second, in a decision of 3 May 1988 relating to practices observed on the salt market,[18] the FCA granted an exemption to an association of salt producers on the basis that this association of undertakings aimed, *inter alia*, at the preservation of the environment.

Third, in an opinion of 14 December 1999 on the organisation and financing of the used battery disposal sector,[19] the FCA stated that environmental protection is a form of "economic progress" referred to in Article L.420-4 (then Article 10 of the Order of 1 December 1986) and Article 101§3 TFEU (then Article 85§3 of the Treaty on European Communities). In this respect, the French Competition Authority affirmed that anti-competitive practices favourable to the environment could benefit from an exemption if all other conditions set in these articles were met. However, the FCA stated that, in most cases, environmental protection is only one of the aspects of economic progress taken into account.

It can be noted, however, that most of these FCA decisions are quite old, and there are no recent examples in the FCA's decision-making practice of anti-competitive agreements being exempted because of their favourable effect on sustainability. Furthermore, there is no FCA decision to date exempting abuses of dominance on this basis.

As regards mergers, in an opinion of 17 May 1994 concerning the proposed creation of a joint subsidiary of the companies Metaleurop and Heubach & Lindgens,[20] the FCA accepted that the proposed merger could be authorised if, in particular, it contributed to economic progress by allowing the abandonment of the most polluting techniques.

However, in an opinion of 9 May 1995 on the acquisition of Compagnie Européenne d'Accumulateurs (CEAC) by Exeide,[21] the FCA issued an unfavourable opinion on the merger on the ground that the economic progress invoked, in particular better environmental protection, consisted of environmental protection, solely compliance with the applicable law.

---

[17] FCA, advice 94-A-31, 6 December 1994.

[18] FCA, decision 88-D-20, 3 May 1988.

[19] FCA, advice 99-A-22, 14 December 1999.

[20] FCA, advice 94-A-18, 17 May 1994.

[21] FCA, advice 95-A-08, 9 May 1995.

6.2.3 Sustainability "as a Trap"

It can easily be imagined that genuine sustainability initiatives serve as a springboard for other anti-competitive behaviour and that the FCA may consider it as a motive or pretext for engaging in anti-competitive behaviour rather than genuine sustainability initiatives.

To date, we are not aware of any decision by the FCA concerning a sustainable development or environmental protection initiative that has led to anti-competitive behaviour.

There is no doubt that in the years to come, decisions of this kind will emerge, particularly with regard to the positions taken by the Competition Authority in this area.

Indeed, the Competition Authority has recently indicated that various behaviours are under scrutiny: "These are behaviours which, under the guise of commitments to environmental or sustainable development objectives, serve to create and conceal an agreement or an abuse, by implementing prohibited practices, such as price fixing, production limitation, market allocation or the eviction of existing or potential competitors."[22]

As an example, we can cite the aforementioned decision[23] regarding practices implemented in the hard-wearing floor covering sector, which was set up by the FCA as one of the first emblematic decisions in this area, even though it cannot be considered to be particularly reflective of the sustainable development trap insofar as, although the individual environmental performance of companies was in question, it was not really a "green" initiative that was in question.

Open Access This chapter is licensed under the terms of the Creative Commons Attribution 4.0 International License (http://creativecommons.org/licenses/by/4.0/), which permits use, sharing, adaptation, distribution and reproduction in any medium or format, as long as you give appropriate credit to the original author(s) and the source, provide a link to the Creative Commons license and indicate if changes were made.

The images or other third party material in this chapter are included in the chapter's Creative Commons license, unless indicated otherwise in a credit line to the material. If material is not included in the chapter's Creative Commons license and your intended use is not permitted by statutory regulation or exceeds the permitted use, you will need to obtain permission directly from the copyright holder.

---

[22] FCA Sustainable development and competition. A https://media.autoritedelaconcurrence.fr/adlc-bilan-activite-2020/en/a-major-and-strategic-concern/. Accessed 27 October 2022.

[23] FCA, decision 17-D-20 of 18 October 2017.

Sustainability and Competition Law in Germany

7

Eckart Bueren and Jennifer Crowder

7.1 Introduction

German competition law has been witnessing a long-standing discussion on how and to what extent non-economic goals can and should play a role in the application of the German Act against Restraints of Competition (GWB).[1] Recently, this discussion has centered on sustainability as a partly economic, partly non-economic goal, which is universally recognised but seems to lack a clear-cut definition (see Sect. 7.2.1). Indeed, promoting sustainability, in particular with regard to climate protection and the protection of human rights in supply chains, is currently one of the most prominent hot topics in economic law in many jurisdictions. In Germany as well as in Europe, not only company law[2] and capital market law[3] but also competition law has come into the focus of this debate. At the outset, it should be stressed that most scholars and practitioners in Germany share the view that sustainability, in particular climate protection, and competition law do not usually collide as competition

---

[1] See e.g. K. Hübner, Außerkartellrechtliche Einschränkungen des Kartellverbots, Carl Heymanns Verlag 1971; J. Fatschek, Die Berücksichtigung außerwettbewerblicher Gesichtspunkte bei Anwendung der Zusammenschlusskontrolle, Lehmann Gerbrunn 1977; J. Hackl, Verbot wettbewerbsbeschränkender Vereinbarungen und nichtwettbewerbliche Interessen, Nomos 2010; Monopolkommission (ed), Politischer Einfluss auf Wettbewerbsentscheidungen, Nomos 2015; L. Fries, Die Berücksichtigung außerwettbewerblicher Interessen in der Fusionskontrolle, Nomos 2020; R. Podszun, Außerwettbewerbliche Interessen im Kartellrecht und ihre Grenzen, in: Kokott, Pohlmann and Polley (eds), Europäisches, deutsches und internationales Kartellrecht: Festschrift für Dirk Schroeder zum 65. Geburtstag, Otto Schmidt 2018, pp. 613-632;

[2] For an overview see H. Fleischer, Klimaschutz im Gesellschafts-, Bilanz- und Kapitalmarktrecht, DB 2022(1), pp. 37-45.

[3] For an overview see E. Bueren, Sustainable Finance, ZGR 2019(5), pp. 813-875.

E. Bueren (✉) · J. Crowder
Georg-August-University Göttingen, Göttingen, Germany
e-mail: eckart.bueren@jura.uni-goettingen.de; jennifer.crowder@jura.uni-goettingen.de

© The Author(s) 2024
P. Këllezi et al. (eds.), *Sustainability Objectives in Competition and Intellectual Property Law*, LIDC Contributions on Antitrust Law, Intellectual Property and Unfair Competition, https://doi.org/10.1007/978-3-031-44869-0_7

promotes an efficient use of resources in line with sustainability targets (see Sect. 7.3.1). However, conflicts may arise, in particular in situations of market failure (see Sect. 7.2.2).[4]

This discussion is not completely new. As a spin-off of the debate and the vast scholarship on (certain) non-economic goals and competition policy, the period between (at least) the early 1970s and the late 1990s saw an intense discussion on German competition law vis-á-vis environmental protection, in particular with regard to so-called self-restraint agreements by undertakings (Selbstbeschränkungsvereinbarungen) and regulation on waste disposal.[5] Besides, since the 1990s, several German monographs have been devoted to the topic, placing a focus on European competition law.[6]

Especially since 2020, the discussion has gained traction again in light of the broader concept of sustainability but is still focussing on environmental aspects, as evidenced by a steady stream of recent articles in German journals and commemorative publications.[7] From a legal perspective, this renewed attention on

---

[4] See Monopolkommission, XXIV. Hauptgutachten, Wettbewerb 2022, BT-Drs. 20/3065, paras 400-401, 409-413; Bundeskartellamt, Open markets and sustainable economic activity – public interest objectives as a challenge for competition law practice, October 2020, pp. 6-7.

[5] See D. Freitag, K. Hansen, K. Markert and V. Strauch, Umweltschutz und Wettbewerbsordnung, Athenäum 1973; G. v. Wallenberg, Umweltschutz und Wettbewerb, V. Florentz 1980; E. Sacksofsky, Wettbewerbliche Probleme der Entsorgungswirtschaft, WuW 1994(4), pp. 320-322; T. Lappe, Zur ökologischen Instrumentalisierbarkeit des Wettbewerbsrechts, WRP 1995(3), pp. 170-180; H. Köhler, Abfallrückführungssysteme der Wirtschaft im Spannungsfeld von Umweltrecht und Kartellrecht, BB 1996(50), pp. 2577-2582; K. Becker-Schwarze, Steuerungsmöglichkeiten des Kartellrechts bei umweltschützenden Unternehmenskooperationen: das Beispiel der Verpackungsverordnung, Nomos 1997; R. Velte, Duale Abfallentsorgung und Kartellverbot: eine Untersuchung zur Zulässigkeit von Umweltschutzkartellen nach deutschem und europäischem Recht am Beispiel des Dualen Systems für Verkaufsverpackungen, Nomos 1999.

[6] On European competition law see e.g. W. F. v. Bernuth, Umweltschutzfördernde Unternehmenskooperationen und das Kartellverbot des Gemeinschaftsrechts, Nomos 1996; K. Rook, Umweltschutz und Wettbewerb im EG-Recht, eine Untersuchung unter besonderer Berücksichtigung der Verwirklichung von Umweltschutzzielen im Kartellrecht nach Article 85 EGV, Institut für Europarecht der Universität Osnabrück 1997; D. Ehle, Die Einbeziehung des Umweltschutzes in das Europäische Kartellrecht: eine Untersuchung zu Article 85 EGV unter besonderer Berücksichtigung kooperativer abfallrechtlicher Rücknahme- und Verwertungssysteme, Heymanns 1997; C. Klados, Umweltschutz als Freistellungsgrund im Gemeinschaftskartellrecht: die Anwendbarkeit der Article 85 Abs. 1 und 3 EGV auf Kooperationen zugunsten des Umweltschutzes am Beispiel der Abfallwirtschaft, LIT 1999; J. Rabus, Die Behandlung von Effizienzvorteilen in der europäischen Fusionskontrolle und in Article 81 Abs. 3 EG, Duncker & Humblot 2008; L. Breuer, Das EU-Kartellrecht im Kraftfeld der Unionsziele, Nomos 2013.

[7] See e.g. T. Lübbig, Nachhaltigkeit als Kartellthematik, WuW 2012(12), pp. 1142-1155; J. Hoffmann et al. (2015), Nachhaltigkeit im Wettbewerb verankern, available at https://library. fes.de/pdf-files/wiso/11440.pdf. Accessed 5 November 2022; R. Podszun, Außerwettbewerbliche Interessen im Kartellrecht und ihre Grenzen, in: Kokott, Pohlmann and Polley (eds), Europäisches, deutsches und internationales Kartellrecht: Festschrift für Dirk Schroeder zum 65. Geburtstag, Otto Schmidt 2018, pp. 613-632; S. Kingston, Greening Competition Law, WuW 2020(6), p. 293; D. Zimmer, Was ist eine Wettbewerbsbeschränkung? Eine Neubestimmung, in: Klose, Klusmann and Thomas (eds), Das Unternehmen in der Wettbewerbsordnung, Festschrift für Gerhard

sustainability-related matters is motivated by international law obligations for climate protection, both by European law and by German constitutional law (see Sect. 7.2.1). Another important driver is the practical need for companies to react to the aforementioned developments: companies must adapt to new laws on supply chain liability[8] and sustainable finance, to increasing societal pressure to stop irresponsible business practices and to growing consumer demand for ESG-compatible business behaviour and products.[9] For these reasons, several companies have already approached the German Federal Cartel Office for advice on whether they can pursue joint sustainability initiatives (see Sect. 7.3). Against this background and in view of recent policy activities in other jurisdictions, such as the Netherlands and on the EU level, the current German government has announced that it will examine whether changes in the GWB to facilitate sustainability initiatives are in order (see Sect. 7.4).

Wiedemann zum 70. Geburtstag, C.H. Beck 2020, pp. 269-284; F. Engelsing, Nachhaltigkeit und Wettbewerbsrecht, DB 2020(10), M4-M5; M. Otto, Gemeinwohlorientierte Absprachen in der Wirtschaft, FAZ 13.02.2021, p. 22; J. Dreyer and E. Ahlenstiel, Berücksichtigung von Umweltschutzaspekten bei der kartellrechtlichen Bewertung von Kooperationen, WuW 2021(2), pp. 76-81; A. Heinemann, Nachhaltigkeitsvereinbarungen, sic! 2021(5), pp. 213-227; W. Frenz, Klimaschutz und Wettbewerb in der digitalen Kreislaufwirtschaft, WRP 2021(8), pp. 995-1003; A. Wambach, Gemeinwohlziele und Wettbewerb, FAZ 20.08.2021, p. 16; D. Seeliger and K. Gürer, Kartellrecht und Nachhaltigkeit, Neue Regeln für Umweltschutzvereinbarungen von Wettbewerbern, BB 2021(6), pp. 2050-2056; D. Seeliger and K. Gürer, ESG und Kartellrecht, neue Chancen für Unternehmen, DB 2021(20), supplement. no. 2, pp. 39-41; A. Bischke and S. Brack, Neuere Entwicklungen im Kartellrecht, NZG 2022(9), pp. 393-396; R. Inderst and S. Thomas (2022), Nachhaltigkeit und Wettbewerb: Zu einer Reform des Wettbewerbsrechts für die Erreichung von Nachhaltigkeitszielen, available at https://safe-frankfurt.de/fileadmin/user_upload/editor_common/Policy_Center/SAFE_Policy_Letter_94.pdf. Accessed 5 November 2022; C. Ritz and F. v. Schreitter, Chain(ed) Reaction? Das Lieferkettengesetz und seine kartellrechtlichen Hürden, NZKart 2022(5), pp. 251-259; J. Hertfelder and D. Drixler, Ein „more sustainable economic approach" – der niederländische Leitlinienentwurf zu Nachhaltigkeitsvereinbarungen im Kartellrecht, BB 2022(22), pp. 1218-1224; E. Wiese (2022), Agenda 2025 - Roadmap to Sustainability, available at https://www.d-kart.de/blog/2022/05/18/agenda-2025-roadmap-to-sustainability/. Accessed 5 November 2022; See also S. Legner, Die Relevanz eines Geschlechteraspekts für das Kartellrecht, ZWeR 2020(2), pp. 289-312; J. Haucap, C. Heldmann and H. Rau, Die Rolle von Geschlechtern für Wettbewerb und Kartellrecht, WuW 2021(7), pp. 408-412; On EU law W. Frenz, Kartellrecht und Umweltschutz im Zeichen der Energiewende, WRP 2012(8), pp. 980-989; C. Mayer, Der Beitrag des Kartellrechts zum Green Deal, WuW 2021(5), pp. 258-260; M. Olthoff and A. v. Bonin, Das „Hinkley Point"-Urteil des EuGH: Berücksichtigung von Nachhaltigkeitsaspekten in der beihilferechtlichen Prüfung sowie allgemein im EU-Kartell- und Fusionskontrollrecht, EuZW 2021(5), pp. 181-187; O. Schley and M. Symann, Article 101 Abs. 3 AEUV goes green (und Nachhaltigkeit), WuW 2022(1), pp. 2-7.

[8] On the ensuing questions for competition law C. Ritz and F. v. Schreitter, Chain(ed) Reaction? Das Lieferkettengesetz und seine kartellrechtlichen Hürden, NZKart 2022(5), pp. 251-259.

[9] D. Seeliger and K. Gürer, Kartellrecht und Nachhaltigkeit, Neue Regeln für Umweltschutzvereinbarungen von Wettbewerbern, BB 2021(6), p. 2050.

7.2 General Framework

7.2.1 Definition of Sustainability

There is no statutory definition of sustainability specifically for German competition law. Neither the German Federal Cartel Office nor competition law scholarship offer clear-cut definitions of what sustainability precisely is. That said, sustainability is generally understood to be a broad concept that should be determined in particular with a view to UN standards and international treaties, such as the Paris Agreement of 2015. The German Federal Cartel office usually stresses that sustainability includes environmental as well as social aspects or standards,[10] sometimes referring to the UN agenda 2030 for sustainable development[11] with its 17 sustainable development goals.[12] In a similar vein, journal articles on competition law briefly state that sustainability includes environmental, social and governance standards.[13]

While it is generally accepted that sustainability comprises all ESG dimensions, the environmental dimension of sustainability, in particular climate protection, is currently at the heart of the case law (see Sect. 7.3) and the academic as well as political discourse in Germany.[14] Secondly, social standards such as the protection

---

[10] See Bundeskartellamt, Open markets and sustainable economic activity – public interest objectives as a challenge for competition law practice, October 2020, p. 5; Bundeskartellamt (2021), Jahresbericht 2020/21, available at https://www.bundeskartellamt.de/SharedDocs/Publikation/DE/Jahresbericht/Jahresbericht_2020_21.pdf?__blob=publicationFile&v=5. Accessed 5 November 2022, p. 46; F. Engelsing and M. Jakobs, Nachhaltigkeit und Wettbewerb, WuW 2019(1), p. 16.

[11] See United Nations General Assembly (2015), Transforming our world: the 2030 Agenda for Sustainable Development, available at https://www.un.org/en/development/desa/population/migration/generalassembly/docs/globalcompact/A_RES_70_1_E.pdf. Accessed 5 November 2022, para 2: "[. . .] sustainable development with its three dimensions – economic, social and environmental [. . .]"

[12] Bundeskartellamt, Open markets and sustainable economic activity – public interest objectives as a challenge for competition law practice, October 2020, p. 5.

[13] A. Bischke and S. Brack, Neuere Entwicklungen im Kartellrecht, NZG 2022(9), p. 393; J. Dreyer and E. Ahlenstiel, Berücksichtigung von Umweltschutzaspekten bei der kartellrechtlichen Bewertung von Kooperationen, WuW 2021(2), pp. 76-77 fn. 1.

[14] Bundeskartellamt, Open markets and sustainable economic activity – public interest objectives as a challenge for competition law practice, October 2020, pp. 7-8; Stellungnahme der Bundesregierung zum Tätigkeitsbericht des Bundeskartellamtes 2019/2020, BT-Drs. 19/30775, pp. III, IV; Monopolkommission, XXIV. Hauptgutachten, Wettbewerb 2022, BT-Drs. 20/3065, para 404; SPD, Bündnis 90/Die Grünen and FDP (2021), Mehr Fortschritt wagen, Koalitionsvertrag 2021-2025, available at https://www.spd.de/fileadmin/Dokumente/Koalitionsvertrag/Koalitionsvertrag_2021-2025.pdf. Accessed 5 November 2022, p. 5: "Die Klimaschutzziele von Paris zu erreichen, hat für uns oberste Priorität. Klimaschutz sichert Freiheit, Gerechtigkeit und nachhaltigen Wohlstand. Es gilt, die soziale Marktwirtschaft als eine sozial-ökologische Marktwirtschaft neu zu begründen."

7 Sustainability and Competition Law in Germany

of human rights in supply chains get attention.[15] By contrast, governance factors as such appear to play a minor role so far.

The lack of a clear explanation or even a precise definition of sustainability in German competition law and practice is partly remedied by the deliberate conformity of the German prohibition of cartels in Sections 1 and 2 GWB with the European prohibition of cartels in Article 101 TFEU, as expressed by the German legislator when amending the GWB in 2005. By consequence, legal terms are generally to be interpreted in parallel, and the legislator intended that the Commission's guidelines be considered in that regard.[16] Adding to that, sustainability has so far been discussed mostly with respect to the prohibition of cartels, not in relation to the prohibition of abuse of a dominant position, which is where German and European Union competition law diverge. Therefore, German competition law can and should draw on definitions of sustainability in European (competition) law whenever possible. This may even be a legal requirement in all cases where the German Federal Cartel Office also enforces Article 101 TFEU (see Article 3 (1)1 reg. 1/2003). In those cases, the application of national competition law may not lead to the prohibition of agreements, decisions by associations of undertakings or concerted practices, which may affect trade between Member States but which do not restrict competition within the meaning of Article 101 (1) TFEU. The same goes for corporate behaviour exempted by Article 101 (3) TFEU or covered by a Regulation for the application of Article 101 (3) TFEU (Article 3 (2)1 reg. 1/2003).

European competition law does not provide a clear-cut definition of sustainability either. With reference to Article 101 TFEU, Article 210a (3) of Regulation 1308/2013 of the European Parliament and of the Council of 17 December 2013 establishing a common organisation of the markets in agricultural products,[17] as amended by Regulation 2021/2117 of the European Parliament and of the Council of 2 December 2021,[18] defines the term "sustainability standard". Again, this definition stresses the environmental dimension of sustainability. Besides, the EU has enacted a taxonomy regulation that defines sustainability for capital market law.[19]

---

[15] Cf. BMWK (2022), Wettbewerbspolitische Agenda des BMWK bis 2025, available at https://www.bmwk.de/Redaktion/DE/Downloads/0-9/10-punkte-papier-wettbewerbsrecht.pdf?__blob=publicationFile&v=6#:~:text=Die%20Kartellbeh%C3%B6rden%20beobachten%20die%20Entwicklungen,und%20auf%20den%20Fernw%C3%A4rmesektor%20ausweiten. Accessed 5 November 2022, p. 3: "Dort, wo Unternehmen über die staatlichen Vorgaben hinaus gemeinsam Nachhaltigkeitsziele oder menschenrechtliche Standards erreichen wollen, muss die Wettbewerbspolitik Rechtssicherheit geben [. . .]."

[16] Gesetzentwurf der Bundesregierung, Entwurf eines Siebten Gesetzes zur Änderung des Gesetzes gegen Wettbewerbsbeschränkungen, BT-Drs. 15/3640, pp. 22-23, 25, 44; but see D. Zimmer, in: Körber, Schweitzer and Zimmer (eds), Immenga/Mestmäcker, Wettbewerbsrecht, Vol. 2, 7th ed, C.H. Beck 2023, sec. 1 GWB paras 15-17.

[17] OJ 2013 L 347, p. 671.

[18] OJ 2021 L 435, p. 262.

[19] OJ 2020 L 198, p. 13; see further E. Bueren, Die EU-Taxonomie nachhaltiger Anlagen, Teil I, WM 2020(35), pp. 1611-1618, Teil II, WM 2020(36), pp. 1659-1664.

However, from a legal perspective, definitions within secondary EU law are inferior to the goals and the provisions associated with sustainability enshrined in the European treaties.

In the EU treaties, sustainability and environmental objectives are enshrined, inter alia, in recital 9 preamble TEU; Article 3 (3) and (5) TEU; Articles 11 and 191 TFEU; and Article 37 CFR. Moreover, there are horizontal clauses on certain aspects of sustainability in the broad ESG sense, such as social protection (Article 9 TFEU), health protection (Article 168 (1), Article 9 TFEU, Article 35 sentence 2 CFR), development policy (Article 208 (1) TFEU), consumer protection (Article 12 TFEU, Article 38 CFR), gender equality (Article 8 TFEU, Article 23 (1) CFR) and culture (Article 167 (4) TFEU).[20] As a general matter, many of these horizontal clauses reflect values integral to the European Union (see Article 2 TEU).

Against this background, the draft Guidelines on the applicability of Article 101 of the Treaty on the Functioning of the European Union to horizontal co-operation agreements (horizontal guidelines) acknowledge a broad concept of sustainability but, just like the discussion in Germany, focus on the environmental dimension. The Commission rightly states that sustainable development is a core principle of the TEU and a priority objective for the Union's policies. At the same time, the Guidelines recall that the Commission committed to implementing the United Nation's sustainable development goals and the growth strategy of the European Green Deal.[21] The draft horizontal guidelines go on to explain:

> in broad terms, sustainable development refers to the ability of society to consume and use the available resources today without compromising the ability of future generations to meet their own needs. It encompasses activities that support economic, environmental, and social (including labour and human rights) development. The notion of sustainability objective therefore includes, but is not limited to, addressing climate change (for instance, through the reduction of greenhouse gas emissions), eliminating pollution, limiting the use of natural resources, respecting human rights, fostering resilient infrastructure and innovation, reducing food waste, facilitating a shift to healthy and nutritious food, ensuring animal welfare, etc.[22]

To sum up, there is a lot of agreement on what sustainability means in abstract due to general European and international rules on sustainability. Nevertheless, there is no concise definition that makes sustainability goals readily operational in practice. This vagueness is circumvented to the extent that current practice, scholarship

---

[20] For an in-depth analysis with a view to competition law, see L. Breuer, Das EU-Kartellrecht im Kraftfeld der Unionsziele, Nomos 2013.

[21] European Commission, Communication from the Commission, Approval of the content of a draft for a Communication from the Commission – Guidelines on the applicability of Article 101 of the Treaty on the Functioning of the European Union to horizontal cooperation agreements, OJ 2022 C 164 para 542.

[22] European Commission, Communication from the Commission, Approval of the content of a draft for a Communication from the Commission – Guidelines on the applicability of Article 101 of the Treaty on the Functioning of the European Union to horizontal cooperation agreements, OJ 2022 C 164 para 543 [footnotes omitted].

and policy discussions on competition law and sustainability in Germany as well as in Europe deal with the environmental dimension of sustainability.

This focus partly distinguishes the discussion on competition law and sustainability from the closely related but more general discussion on non-economic goals in competition law. While sustainability is often considered a non-economic goal and while joint sustainability initiatives can conflict with the cartel prohibition, environmental problems are often caused by external effects, i.e. a kind of market failure that impedes workable competition.[23] From that perspective, the promotion of sustainability can actually promote efficiency, workable competition and long-term consumer welfare, a goal central to competition law. However, conventional competition law only involves the welfare of the consumers in the markets affected by the restriction of competition. In western economies, these consumers will often be the ones who produce and take advantage of external effects rather than the one exposed to them (see Sects. 7.2.2.2.2 and 7.2.2.4).[24]

7.2.2 Role of Sustainability in Competition Law

7.2.2.1 General Framework: Constitutional Law, European Law, Climate Protection Act (KSG)

As a general matter, the German constitution (Grundgesetz (GG)), similarly to European primary law, obliges the state to protect certain dimensions of sustainability, in particular natural livelihoods and human rights, see Articles 1, 2 (1) and 20a GG. Article 20a GG has recently come into focus as an obligation for the legislator to fight global warming in order to protect future generations.[25] It is generally acknowledged that Article 20a GG, while predominantly addressing the legislator, is also binding for public authorities. When applying and interpreting a general clause, i.e. a statute that grants an authority discretion, the authority must take the goals of Article 20a GG into account.[26] In the same vein, the interpretation

---

[23] D. Freitag, K. Hansen, K. Markert and V. Strauch, Umweltschutz und Wettbewerbsordnung, Athenäum 1973, pp. 50-52; Bundeskartellamt, Open markets and sustainable economic activity – public interest objectives as a challenge for competition law practice, October 2020, p. 5; F. Wagner v. Papp, in: Säcker and Meier-Beck (eds), Münchener Kommentar zum Wettbewerbsrecht, Vol. 2, 4th ed, C.H. Beck 2022, sec. 1 GWB para 362.

[24] Monopolkommission, XXIV. Hauptgutachten, Wettbewerb 2022, BT-Drs. 20/3065, paras 434-435.

[25] Bundesverfassungsgericht, BVerfGE 157, 30.

[26] See S. Huster and S. Rux, in: Epping and Hillgruber (eds), BeckOK Grundgesetz, 51st ed, C.H. Beck 15.5.2022, GG Article 20a paras 27, 32-33; H. Schulze-Fielitz, in: Dreier (ed), Grundgesetz Kommentar, Vol. 2, 3rd ed, Mohr Siebeck 2015, GG Article 20a paras 67, 74-75; R. Scholz, in: Dürig, Herzog and Scholz (eds), Grundgesetz-Kommentar, 96th supplement November 2021, C.H. Beck 2021, GG Article 20a paras 46, 56; H. Jarass, in: Jarass and Pieroth (eds), Grundgesetz für die Bundesrepublik Deutschland, 17th ed, C.H. Beck 2022, GG Article 20a paras 18, 21; K.-P. Sommermann, in: v. Münch and Kunig (eds), Grundgesetz-Kommentar, Vol. 1, 7th ed, C.H. Beck 2021, GG Article 20a, paras 41, 46-48.

of general clauses must incorporate the protection of fundamental rights (interpretation in conformity with fundamental rights or other constitutional law).

This applies to competition law, too, as both the prohibition of anti-competitive agreements (Section 1 GWB) and the exemption pursuant to Section 2 GWB as well as the prohibition of an abuse of a dominant position (Section 19 GWB) are general clauses. By consequence, the German Constitutional Court as well as the German Federal Court (Bundesgerichtshof (BGH)) have repeatedly held that Section 19 GWB must be interpreted in conformity with fundamental rights.[27]

Furthermore, since 2019, Section 13 (1)1 of the Climate Protection Act (Klimaschutzgesetz (KSG)) has spelled out and clarified that (inter alia) federal state authorities must interpret and apply general clauses with regard to climate protection when they make discretionary decisions. The purpose of the KSG is to soften the effects of global climate change by ensuring that national and European climate protection targets are met. Section 13 (1)1 KSG requires that wherever substantive federal law uses legal terms that require interpretation or accord assessment or discretionary powers, the purpose and objectives of the KSG need to be included in the considerations as (co-)decisive aspects. This can significantly influence the application practice of other federal laws.[28]

Notably, however, this requirement is restricted to discretionary decision-making within the framework of the statutory requirements.[29] Moreover, Section 13 (1) KSG only requires that climate protection be taken into account. Section 13 (1) KSG does not postulate a strict obligation to effectively take account of climate protection targets, nor does it give them a special weight in the sense of an optimisation requirement. This means that climate protection concerns can be set aside in favour of other concerns.[30] Therefore, it seems somewhat doubtful whether Section 13 (1) KSG goes beyond the requirements of Article 20a GG.[31] Regarding competition law, the discussion so far has been mostly about whether and under which conditions the GWB allows the Federal Cartel Office to apply more lenient standards to (potentially) anti-competitive behaviour at all (see Sects. 7.2.2.2 and 7.2.2.3). If the GWB does not grant permission to do so, there will arguably be no scope for

---

[27] See recently Bundesverfassungsgericht, 3 June 2022, 1 BvR 2103/16, ECLI:DE:BVerfG:2022:rk20220603.1bvr210316, paras 35, 37; BGH, 23 June 2020, KVR 68/19, ECLI:DE:BGH:2020:230620BKVR69.19.0, para 105; critical A. Fuchs, in: Körber, Schweitzer and Zimmer (eds), Immenga/Mestmäcker, Wettbewerbsrecht, Vol. 2, 7th ed, C.H. Beck 2023, sec. 19 GWB paras 211c-211d.

[28] S. Klinski et al., Das Bundes-Klimaschutzgesetz, NVwZ 2020(1-2), p. 6.

[29] See the government explanatory memorandum accompanying the draft act, BT-Drs. 19/14337, p. 36; M. Wickel, Das Bundes-Klimaschutzgesetz und seine rechtlichen Auswirkungen, ZUR 2021(6), p. 337.

[30] M. Wickel, Das Bundes-Klimaschutzgesetz und seine rechtlichen Auswirkungen, ZUR 2021(6), p. 337; M. Kment, Klimaschutzziele und Jahresemissionsmengen – Kernelemente des neuen Bundes-Klimaschutzgesetzes, NVwZ 2020(21), p. 1544.

[31] See S. Schlacke, Klimaschutzrecht im Mehrebenensystem, NVwZ 2022(13), p. 911.

7 Sustainability and Competition Law in Germany

(discretionary) decision-making with regard to climate protection, and Section 13 KSG may not come into play as far as substantive competition law is concerned.

To summarise, for the reasons explained above, from a theoretical point of view, Section 13 KSG does not necessarily affect the positions in the discussion on German competition law and sustainability at all. Nevertheless, the legislative intention of Section 13 (1) KSG ought to get more attention in German competition law scholarship. So far, it seems to have been completely ignored.

Another source of conflict between competition law and sustainability have been the legislator's efforts to encourage cooperations aiming to achieve sustainability goals, even though their agreements were likely to contravene competition law. This scenario can put authorities in an enforcement dilemma (see Sect. 7.2.2.2).

7.2.2.2 Cartel Prohibition

7.2.2.2.1 Section 1 GWB (Equivalent to Article 101 TFEU)

The obligation to interpret general clauses in conformity with the constitution is not to say, however, that Sections 1 and 19 GWB can be narrowed or even set aside to account for fundamental rights or sustainability according to Article 20a GG. Whether this is possible is in dispute. If it is possible by law, the KSG (see Sect. 7.2.2.1) requires a climate change impact assessment.[32] The position of the Federal Cartel Office regarding the latter has changed over time.

Until 1995, the German Federal Cartel Office used to apply a balancing test to justify restrictions of competition for public interest reasons, along two lines of reasoning: sometimes the balancing test was meant to restrict the scope of application of Section 1 GWB; sometimes it was presented as a part of prosecutorial discretion. Arguably, the Federal Cartel Office has applied this policy mostly to issues of health protection[33] but also to agreements in the interest of environmental protection.[34]

---

[32] BT-Drs. 19/14337, p. 36; S. Schlacke, Klimaschutzrecht im Mehrebenensystem, NVwZ 2022(13), p. 911.

[33] Bericht des Bundeskartellamtes über seine Tätigkeit im Jahre 1966 sowie über Lage und Entwicklung auf seinem Aufgabengebiet (§ 50 GWB), BT-Drs. V/1950, p. 58: restrictions of cigarette advertising. In this report, it is unclear whether the balancing test was meant to restrict the scope of application of sec. 1 GWB or whether it was considered part of prosecutorial discretion. With regard to a subsequent agreement of the cigarette industry amending the aforementioned one, the Federal Cartel Office made clear that it considered the balancing test to restrict the scope of the cartel prohibition, see Bericht des Bundeskartellamtes über seine Tätigkeit im Jahr 1976 sowie über die Lage und Entwicklung auf seinem Aufgabengebiet (§ 50 GWB), BT-Drs. 8/704, pp. 9, 79; also clearly arguing for a restriction of the scope of the cartel prohibition Bundeskartellamt, 20 February 1960 - *Doppelstecker*, WuW/E Bundeskartellamt pp. 145, 149-152, concerning an agreement to build in protective contacts for plugs.

[34] Bericht des Bundeskartellamtes über seine Tätigkeit in den Jahren 1985/1986 sowie über Lage und Entwicklung auf seinem Aufgabengebiet (§ 50 GWB), BT-Drs. 11/554, p. 70, as part of prosecutorial discretion.

Following a conference with scholars on packaging regulation,[35] the Federal Cartel Office in 1995 changed its policy in view of scholarly critique[36] and backed by critical studies of the Centre for European Economic Research (ZEW) as well as the German Monopolies Commission:[37] The agency gave up the approach of a balancing test and moved instead to exclusive reliance on opportunity decisions (discretionary prosecution principle – Opportunitätsprinzip, Section 47 (1) OWiG), i.e. toleration (non-prosecution) of certain kinds of sustainability agreements even if they were considered to violate competition law (for particular cases, see Sect. 7.3.1).[38]

The Federal Cartel Office may have had different reasons for its decision. First, as a general matter, the discretionary prosecution principle shall enable the authority to avoid spending resources on infringements that, although (potentially) illegal, are thought not to produce meaningful harm, for instance because the affected business volume is (very) small.

Second, in the 1990s, the Federal Cartel Office used this strategy to avoid an enforcement dilemma with sustainability goals. In particular, environmental law on European as well as on German level has occasionally pursued the strategy to encourage (as opposed to require) companies to cooperate in order to achieve certain regulatory goals, for instance with regard to the reduction or recycling of waste or the avoidance of CFC.[39] This tendency can give rise to a conflict with competition law. Most notably, when enacting the packaging ordinance (Verpackungsverordnung) in 1991, the German legislator had been warned beforehand by the German Federal Cartel Office of a conflict with competition law but decided to leave the regulatory framework unchanged. This presented the competition authority with the dilemma that strict competition law enforcement could have rendered the regulatory concept

---

[35] Bundeskartellamt, Arbeitsunterlage für die Sitzung des Arbeitskreises Kartellrecht am 4. und 5. Oktober 1993, Wettbewerbspolitische und kartellrechtliche Probleme der deutschen Entsorgungswirtschaft, September 1993.

[36] For an overview, see K. Krauß, in: Bunte (ed), Kommentar zum Deutschen und Europäischen Kartellrecht, Vol. 1, 14th ed, Luchterhand 2022, sec. 1 GWB para 162; H. Köhler, Abfallrückführungssysteme der Wirtschaft im Spannungsfeld von Umweltrecht und Kartellrecht, BB 1996(50), p. 2579; T. Lappe, Zur ökologischen Instrumentalisierbarkeit des Wettbewerbsrechts, WRP 1995(3), p. 174; D. Freitag, K. Hansen, K. Markert and V. Strauch, Umweltschutz und Wettbewerbsordnung, Athenäum 1973, pp. 73-75, 79-80.

[37] Monopolkommission, XI. Hauptgutachten 1994/1995, Wettbewerbspolitik in Zeiten des Umbruchs, 1996, BT-Drs. 13/5309, paras 75-98.

[38] Bericht des Bundeskartellamtes über seine Tätigkeit in den Jahren 1995/1996 sowie über Lage und Entwicklung auf seinem Aufgabengebiet (§ 50 GWB), BT-Drs. 13/7900, pp. 39-40; Bundeskartellamt, Arbeitsunterlage für die Sitzung des Arbeitskreises Kartellrecht am 4. und 5. Oktober 1993, Wettbewerbspolitische und kartellrechtliche Probleme der deutschen Entsorgungswirtschaft, September 1993, pp. 28, 40.

[39] I. Pernice, Rechtlicher Rahmen der europäischen Unternehmenskooperation im Umweltbereich unter besonderer Berücksichtigung von Article 85 EWGV, EuZW 1992(5), pp. 139-140; Bundeskartellamt, Arbeitsunterlage für die Sitzung des Arbeitskreises Kartellrecht am 4. und 5. Oktober 1993, Wettbewerbspolitische und kartellrechtliche Probleme der deutschen Entsorgungswirtschaft, September 1993, pp. 10-12.

7 Sustainability and Competition Law in Germany

vacuous. The authority therefore decided to tolerate cooperation under certain circumstances (see also Sect. 7.3.1).[40]

Such a decision does not prevent the Federal Cartel Office from changing its conclusion and finding an infringement in the future, nor does it affect the civil law consequences of a possible infringement of the cartel prohibition. Hence, there remains legal uncertainty for the undertakings concerned.[41] Inter alia for this reason and in fear of a possible influence of political and social interests, this newer line of reasoning has been subject to criticism.[42]

Today, the German Federal Cartel Office and the leading opinion in German competition law scholarship continue to hold the view that it is not possible to deny a violation of Section 1 GWB, the prohibition of anti-competitive agreements, by means of a balancing test. Weighing the benefits of competition via-á-vis other legal interests[43] can therefore not justify a sustainability or an ecological cartel privilege, unless explicitly stipulated by law (on such provisions see Sect. 7.2.3). If no such exemptions exist, restrictions of competition that serve a public interest do not escape the prohibition of cartels. In particular, it is considered impossible to tailor the cartel prohibition by interpretation for it to protect only competition on the merits (Leistungswettbewerb) as this would give rise to enormous problems of delimination.[44]

A strong argument in support of this position is that previous versions of the GWB explicitly allowed for the exceptional permission of cartels due to overriding reasons of the overall economy and of public interest (Section 8 GWB former version; see Sect. 7.2.3). The legislator has abolished these provisions so that the current GWB lacks a (codified) public interest exception. Any exemptions are deliberately left to Section 2 GWB, the German equivalent of Article 101 (3) TFEU. One can therefore argue that acknowledging an uncodified balancing

---

[40] Bundeskartellamt, Arbeitsunterlage für die Sitzung des Arbeitskreises Kartellrecht am 4. und 5. Oktober 1993, Wettbewerbspolitische und kartellrechtliche Probleme der deutschen Entsorgungswirtschaft, September 1993, pp. 28, 40; see also Monopolkommission, XI. Hauptgutachten 1994/1995, Wettbewerbspolitik in Zeiten des Umbruchs, 1996, BT-Drs. 13/5309, pp. 30-31, paras 75-76.

[41] H. Köhler, Abfallrückführungssysteme der Wirtschaft im Spannungsfeld von Umweltrecht und Kartellrecht, BB 1996(50), p. 2578.

[42] For example by F. Säcker and A. Zorn, in: Säcker and Meier-Beck (eds), Münchener Kommentar zum Kartellrecht, Vol. 2, 4th ed, C.H. Beck 2022, sec. 1 GWB para 185; J. Heyers, in: Jaeger et al. (eds), Frankfurter Kommentar zum Kartellrecht, Vol. 4, 91st supplement August 2018, Otto Schmidt 2018, sec. 2 GWB para 72.

[43] F. Säcker and A. Zorn, in: Säcker and Meier-Beck (eds), Münchener Kommentar zum Kartellrecht, Vol. 2, 4th ed, C.H. Beck 2022, sec. 1 GWB para 185.

[44] D. Freitag, K. Hansen, K. Markert and V. Strauch, Umweltschutz und Wettbewerbsordnung, Athenäum 1973, pp. 67-68. Note, however, that performance-based competition can be the object of competition rules by business and trade associations, sec. 24 GWB. These rules can be recognised by the Federal Cartel Office. But a recognition only means that the competition authority will not exercise its power to intervene.

test would circumvent the abolishment of the old exemptions and thereby contradict the intention of the legislator.[45]

Besides, the German Federal Cartel Office and German scholarship argue that it is good policy to separate objectives related to competition from those related to non-economic goals, such as sustainability. The Federal Cartel Office considers itself well positioned to safeguard the former but argues that the democratically legitimised legislator is best positioned to protect the latter and, when doing so, to strike a balance with the interest in workable competition.[46] This task should not be imposed on companies in their self-assessment nor on the Federal Cartel Office.

The Federal Court of Justice has also held on several occasions—which did not concern sustainability agreements—that a restriction of Section 1 GWB by means of a balancing test is impossible.[47] The German Federal Court as well as the German Constitutional Court have so far adhered to this principle when employing a so-called *materialisation* in their interpretation of competition law (Pechstein, Facebook; see Sect. 7.2.2.1 with fn. 27): as indicated by recent judgments, the courts have resorted to fundamental rights only as an argument for applying competition law, i.e. a strict interpretation, but, to our knowledge, never as an argument to effectively weaken it.

7.2.2.2.2 Section 2 GWB (Equivalent to Article 101 (3) TFEU)

It follows that according to the Federal Cartel Office and the majority view in scholarship, sustainability or other political objectives (such as "fair trade") may only be taken into account as a consumer benefit within the scope of the exemption requirements of Section 2 GWB, the German equivalent of Article 101 (3) TFEU. Section 2 (1) GWB was aligned with Article 101 (3) TFEU in 2005, and Section 2 (2) GWB has since then contained a dynamic reference to the European block exemption regulations.[48] By consequence, the question to what extent sustainability can be a benefit offsetting a restriction of competition in German competition law should generally be answered in the same way as in European competition law.[49] However, European competition law is currently witnessing an intense legal and policy discussion on the matter. At present, the Commission is of the opinion that

---

[45] See Unabhängige Sachverständigenkommission zum Umweltgesetzbuch, in: Bundesministerium für Umwelt, Naturschutz und Reaktorsicherheit (ed), Umweltgesetzbuch (UGB-KomE), Duncker & Humblot 1998, p. 504: a general privilege under competition law for anti-competitive agreements that protect the environment requires a legislative decision.

[46] Bundeskartellamt, Open markets and sustainable economic activity – public interest objectives as a challenge for competition law practice, October 2020, pp. 13-15.

[47] BGH, 9 March 1999, KVR 20/97 - *Lottospielgemeinschaft*, WuW/E DE-R 289, 293, 295; BGH, 11 December 1997, KVR 7/96 - *Europapokal-Heimspiele*, NJW 1998, pp. 756, 759-760.

[48] On the object and consequences of this dynamic reference, see J. Heyers, in: Jaeger et al. (eds), Frankfurter Kommentar zum Kartellrecht, Vol. 4, 91st supplement August 2018, Otto Schmidt 2018, sec. 2 GWB paras 36-40a.

[49] See J. Heyers, in: Jaeger et al. (eds), Frankfurter Kommentar zum Kartellrecht, Vol. 4, 91st supplement August 2018, Otto Schmidt 2018, sec. 2 GWB para 72.

sustainability agreements are to be examined according to the usual criteria for Article 101 (3), i.e. they do not receive preferential treatment (see Sect. 7.2.2.5). In line with this, the Federal Cartel Office and the majority opinion in Germany require that consumers consider the relevant feature (such as sustainable production) to be a qualitative value of the product.[50] Additionally, the benefit needs to accrue essentially to the same group of consumers negatively affected by the restriction of competition.[51] Such collective advantages do not require every single consumer who is negatively affected to profit from the advantages. Nevertheless, this prerequisite does not seem to allow for restrictions of competition that would internalise external effects, e.g. in situations where consumers in Western Europe produce emissions that cause damage (predominantly) in other parts of the world. Again, this issue is in dispute. The opposing view advocates a more generous inclusion of the non-economic goals that are enshrined in the constitution.[52] Besides, the Federal Cartel Office as well as German competition law scholarship show great interest in pioneering approaches in other jurisdictions to broaden the scope for efficiency exemptions, e.g. by the Dutch Competition Authority.[53] Finally, the Commission has put forward a partly more generous concept of collective advantages in the draft horizontal guidelines.[54]

7.2.2.3 Prohibition of Abuse of a Dominant Position (Section 19 GWB)

While there is an intense discussion on sustainability with regard to the prohibition of cartels in Section 1 GWB, sustainability has not played a considerable role so far with regard to Sections 19–20 GWB, the prohibitions of abuse of a dominant

---

[50] K. Krauß, in: Bunte (ed), Kommentar zum Deutschen und Europäischen Kartellrecht, Vol. 1, 14th ed, Luchterhand 2022, sec. 1 GWB paras 161, 164; J. Nordemann and C. Grave, in Loewenheim et al. (eds), Kartellrecht, 4th ed, C.H. Beck 2020, sec. 2 GWB para 14; R. Ellger and A. Fuchs, in: Körber, Schweitzer and Zimmer (eds), Immenga/Mestmäcker, Wettbewerbsrecht, Vol. 2, 7th ed, C.H. Beck 2023, sec. 2 GWB para 72; D. Zimmer, in: Körber, Schweitzer and Zimmer (eds), Immenga/Mestmäcker, Wettbewerbsrecht, Vol. 2, 7th ed, C.H. Beck 2023, sec. 1 GWB para 49.

[51] Bundeskartellamt, Open markets and sustainable economic activity – public interest objectives as a challenge for competition law practice, October 2020, p. 28; F. Wagner v. Papp, in: Säcker and Meier-Beck (eds), Münchener Kommentar zum Wettbewerbsrecht, Vol. 2, 4th ed, C.H. Beck 2022, sec. 1 GWB paras 368, 372; Monopolkommission, XXIV. Hauptgutachten, Wettbewerb 2022, BT-Drs. 20/3065, paras 435, 445-449.

[52] See for this view e.g. J. Heyers, in: Jaeger et al. (eds), Frankfurter Kommentar zum Kartellrecht, Vol. 4, 91st supplement August 2018, Otto Schmidt 2018, sec. 2 GWB paras 69-74 with further references; J. Hackl, Verbot wettbewerbsbeschränkender Vereinbarungen und nichtwettbewerbliche Interessen, Nomos 2010, pp. 245-247.

[53] Authority for Consumers & Markets (2021), Guidelines, Sustainability Claims, available at https://www.acm.nl/sites/default/files/documents/guidelines-sustainability-claims.pdf. Accessed 5 November 2022.

[54] European Commission, Communication from the Commission, Approval of the content of a draft for a Communication from the Commission – Guidelines on the applicability of Article 101 of the Treaty on the Functioning of the European Union to horizontal cooperation agreements, OJ 2022 C 164 paras 601-607; Monopolkommission, XXIV. Hauptgutachten, Wettbewerb 2022, BT-Drs. 20/3065, paras 446-448.

position and of relative market power, respectively. The conventional leading opinion argues that the German law on abuse of a dominant position is essentially competitive in nature. Therefore, according to this view, non-competitive considerations can only be taken into account in the interpretation if this is permitted by a special provision. Otherwise, the inclusion of non-competitive considerations would generally counteract the basic legislative intention according to which the control of abusive behaviour is to promote the regulatory principles of free competition.[55]

This strict position may seem somewhat surprising because, as far as constitutional law and European primary law are concerned, the requirements for the interpretation of general clauses are the same for the cartel prohibition and for the prohibition of abuse. When applying and interpreting a general clause, the authority and the courts must take the fundamental rights and the goals of Article 20a GG into account (Sect. 7.2.2.1). Section 19 GWB seems even more prone to the influence of non-economic goals, such as sustainability, than Section 1 GWB because the application of Section 19 GWB always requires a balancing test. This can incorporate, apart from the interest in the protection of free competition, certain value judgments that the legislator has made in constitutional law and public law.[56] Indeed, the German Constitutional Court as well as the BGH have repeatedly interpreted Section 19 GWB with regard to fundamental rights (Sect. 7.2.2.1) and with regard to value judgments of other laws, e.g. of social security law.[57] In principle, the same holds with regard to the horizontal clauses in European primary law. To the extent that they apply, they can affect the interpretation of the prohibition of abuse.[58] If there is scope to consider environmental sustainability, Section 13 (1) KSG requires the Federal Cartel Office to perform the above-mentioned climate change impact assessment (see Sect. 7.2.2.2.1).

There are, however, also reservations against taking value judgments of other laws into account, in particular with regard to a line of reasoning according to which a violation of another law can amount to or contribute significantly to finding an

---

[55] W. Wurmnest, Marktmacht und Verdrängungsmissbrauch, 2nd ed, Mohr Siebeck 2012, pp. 100 f.; K. Markert and R. Podszun, in: Körber, Schweitzer and Zimmer (eds), Immenga/Mestmäcker, Wettbewerbsrecht, Vol. 2, 7th ed, C.H. Beck 2023, sec. 20 GWB para 9; see also J. Nothdurft, in: Bunte (ed), Kommentar zum Deutschen und Europäischen Kartellrecht, Vol. 1, 14th ed, Luchterhand 2022, sec. 19 GWB para 369 at the end.

[56] J. Nothdurft, Bunte (ed), Kommentar zum Deutschen und Europäischen Kartellrecht, Vol. 1, 14th ed, Luchterhand 2022, sec. 19 GWB para 369.

[57] See M. Kubiciel, Verhaltensbeschränkungen marktbeherrschender Unternehmen durch außerwettbewerbliche Zielsetzungen, WuW 2004(2), pp. 163-164 with regard to BGHZ 129, 53 which relied on the legislator's conception of competition in the pharmaceutical wholesale market inherent in sec. 129 SGB V.

[58] With respect to EU law, W. Wurmnest, Marktmacht und Verdrängungsmissbrauch, 2nd ed, Mohr Siebeck 2012, p. 92.

7 Sustainability and Competition Law in Germany

infringement of Section 19 GWB.[59] A popular argument for a restrictive reception of other laws in the balancing test for Section 19 GWB is that Section 19 GWB should not become the repair kit for enforcement problems in other parts of the law unrelated to competition. This topic created great controversy on the occasion of the Facebook case. The German Federal Cartel Office argued that a dominant company violating data protection law at the same time abuses its market power vis-à-vis consumers, irrespective of whether there is a causal link between the company's market power and the violation of data protection law.[60] This gave rise to the question of whether and to what extent the violation of other laws, inter alia environmental laws, can amount to an abuse of a dominant position.[61] After the Higher Regional Court of Düsseldorf had halted the decision of the Federal Cartel Office,[62] the BGH upheld it with a modified line of reasoning.[63]

As a reaction to the fierce dispute about the case, the German legislator modified Section 19 GWB to support the position of the German Federal Cartel Office.[64] However, the legislator also intended to restrict the scope of provisions whose violation can amount to an abuse of a dominant position. The legislator did not want to include any kind of illegal conduct by dominant companies. This shall only be the case if a statute is market related, i.e. if it regulates a supply-demand relationship, and if it serves to protect market participants. According to the government's explanatory memorandum to the tenth amendment of the GWB, data protection law fulfils these criteria. Conversely, according to the government's memorandum, violations by dominant companies of other norms do not suffice to constitute a violation of competition law. The government's explanatory memorandum mentions infringements of tax, labour or environmental law as examples.[65]

However, not all of the negative examples mentioned by the government memorandum are self-explanatory or readily apparent from a competition law

---

[59] On the case law and the disputes associated with this line of reasoning, see M. Wolf, in: Säcker and Meier-Beck (eds), Münchener Kommentar zum Wettbewerbsrecht, Vol. 2, 4th ed, C.H. Beck 2022, sec. 19 GWB paras 35d-35f.

[60] Bundeskartellamt, 6 February 2019 – B6–22/16, paras 522-523, 573, 630, 872-873, 876.

[61] See i.a. with regard to environmental law T. Körber, „Ist Wissen Marktmacht?" Überlegungen zum Verhältnis von Datenschutz, „Datenmacht" und Kartellrecht – Teil 2, NZKart 2016(8), p. 354; F. Bien, in: Säcker et al. (eds), Münchener Kommentar zum Wettbewerbsrecht, 3rd ed, C.H. Beck 2020, Article 101 TFEU para 265; M. Wolf, in: Säcker and Meier-Beck (eds), Münchener Kommentar zum Wettbewerbsrecht, Vol. 2, 4th ed, C.H. Beck 2022, sec. 19 GWB paras 35d-35f.

[62] OLG Düsseldorf, 26 August 2019, VI-Kart 1/19 (V) – *Facebook I*, ECLI:DE:OLGD:2019:0826. KART1.19V.00.

[63] BGH, 23 June 2020 – KVR 69/19, NZKart 2020(9), p. 473; for a detailed analysis, see E. Bueren, Die Neufassung der Missbrauchsgeneralklausel durch die 10. GWB-Novelle im Lichte des Facebook-Beschlusses, ZHR 2021(4), pp. 569-574.

[64] For a detailed analysis, see E. Bueren, Die Neufassung der Missbrauchsgeneralklausel durch die 10. GWB-Novelle im Lichte des Facebook-Beschlusses, ZHR 2021(4), pp. 577-594.

[65] Government draft 10th amendment of the GWB, BT-Drs. 19/23492, p. 71.

perspective.[66] As far as sustainability is concerned, environmental law, though it will generally not be relevant,[67] could at least selectively have an impact on market relations, especially in times of increasing emphasis on CSR and sustainable finance. The outcome of individual cases will depend on whom the laws in question are intended to protect. With this in mind, it does seem possible to meet the requirements spelled out in the government's explanatory memorandum even for areas of the law that the government has explicitly excluded. Ultimately, only the violation of a norm can reveal its competitive relevance because strategies that conform to competition law as well as those that do not are the result of a development process. The decisive test should therefore rather be whether a violation can negatively affect the objectives of Section 19 GWB, not a formal grouping of statutes that are considered per se relevant or irrelevant for competition.[68]

To sum up, even though it does not seem inconceivable that a violation of certain environmental laws by a dominant firm may, depending on the circumstances of the case, be a strategy to strengthen the dominant position or exploit the opposite market side, amounting to a violation of competition law, such a line of reasoning has not yet been accepted by the majority view in Germany.

7.2.2.4 Merger Control

Similar to the prohibition of abuse of a dominant position, German merger control has so far only played a peripheral role in the discussion on competition law and sustainability.

A concentration that would significantly impede effective competition, in particular a concentration expected to create or strengthen a dominant position, shall be prohibited by the Federal Cartel Office, Section 36 (1)1 GWB. Importantly, the Federal Cartel Office has no discretion in that regard.[69] It follows that, in contrast to horizontal or vertical agreements between undertakings (see Sect. 7.2.2.2.1), the Cartel Office cannot tolerate mergers that do not pass the legal test but would be beneficial to sustainability. This means that the Federal Cartel Office's current main tool with respect to sustainability concerns is not available in merger control.

By consequence, the Federal Cartel Office can only consider sustainability in merger control if the substantive merger test allows for this. According to the majority view, the extent to which this is possible is broadly similar to Section 1 GWB. The merger control test is generally taken to focus predominantly on

---

[66] E. Bueren, Die Neufassung der Missbrauchsgeneralklausel durch die 10. GWB-Novelle im Lichte des Facebook-Beschlusses, ZHR 2021(4), pp. 590-591.

[67] See T. Körber, „Ist Wissen Marktmacht?" Überlegungen zum Verhältnis von Datenschutz, „Datenmacht" und Kartellrecht – Teil 2, NZKart 2016(8), p. 354; J. Nothdurft, in: Bunte (ed), Kommentar zum Deutschen und Europäischen Kartellrecht, Vol. 1, 14th ed, Luchterhand 2022, sec. 19 para 229, arguing that environmental law be irrelevant per se.

[68] M. Wolf, in: Säcker and Meier-Beck (eds), Münchener Kommentar zum Wettbewerbsrecht, Vol. 2, 4th ed, C.H. Beck 2022, sec. 19 GWB paras 35-36; Monopolkommission, Wettbewerbspolitik: Herausforderungen für digitale Märkte, BT-Drs 18/5080, paras 523-524.

[69] M. Dreher and M. Kulka, Wettbewerbs- und Kartellrecht, 11th ed, C.H. Beck 2021, para 1638.

competitive concerns, as does Section 1 GWB.[70] Therefore, sustainability can only be considered as part of the competitive process, for instance if markets for certain sustainable products exist and/or if consumers value sustainability as a product quality.

As with Section 2 GWB, the Federal Cartel Office can take efficiencies into account, which might allow for the promotion of sustainability to come into play.

First, it might be possible to consider efficiencies in the competitive assessment of the merger as part of the SIEC test. As the GWB does not allow for this explicitly, it is in dispute whether this is possible; the question has not yet been clarified by the BGH.[71] The Federal Cartel Office has already examined efficiency arguments raised by the merging parties.[72] However, the standard of proof that the parties face concerning efficiencies seems very high. So far, the Federal Cartel Office has always held that the submissions of the parties to the merger did not meet the requirements formulated by the Commission in this regard.[73] Furthermore, the efficiencies must be passed on to the consumers, that is the opposite market side,[74] which implies that merger efficiencies do not usually contribute to internalising external effects. In principle, the discussion concerning Section 2 GWB, Article 101 (3) TFEU, whether and how the concept of (collective) efficiencies should be broadened, applies here mutatis mutandis (see Sects. 7.2.1 and 7.2.2.2.2). However, as German merger control already offers another solution to account for non-economic goals (ministerial approval, see Sect. 7.2.3.1), it seems more difficult to argue that the current approach must be changed to promote sustainability.

Second, German merger control explicitly allows for an examination of efficiencies and certain other advantages under the so-called balancing clause, Section 36 (1) 2 No. 1 GWB. This clause allows for a merger to be cleared if it improves the competitive conditions—which, according to the majority view, must refer to other markets[75]—and if these improvements outweigh the restriction of competition caused by the merger. The "other markets" must at least partly extend to

---

[70] J. Dreyer and E. Ahlenstiel, Berücksichtigung von Umweltschutzaspekten bei der kartellrechtlichen Bewertung von Kooperationen, WuW 2021(2), pp. 76-77; Bundeskartellamt, Open markets and sustainable economic activity – public interest objectives as a challenge for competition law practice, October 2020, p. 42.

[71] M. Dreher and M. Kulka, Wettbewerbs- und Kartellrecht, 11th ed, C.H. Beck 2021, para 1640; S. Thomas, in: Körber, Schweitzer and Zimmer (eds), Immenga/Mestmäcker, Wettbewerbsrecht, Vol. 2, 6th ed, C.H. Beck 2020, sec. 36 GWB paras 488-491.

[72] S. Thomas, in: Körber, Schweitzer and Zimmer (eds), Immenga/Mestmäcker, Wettbewerbsrecht, Vol. 2, 6th ed, C.H. Beck 2020, sec. 36 GWB para 510.

[73] Monopolkommission, XXIV. Hauptgutachten, Wettbewerb 2022, BT-Drs. 20/3065, para 427.

[74] M. Dreher and M. Kulka, Wettbewerbs- und Kartellrecht, 11th ed, C.H. Beck 2021, para 1641 S. Thomas, in: Körber, Schweitzer and Zimmer (eds), Immenga/Mestmäcker, Wettbewerbsrecht, Vol. 2, 6th ed, C.H. Beck 2020, sec. 36 GWB para 489.

[75] Efficiencies in the markets affected by the merger are, according to the majority view in Germany, already part of the SIEC test, M. Dreher and M. Kulka, Wettbewerbs- und Kartellrecht, 11th ed, C.H. Beck 2021, para 1641, 1657; A. Christiansen and J. Knebel, in: Säcker and Meier-Beck (eds), Münchener Kommentar zum Wettbewerbsrecht, Vol. 2, 4th ed, C.H. Beck 2022, sec. 36 GWB para 245; for a slightly different view see S. Thomas, in: Körber, Schweitzer and Zimmer (eds), Immenga/Mestmäcker, Wettbewerbsrecht, Vol. 2, 6th ed, C.H. Beck 2020, sec. 36 GWB para 626.

Germany—purely foreign markets are irrelevant.[76] According to the wording of Section 36 (1) 2 No. 1 GWB, only positive effects on the conditions of competition are relevant. Whether these improvements can include efficiency-related advantages is in dispute; the majority view seems to answer in the affirmative.[77] By contrast, non-economic advantages are clearly excluded.[78]

7.2.2.5 Reference Point: European Competition Law

When authorities of the member states, such as the German Federal Cartel Office, enforce European competition law, they must comply with Article 51 (2) CFR which protects several dimensions of sustainability. In particular, it requires a high level of environmental protection (Article 37 CFR[79]) (see Sect. 7.2). Furthermore, though the matter is controversial, it is argued that member states must also respect the horizontal clauses,[80] which include several aspects of sustainability, in particular environmental protection, when applying or enforcing European (competition) law (see Sect. 7.2).

As a general matter, both fundamental rights and horizontal clauses can influence the application of European competition law. With regard to sustainability, in

---

[76] S. Thomas, in: Körber, Schweitzer and Zimmer (eds), Immenga/Mestmäcker, Wettbewerbsrecht, Vol. 2, 6th ed, C.H. Beck 2020, sec. 36 GWB para 635; A. Christiansen and J. Knebel, in: Säcker and Meier-Beck (eds), Münchener Kommentar zum Wettbewerbsrecht, Vol. 2, 4th ed, C.H. Beck 2022, sec. 36 GWB para 247.

[77] S. Thomas, in: Körber, Schweitzer and Zimmer (eds), Immenga/Mestmäcker, Wettbewerbsrecht, Vol. 2, 6th ed, C.H. Beck 2020, sec. 36 GWB para 619; M. Dreher and M. Kulka, Wettbewerbs- und Kartellrecht, 11th ed, C.H. Beck 2021, para 1657; A. Christiansen and J. Knebel, in: Säcker and Meier-Beck (eds), Münchener Kommentar zum Wettbewerbsrecht, Vol. 2, 4th ed, C.H. Beck 2022, sec. 36 GWB para 235. Besides, it is not completely clear if improvements must be structural in nature, see A. Christiansen and J. Knebel, in: Säcker and Meier-Beck (eds), Münchener Kommentar zum Wettbewerbsrecht, Vol. 2, 4th ed, C.H. Beck 2022, sec. 36 GWB para 235; Monopolkommission, XXIV. Hauptgutachten, Wettbewerb 2022, BT-Drs. 20/3065, para 427.

[78] M. Dreher and M. Kulka, Wettbewerbs- und Kartellrecht, 11th ed, C.H. Beck 2021, para 1656.

[79] It is generally acknowledged that Article 37 CFR, despite its somewhat ambiguous wording, is binding also for the authorities of the member states, see H. Jarass, Charta der Grundrechte der EU, 4th ed, C.H. Beck 2021, Article 37 para 4; D. Winkler, in: Stern and Sachs (eds), Europäische Grundrechte-Charta, 1st ed, C.H. Beck 2016, Article 37 para 13; A. Schwerdtfeger, in: Meyer and Hölscheidt (eds), Charta der Grundrechte der Europäischen Union, 5th ed, Nomos 2019, Article 37 para 32.

[80] For an in depth analysis with a view to competition law, see L. Breuer, Das EU-Kartellrecht im Kraftfeld der Unionsziele, Nomos 2013, pp. 166-179; with regard to Article 11 TFEU J. Nowag, Environmental Integration in Competition and Free-Movement Laws, Oxford University Press 2016, pp. 21-24.

7 Sustainability and Competition Law in Germany

particular environmental protection, this can occur in two forms:[81] first, competition law can be narrowed or restricted to the benefit of sustainability. In order to achieve this, one can interpret competition law in a way that allows for sustainability-friendly measures, or beyond that, one could even set competition law aside to the benefit of other non-economic goals related to sustainability. Second, in other circumstances, one can tighten the application of competition law in order to prevent activities that are harmful to sustainability.

There is long-standing and wide-ranging case law of the Commission and by the European courts on non-economic goals in the application of Article 101 (1) and (3) TFEU.[82] Several aspects of this case law are in dispute or somewhat unclear.[83] A detailed analysis would surpass the scope of this chapter, which focuses on Germany and includes EU law only as an important reference point for the application of German competition law. Therefore, a rough sketch shall suffice.

The case law of the European Court of Justice (ECJ) on Article 101 (1) TFEU recognises that certain non-economic objectives enshrined in European primary law can exempt anti-competitive arrangements from the cartel prohibition if the associated restriction of competition is necessary to achieve the non-economic objective.[84] On some occasions, the ECJ has restricted the ban on cartels with such teleological considerations, such as in Albany[85] (collective bargaining autonomy) and Wouters[86] (liberal professions). In essence, this is an uncodified restriction of the cartel prohibition clause, by which the objectives established in particular in the horizontal clauses of Article 8 ff. TFEU are brought to a balance with the interests protected by competition law. This provides a solution for those cases where the exemption clause of Article 101 (3) TFEU seems unsuitable to achieve such a balance.[87] However, so far, the ECJ has not extended this line of reasoning to sustainability agreements.[88]

To date, only the case law of the Commission has explicitly dealt with environmental issues. However, its approach has changed over time and does not always

---

[81] See J. Nowag, Environmental Integration in Competition and Free-Movement Laws, Oxford University Press 2016, pp. 3-4, 53, 139, who uses the terms "supportive integration" and "preventive integration".

[82] For an in depth analysis with a view to competition law, see L. Breuer, Das EU-Kartellrecht im Kraftfeld der Unionsziele, Nomos 2013, pp. 319-568.

[83] D. Zimmer, in: Körber, Schweitzer and Zimmer, Immenga/Mestmäcker, Wettbewerbsrecht, Vol. 1, 6th ed, C.H. Beck 2020, Article 101 AEUV para 164.

[84] W.-H. Roth and T. Ackermann, in: Jaeger et al. (eds), Frankfurter Kommentar zum Kartellrecht, Vol. 4, 73rd supplement January 2011, Otto Schmidt 2011, sec. 1 GWB para 105.

[85] ECJ, case C-67/96, *Albany International BV v Stichting Bedrijfspensioenfonds Textielindustrie,* ECLI:EU:C:1999:430.

[86] ECJ, case C-309/99, *J. C. J. Wouters, J.W. Savelbergh, Price Waterhouse Belastingadviseurs BV v. Algemene Raad van de Nederlandse Orde van Advocaten,* ECLI:EU:C:2002:98.

[87] W.-H. Roth and T. Ackermann, in: Jaeger et al. (eds), Frankfurter Kommentar zum Kartellrecht, Vol. 4, 73rd supplement January 2011, Otto Schmidt 2011, sec. 1 GWB para 105.

[88] Monopolkommission, XXIV. Hauptgutachten, Wettbewerb 2022, BT-Drs. 20/3065, para 422.

appear to be clear and consistent.[89] The Commission's draft horizontal guidelines (see Sect. 7.2.1) refuse to set competition law aside, referring to the aforementioned case law.[90] In a nutshell, the draft horizontal guidelines examine sustainability agreements broadly along the standard lines of competition law. The Commission, while open to a favourable treatment of sustainability agreements, does not acknowledge more lenient competition law standards for them. Several problems in that regard are currently unresolved, in particular how to verify and quantify qualitative and quantitative environmental efficiencies for Article 101 (3) TFEU.[91]

All in all, German and European competition law thus seem mostly aligned in their approach to sustainability agreements. That said, there also appear to be two divergent tendencies.

On the one hand, the fact that the ECJ has at least occasionally restricted the ban on cartels due to teleological considerations seems to be an important deviation from German competition law, where the majority view rejects such sweeping exemptions. This majority opinion is in dispute. The opposite view in scholarship argues that since German competition law was aligned with European law in 2005,[92] German competition law ought to embrace the case law of the ECJ.[93] Regardless, for the time being, it does not seem realistic that broad exemptions will be justified in this way.

On the other hand, there is so far no indication for a stricter interpretation of European competition law when activities harmful to sustainability are at stake.[94] This could only influence the level of the fines imposed on undertakings. By contrast, in Germany, the interpretation of competition law general clauses in conformity with fundamental rights has so far served to advocate a strict application of competition law (see Sect. 7.2.2.3). While the relevant cases did not concern pure sustainability cases, it seems possible to make similar arguments with regard to environmental sustainability, drawing on Articles 2 (1), 1 (1) and Article 20a GG.[95]

---

[89] For a detailed review see L. Breuer, Das EU-Kartellrecht im Kraftfeld der Unionsziele, Nomos 2013, pp. 322-370; for a concise overview of the current discussion, see F. Wagner v. Papp, in: Säcker and Meier-Beck (eds), Münchener Kommentar zum Wettbewerbsrecht, Vol. 2, 4th ed, C.H. Beck 2022, sec. 1 GWB paras 361-372.

[90] European Commission, Communication from the Commission, Approval of the content of a draft for a Communication from the Commission – Guidelines on the applicability of Article 101 of the Treaty on the Functioning of the European Union to horizontal cooperation agreements, OJ 2022 C 164 para 548 with fn. 315.

[91] See Monopolkommission, XXIV. Hauptgutachten, Wettbewerb 2022, BT-Drs. 20/3065, paras 431-432, 439-443.

[92] See fn. 16.

[93] W.-H. Roth and T. Ackermann, in: Jaeger et al. (eds), Frankfurter Kommentar zum Kartellrecht, Vol. 4, 73rd supplement January 2011, Otto Schmidt 2011, sec. 1 GWB para 105.

[94] See on environmental protection J. Nowag, Environmental Integration in Competition and Free-Movement Laws, Oxford University Press 2016, pp. 140-142.

[95] Cf. F. Wagner v. Papp, in: Säcker and Meier-Beck (eds), Münchener Kommentar zum Wettbewerbsrecht, Vol. 2, 4th ed, C.H. Beck 2022, sec. 1 GWB para 365.

7.2.3 Specific Legislative Provisions on or with Relevance to Sustainability and Competition Law

7.2.3.1 Exemptions in Force

At the time of writing, German competition law does no longer contain specific legislative provisions that focus on sustainability as such. This has been different in the past (see Sect. 7.2.3.2) and is likely to change again with the next amendment to the GWB (see Sect. 7.4.2).

With that said, four provisions or groups of provisions with relevance for sustainability in competition law deserve mention.

First, concerning the cartel prohibition, Section 2 GWB, the German equivalent of Article 101(3) TFEU, can exempt an agreement that restricts competition if it promotes sustainability. This requires that, in the case at hand, sustainability be a quality that consumers appreciate or that the promotion of sustainability result in a more efficient use of resources that in turn creates cost savings. Furthermore, these savings need to be passed on to the consumers affected by the restriction of competition. More far-reaching approaches are currently being discussed but have not yet been accepted by the majority view in Germany (for more detail, see Sect. 7.2.2.2.2).

Second, concerning the cartel prohibition and the prohibition of abuse of a dominant position, the GWB (inter alia) contains special provisions for agriculture (Section 28 GWB) and a sectoral exemption for the water industry. The latter one applies to companies involved in the public supply of water, in particular municipal water supply companies (Sections 31–31b GWB).[96] None of these provisions are tailored to sustainability, but one might argue that they can, in principle, promote at least some aspects of sustainability: the exemption for the agricultural sector in Section 28 GWB corresponds in essence to the regulation at European level by Council Regulation (EC) No 1184/2006 of 24 July 2006 applying certain rules of competition to the production of, and trade in, agricultural products.[97] It declares Section 1 GWB inapplicable to certain agreements by agricultural producers, inter alia about production. This allows for agreements about sustainable production, irrespective of consumer preferences.[98] Section 28 (2) GWB adds a limited exemption for vertical price maintenance. Besides, there is a sectorial exemption in the German law on forestry (Section 46 Bundeswaldgesetz – BWaldG), inter alia concerning the planning and execution of silvicultural measures. Again, this could pave the way for agreements on sustainable measures. However, neither of these exemptions have been introduced with the goal of promoting sustainability.

---

[96] V. Emmerich and K. Lange, Kartellrecht, 15th ed. C.H. Beck 2021, p. 204.

[97] OJ 2006, L 214, p. 7.

[98] Cf. H. Schweitzer, in: Körber, Schweitzer and Zimmer (eds), Immenga/Mestmäcker, Wettbewerbsrecht, Vol. 2, 7th ed, C.H. Beck 2023, sec. 28 GWB para 34.

Section 46 BWaldG is rather heavily criticised for being the result of successful lobbying to avoid effective competition law.[99]

Section 31 (1) GWB stipulates that the prohibition of cartels under Section 1 GWB does not apply to certain anti-competitive contracts that are customary in the water industry sector. It forms the basis of the territorial monopolies of the municipal water supply companies. To compensate for the permission of these territorial monopolies, Section 31 (3)–(4) provides for control of abuse conducted by the Cartel Offices. However, Section 185 (1) 2 GWB determines that Sections 19–20 and 31b (5) GWB do not apply to fees and contributions under public law. This serves the purpose of preventing competition law control of water charges. These rules are considered by some to be a reason for the exceptionally high water prices in Germany.[100] However, one could argue that this result promotes sustainability because, ceteris paribus, high water prices will make consumers more frugal in their use of water.[101]

Third, with regard to merger control, German law grants companies the option to apply for ministerial approval after the Federal Cartel Office has prohibited a merger project. Pursuant to Section 42 GWB, the Federal Minister for Economic Affairs may, upon application, permit a merger prohibited by the Federal Cartel Office if the overall economic benefits of the merger outweigh the restrictions on competition in the case at hand or if the merger is justified by an overriding public interest. According to the Ministry's legal application of Section 42 GWB and the explanatory memorandum to the second amendment to the GWB, reasons of public interest can only be taken into account if they are of great weight in the individual case, demonstrated by concrete evidence, and if there are no alternative state-induced remedies that are better in line with competition law.[102]

The public benefits recognised in the ministerial approval may overlap with the efficiencies examined under the efficiency defence in the proceedings of the Federal Cartel Office. However, these are not usually cost savings that can be quantified but

---

[99] See R. Podszun, Außerwettbewerbliche Interessen im Kartellrecht und ihre Grenzen, in: Kokott, Pohlmann and Polley (eds), Europäisches, deutsches und internationales Kartellrecht: Festschrift für Dirk Schroeder zum 65. Geburtstag, Otto Schmidt 2018, pp. 613, 618-619; V. Emmerich and K. Lange, Kartellrecht, 15th ed. C.H. Beck 2021, p. 205.

[100] V. Emmerich and K. Lange, Kartellrecht, 15th ed. C.H. Beck 2021, p. 217; see also T. Reif, in: Säcker and Meier-Beck (eds), Münchener Kommentar zum Wettbewerbsrecht, Vol. 2, 4th ed, C.H. Beck 2022, sec. 31 GWB paras 18-19, 26, 39, 40; J. Scholl, in: Körber, Schweitzer and Zimmer (eds), Immenga/Mestmäcker, Wettbewerbsrecht, Vol. 2, 7th ed, C.H. Beck 2023, sec. 31 GWB para 10.

[101] But see T. Reif, in: Säcker and Meier-Beck (eds), Münchener Kommentar zum Wettbewerbsrecht, Vol. 2, 4th ed, C.H. Beck 2022, sec. 31 GWB para 33, who argues that concerns about the competition authorities neglecting environmental policy objectives are unfounded.

[102] BMWi, 19 August 2019, I B 2 - 20302/14-02 – Miba/Zollern, para 163 referring to BMWi, 22 May 2006, I B 2 – 221410/02 - Rhön/Grabfeld, para 63; BMWi, 17 April 2008, I B 1 - 221410/03 – Universitätsklinikum Greifswald/Kreiskrankenhaus Wolgast, para 58; BT-Drs. VI/2520, Änderung des Gesetzes gegen Wettbewerbsbeschränkungen (Gesetzentwurf der Bundesregierung), p. 31.

7 Sustainability and Competition Law in Germany

rather benefits that should be assessed more qualitatively. According to the Ministry, fields of public interests potentially relevant to Section 42 GWB include economic policy, environmental concerns, employment, social and educational aspects, academic research, defence policy and health policy.[103]

Moreover, public welfare benefits may be, and will usually be, benefits that (also) have an impact outside of the affected markets. Against this background, environmental protection in general and climate protection in particular are generally accepted as important public interests that can justify ministerial approval.[104] As concerns the legal-economic peculiarities of climate protection—so far at the heart of the sustainability debate – the Ministry set out the requirements of the public benefit exemption of Section 42: with regard to future scenarios for which no experience rates (*Erfahrungssätze*) exist, the public benefit prognosis has to rely on reliable facts, satisfy the general rules of logic and occur with sufficient probability.[105]

Nevertheless, the aspect has not been very important in the case law so far.[106] The only case in which it played a major role was the Miba/Zollern merger (see Sect. 7.3.2.1.2). In the E.ON/Ruhrgas merger, the parties put forward environmental advantages as a public interest, but neither the Monopolies Commission[107] nor the Minister for Economic Affairs[108] followed this line of reasoning. The Minister granted approval for other reasons (see Sect. 7.3.2.1.2).

Fourth, Sections 24–27 GWB allow business and trade associations and professional organisations to establish competition rules within their area of business and to have these, upon examination, recognised by the Federal Cartel Office. Competition rules are provisions that regulate the conduct of undertakings and that aim to safeguard and encourage fair competition on the merits, Section 24 (1)–(2) GWB. It has been discussed whether the target of effective competition on the merits can

---

[103] BMWi, 19 August 2019, I B 2 - 20302/14-02 – *Miba/Zollern*, para 164.

[104] L. Fries, Die Berücksichtigung außerwettbewerblicher Interessen in der Fusionskontrolle, Nomos 2020, pp. 166-168; S. Thomas, in: Körber, Schweitzer and Zimmer (eds), Immenga/Mestmäcker, Wettbewerbsrecht, Vol. 2, 6th ed, C.H. Beck 2020, sec. 42 GWB para 84; E. Bremer and F. Scheffczyk, in: Säcker and Meier-Beck (eds), Münchener Kommentar zum Wettbewerbsrecht, Vol. 2, 4th ed, C.H. Beck 2022, sec. 42 GWB para 29.

[105] „Liegen keine derartigen Erfahrungssätze vor, genügen plausible und hinreichend wahrscheinliche Prognosen auf Grundlage gegenwärtig gesicherter Tatsachen. Beziehen sich die Gemeinwohlvorteile auf künftige Sachverhalte, genügt es, wenn die Prognosen auf konkreten Tatsachen aufbauen und nach logischen Denkgesetzen mit einer hinreichenden Wahrscheinlichkeit eintreten", BMWi, 19 August 2019, I B 2 -20302/14-02 – *Miba/Zollern,* para 165 with reference to BMWi, 5 July 2002 – I B 1 – 220840/129 – *E.ON/Ruhrgas,* WuW/E DE-V 573.

[106] But see ; E. Bremer and F. Scheffczyk, in: Säcker and Meier-Beck (eds), Münchener Kommentar zum Wettbewerbsrecht, Vol. 2, 4th ed, C.H. Beck 2022, sec. 42 GWB para 29, who note a general tendency that climate protection and environmental protection are becoming more important.

[107] Monopolkommission, Sondergutachten 34, Zusammenschlussvorhaben der E.ON AG mit der Gelsenberg AG und der E.ON AG mit der Bergemann GmbH, Nomos 2002, para 212.

[108] BMWi, 5 July 2002 – I B 1 – 220840/129 – *E.ON/Ruhrgas,* WuW/E DE-V 573.

capture agreements on sustainability.[109] In current practice and in leading commentaries, however, the topic does not seem to play a role. In principle, it seems conceivable to argue that effective competition on the merits should not create external effects. Accordingly, effective competition should cover sustainability agreements that internalise external effects. In any event, however, a recognition of competition rules by the Federal Cartel Office only implies that the authority will not exercise its enforcement powers with regard to the rules. Private plaintiffs and the courts are not bound by this and may consider certain competition rules an infringement of competition law.

7.2.3.2 Former Exemptions

In the past, German competition law used to contain more exemptions, which have gradually been abolished. Sometimes this was intended to effectively restrict the scope for exemptions; at other times, the deletions were not considered to bring about a change in substance. While none of the former exemptions focused on sustainability as such, several of them captured or at least were able to capture certain aspects of sustainability.

First, from the oldest version of the GWB in 1957 until the seventh reform in 2005, the GWB contained an explicit exemption for agreements on standards and types in Section 5 GWB old version. This exemption could also apply to agreements on sustainability standards.[110] Today, such agreements are examined under Section 2 GWB. Over the same period, the GWB contained an explicit exemption for agreements on rationalisation in Section 5 GWB old version. Somewhat surprisingly, however, this did not seem to apply to agreements that promote sustainability because the provision required that the cost/income ratio of the undertakings concerned be improved. Therefore, the internalisation of external effects, i.e. of costs that were caused by the producing companies but borne by others, was not covered.[111]

Again from 1957 until the seventh amendment in 2005, Section 8 GWB provided that if the requirements of the exemptions pursuant to Sections 2-7 GWB were not met, the Federal Minister for Economic Affairs could, upon application, grant permission for a contract or decision within the meaning of Section 1 GWB. This exception required the restriction of competition to be necessary for overriding reasons related to the economy as a whole and the public interest. It was in dispute whether "overriding reasons for the economy as a whole" and "the public interest" were cumulative criteria or if one or the other sufficed to grant permission. In any event, arguably the majority view, shared in particular by officials of the Federal

---

[109]D. Freitag, K. Hansen, K. Markert and V. Strauch, Umweltschutz und Wettbewerbsordnung, Athenäum 1973, pp. 85-87.

[110]D. Freitag, K. Hansen, K. Markert and V. Strauch, Umweltschutz und Wettbewerbsordnung, Athenäum 1973, pp. 81-82.

[111]D. Freitag, K. Hansen, K. Markert and V. Strauch, Umweltschutz und Wettbewerbsordnung, Athenäum 1973, pp. 82-84.

7 Sustainability and Competition Law in Germany

Cartel Office, considered Section 8 GWB suitable to legalise sustainability agreements.[112]

Third, in 1999, as part of the sixth amendment to the GWB, the legislator passed Section 7 GWB old version, which enabled the Federal Cartel Office to exempt agreements between companies that contributed to an improvement, inter alia, in the take back or disposal of goods or services. This required that consumers receive a fair share of the resulting benefit and that the improvement could not be achieved by the undertakings in any other way. Furthermore, the benefits had to be proportionate to the restriction of competition, and the latter could not create or strengthen an undertaking's dominant position.

The requirement of "take back or disposal" was supposed to exempt agreements from Section 1 GWB that aimed at ensuring compliance with the obligations arising from the Recycling and Waste Management Act (Kreislaufwirtschafts- und Abfallgesetz).[113] The goal of this act (and thus of the aforementioned elements of Section 7 GWB old version[114]) was to protect natural resources and make waste disposal more environmentally friendly.[115] Besides, the legislator argued that European law required competition policy to account for environmental protection.[116]

Section 7 GWB was removed in 2005 as part of the seventh amendment to the GWB. The legislator did not intend to bring about a change in substance. Rather, the deletion intended to align German with European competition law (Article 101 (3) TFEU).[117] Aspects that were previously relevant to Section 7 GWB should henceforth be covered by Section 2 GWB.[118]

7.3 Case Law

7.3.1 Sustainability as a Sword

As already explained in Sect. 7.1, it is generally acknowledged that the objectives promoted by competition, such as the effective use and allocation of resources tend to go hand in hand with sustainability objectives.[119] Despite some constellations in

---

[112] D. Freitag, K. Hansen, K. Markert and V. Strauch, Umweltschutz und Wettbewerbsordnung, Athenäum 1973, pp. 88-92.

[113] BT-Drs. 13/9720, p. 33.

[114] BT-Drs. 13/9720, p. 33.

[115] See Recycling and Waste Management Act (Kreislaufwirtschafts- und Abfallgesetz), sec. 1, former version.

[116] BT-Drs. 13/9720, p. 33.

[117] BT-Drs. 15/3640, pp. 24-25.

[118] BT-Drs. 15/3640, p. 27.

[119] Bundeskartellamt, Open markets and sustainable economic activity – public interest objectives as a challenge for competition law practice, October 2020, p. 8.

which public interest objectives and the objective to protect competition do in fact collide, competition law enforcement is typically designed to foster innovation, and innovation includes the realm of sustainability. Against this backdrop, the German Federal Cartel Office in several cases ended up promoting sustainability by intervening against anti-competitive practices.[120] The fact that the agency did not explicitly mention possible concerns about sustainability in its reasoning raises the question of whether the beneficial consequences to sustainability were in fact intended or merely coincidental in nature.

7.3.1.1 DSD

There seems to be widespread agreement that the primary responsibility to reach sustainability targets resides with the legislator, resulting in no more than a subordinate role for the executive branch – including the competition law agencies – with regard to sustainability matters (see Sect. 7.2.2.2.1). However, as indicated by the case against Duales System Deutschland,[121] this division of responsibility is by no means linear in nature: instead, in this particular case, it was precisely compliance with the obligations created by the legislative branch in its Packaging Ordinance (Verpackungsverordnung[122]) that caused the competition law agencies to become active (see already Sect. 7.2.2.2.1).

Duales System Deutschland (DSD, "The Green Dot") used to be the sole nation-wide system for the return and disposal of packaging. Operating through a net of service contracts with waste disposal firms, DSD took on the disposal obligations of manufacturers and dealers under the Packaging Ordinance against the payment of a licence fee. The resulting bundling of the demand constituted an appreciable restriction of competition, making DSD a monopoly player in the market for disposal services. After the Federal Cartel Office had been tolerating the DSD contract system within its discretion to take up a case, it initiated a competition law inquiry in

---

[120] See for example the DSD case discussed below.

[121] Even before the inquiry by the German Federal Cartel Office, the EU Commission had investigated DSD's practices and in 2001 found its payment system to violate former Article 82 TEC (now Article 102 TFEU), case T-151/01 (available at https://ec.europa.eu/commission/presscorner/detail/null/IP_01_584. Accessed 5 November 2020), upheld by the CFI, case T-151/01, *Der Grüne Punkt - Duales System Deutschland v Commission,* ECLI:EU:T:2007:154 and ECJ, case C-385/07 P, ECLI:EU:C:2009:456. Interestingly enough, concluding investigations under Article 81 TEC (now Article 101 TFEU), the DSD's was granted an exemption under Article 81 (3), European Commission, case T-289-01 (available at https://ec.europa.eu/commission/presscorner/detail/en/IP_01_1279. Accessed 5 November 2020). This outcome was highly motivated by the fact that DSD had reduced the duration of its exclusivity provisions to end in 2003. Also, the exemption was subject to two obligations regarding DSD's exclusivity clauses that were later upheld in CFI, case T-289-01, *Duales System Deutschland v Commission,* ECLI:EU:T:2007:155.

[122] Verordnung über die Vermeidung und Verwertung von Verpackungsabfällen, Federal Law Gazette 1991 I, p. 1234, replaced by the Verpackungsgesetz (Gesetz über das Inverkehrbringen, die Rücknahme und die hochwertige Verwertung von Verpackungen) in 2019, Federal Law Gazette 2017 I, p. 2234.

2002,[123] motivated by the aforementioned sixth amendment to the GWB and the introduction of the new exemption in Section 7 GWB.

Following an announcement by the Federal Cartel Office that from 2006 on it would no longer tolerate the contract system without intervention, DSD decided to dismantle its cartel-like structure.[124] Contributing to the "liberalization of the market for disposal services the Commission strives for",[125] the Federal Cartel Office successfully refuted the sustainability claims previously made by DSD. According to these, competition in the disposal service market would defeat the objective of the Packaging Ordinance and negatively affect environmental protection. Instead, in its sector inquiry published in 2012, the Federal Cartel Office observed substantial cost savings and improvements in the quality of recycling due to the opening up of the market.[126] In fact, contrary to the concerns by DSD, a wave of innovation in technology for sorting the mix of waste material led to higher quality recycling. Disregarding these improvements, municipal waste management providers and the private waste management industry continued to argue for the elimination of competition between compliance schemes to the legal framework, thereby reflecting their entrepreneurial interests.[127]

The German DSD case underscores concerns that sustainability cooperations more often than not turn out to be attempts at greenwashing what turn out to be purely economic interests of market participants. Moreover, the move from a general exemption outside the realm of competition law to a consideration within the competition law framework reflects the overall development in the approach by the German Federal Cartel Office (see Sect. 7.2.2.2.1).

---

[123] Bundeskartellamt (2002), Bundeskartellamt examines whether DSD is compatible with competition law, available at https://www.bundeskartellamt.de/SharedDocs/Meldung/EN/Pressemitteilungen/2002/23_08_2002_DSD_eng.html;jsessionid=7B56D1A14EF4BDC4A9EEF1D460FAB063.1_cid378?nn=3591568. Accessed 5 November 2022.

[124] Bundeskartellamt (2004), Bundeskartellamt welcomes opening-up of DSD to more competition, available at https://www.bundeskartellamt.de/SharedDocs/Meldung/EN/Pressemitteilungen/2004/12_10_2004_DSD_eng.html?nn=3591568. Accessed 5 November 2022.

[125] Bundeskartellamt (2002), Bundeskartellamt examines whether DSD is compatible with competition law, available at https://www.bundeskartellamt.de/SharedDocs/Meldung/EN/Pressemitteilungen/2002/23_08_2002_DSD_eng.html;jsessionid=7B56D1A14EF4BDC4A9EEF1D460FAB063.1_cid378?nn=3591568. Accessed 5 November 2022.

[126] Bundeskartellamt (2012), Bundeskartellamt presents results of its sector inquiry into compliance schemes, available at https://www.bundeskartellamt.de/SharedDocs/Meldung/EN/Pressemitteilungen/2012/03_12_2012_SU-duale-Systeme.html. Accessed 5 November 2022. The full version of the sector inquiry in German language is available at https://www.bundeskartellamt.de/SharedDocs/Publikation/DE/Sektoruntersuchungen/Sektoruntersuchung%20Duale%20Systeme%20-%20Abschlussbericht.pdf?__blob=publicationFile&v=7. Accessed 5 November 2022.

[127] Bundeskartellamt (2012), Bundeskartellamt presents results of its sector inquiry into compliance schemes, available at https://www.bundeskartellamt.de/SharedDocs/Meldung/EN/Pressemitteilungen/2012/03_12_2012_SU-duale-Systeme.html. Accessed 5 November 2022.

7.3.1.2 DSD/Remondis

Furthermore, DSD was the subject of yet another competition law case relevant to the field of sustainability: in 2019, the German Federal Cartel Office prohibited the acquisition of DSD by REMONDIS SE & Co. KG, stating that the merger would have significantly impeded competition between the dual systems for packaging recycling.[128] Remondis is Germany's largest waste management company and among the contract partners of DSD. Due to the increase in concentration achieved by the merger, Remondis would have had an incentive to charge DSD's competitors higher prices.[129] In a way, the prohibition thus served the purpose of maintaining the benefits achieved by opening up the market sector in the previous DSD proceedings. Additionally, the Federal Cartel Office points to sustainability concerns regarding the related market of glassworks: the expected decrease in the supply of glass cullet due to the proposed merger would cause glassworks to resort to alternative raw materials where obtainable. Severe increases in production costs aside, this adjustment would result in greater energy consumption, contradicting the overall aims of the recycling economy.[130] Even though this implication was not at the centre of the Federal Cartel Office's reasoning for the prohibition of the merger, the line of thought reveals the authority's awareness of sustainability effects in the field of disposal services.

7.3.2 Sustainability as a Shield

7.3.2.1 Merger Control Cases

On other more recent occasions, German competition law enforcement dealt with sustainability as a potential justification for non-intervention. As explained in Sect. 7.2.3.1, Section 42 GWB allows the Federal Minister for Economic Affairs to override a prohibition by the Federal Cartel Office if the restrictions on competition are offset by public benefits of a certain weight. Three recent cases indicate how environmental concerns increasingly feed into the consultation process required by

---

[128] Bundeskartellamt (2019), Bundeskartellamt prohibits Remondis/DSD merger, available at https://www.bundeskartellamt.de/SharedDocs/Meldung/EN/Pressemitteilungen/2019/11_07_2019_Remondis_DSD.html;jsessionid=2ED3D994A5FC35EA94B18CEC544478BF.1_cid378?nn=3591568. Accessed 5 November 2022. Confirmed by the Superior Court of Düsseldorf in OLG Düsseldorf, 24 April 2020 – VI-Kart 3/19 (V) – *Remondis/DSD*, ECLI:DE:OLGD:2020:0422. KART3.19V.00.

[129] For more details see Bundeskartellamt, 11 July 2019, B4-21/19, available at https://www.bundeskartellamt.de/SharedDocs/Entscheidung/DE/Entscheidungen/Fusionskontrolle/2019/B4-21-19.pdf;jsessionid=B2C1593C5F903CA2353E8C3D0DB97A7B.2_cid378?__blob=publicationFile&v=2. Accessed 5 November 2022; OLG Düsseldorf, 22 April 2020, VI-Kart 3/19 (V) – *Remondis/DSD*, ECLI:DE:OLGD:2020:0422.KART3.19V.00.

[130] Bundeskartellamt, 11 July 2019, B4-21/19, https://www.bundeskartellamt.de/SharedDocs/Entscheidung/DE/Entscheidungen/Fusionskontrolle/2019/B4-21-19.pdf;jsessionid=B2C1593C5F903CA2353E8C3D0DB97A7B.2_cid378?__blob=publicationFile&v=2. Accessed 5 November 2022, pp. 190-191.

7 Sustainability and Competition Law in Germany 111

Section 42 GWB, but they also reveal the shortcomings of this solution.[131] The consideration within the procedure for ministerial approval is subject to similar economic realities and constraints as is the political debate about climate change in general: in E.ON/Ruhrgas (Sect. 7.3.2.1.1), the Ministry clarified that competitive constraints generally occur right after the implementation of a merger, whereas the overall economic benefits can only be determined in the long term.[132] This is exactly one of the difficulties with climate change protection and a major obstacle to political solutions to the problem. This starting position provokes calls for an examination of sustainability concerns irrespective of their inclusion in consumers' preferences. Whether or not competition law is the right place for this remains an open question. For the moment, the general opinion within German competition law is doubtful in this regard.

7.3.2.1.1 E.ON/Ruhrgas

In the merger case of E.ON/Ruhrgas, the Ministry allowed for the merger of the two energy providers to proceed in accordance with additional requirements,[133] contradicting the previous prohibition by the Federal Cartel Office.[134] Besides the procedural particularities of the case,[135] both the Monopolies Commission's opposition and the ministerial approval examined the macro-economic consequences of the intended vertical integration.[136] Highlighting the significance of natural gas for the German energy supply, the companies in their application relied on the joint venture's international competitiveness, their contribution to securing Germany's energy supply and the preservation of jobs.[137] Furthermore, they claimed that the merger would support the government's climate protection efforts and

---

[131] Cf. inter alia S. Roth and M. Voigtländer, Die Ministererlaubnis für den Zusammenschluss von Unternehmen – ein Konflikt mit der Wettbewerbsordnung ZfW 2002(2), p. 239.

[132] BMWi, 5 July 2002 – I B 1 – 220840/129 – *E.ON/Ruhrgas,* WuW/E DE-V p. 573, para 140.

[133] BMWi, 5 July 2002 – I B 1 – 220840/129 – *E.ON/Ruhrgas,* WuW/E DE-V p. 573.

[134] Bundeskartellamt (2002), Bundeskartellamt prohibits E.ON/Gelsenberg (Ruhrgas) merger, available at https://www.bundeskartellamt.de/SharedDocs/Meldung/EN/Pressemitteilungen/2002/21_01_2002_EON_eng.html. Accessed 5 November 2022. According to the agency, the merger would have strengthened dominant positions in the gas and electricity sales markets.

[135] Since the Minister for Economic Affairs at the time, Werner Müller, was legally prevented from deciding the case due to possible conflicts of interest, the decision was issued by state secretary Alfred Tacke, whose responsibility had been under debate, see BMWi, 5 July 2002 – I B 1 – 220840/129 – *E.ON/Ruhrgas,* WuW/E DE-V 573, paras 87-91.

[136] Monopolkommission, Sondergutachten 34, Zusammenschlussvorhaben der E.ON AG mit der Gelsenberg AG und der E.ON AG mit der Bergemann GmbH, Nomos 2002, paras 137 ff; BMWi, 5 July 2002 – I B 1 – 220840/129 – *E.ON/Ruhrgas,* WuW/E DE-V 573.

[137] Application by E.ON as reported by Monopolkommission, Sondergutachten 34, Zusammenschlussvorhaben der E.ON AG mit der Gelsenberg AG und der E.ON AG mit der Bergemann GmbH, Nomos 2002, paras 84-95.

environmental policy.[138] They predicted that gas would replace a major part of the energy supplied by oil and nuclear energy and thus help achieve the government's climate policy goals, fulfilling the public interest criteria of Section 42 GWB. The Monopolies Commission in turn highlighted the primacy of functioning competition for innovation and the efficient use of resources when opposing the merger.[139] The expert panel also criticised the idea that the government could depend on a single company when trying to achieve its environmental objectives.[140] On a side note, in view of the current energy crisis, the Monopolies Commission's concerns regarding the diversification of energy sources turned out to be very accurate.[141]

The E.ON/Ruhrgas case was by no means the first merger in which German's security of supply had been invoked as an area of public interest: Already in its first approval case in 1974[142] and again in 1978,[143] the Ministry overruled the Federal Cartel Office's prohibitions based on similar policy considerations.[144] Criticising its own previous work, the Ministry recognised the problems related to energy policy considerations within the scope of Section 42 GWB.[145] Whether or not the uncertainty of forecast decisions is a problem peculiar to policy considerations or even to the ministry approval procedure is a different issue altogether.

Interestingly enough, irrespective of the environmental implications of natural gas as a fossil fuel, the German government subscribed to safeguarding the reliability of supply and the competitiveness of the energy supply sector as some of its major

---

[138] Application by E.ON as reported by Monopolkommission, Sondergutachten 34, Zusammenschlussvorhaben der E.ON AG mit der Gelsenberg AG und der E.ON AG mit der Bergemann GmbH, Nomos 2002, para 95.

[139] Application by E.ON as reported by Monopolkommission, Sondergutachten 34, Zusammenschlussvorhaben der E.ON AG mit der Gelsenberg AG und der E.ON AG mit der Bergemann GmbH, Nomos 2002, paras 183, 190, 236.

[140] Application by E.ON as reported by Monopolkommission, Sondergutachten 34, Zusammenschlussvorhaben der E.ON AG mit der Gelsenberg AG und der E.ON AG mit der Bergemann GmbH, Nomos 2002, para 211: "Davon abgesehen erscheint es wenig sinnvoll, dass sich die Bundesregierung bei der Erreichung ihrer umwelt- und klimapolitischen Zielsetzungen im Wesentlichen von einem Unternehmen abhängig macht."

[141] Application by E.ON as reported by Monopolkommission, Sondergutachten 34, Zusammenschlussvorhaben der E.ON AG mit der Gelsenberg AG und der E.ON AG mit der Bergemann GmbH, Nomos 2002, para 162. Nonetheless, the Monopolies Commission in its statement underestimated the consequences of Germany's high import dependency in the energy sector, cf. para 173.

[142] BMWi, 1 February 1974 – I B 5 – 810607 – *VEBA/Gelsenberg*, WuW/E BWM p. 147.

[143] BMWi, 5 March 1979 – I B 6 – 220840/15 – *VEBA/BP*, WuW/E BWM p. 165.

[144] For more details, see L. Fries, Die Berücksichtigung außerwettbewerblicher Interessen in der Fusionskontrolle, Nomos 2020, pp. 169-170.

[145] "Durchgesetzt hat sich die Einsicht, daß die Ministererlaubnis kein Instrument zur wirtschaftspolitischen Steuerung ist. Wie wenig die Ministererlaubnis hierfür geeignet ist, zeigen die Fälle VEBA/Gelsenberg (1974) und VEBA/BP (1979). Die dabei zugrunde gelegten energiepolitischen Prognosen haben sich im Rückblick als recht problematisch erwiesen", Erfahrungsbericht des Bundeswirtschaftsministeriums über Ministererlaubnis-Verfahren bei Firmen-Fusionen, WuW 42(11), 1992, p. 928.

7 Sustainability and Competition Law in Germany

political targets.[146] Anticipating the market pressure caused by the EU's efforts to open up the European gas market, the Ministry believed that the merger would contribute to achieving the above-mentioned target while at the same time ensuring the international competitiveness of Ruhrgas.[147]

Moreover, the Ministry explicitly went on to discuss contributions to climate protection as potential public benefits within the realm of Section 42 GWB.[148] In accordance with the Monopolies Commission, it observed that, generally, the transition from emission-heavy oil to gas might indeed meet the threshold of Section 42. Nevertheless, it concluded that the reduction in emissions remained uninfluenced by the merger and hence did not fulfill the causality requirement.[149] Even though the E.ON/Ruhrgas case is said to have opened up Section 42 GWB to environmental policy goals,[150] the ministerial approval was granted irrespective of sustainability aspects.[151] Instead, the Ministry questioned the R&D intentions of the merging parties, maybe even suggesting that the stated benefits due to improvements in fuel cell technology may have been a sign of greenwashing.[152]

7.3.2.1.2 Miba/Zollern

The Ministry for Economic Affairs further specified the E.ON/Ruhrgas ground rules in the more recent Miba/Zollern merger.[153] Initially, the Federal Cartel Office had prohibited the proposed merger in the bearing production sector vital to mechanical and plant engineering as well as engine construction. The joint venture would have increased concentration in a market with only a few suppliers and high barriers to entry.[154] According to the Federal Cartel Office, the parties had not sufficiently

---

[146] Bundesregierung (2017), Die Energie der Zukunft – Langfassung, Zweiter Fortschrittsbericht zur Energiewende, available at https://www.bmwk.de/Redaktion/DE/Publikationen/Energie/fortschrittsbericht-monitoring-energiewende-kurzfassung.pdf?__blob=publicationFile&v=8. Accessed 5 November 2022, p. 21.

[147] BMWi, 5 July 2002 – I B 1 – 220840/129 – *E.ON/Ruhrgas,* WuW/E DE-V 573, paras 114-121.

[148] BMWi, 5 July 2002 – I B 1 – 220840/129 – *E.ON/Ruhrgas,* WuW/E DE-V 573, para 136.

[149] BMWi, 5 July 2002 – I B 1 – 220840/129 – *E.ON/Ruhrgas,* WuW/E DE-V 573, paras 137-138.

[150] Quoted for example by BMWi, 19 August 2019, I B 2 - 20302/14-02 – *Miba/Zollern*, para 172.

[151] In support of this conclusion, see L. Fries, Die Berücksichtigung außerwettbewerblicher Interessen in der Fusionskontrolle, Nomos 2020, p. 166.

[152] BMWi, 5 July 2002 – I B 1 – 220840/129 – *E.ON/Ruhrgas,* WuW/E DE-V 573, para 139. The Monopolies Commission had arrived at a similar conclusion, pointing to E.ON's past R&D behavior, Monopolkommission, Sondergutachten 34, Zusammenschlussvorhaben der E.ON AG mit der Gelsenberg AG und der E.ON AG mit der Bergemann GmbH, Nomos 2002, paras 209-210. The mere declarations of intent did not meet the burden of proof required by sec. 42 GWB, cf. para 212.

[153] Bundeskartellamt, 17 January 2019, B 5-29/18, BMWi, 19 August 2019, I B 2 - 20302/14-02 – *Miba/Zollern.*

[154] Bundeskartellamt (2019), Bundeskartellamt prohibits merger between Miba and Zollern in bearing production sector, available at https://www.bundeskartellamt.de/SharedDocs/Meldung/EN/Pressemitteilungen/2019/17_01_2019_Miba_Zollern.html;jsessionid=6321A35D69BB2506F4AF94C49DF15AAB.2_cid378?nn=3591568. Accessed 5 November 2022.

substantiated that the merger would result in the claimed efficiencies and synergies.[155] They had not met the requirements set by the balancing clause in Section 36 (1) 2 No. 1 GWB (see Sect. 7.2.2.4).[156] Once again, the agency refrained from deciding whether efficiencies can be considered within the SIEC test, pointing out that the parties' submissions were insufficient in this regard (see Sect. 7.2.2.4).[157]

The parties to the proposed merger then filed for ministerial approval under Section 42 GWB. In addition to the preservation of jobs—a frequently tried argument in merger cases[158]—the companies' application pointed to the "know-how and innovation potential for energy transition and sustainability" and the "preservation of the technological lead of Germany and the EU".[159] They argued that their pooled production activities would contribute significantly to the desired energy transition and reduction of emissions.[160] The combination of unique expertise, (intellectual) property rights and resources would (re-)establish and ensure the undertaking's international competitiveness in the highly specialised industry of the bearing production sector.[161] Disregarding the contrary recommendation by the German Monopolies Commission,[162] Federal Minister Peter Altmaier granted authorisation for the Miba/Zollern merger to proceed subject to conditions and obligations.

Even though the German legislator made a conscious decision when separating the areas of responsibility of the Federal Cartel Office on the one hand and of the

---

[155] Bundeskartellamt, 17 January 2019, B 5-29/18, paras 364-367, 372-377.

[156] Bundeskartellamt, 17 January 2019, B 5-29/18, paras 416-435.

[157] Bundeskartellamt, 17 January 2019, B 5-29/18, para 372.

[158] Cf. Monopolkommission, Sondergutachten 81, Zusammenschlussvorhaben der Miba AG mit der Zollern GmBH & Co. KG, Nomos 2019, para 129.

[159] BMWi, 19 August 2019, I B 2 - 20302/14-02 – *Miba/Zollern,* paras 24-30.

[160] The special report issued by the German Monopolies Commission refers to ongoing research and development in the area of wind turbine technology as one of the examples states by the applicants, Monopolkommission, Sondergutachten 81, Zusammenschlussvorhaben der Miba AG mit der Zollern GmBH & Co. KG, Nomos 2019, paras 37, 41.

[161] Application by Miba and Zollern as reported by Monopolkommission, Sondergutachten 81, Zusammenschlussvorhaben der Miba AG mit der Zollern GmBH & Co. KG, Nomos 2019, paras 43-45.

[162] Monopolkommission, Sondergutachten 81, Zusammenschlussvorhaben der Miba AG mit der Zollern GmBH & Co. KG, Nomos 2019. While recognizing the maintenance of technological know-how, of the potential for innovation and the technological lead to generally be in the public interest, the Monopolies Commission argued that the companies had not met the high burden of proof required for public benefits in the realm of sec. 42 GWB, cf. paras 147-148. In continuation of its restrictive interpretation, the Monopolies Commission underscored the companies' need to show that the claimed public benefits are merger specific in nature and would not have occurred otherwise, cf. paras 80, 147, 152. The Commission also pointed out that European public interests – as compared to German public interests – are not eligible for consideration within sec. 42 GWB. Balancing impediments to competition and public benefits, sec. 42 GWB calls for a uniform reference point. Since the GWB's aim is to protect competition in the domestic market, the public benefits are confined to those with impact on the German market as well. The Monopolies Commission supports this view by recalling that the German Minister for Economic Affairs's range of authority is restricted to German territory, cf. paras 69-79.

7 Sustainability and Competition Law in Germany

Federal Ministry for Economic Affairs and Energy on the other hand,[163] the balancing process pursuant to Section 42 GWB might require a re-evaluation of the competitive restraints in individual cases. This balancing process necessitates the weighting of the competitive constraints found by the Federal Cartel Office[164] in order to determine the exact requirements for the public benefits.[165] In principle, the division of responsibility keeps the Federal Cartel Office from taking into account non-competition-related interests in the scope of its merger assessment.[166] Accordingly, the Federal Ministry is bound by the findings of the Federal Cartel Office as far as they concern the restriction of competition.[167] This notwithstanding, the Minister may deviate from this principle when new facts relevant to the ruling demand a re-evaluation of the circumstances.[168] Contradicting the assessment by the Monopolies Commission, the Federal Minister—based on a competitor's pleading in the oral proceedings—gave greater importance to the remaining competition (*Restwettbewerb*) in the sector.[169] This in turn caused the Minister to diverge from both the Federal Cartel Office's and the Monopolies Commission's evaluations. As a result, the Minister saw only a minor impairment to competition due to the proposed merger.[170]

While acknowledging the relation of rule and exception within the framework of the GWB,[171] Minister Altmaier concluded that the obligations imposed on the merging parties would safeguard the realisation of the sustainability benefits. Drawing on Section 20a GG, the Ministry set forth its practice from the E.ON/Ruhrgas case:[172] through their joint venture, Miba and Zollern would contribute significantly to the desired energy transition and thus help implement the "leading principle of sustainability" (*Leitprinzip der Nachhaltigkeit*).[173] Altmaier based his approval on the public interest in the "know-how and innovation potential for energy

---

[163] BT-Drs. 13/9720, p. 44.

[164] BMWi, 5 July 2002 – I B 1 – 220840/129 – *E.ON/Ruhrgas,* WuW/E DE-V 573 para 140.

[165] BMWi, 19 August 2019, I B 2 - 20302/14-02 – *Miba/Zollern,* para 155; see also KG, WuW/E OLG 1937, 1939 – *Thyssen/Hüller-Hille*.

[166] Bundeskartellamt, Open markets and sustainable economic activity – public interest objectives as a challenge for competition law practice, October 2020, pp. 42-43.: "Due to the clear separation of the areas of responsibility of the Bundeskartellamt and those of the Federal Ministry for Economic Affairs and Energy within the context of ministerial authorisation proceedings, it could hardly be justified to take into account non-competition related interests opposing the clearance of a merger project in the scope of the competition authority's merger assessment."

[167] BMWi, 19 August 2019, I B 2 - 20302/14-02 – *Miba/Zollern*, para 151.

[168] BMWi, 19 August 2019, I B 2 - 20302/14-02 – *Miba/Zollern*, para 154.

[169] BMWi, 19 August 2019, I B 2 - 20302/14-02 – *Miba/Zollern*, paras 157-158.

[170] BMWi, 19 August 2019, I B 2 - 20302/14-02 – *Miba/Zollern,* para 159.

[171] BMWi, 19 August 2019, I B 2 - 20302/14-02 – *Miba/Zollern,* para 161: "Der Schutz des Wettbewerbs ist die Regel, während die Erteilung einer Ministererlaubnis nur die Ausnahme sein kann, wie die Worte „im Einzelfall" in § 42 Abs. 1 S. 1 GWB verdeutlichen."

[172] BMWi WuW/E DE-V 573, 593 – E.ON/Ruhrgas.

[173] BMWi, 19 August 2019, I B 2 - 20302/14-02 – *Miba/Zollern*, para 180.

transition and sustainability"[174] while rejecting the more standard arguments along the lines of general international competitiveness, job protection, armaments sector security and defence policy.[175] To secure the promotion of the public benefit at the heart of the case, the ministerial approval demanded the creation of an escrow account containing a total investment sum of EUR 50 million.[176]

7.3.2.1.3 Beretta/Ammotec

Without having to undergo the procedure of Section 42 GWB, Beretta, an Italian manufacturer of firearms, was granted permission to acquire ammunition specialist Ammotec earlier this year. The Federal Cartel Office cleared the merger plans, observing only minor market share additions in the area for ammunition for civilian purposes.[177] It elaborated that Ammotec's supply of low-pollutant ammunition is in great demand by the German army (Bundeswehr). Collaboration with firearms manufacturers would facilitate compliance with goals for environmental protection and occupational safety.[178] The Federal Cartel Office saw no indication that competitors' access to low-polluting ammunition or its components could be restricted due to the merger. Whether the sustainability implications served as a shield in this particular case is hard to judge, given the little insight that the agency gave in its press release. The Federal Cartel Office's short remarks give the impression that there was no conflict between competition law and sustainability concerns in the case at hand.

7.3.2.2 Individual Guidance

The Federal Cartel Office further specified its stance towards sustainability cooperations through its individual guidance in constellations relevant to competition. The authority explicitly acknowledged the increasing frequency and significance of sustainability initiatives.[179]

---

[174] BMWi, 19 August 2019, I B 2 - 20302/14-02 – *Miba/Zollern*, paras 149, 167-168.

[175] BMWi, 19 August 2019, I B 2 - 20302/14-02 – *Miba/Zollern*, paras 202-233.

[176] BMWi, 19 August 2019, I B 2 - 20302/14-02 – *Miba/Zollern*, paras 1.2-1.4.

[177] Bundeskartellamt (2022), Small arms and ammunition sector: Merger between Beretta and Ammotec cleared, available at https://www.bundeskartellamt.de/SharedDocs/Meldung/EN/Pressemitteilungen/2022/11_05_2022_Beretta_Ammotec.html?nn=3599398. Accessed 5 November 2022.

[178] Special requirements in tender procedures increase the importance of non-toxic ammunition for the German Bundeswehr, cf. Bundeskartellamt (2022), Small arms and ammunition sector: Merger between Beretta and Ammotec cleared, available at https://www.bundeskartellamt.de/SharedDocs/Meldung/EN/Pressemitteilungen/2022/11_05_2022_Beretta_Ammotec.html?nn=3599398. Accessed 5 November 2022.

[179] Bundeskartellamt, Tätigkeitsbericht 2015/16, BT-Drs. 18/12760, p. 54.

7 Sustainability and Competition Law in Germany

Making use of its discretion to take up a case (see Sect. 7.2.2.2.1), the Federal Cartel Office in "Fairtrade",[180] "Living Wages"[181] and "Initiative Tierwohl"[182] clarified the ground rules for private sustainability initiatives. The Fairtrade Labelling Organisations International, e.V., is an internationally recognised, non-profit organisation that aims for fairer trade terms for farmers and workers in developing countries. Participation in the labelling process restricted to goods produced in non-EU countries is voluntary and non-exclusive. Despite the price elements of the agreement (agreement of a minimum price for certain import goods and the payment of the so-called Fairtrade Premium, irrespective of production costs[183]), the Federal Cartel Office decided not to initiate proceedings against Fairtrade. Drawing on the experience with quality mark associations (*Gütezeichengemeinschaften*) within German competition law,[184] it seems worth considering if parts of the Fairtrade mechanism could potentially pass as "quality features"[185] and might therefore qualify for an expanded interpretation of the exemptions granted to quality mark associations.[186] Others discuss the application of ancillary restraints or the rule of reason (*Immanenztheorie*)[187] with regard to the social standards implemented through private initiatives, such as the Fairtrade label.[188] According to this doctrine, competitive restraints necessary to ensure a main purpose can be exempted from

---

[180] Bundeskartellamt, Tätigkeitsbericht 2017/18, BT-Drs. 19/10900, p. 52.

[181] Bundeskartellamt, 25 November 2021, B2-90/21; Bundeskartellamt (2022), Achieving sustainability in a competitive environment – Bundeskartellamt concludes examination of sector initiatives, available at https://www.bundeskartellamt.de/SharedDocs/Meldung/EN/Pressemitteilungen/2022/18_01_2022_Nachhaltigkeit.html;jsessionid=2E99AFE29AA5859CB0CA98F0FB1D6331.2_cid381?nn=3591568. Accessed 5 November 2022.

[182] Bundeskartellamt, Tätigkeitsbericht 2013/14, BT-Drs. 18/5210, pp. 53-54.; Bundeskartellamt (2017), Bundeskartellamt calls for more consumer transparency in animal welfare initiative, available at https://www.bundeskartellamt.de/SharedDocs/Meldung/EN/Pressemitteilungen/2017/28_09_2017_Tierwohl.html;jsessionid=2B4801015737BFBD09A5AE315F90D570.1_cid390?nn=3591568. Accessed 5 November 2022; Bundeskartellamt (2022), Achieving sustainability in a competitive environment – Bundeskartellamt concludes examination of sector initiatives, available at https://www.bundeskartellamt.de/SharedDocs/Meldung/EN/Pressemitteilungen/2022/18_01_2022_Nachhaltigkeit.html;jsessionid=2E99AFE29AA5859CB0CA98F0FB1D6331.2_cid381?nn=3591568. Accessed 5 November 2022.

[183] Fairtrade International, available at https://www.fairtrade.net/about/how-fairtrade-works. Accessed 5 November 2022. For a more detailed discussion of the particularities, cf. F. Engelsing and M. Jakobs, Nachhaltigkeit und Wettbewerb, WuW 2019(1), p. 18.

[184] U. Loewenheim, in: Loewenheim et al. (eds), Kartellrecht, 4th ed, C.H. Beck 2020, sec. 20 GWB para 93; Krauß, in: Bunte (ed), Kommentar zum Deutschen und Europäischen Kartellrecht, Vol. 1, 14th ed, Luchterhand 2022, sec. 1 GWB para 272.

[185] K. Krauß, in: Bunte (ed), Kommentar zum Deutschen und Europäischen Kartellrecht, Vol. 1, 14th ed, Luchterhand 2022, sec. 1 GWB paras 161, 164.

[186] F. Engelsing and M. Jakobs, Nachhaltigkeit und Wettbewerb, WuW 2019(1), p. 19.

[187] D. Zimmer, in: Körber, Schweitzer and Zimmer (eds), Immenga/Mestmäcker, Wettbewerbsrecht, Vol. 2, 7th ed, C.H. Beck 2023, sec. 1 GWB para 50.

[188] F. Engelsing and M. Jakobs, Nachhaltigkeit und Wettbewerb, WuW 2019(1), p. 19; see also Bundeskartellamt, Open markets and sustainable economic activity – public interest objectives as a

competition law as long as that main purpose does not raise any concerns under competition law. Commonly invoked in partnership and acquisition agreements, the same line of thought could extend to environmental protection.[189] This approach is broadly similar to the teleological considerations made by the ECJ in *Wouters*[190] and *Albany*[191] (see Sect. 7.2.2.3).[192] Nonetheless, it would contradict recent efforts to embed sustainability concerns within the efficiency considerations of Article 101 (3) TFEU/Section 2 GWB (Sect. 7.2.2.2.2).

In its decision regarding Fairtrade, the Federal Cartel Office points to the initiative's relatively low market coverage within the EU as well as the existence of competing sustainability labels in Germany.[193] It also stressed the importance of a voluntary and discrimination-free participation in the certifications[194] as well as transparency for consumers.[195] Voluntary labels primarily increase customer choice and thus tend to be less objectionable under competition law.[196] For the same reason, the Federal Cartel Office announced that it has no competition concerns about "Living Wages", a voluntary commitment to achieving common standards for wages in the Ecuadorian banana sector.[197] Among its objectives are the introduction of responsible procurement practices and the development of monitoring transparent wages. The cooperation operates without any price-fixing arrangements and up to this point seems to comply with competition law requirements.

The guidance issued regarding the Initiative Tierwohl underscored the transparency requirement generally set out by the Federal Cartel Office. Initiative Tierwohl is an industry alliance of agriculture, the meat industry and food retailers. Its purpose is to reward livestock farmers for improving the conditions in which they keep their animals. To do so, the participants (major food retailers) agree to the payment of a uniform surcharge via the participating slaughterhouses. Employing labels for meat produced in line with animal welfare criteria, the initiative wants to create

---

challenge for competition law practice, October 2020, p. 20; L. Breuer, Das EU-Kartellrecht im Kraftfeld der Unionsziele, Nomos 2013, pp. 529-568.

[189] For an EU perspective of this line of thought, cf. S. Holmes, Climate Change, sustainability, and competition law, J. Antitrust Enforc. 2020 (2), pp. 370-371.

[190] ECJ, case C-309/99, *J. C. J. Wouters, J.W. Savelbergh, Price Waterhouse Belastingadviseurs BV v. Algemene Raad van de Nederlandse Orde van Advocaten,* ECLI:EU:C:2002:98.

[191] ECJ, case C-67/96, *Albany International BV v Stichting Bedrijfspensioenfonds Textielindustrie,* ECLI:EU:C:1999:430.

[192] F. Engelsing and M. Jakobs, Nachhaltigkeit und Wettbewerb, WuW 2019(1), p. 19.

[193] Bundeskartellamt, Tätigkeitsbericht 2017/18, BT-Drs. 19/10900, p. 52.

[194] Bundeskartellamt, Tätigkeitsbericht 2013/14, BT-Drs. 18/5210, pp. 53-54.

[195] Bundeskartellamt, Tätigkeitsbericht 2017/18, BT-Drs. 19/10900, p. 52.

[196] F. Engelsing and M. Jakobs, Nachhaltigkeit und Wettbewerb, WuW 2019(1), p. 20.

[197] Bundeskartellamt, 25 November 2021, B2-90/21; Bundeskartellamt (2022), Achieving sustainability in a competitive environment – Bundeskartellamt concludes examination of sector initiatives, available at https://www.bundeskartellamt.de/SharedDocs/Meldung/EN/Pressemitteilungen/2022/18_01_2022_Nachhaltigkeit.html;jsessionid=2E99AFE29AA5859CB0CA98F0FB1D6331.2_cid381?nn=3591568. Accessed 5 November 2022.

7 Sustainability and Competition Law in Germany

transparency and consumer awareness. Due to difficulties in the implementation of a monitoring process for pork products, the Initiative Tierwohl initially restricted the use of its label to poultry. Again stressing the importance of consumer choice, the Federal Cartel Office demanded an adjustment for the label to include pork products.[198] Later on, it also urged the parties to revise the structure of the financing model by gradually introducing competitive elements. Working towards a reconciliation of competition law requirements and sustainability efforts, the Federal Cartel Office suggested for the participants to part ways with the standard premium and to consider compensation for animal welfare costs instead.[199] Apart from this, the authority early on clarified the need for the participants to avoid the exchange of any information relevant to competition, thereby trying to keep sustainability initiatives from becoming a coincidental facilitator of illegal conduct.[200]

This detailed advice shows that in order to support companies in their sustainability efforts, the Federal Cartel Office focuses in particular on assisting companies in structuring their respective agreements.[201] There are further cases along these lines that cannot be analysed in detail here.[202] At the same time, it should be noted that there is only limited information publicly available on the cases that are the basis for the Federal Cartel Office's guidance. This implies a certain degree of opacity and a lack of legal certainty, which constitutes a weakness of the solution via the authorities' prosecutorial discretion (see also Sect. 7.4.1).

---

[198] Bundeskartellamt (2017), Bundeskartellamt calls for more consumer transparency in animal welfare initiative, available at https://www.bundeskartellamt.de/SharedDocs/Meldung/EN/Pressemitteilungen/2017/28_09_2017_Tierwohl.html;jsessionid=2B4801015737BFBD09A5AE315F90D570.1_cid390?nn=3591568. Accessed 5 November 2022.

[199] Bundeskartellamt (2022), Achieving sustainability in a competitive environment – Bundeskartellamt concludes examination of sector initiatives, available at https://www.bundeskartellamt.de/SharedDocs/Meldung/EN/Pressemitteilungen/2022/18_01_2022_Nachhaltigkeit.html;jsessionid=2E99AFE29AA5859CB0CA98F0FB1D6331.2_cid381?nn=3591568. Accessed 5 November 2022.

[200] Bundeskartellamt, Tätigkeitsbericht 2013/14, BT-Drs. 18/5210, p. 54.

[201] Bundeskartellamt (2022), Achieving sustainability in a competitive environment – Bundeskartellamt concludes examination of sector initiatives, available at https://www.bundeskartellamt.de/SharedDocs/Meldung/EN/Pressemitteilungen/2022/18_01_2022_Nachhaltigkeit.html;jsessionid=2E99AFE29AA5859CB0CA98F0FB1D6331.2_cid381?nn=3591568. Accessed 5 November 2022; the Federal Cartel office reports a certain expectation on the part of politicians and the public not to obstruct such agreements, see Bundeskartellamt, Tätigkeitsbericht 2015/2016, BT-Drs. 18/12760, p. 54.

[202] Bündnis für nachhaltige Textilien ("Grüner Knopf") and Initiative zur Reduzierung von Fett, Zucker und Salz in Fertigprodukten und Getränken, see Bundeskartellamt (2021), Jahresbericht 2020/21, available at https://www.bundeskartellamt.de/SharedDocs/Publikation/DE/Jahresbericht/Jahresbericht_2020_21.pdf?__blob=publicationFile&v=5. Accessed 5 November 2022, p. 46; Initiative des Handels zur Verringerung des Verbrauchs von Kunststofftragetaschen, see Bundeskartellamt, Tätigkeitsbericht 2015/16, BT-Drs. 18/12760, p. 54.

7.3.3 Greenwashing Cartels: Sustainability Initiatives as a Cover for Illegal Conduct

7.3.3.1 GGA

Whether ultimately intended or not, competition law enforcement turned out to promote sustainability by breaking up the DSD cartel (see Sect. 7.3.1 above). Adjusting its policy to European standards, the Federal Cartel Office rejected the suggested pretext that the undertaking's structure was the only feasible way to meet the requirements imposed by the Packaging Ordinance. Following a similar path, the Federal Cartel Office in 2007 took action against a cooperation of German container glass manufacturers.[203] The participants had set up the "Gesellschaft für Glasrecycling und Abfallvermeidung" (GGA) to jointly purchase waste glass recovered from household collections. Similar to the line of argument in the DSD case, the parties in GGA claimed that the purchasing cartel was the only way to guarantee high recycling quotas for waste glass. Going beyond the agency's stance in the DSD proceedings, the Federal Cartel Office's president accused the companies of greenwashing: "We will not allow anti-competitive cartel agreements to be made under the guise of environmental protection."[204] He went on by referring to end consumers who are denied their share of potential sale proceeds due to the elimination of competition in the sector. The Higher Regional Court sided with the Federal Cartel Office when denying GGA's motion to suspend the decision's implementation.[205] The purchasing cartel served the individual economic interests of the participants while neglecting the overall public interest in opening up the market.

7.3.3.2 Agrardialog Milch

Another case that helped draw boundaries for the treatment of cooperative initiatives within German competition law enforcement was the individual guidance provided to Agrardialog Milch.[206] This agricultural policy project had suggested a financing concept in favor of raw milk producers whose costs oftentimes exceed the price paid for milk. The Federal Cartel Office heavily criticised the idea of a surcharge

---

[203] Bundeskartellamt (2007), Bundeskartellamt prohibits purchasing cartel in the container glass industry, available at https://www.bundeskartellamt.de/SharedDocs/Meldung/EN/Pressemitteilungen/2007/01_06_2007_Beh%C3%A4lterglasindustrie.html. Accessed 5 November 2022; Bundeskartellamt, 31 May 2007, B4-1006-06, confirmed by OLG Düsseldorf, 14 June 2007 - VI-Kart 9/07 (V), ECLI:DE:OLGD:2007:0614.VI.KART9.07V.00.

[204] Bundeskartellamt (2007), Bundeskartellamt prohibits purchasing cartel in the container glass industry, available at https://www.bundeskartellamt.de/SharedDocs/Meldung/EN/Pressemitteilungen/2007/01_06_2007_Beh%C3%A4lterglasindustrie.html. Accessed 5 November 2022.

[205] OLG Düsseldorf, 14 June 2007 - VI-Kart 9/07 (V), ECLI:DE:OLGD:2007:0614.VI.KART9.07V.00.

[206] Bundeskartellamt, 8 March 2022, B2-87/21; Bundeskartellamt (2022), Surcharges without improved sustainability in the milk sector: Bundeskartellamt points out limits of competition law, available at https://www.bundeskartellamt.de/SharedDocs/Meldung/EN/Pressemitteilungen/2022/25_01_2022_Agrardialog.html. Accessed 5 November 2022.

independent of sustainability efforts by the producers. It pointed out the lack of specific criteria for milk production that would take into account sustainability concerns. Reinforcing its stance from the Initiative Tierwohl case, the agency advised the participants to link the intended surcharge to consumer demand.[207] The current version proved to be a price agreement to the disadvantage of consumers, rendering the project "not acceptable under competition law".[208] The Federal Cartel Office rejected the claim that the initiative would help finance the transformation of the domestic agricultural sector. Bearing in mind that the Federal Cartel Office in previous decisions attached importance to the market coverage of sustainability cooperatives, the intended sector-wide implementation of Agrardialog Milch certainly did not help its case.

7.4 Policy and Reform

7.4.1 General Guidance

The Federal Cartel Office is sympathetic to the goal of sustainability, acknowledging the importance of sustainability for society as a whole[209] and for consumers in particular.[210] Nevertheless, it has not provided general guidance for the consideration of sustainability in competition law yet. This is explained by four reasons: first, there has been insufficient case law on the matter up to now. Second, the effects of the specific conduct in question largely depend on the individual case at hand. Third, the formulation of general guidance would entail the risk of false negatives, i.e. the inclusion of agreements that are on balance harmful to competition and consumers.[211] Fourth, the Federal Cartel Office points out that its primary task is to protect competition; therefore, it wants to consider non-economic goals only in atypical cases.[212] In the view of the Federal Cartel Office, it is the legislator's task to

---

[207] Cf. F. Engelsing and M. Jakobs, Nachhaltigkeit und Wettbewerb, WuW 2019(1), p. 22.

[208] Bundeskartellamt (2022), Surcharges without improved sustainability in the milk sector: Bundeskartellamt points out limits of competition law, available at https://www.bundeskartellamt. de/SharedDocs/Meldung/EN/Pressemitteilungen/2022/25_01_2022_Agrardialog.html. Accessed 5 November 2022.

[209] Bundeskartellamt, Tätigkeitsbericht 2019/2020, BT-Drs. 19/30775, p. 13.

[210] Bundeskartellamt (2022), Achieving sustainability in a competitive environment – Bundeskartellamt concludes examination of sector initiatives, available at https://www. bundeskartellamt.de/SharedDocs/Meldung/EN/Pressemitteilungen/2022/18_01_2022_ Nachhaltigkeit.html;jsessionid=2E99AFE29AA5859CB0CA98F0FB1D6331.2_cid381?nn=3591 568.

[211] E. Wiese (2022), Agenda 2025 - Roadmap to Sustainability, available at https://www.d-kart.de/ blog/2022/05/18/agenda-2025-roadmap-to-sustainability/. Accessed 5 November 2022.

[212] Bundeskartellamt, Tätigkeitsbericht 2015/2016, BT-Drs. 18/12760, p. 54; Bundeskartellamt, Tätigkeitsbericht 2019/2020, BT-Drs. 19/30775, p. 13.

determine which goals to pursue and how much weight to attach to them.[213] Accordingly, the Federal Cartel Office so far has been acting on a case-by-case basis (see Sects. 7.2.2.2.1 and 7.3.3.2).[214] This lack of general guidance is subject to criticism since it can discourage companies from implementing sustainability initiatives.[215] The government has already announced its intention to address this problem; see Sect. 7.4.2.

7.4.2 Reform Proposals

7.4.2.1 Monopolies Commission
In 2022, the Monopolies Commission published a report, inter alia, on the consideration of sustainability goals in competition law, focusing on climate and environmental protection. The expert panel shares the majority view in German competition law scholarship that sustainability and competition law tend to go hand in hand, yet it also acknowledges a first-mover disadvantage[216] for sustainability initiatives and discusses possible remedies. It rejects the idea of exempting cooperations for environmental and climate protection from the cartel prohibition (Section 1 GWB) through teleological interpretation as this would reduce legal certainty.[217] Instead, the Monopolies Commission suggests further elaboration on Section 2 GWB, Article 101 (3) TFEU, the existing efficiency defence, which requires balancing efficiencies with restrictions of competition case by case.[218] In a similar vein, the Monopolies Commission is of the opinion that in merger control, the—currently unwritten—efficiency defence provides a suitable framework for sustainability considerations.[219] It contemplates clarifying this defence in the GWB[220] as this could encourage notifying parties to assert and substantiate a positive impact of the proposed merger in terms of efficiencies with respect to sustainability.[221] However, somewhat paradoxically, the Monopolies Commission recommends the

---

[213] Bundeskartellamt, Tätigkeitsbericht 2019/2020, BT-Drs. 19/30775, p. 13.

[214] Bundeskartellamt, Tätigkeitsbericht 2015/2016, BT-Drs. 18/12760, p. 54; Tätigkeitsbericht 2019/2020, BT-Drs. 19/30775, p. 13.

[215] Cf. Monopolkommission, XXIV. Hauptgutachten, Wettbewerb 2022, BT-Drs. 20/3065, para 414; critical of this approach also E. Wiese (2022), Agenda 2025 - Roadmap to Sustainability, available at https://www.d-kart.de/blog/2022/05/18/agenda-2025-roadmap-to-sustainability/. Accessed 5 November 2022.

[216] Monopolkommission, XXIV. Hauptgutachten, Wettbewerb 2022, BT-Drs. 20/3065, paras 408-413.

[217] Monopolkommission, XXIV. Hauptgutachten, Wettbewerb 2022, BT-Drs. 20/3065, para 424.

[218] Monopolkommission, XXIV. Hauptgutachten, Wettbewerb 2022, BT-Drs. 20/3065, paras 418, 424.

[219] Monopolkommission, XXIV. Hauptgutachten, Wettbewerb 2022, BT-Drs. 20/3065, para 462.

[220] Monopolkommission, XXIV. Hauptgutachten, Wettbewerb 2022, BT-Drs. 20/3065, para 475.

[221] Monopolkommission, XXIV. Hauptgutachten, Wettbewerb 2022, BT-Drs. 20/3065, para 463; J. Welsch (2022), Agenda 2025: Mehr Nachhaltigkeit in der Fusionskontrolle?, available at https://

7 Sustainability and Competition Law in Germany

discussed amendment only after more companies have invoked the efficiency defence with regard to sustainability.[222]

At the same time, the Monopolies Commission acknowledges several problems for the efficiency defence in its current form: first, it takes note of the fact that especially in merger control, it is difficult for the notifying parties to meet the strict requirements of the efficiency defence as well as the burden of proof, in particular with regard to a positive impact of sustainability issues on consumers.[223] Second, it raises the question of whether the majority view that excludes out-of-market-efficiencies from Section 2 GWB has to adjust in order to take all efficiencies of climate protection into account.[224] Third, a similar problem arises with regard to the timeline in which efficiencies materialise[225] and to the territorial restriction for efficiencies in merger control, i.e. the fact that only those efficiencies are eligible for consideration that manifest within Germany.[226]

In any event, the Monopolies Commission recommends providing the Federal Cartel Office with more resources to perform sustainability assessments, especially experts on environmental issues and climate economics.[227]

7.4.2.2 Legislative Activity

In its coalition agreement, the current German government made climate protection an objective that it will pursue in all policy areas. Accordingly, the social market economy should transform to a social-ecological market economy.[228] While the coalition agreement announced that the 11th amendment to the GWB would focus on digital markets, ministerial approval and consumer protection,[229] the German Ministry for Economic Affairs and Climate Action (BMWK) put forward ten points for making sustainable competition a cornerstone of the socioecological market economy, which address sustainability, social justice and innovation. Inter alia, the BMWK argues that where companies want to cooperate to achieve sustainability

---

www.d-kart.de/blog/2022/08/08/agenda-2025-mehr-nachhaltigkeit-in-der-fusionskontrolle/. Accessed 5 November 2022.

[222] Monopolkommission, XXIV. Hauptgutachten, Wettbewerb 2022, BT-Drs. 20/3065, para 462.

[223] Monopolkommission, XXIV. Hauptgutachten, Wettbewerb 2022, BT-Drs. 20/3065, paras 440, 444, 461-462.

[224] Monopolkommission, XXIV. Hauptgutachten, Wettbewerb 2022, BT-Drs. 20/3065, paras 434-436

[225] Monopolkommission, XXIV. Hauptgutachten, Wettbewerb 2022, BT-Drs. 20/3065, paras 437-438.

[226] Monopolkommission, XXIV. Hauptgutachten, Wettbewerb 2022, BT-Drs. 20/3065, para 445.

[227] Monopolkommission, XXIV. Hauptgutachten, Wettbewerb 2022, BT-Drs. 20/3065, paras 472, 475.

[228] SPD, Bündnis 90/Die Grünen and FDP (2021), Mehr Fortschritt wagen, Koalitionsvertrag 2021-2025, available at https://www.spd.de/fileadmin/Dokumente/Koalitionsvertrag/Koalitionsvertrag_2021-2025.pdf. Accessed 5 November 2022, pp. 5, 43.

[229] SPD, Bündnis 90/Die Grünen and FDP (2021), Mehr Fortschritt wagen, Koalitionsvertrag 2021-2025, available at https://www.spd.de/fileadmin/Dokumente/Koalitionsvertrag/Koalitionsvertrag_2021-2025.pdf. Accessed 5 November 2022, p. 25.

goals or human rights standards beyond government requirements, competition policy must provide legal certainty as regards conformity with competition law. The BMWK is therefore examining whether and how the competition law framework can be adapted. Its aim is to provide companies with a clear legal framework for sustainability cooperation while eliminating "greenwashing" cartels or other forms of disguised restrictions on competition.[230] Up until now, it is still unclear how exactly the government intends to achieve these goals. The government has recently announced that the upcoming 12th amendment to the GWB will cover these issues.[231]

Some scholarly proposals in a series of blog articles have tried to push the debate regarding the government's reform plans forward. They have advocated going beyond comfort letters by establishing an effective exemption for sustainability agreements, pointing to the Austrian example of Section 2 (1) KartG. Stressing the importance of legal certainty, these authors urge the legislator to focus on contributing to sustainability rather than on prioritizing the prevention of greenwashing.[232] They also highlight a possible supportive effect for reforms on the European level.[233] Furthermore, the same scholars have proposed strengthening sustainability in competition law by way of private enforcement. If cartels agree to refrain from producing more eco-friendly alternatives, it will be difficult to allocate damages produced by the external effects of the more harmful production. As a remedy, the authors suggest appointing a public interest representative who can claim these damages to the environment.[234]

---

[230] BMWK (2022), Wettbewerbspolitische Agenda des BMWK bis 2025, available at https://www.bmwk.de/Redaktion/DE/Downloads/0-9/10-punkte-papier-wettbewerbsrecht.pdf?__blob=publicationFile&v=6#:~:text=Die%20Kartellbeh%C3%B6rden%20beobachten%20die%20Entwicklungen,und%20auf%20den%20Fernw%C3%A4rmesektor%20ausweiten. Accessed 5 November 2022, pp. 2-3.

[231] Bundesministerium für Wirtschaft und Klimaschutz (2022), Bundeswirtschaftsministerium legt Entwurf zur Verschärfung des Wettbewerbsrechts vor, available at https://www.bmwk.de/Redaktion/DE/Meldung/2022/20220920-bmwk-legt-entwurf-zur-verscharfung-des-wettbewerbsrechts-vor.html. Accessed 5 November 2022.

[232] E. Wiese (2022), Agenda 2025 - Roadmap to Sustainability, available at https://www.d-kart.de/blog/2022/05/18/agenda-2025-roadmap-to-sustainability/. Accessed 5 November 2022.

[233] P. Hauser (2022), Agenda 2025: Nachhaltigkeitskooperationen & Private Enforcement, available at https://www.d-kart.de/blog/2022/06/07/agenda-ZOZS-nachhaltigkeitskooperationen-private-enforcement/. Accessed 5 November 2022.

[234] P. Hauser (2022), Agenda 2025: Nachhaltigkeitskooperationen & Private Enforcement, available at https://www.d-kart.de/blog/2022/06/07/agenda-ZOZS-nachhaltigkeitskooperationen-private-enforcement/. Accessed 5 November 2022.

7 Sustainability and Competition Law in Germany 125

Open Access This chapter is licensed under the terms of the Creative Commons Attribution 4.0 International License (http://creativecommons.org/licenses/by/4.0/), which permits use, sharing, adaptation, distribution and reproduction in any medium or format, as long as you give appropriate credit to the original author(s) and the source, provide a link to the Creative Commons license and indicate if changes were made.

The images or other third party material in this chapter are included in the chapter's Creative Commons license, unless indicated otherwise in a credit line to the material. If material is not included in the chapter's Creative Commons license and your intended use is not permitted by statutory regulation or exceeds the permitted use, you will need to obtain permission directly from the copyright holder.

Sustainability and Competition Law in Hungary

8

András M. Horváth

8.1 Introduction

8.1.1 Definition of Sustainability

The concept of sustainable yield—serving as the origin of the concept of sustainable development – was introduced in response to dwindling forest resources in the seventeenth and eighteenth centuries.[1] This suggests that the concept of sustainability is inherently economical, not alien from competition law, but having benefits going beyond simple economic gains.

The members of the working group have contributed to this report as follows: Eszter Ritter, Iván Sólyom, Zsolt Gyebrovszki and Ákos Réger to Sect. 8.2, Álmos Papp to Sect. 8.3, Aranka Nagy, Anikó Keller and Emil Szabó to Sect. 8.4, Tamás Polauf, Márton Kocsis, Boglárka Priskin and Anna Pintér to Sect. 8.5, Bálint Bassola and Kristóf Csillik to Sect. 8.6, as well as Csaba Kovács to Sect. 8.7. The author is grateful to the members of the working group for their invaluable contribution; all errors are those of the author. The views expressed here are not necessarily identical to the official position of the Hungarian Competition Authority.

[1] B. Purvis, Y. Mao and D. Robinson, Three pillars of sustainability: in search of conceptual origins, *Sustainability Science,* 2019(14), pp. 681-695.

A. M. Horváth (✉)
Hegymegi-Barakonyi and Fehérváry Baker & McKenzie, Budapest, Hungary
e-mail: andras.horvath@bakermckenzie.com

© The Author(s) 2024
P. Këllezi et al. (eds.), *Sustainability Objectives in Competition and Intellectual Property Law*, LIDC Contributions on Antitrust Law, Intellectual Property and Unfair Competition, https://doi.org/10.1007/978-3-031-44869-0_8

Although Article Q(1) of the Fundamental Law of Hungary mentions sustainable development,[2] it does not provide a definition. This is separate from the right to a healthy environment provided in Article XXI of the Fundamental Law of Hungary.

There is no exact definition of sustainability in any Hungarian act or regulation. However, Sect. 8.3.1 of the Annex of Parliamentary Decree No. 18/2013 (III. 28.) on the Framework Strategy of National Sustainable Development provides the following description: "By sustainability, we mean that the generation that creates its own well-being at a given moment in time does not exhaust its resources but preserves and expands them in sufficient quantity and quality for future generations. The interests of those who are not yet born i.e., those who do not yet have the right to vote, can be protected by imposing value, constitutional or other institutional limits on the freedom of movement of those who are alive today. They clarify the boundaries beyond which they will not, or cannot, take certain steps, and to resist the temptation to do so, they put up barriers in advance."

There is an exact definition for sustainable development, though in Section 4 point 29 of the Environmental Protection Act, it is stated:[3] "a system of social and economic conditions and activities that preserves the natural values for the current and future generations, uses natural resources economically and expediently and, in ecological terms, ensures the improvement of the quality of life and the preservation of diversity in the long run".

The Parliamentary Decree refers to the Brundtland report.[4] It is considered that the Brundtland report introduced sustainable development into public discourse. A simplified definition of sustainable development is the following: development, which satisfies the needs of today without jeopardizing the satisfaction of the needs of future generations. The three pillars of sustainability in the Brundtland report were environmental, economic and social.[5] Notably, the Parliamentary Decree accepted these pillars and suggested the addition of a fourth pillar: human resources.

---

[2] "In order to establish and maintain peace and security, and to ensure the sustainable development of humanity, Hungary shall endeavour to live in harmony with all the peoples and countries of the world."

[3] Act LIII of 1995 on the General Rules of Environmental Protection.

[4] Report of the World Commission on Environment and Developments: Our Common Future, United Nations, 1987.

[5] I. Gyulai, Fenntartható fejlődés és fenntartható növekedés, *Statisztikai Szemle*, 2013(8-9), p. 797.

Both the Draft Revised Horizontal Guidelines[6] and the Sustainability Expert Advice[7] refer to UN Resolution 66/288,[8] whereby the UN General Assembly decided to renew its commitment to sustainable development and "ensuring the promotion of an economically, socially and environmentally sustainable future for our planet and for present and future generations". This confirms the three pillars of sustainability suggested in the Brundtland report.

Due to the three pillars, sustainability covers numerous issues, not just environmental protection. This is exemplified by paragraph 543 of the Draft Revised Horizontal Guidelines, which—non-exhaustively—lists the following issues: "addressing climate change (for instance, through the reduction of greenhouse gas emissions), eliminating pollution, limiting the use of natural resources, respecting human rights, fostering resilient infrastructure and innovation, reducing food waste, facilitating a shift to healthy and nutritious food, ensuring animal welfare".

However, sustainability is not limited to environmental protection; for the purposes of this report, we focussed on this pillar.

8.1.2 Sustainability in Competition Law Rules

With respect to the above, it is not surprising that sustainability is not mentioned explicitly in Hungarian competition rules. Nevertheless, Section 17 a) of the Competition Act[9]—regulating the criteria of individual exemption from the prohibition of anti-competitive agreements (Section 11 of the Competition Act)—contains an explicit reference to environmental protection.[10] This provision is the equivalent of Article 101(3) TFEU. As described below, this formed the basis for defensive

---

[6] Annex to the Communication from the Commission – Approval of the content of a draft for a Communication from the Commission – Guidelines on the applicability of Article 101 of the Treaty on the Functioning of the European Union to horizontal co-operation agreements, Brussels, 1 March 2022 C(2022) 1159 final.

[7] Incorporating Sustainability into an Effects-Analysis of Horizontal Agreements – Expert advice on the assessment of sustainability benefits in the context of the review of the Commission Guidelines on horizontal cooperation agreements, Author: Roman Inderst, 2022.

[8] Resolution adopted by the General Assembly on 27 July 2012 (66/288).

[9] Act LVII of 1996 on the Prohibition of Unfair Market Practices and the Restriction of Competition.

[10] "The prohibition defined in Section 11 shall not apply to an agreement if:

(a) it contributes to the improvement of efficiency of production or distribution, or the promotion of technical or economic development, or the improvement of means of environmental protection or competitiveness;
(b) a fair share of the benefits arising from the agreement is conveyed to business partners who are not parties to the agreement;
(c) the concomitant restriction or exclusion of economic competition does not exceed the extent required for attaining the economically justified common goals;
(d) it does not provide for the exclusion of competition in connection with a considerable part of the goods concerned."

arguments in various procedures before the Hungarian Competition Authority (*GVH*). Notably, it could be argued that sustainability benefits not falling under the environmental pillar of sustainability could still fall under other types of efficiency gains mentioned in the Competition Act (technical development, competitiveness).

8.1.3 Sustainability Rules Relevant from a Competition Law Perspective

We understand that practically any environmental protection rule can be relevant from a competition law perspective because undertakings compete directly (e.g. by advertising their products as more sustainable) or indirectly (e.g. achieving cost reductions by more sustainable production technology) on sustainability. Also, there are secondary markets, which were explicitly created by environmental protection rules. For example, the Deposit Regulation[11] enabled companies to introduce a deposit fee on their packaging to facilitate recycling/recollection. On the one hand, companies compete in the amount of the deposit fee (included in the retail price). On the other hand, companies compete to obtain more recycled raw materials at a lower cost than new raw materials.

The main rules regulating companies' environmental protection obligations are the following:

- Environmental Protection Act: providing for an obligation to use the environment in such a manner that results in the smallest degree of environmental loading and utilization, prevents environmental pollution and precludes damage to the environment
- Environmental Pollution Charge Act:[12] providing for an obligation to pay a charge in the case of releasing certain substances into the air, waters or soil in excess of the environmental protection thresholds
- Environmental Protection Product Charge Act:[13] providing for an obligation to pay a surcharge in case of certain products (batteries, packaging material, electric and electronic equipment, tyres and motor oil) to facilitate the recollection/ recycling of the given products
- Waste Management Act:[14] providing for producers' responsibility for disposing waste stemming from their products and their packaging and – as of 1 March 2021 – providing for extended producer responsibility, creating an obligation to select production technologies, raw material, etc. most favourable for the environment

---

[11] Government Regulation No. 209/2005 (X. 5.) on the Application of Deposit Fees.

[12] Act LXXXIX of 2003 on Environmental Pollution Charge.

[13] Act LXXXV of 2011 on Environmental Protection Product Charge.

[14] Act CLXXXV of 2012 on Waste.

− Deposit Regulation: providing for the possibility for producers to apply a deposit fee on the packaging of products

8.2 Cartels and Sustainability

In this section, we examine the role of sustainability in the assessment of anti-competitive agreements, decisions and concerted practices in the context of Hungarian regulation. We examine two scenarios: (i) *ex post*, where an agreement is exempted on the basis of sustainability considerations, and (ii) *ex ante*, where an agreement is not unlawful due to sustainability considerations.

8.2.1 Use of Sustainability *Ex Post* (Exempted Agreements)

8.2.1.1 Legislative Framework
As mentioned above, the rules of the Competition Act on anti-competitive agreements take into account sustainability considerations as part of the assessment of individual exemption. Examples of potentially exempted agreements could be the following:

− An agreement between competitors to jointly develop a production technology that reduces energy consumption
− An agreement between competitors to share infrastructure with a view to reducing the environmental footprint of a production process
− An agreement between competitors to jointly purchase products having a limited environmental footprint as an input for their production
− An agreement between competitors to only purchase from suppliers observing certain sustainability principles.

As confirmed by the courts, all four conditions of individual exemption must be fulfilled simultaneously for a given anti-competitive agreement to be exempted.[15] Therefore, these agreements could only be exempted if the undertakings can prove other conditions too (particularly, that the sustainability benefits cannot be achieved without the anti-competitive agreement).

Pursuant to Section 75 (1) of the Competition Act, the GVH can accept commitments instead of finding an infringement and imposing a fine if the commitment decision can ensure the effective protection of public interests. Pursuant to the

---

[15] See the judgments of the Hungarian Supreme Court (*Kúria*) no. Kf.II.40.072/2000/5 and the Metropolitan Court of Appeal no. 2.Kf.27.314/2005/6.

Commitment Notice of the GVH,[16] the effective protection of the public interest means that the commitment will result in an – even indirect – benefit, perceptible for the market and a wide range of consumers. As of 1 January 2021, point h) of paragraph 19 of the Commitment Notice explicitly mentions as an example of such benefits commitments that contribute to sustainability or environmental protection and thereby to consumer welfare. Notably, paragraph 12 of the Commitment Notice suggests that hard-core cartels are typically unsuitable for commitments (unless it is a novel type or carried out by small and medium-sized enterprises (SMEs)).

Pursuant to Section 78 (3) of the Competition Act, the fine shall be determined taking into account all applicable circumstances (including any cooperation with the investigation). Pursuant to the Fining Notice of the GVH,[17] active reparation is considered as a form of cooperation. As a result of (voluntary) active reparation, the GVH may grant immunity from or a reduction of the fine. Under the Fining Notice, conduct is considered to constitute active reparation where the undertaking subject to the infringement—in whole or in part—remedies the negative effects of the infringement (compensation). As of 1 January 2021, paragraph 33 point iv) of the Fining Notice explicitly mentions as an example of such benefits conduct that contributes to sustainability or environmental protection and thereby to consumer welfare. Notably, active reparation must be supported by data and analysis beforehand and verified by audited results afterwards.

Finally, the GVH can also apply EU competition law. According to the case law,[18] in the case of proceedings with a dual legal basis, Article 101 TFEU takes precedence over the provisions of Hungarian law. This also means that if an anti-competitive agreement is not prohibited under Article 101 (3) TFEU, there is no longer a need for any justification for a separate exemption under Hungarian law. On the one hand, the GVH could apply the Article 101 (3) Guidelines,[19] which clarifies that cost efficiencies and qualitative efficiencies (though certain sustainability benefits would clearly qualify as qualitative efficiencies) are not the only types of efficiencies acceptable under Article 101 (3) TFEU; rather, the individual exemption applies to all objective economic efficiencies (paragraph 59). On the other hand, the GVH could apply the Horizontal Guidelines,[20] or its likely successor, the Draft Revised Horizontal Guidelines, which dedicates a whole chapter to sustainability

---

[16] Notice no. 1/2018 of the President of the Hungarian Competition Authority and the President of the Competition Council of the Hungarian Competition Authority on the application of Section 75 of the Competition Act on commitments.

[17] Notice no. 1/2020 of the President of the Hungarian Competition Authority and of the President of the Competition Council of the Hungarian Competition Authority on setting the amount of the fine in the case of infringements of the prohibitions on antitrust infringements.

[18] VJ-180/2004 *Hungarian Bar Association* case (14 June 2006).

[19] Guidelines on the application of Article 81(3) of the Treaty, OJ 2004 C 101, p. 97-118.

[20] Guidelines on the applicability of Article 101 of the Treaty on the Functioning of the European Union to horizontal co-operation agreements, OJ 2011 C 11, p. 1-72.

8 Sustainability and Competition Law in Hungary

agreements. This is a likely scenario, particularly, because the GVH does not have equivalent guidance at the national level.

8.2.1.2 Sustainability Agreements in Hungarian Case Law

Although the GVH has rejected applications for individual exemption based on sustainability in all cases so far, it is important to highlight that the GVH is hearing sustainability defences. In the next sections, we demonstrate this relevant case law.

8.2.1.2.1 Portable Batteries Case

The most relevant case for the use of sustainability as a shield is the *Portable Batteries* case.[21] In Hungary, the obligation of recollection of portable batteries and accumulators was defined by the Waste Batteries Regulation.[22] The manufacturers could delegate their obligation to a coordinating/intermediary organisation in return for a fee. The coordinating organisation set up a collection system on behalf of its contractual partners, placing common collection containers in shops, schools and other community places and coordinating the emptying of full containers and the disposal of used batteries. To ensure effective recycling, the three investigated companies (Energizer, Procter & Gamble, Spectrum) established a non-profit organization (RE'LEM) to fulfil their obligation. The representatives of the companies as board members of RE'LEM agreed to set a waste management fee that RE'LEM collected from the companies on a per kilogram basis. Being an industry standard, the GVH confirmed that the way the companies set RE'LEM's waste management fee was not anti-competitive (although RE'LEM was held liable as a facilitator). What the GVH found anti-competitive, however, was that the manufacturers coordinated the uniform passing on of the waste management fee to retailers (and thereby to consumers).[23] The GVH considered this a restriction by object. Notably, the waste management fee was only a negligible percentage of the wholesale price of portable batteries and accumulators; however, this was only assessed in connection with the calculation of the fine (gravity of the infringement).

P&G and Spectrum argued that passing on did not affect their wholesale prices and that passing on (and the implementation of the agreement) was not complete. It was RE'LEM that put forward environmental protection considerations, in particular that the agreement followed environmental protection goals and that the passing on was a waste management industry standard. These considerations were assessed by

---

[21] VJ-43/2015 *Portable Batteries* case (15 April 2019). There was no judicial review in this case (Energizer was granted immunity, and P&G was granted a fine reduction due to its leniency application, while P&G and Spectrum were granted a fine reduction due to a settlement statement).

[22] At the start of the infringement (2005) it was Government Regulation No. 109/2005 (VI. 23.) on the Recollection of Waste Batteries and Accumulators, from 2008 until 2012 it was Government Regulation No. 181/2008 (VII. 8.) on the Recollection of Waste Batteries and Accumulators. At the time of the termination of the infringement (2015), it was Government Regulation No. 445/2012 (XII. 29.) on Waste Management Activities in relation to Waste Batteries and Accumulators.

[23] The companies applied the same notice template addressed to retailers on passing on. Also, the companies indicated the fee separately on their wholesale invoices.

the GVH in connection with the calculation of the fine as a mitigating factor concerning RE'LEM's fine. RE'LEM only alluded to the possibility of an individual exemption. This unsupported argument was dismissed by the GVH. Notably, the GVH mentioned that it would have been unlikely to exempt the agreement – even in the case of verified benefits – due to the hard-core nature of the infringement.

8.2.1.2.2 Lead-Acid Accumulators Case

Another relevant cartel case for the sustainability defence is the *Lead-Acid Accumulators* case.[24] The three investigated companies (Fe-Group, Alcufer, Jász-Plasztik) were all active in the recollection of waste lead-acid accumulators. Jász-Plasztik was also a manufacturer of new lead-acid accumulators. When Jász-Plasztik decided to construct a new recycling plant, the three companies started negotiations to create a consortium. As the capacity of the new plant covered the entire supply of waste lead-acid accumulators in Hungary, Fe-Group and Alcufer agreed to sell all or at least most of their recollection to Jász-Plasztik. To make recollection more efficient, the parties agreed to allocate regions in Hungary. On the one hand, Jász-Plasztik guaranteed Alcufer and Fe-Group a favourable take-over price (by passing on a part of the savings on transport costs). On the other hand, the parties agreed on a maximum price for the purchase (recollection) of waste lead-acid accumulators. The GVH considered this a restriction by object. Notably, the GVH found that the relevant geographic market was wider than national because—in the absence of a Hungarian recycling plant at the time—waste lead-acid accumulators recollected in Hungary were sold to processors in other EU Member States, including Austria, the Czech Republic, Slovakia, Slovenia, Bulgaria and Romania

In the course of the procedure, two companies used the sustainability defence in their arguments. Relying on individual exemption, Alcufer argued that (i) Section 17 of the Competition Act mentions environmental benefits, but coordinating the transport of waste to the recycling plant also creates efficiency gains; (ii) the cost savings are passed on to sellers, while consumers enjoy environmental benefits by lowering emissions from more efficient transport; (iii) the related restriction of competition was proportionate; and (iv) the agreement did not eliminate competition due to the high number of competitors in the market. Jász-Plasztik invoked the proximity principle as an environmental argument: if the capacity of the recycling plant is covered by Hungarian recollection, there is no need for waste import. Jász-Plasztik also argued that efficiency could have been only achieved through coordinated collection.

The GVH dismissed these arguments. The GVH emphasised that the environmental objectives of the agreement do not eliminate liability for an anti-competitive agreement. The GVH stated that the companies failed to demonstrate that the efficiencies outweigh the restriction of competition. The company only relied on

---

[24] VJ-2/2015 *Lead-acid Accumulators* case (26 July 2016). The decision was ultimately upheld by the Kúria in case No. Kf.II.37.959/2018 (4 December 2019). Although, the Kúria annulled all fines, due to the GVH exceeding the procedural deadline.

8 Sustainability and Competition Law in Hungary

general statements concerning the benefits being shared with consumers. The GVH considered the hard-core restrictions (market allocation, price fixing) to exclude individual exemptions. Nevertheless, the GVH assessed these arguments in connection with the calculation of the fine.

In judicial review, the Kúria—upholding the dismissal of the sustainability defence—stated that for an individual exemption, the benefits must significantly outweigh the harm caused by the restriction. The Kúria also emphasised that it is the obligation of the companies to present and prove this.

8.2.1.2.3 Beer Case

The sustainability defence was also used in the *Beer* case.[25] The GVH investigated the exclusive purchase and sale clauses in the contracts of the four Hungarian breweries (Heineken, Borsodi, Dreher and Pécsi) concluded with HoReCa (hotels, restaurants, catering) customers. Ultimately, the GVH considered that (a) due to its market share, Pécsi qualified for *de minimis*, and the procedure was terminated against Pécsi; (b) due to their market shares, the three other breweries qualified for block exemption, but due to the cumulative effects of the agreements, the GVH withdrew the block exemption; (c) the GVH accepted the commitments of the three other breweries to limit the proportion of "reserved quantities" by 19%.

All breweries referred to both block exemption and individual exemption, but it was Borsodi that submitted a sustainability defence. It argued that close cooperation between breweries and HoReCa units contributed to increasing distribution and production efficiency, allowing long-term sales planning, which is particularly important in the case of perishable products like beer, with a relatively short shelf-life. Borsodi stressed that by preventing the production of surplus quantities, the agreements have a positive environmental impact due to the more efficient use of resources and the possibility of avoiding the destruction of surplus. Borsodi also relied on arguments concerning (i) intensive competition in the market (strong competitors, price-sensitive consumers), ensuring the fair share of consumers; (ii) the difficulty of long-term planning without the investigated contractual clauses, making the restriction indispensable; and (iii) the low market share of its HoReCa partners, suggesting no elimination of competition.

The GVH accepted that efficiency gains could arise (without mentioning sustainability benefits as such) but considered that other conditions of individual exemption were not met.

8.2.1.2.4 Lubricants Case

Finally, in the *Lubricants* case,[26] the GVH examined whether clauses restricting resale below the gross retail price and exclusivity clauses that are longer than 5 years,

---

[25] VJ-49/2011 *Beer* case (22 July 2015). There was no judicial review in this case because it was a commitment decision.

[26] VJ-7/2008 *Lubricants* case (29 March 2010). There was no judicial review in this case.

set by Castrol in its contracts with car repairers for the sale of lubricants, could restrict economic competition.

Castrol submitted detailed arguments concerning individual exemption and also referred to the environmental protection goals mentioned in Section 17 of the Competition Act. On the one hand, in Castrol's view, the exclusivity clause contributed significantly to production and distribution efficiencies and increases the cost-effectiveness of production. Castrol emphasised that a professionally maintained car is demonstrably less polluting and consumes less fuel when used properly, which is a benefit felt both locally (at the user level) by stakeholders and globally (at the community level). To secure this benefit, it is essential for Castrol to exclude free riding. On the other hand, in Castrol's view, the use of exclusivity indirectly contributed to the achievement of environmental objectives. Castrol – via a specialised company but at its own expense – carried out the treatment of waste lubricants for the repairers. If repairs were to distribute and use other lubricants, competing companies would benefit from the advantages offered by the collection scheme of Castrol.

Ultimately, the GVH established the infringement; however, it did not impose a fine on Castrol because it has not yet enforced the investigated clauses. The GVH only obligated Castrol to remove the RPM clause and limit exclusivity to 5 years. The sustainability defence of Castrol was not addressed by the GVH.

8.2.1.2.5 Comments to the Case Law

It is clear from the case law that the GVH is open to assessing sustainability-related arguments; however, the benefits must be clearly demonstrated, and quantified, and other conditions of individual exemption must also be met.

The general consideration of sustainability-related literature in environmental law is the promotion of the polluter pays principle.[27] In simple terms, the most environmentally friendly solution to non-sustainable production is that the polluter should bear all costs related to the ecological impact. By obliging manufacturers to organise the effective collection and recycling of their products, the legislator decided to follow the polluter pays principle (at least in relation to recycling). Of course, depending on the level of competition in the market, some of the costs related to recycling are expected to be passed on to customers. In *Portable Batteries* case, the GVH was correct to expect companies to compete in the level of passing on the waste management fee, similarly to how they compete in the level of passing on other types of costs.[28]

Agreeing to pass on these costs in the form of a separate item in the invoice, however, resulted in the consumer paying the costs of recycling fully. Although not

---

[27] United Nations, A/CONF.151/26 (Vol. I), Report of the United Nations Conference on Environment and Development, 12 Aug 1992 (1992 Rio Declaration), Principle 16.

[28] Interestingly, the more competitive a market is, the higher the expected passing on – according to economic theories of passing on. In a perfectly competitive market, consumers would bear the full cost of the environmental impact.

an enforcement priority of the European Commission (*EC*), one could argue that the most environmental-friendly solution is when the environmental costs are fully reflected in consumer prices.[29] Such an approach may be beneficial if the purpose of the legislation was, for example, to reduce the consumption of unsustainable batteries or encourage more environmental-friendly manufacturing and purchases. However, for a battery recycling system only, a passing-on agreement does not seem indispensable to achieve the goals set by the legislator. The shared collection and recycling in exchange for a fee paid by battery sellers seems sufficient for the system to work well; a market-wide passing on of these fees to consumers is not a necessary element.

In Hungary, environmental regulations on battery recycling require that a certain percentage of all batteries sold is collected and recycled. Such regulations were in place throughout the entire infringement period of the *Portable Batteries* case. However, regulation did not prevent battery manufacturers from collecting and recycling more batteries than what was required by law. This could have clearly been an environmental-friendly initiative, benefitting society. According to the Draft Revised Horizontal Guidelines, the willingness to pay of battery consumers should be considered.[30] The Draft Revised Horizontal Guidelines suggest that Article 101(3) may be applied if battery consumers were willing to pay a higher price for a more efficient battery recycling system than the cost of recycling (because they benefit from reduced environmental pollution). Measuring the difference can be done by analysing battery purchasing data or by carrying out surveys involving a pool of representative consumers.[31] Therefore, if the Draft Revised Horizontal Guidelines would have been applicable, the companies involved in the infringement should have first demonstrated that the efficiency, territorial coverage or other aspects of the recycling procedure would have been enhanced by the agreement of passing on those extra costs. Moreover, they should have also shown that battery consumers were willing to pay for the cost of more efficient recycling. Notably, the Draft Revised Horizontal Guidelines also accept an assessment of collective benefits.[32] Such benefits have been invoked in the *Beer* and *Lubricants* cases. However, this may result in a lower threshold for individual exemptions. In the *Portable Batteries* and the *Lead-Acid Accumulators* case, it may have been said that toxic chemicals of non-recycled batteries may poison the topsoil of agricultural land.

---

[29] More general argumentations about the use of "true costs" in sustainability assessments have been emerging in the EC's public consultation on the Draft Revised Horizontal Guidelines (see, for example, the submission of Frontier Economics, available at https://www.frontier-economics.com/media/5170/ec-draft-revised-guidelines-on-horizontal-agreements-section-9-sustainability-agreements.pdf). Accessed 28 October 2022.

[30] Draft Revised Horizontal Guidelines, paragraph 597.

[31] OECD Measuring environmental benefits in competition cases – Note by Nadine Watson, Roundtable on Environmental Considerations in Competition Enforcement, 1 Dec 2021, DAF/COMP(2021)14.

[32] As long as there is a "substantial overlap" between purchasing consumers and the beneficiaries of the agreement (Draft Revised Horizontal Guidelines, paragraph 606).

Therefore, an agreement resulting in better recycling activities benefits society. However, estimating out-of-market benefits could have been challenging.

8.2.2 Use of Sustainability *Ex Ante* (Preventive Integration)[33]

Hungarian competition law is not characterised by preventive integration in the assessment of anti-competitive agreements, i.e. interpreting competition law in a way that leads to, for example, the prevention of environmental degradation. The reason for this is that within the institutional framework of Hungarian competition law, the use of preventive integration instruments is subject to certain limitations. The use of preventive integration may exceed the scope of competition law, and the GVH may exceed its own powers, thus risking the separation of powers. For this reason, the Hungarian competition rules on anti-competitive agreements do not contain environmental standards as such environmental standards are set by the relevant legislator. For the above reason, the use of preventive integration tools does not arise in Hungarian competition proceedings concerning cartels.

However, theoretically, the violation of sustainability by an anti-competitive agreement could be considered by the GVH. If a restriction by object cannot be established, the effects of the agreement need to be examined, including whether the agreement had a negative effect on innovation.[34] An anti-competitive agreement leading to suppression of unsustainable production or distribution could be considered harmful to innovation. Also, the violation of sustainability could be considered an aggravating factor in the imposition of fines in the case of a restriction by object. Such an approach has not been reflected yet in the practice of the GVH.

However, this approach could change. The GVH follows the practice in the EC. In the *Car Emissions* case,[35] the EC's competition law analysis showed that a certain level of preventive integration is easily achievable. The EC is focusing on harm to innovation, which in turn has led to environmental damage. Such an approach seems to be a viable way to indirectly enforce sustainability through competition law instruments.

8.3 Abuse of Dominance and Sustainability

In this section, we examine the role of sustainability in the assessment of abusive practices under the Competition Act.

---

[33] Preventive integration (sword) is determining the question of how the scope of the competition provisions can be interpreted possibly more extensively to subject such measures to competition assessment.

[34] VJ-180/2004 *Hungarian Bar Association* case (14 June 2006).

[35] AT.40178 *Car Emissions* case (8 July 2021).

8.3.1 Legislative Framework

The prohibition of abuse of a dominant position is regulated by Section 21 of the Competition Act, equivalent to Article 102 TFEU, with some more examples of the prohibited abusive behaviours.[36] The definition of a dominant position is regulated in Section 22 of the Competition Act, equivalent to the case law of the Court of Justice of the European Union.[37]

8.3.2 Use of Sustainability as a Sword

The rules of the Competition Act on abusive practices fail to directly mention sustainability (or environmental) considerations. This makes it more difficult to use Section 21 of the Competition Act as a sword to achieve sustainability-related objectives.

Notably, Section 21 of the Competition Act defines abuse so broadly that it may be interpreted to imply a prohibition of measures aiming to, or otherwise resulting in, harm to the environment if it also involves an appreciable restriction on competition. As an example, Section 21 (b) of the Competition Act (equivalent to Article 102 (b) TFEU) prohibits the limitation of technical development to the detriment of consumers. The term "technical development" may be construed to include, *inter alia*, the more sustainable production and distribution. However, the GVH's practice on abusive practices has not endorsed this broad interpretation yet.

---

[36] Additional examples are the following:

- to influence the other party's business decisions for the purpose of gaining unjustified advantages;
- to withdraw goods from general circulation or to withhold goods without justification prior to price increases or for the purpose of causing prices to rise, or by means otherwise capable of securing unjustified advantages or causing a disadvantage in competition;
- to force competitors off the relevant market, or to use excessively low prices which are based not upon better efficiency in comparison to that of the competitors, to prevent competitors from entering the market;
- to hinder competitors from entering the market in any other unjust manner; or
- to create a market environment that is unreasonably disadvantageous for the competitors or to influence their business decisions for the purpose of gaining unjustified benefits.

[37] ECJ, case 2/76 *United Brands Company and United Brands Continental BV v Commission of the European Communities*, ECR 1978 207, pt 65: "a position of economic strength enjoyed by an undertaking which enables it to prevent effective competition being maintained on the relevant market by affording it the power to behave to an appreciable extent independently of its competitors, its customers and ultimately of its consumers."

8.3.2.1 Practice

In the *Budapest Waste Collection* case,[38] the GVH investigated a Consortium Agreement among ten waste collecting/management companies, and the Municipality of Budapest in 1998 aimed at collecting communal and other mixed waste in the territory of Budapest. The Consortium carried out its activities exclusively as a public service against fees capped by the municipality. In 2001, the waste collection regulatory regime was amended to require business or industrial polluters to collect (and transport) their non-communal waste either by themselves or through a third-party service. While communal waste collection remained in the public service domain reserved for the Consortium, non-communal waste collection became a liberalised activity, free of regulatory constraints. The new market entrants collected both the communal and the non-communal waste at unified prices, significantly lower than those the Consortium charged for the collection of the communal waste. Therefore, the Consortium offered even lower fees. This measure was investigated by the GVH as a possible collective abuse of dominant position by the Consortium aimed at foreclosing competitors through low prices (predatory pricing). The GVH concluded that because the customers failed to adequately separate their non-communal from their communal waste before collection, the mixed waste had to be treated generally as communal waste, the collection of which was also a public service. The only activity that remained under the scope of the Competition Act was the collection of purely non-communal waste, which accounted at that time for a small fraction of the waste collection activities. While the GVH emphasised in its reasoning the importance of the sustainability objectives, its conclusion may be seen as an overly broad interpretation of public services and an inadequate limitation of the scope of competition law.

Ultimately, the GVH did not assess whether sustainability could support the finding of an infringement, and it terminated the procedure without the finding of an infringement. The GVH concluded that it had no jurisdiction to assess the effects of the Consortium's behaviour even on the only market where the Competition Act remained applicable (the collection of purely non-communal waste).

8.3.3 Use of Sustainability as a Shield

As explained in connection with the cartels in Sect. 8.2 above, competition law can be used as a shield (defence) to permit measures targeting sustainability objectives to offset specific anti-competitive effects otherwise stemming from the same practice. In the realm of the abusive cases, we see only a few examples where the reason (or notion) of sustainability was invoked as a defensive argument.

---

[38] VJ-43/2005 *Budapest Waste Collection* case (24 August 2006).

8.3.3.1 Practice

8.3.3.1.1 Children's Safety Service Case

In the *Children's Safety Service* case,[39] the GVH investigated whether the Municipality of Budapest—as the owner of the largest recreational park of Budapest—abused its dominant position when it refused amusement park services from the International Children's Safety Service[40] (ICCS), which intended to organise an international children's day event.

The Municipality of Budapest argued that the refusal was based on environmental protection considerations: the protection of the plants and vegetation in the recreational park, which had a unique vegetable environment. It further argued that its decision was based on the local regulation adopted by the municipality to protect the climate and vegetation of Budapest. According to the municipality, the amusement park services provided to ICCS in the previous years resulted in major deterioration in the park's plants, vegetation and grass, which was no longer tolerable.

The reason for the lack of competence was that the refusal of the municipality was adopted based on local regulation, i.e. the decision was of a public law enforcement nature rather than a market behaviour.

Ultimately, the GVH did not assess the sustainability argument because it terminated the procedure without the finding of an infringement. The GVH concluded that it had no jurisdiction to assess the municipality's decision because it was not a commercial act (market practice) but rather an administrative act (public law decision).

8.3.3.1.2 Rail Freight Case

In the *Rail Freight* case,[41] the GVH investigated a complaint that Hungarocombi, as a retaliatory measure, refused to deal with some of its customers who had previously cancelled their orders too often. Hungarocombi was a joint venture established to facilitate, organise and manage the transportation of loaded commercial trucks via railway through Hungary and Austria (Rolling Road or RO-LA services).

In support of the lawfulness of its practices, Hungarocombi argued, *inter alia*, that RO-LA services were based on environmental protection considerations as they intended to mitigate environmental damage by lessening the commercial road transportation traffic.

Ultimately, the GVH did not assess the sustainability argument because it terminated the procedure without the finding of an infringement. The GVH concluded that the dominant position of Hungarocombi cannot be excluded; however, the investigation showed no abuse. The Reservation System of Hungarocombi was in line with the nature of the RO-LA services, and there was no evidence of

---

[39] VJ-126/2001 *Children's Safety Service* case (19 March 2002).

[40] Website of the ICCS available at https://gyermekmento.hu/en/Contents. Accessed 28 October 2022.

[41] VJ-80/2002 *Rail Freight* case (14 November 2002).

Hungarocombi deviating from the system, which would suggest an unjustified refusal to deal.

8.4 Mergers and Sustainability

In this section, we examine the role of sustainability in the assessment within the framework of merger control.

8.4.1 Legislative Framework

To date, there have been a few explicit references to sustainability in the regulation of merger control. This raises the question of whether there is a need to take legislative actions to accommodate such considerations. This section will explore the currently existing institutions that could hypothetically devise and implement sustainability considerations into merger control rules.

The most obvious basis from the currently existing toolbox is the public interest considerations in merger control. Including sustainability considerations in public interest considerations has its pros and cons. On the one hand, such an inclusion would expand the application of the already highly debated[42] public interest instrument in merger control. On the other hand, however, it would enable local enforcement agencies to include sustainability considerations in their merger control analysis through (general) public interest considerations, without the need for a long legislative process.

From a European viewpoint, where the Hungarian regime stands, a merger either has a community dimension under the EUMR[43] or is subject to one (or more) national jurisdiction(s). Thus, we will approach the legislative background review from these two angles.

8.4.1.1 Mergers with a Community Dimension

In line with Articles 1(2) and 1(3) of the EUMR, the EC has an exclusive right over mergers that have a community dimension. However, Article 21(4) of the EUMR allows Member States to take appropriate measures to protect legitimate interests other than those taken into consideration by the EUMR. The EUMR specifies three such considerations that are regarded as legitimate interests: public security, plurality of the media and prudential rules. Further, other elements can be taken into

---

[42] See BIAC's Summary of Discussion Points presented by the Business and Industry Advisory Committee to the OECD Competition Committee Working Party No. 3 on Co-operation and Enforcement Public Interest Considerations in Merger Control (June 2016), which underlines that *introducing public interest considerations into the merger review analysis is unnecessary and potentially counter-productive.*

[43] Council Regulation 139/2004 of 20 January 2004 on the control of concentrations between undertakings, OJ 2004 L 24, p. 1-22.

consideration as well, but those must be communicated to the EC at the outset and shall be recognised by the EC after an assessment of its compatibility with the general principles and other provisions of EU law.

Some authors[44] already argue that sustainability considerations can fall within one of the recognised legitimate interests of the EUMR, most likely falling under public security interests, which can be interpreted in multiple ways. Another possibility for sustainability considerations is that they may fall into the category of any other public interest, also mentioned under Article 21(4) of the EUMR, which needs to be communicated and approved by the EC before the measure of the Member State has taken place.

8.4.1.2 Mergers with No Community Dimension

Mergers that do not have a community dimension are subject to national merger regimes. As we mentioned before, there are very limited examples of explicit references to sustainability considerations. One such example is Spain, where Article 10 of the Competition Law[45] contains a non-exhaustive list of grounds for public interest, including the protection of the environment. But this is the rare exception to the more common situation where national merger regimes make no such references.

As for Hungary, pursuant to Section 24/A of the Competition Act, "The Government may, in the public interest, and in particular to preserve jobs and to assure the security of supply, declare a concentration of undertakings to be of strategic importance at the national level". The public interest exemption does not explicitly refer to sustainability, but just as with the above-described EUMR, it may be covered by the notion of public interest for the purposes of this provision. Notably, to date, the public interest exemption in Hungary has never been used to exempt mergers based on sustainability considerations, but we cannot exclude that it will be used to serve that purpose in the future.

We can already see examples from other Member States, such as the *Miba/ Zollern* merger in Germany. This merger was initially prohibited by the German Competition Authority, but the decision was ultimately overridden by the German Federal Minister for Economic Affairs and Energy, who granted ministerial authorisation to the parties for environmental policy reasons.[46]

Additionally, while the Hungarian merger control regulation is based on the SIEC test, there is one aspect of Hungarian merger control that bears the hallmarks of the public interest test. Section 67 (4) b)of the Competition Act refers to Section 171

---

[44] S. Holmes, Climate change, sustainability, and competition law, *Journal of Antitrust Enforcement*, 2020(8) pp. 354–405, p. 396.

[45] Law 15/2007 of 3 July on the Defence of Competition (*Ley 15/2007, de 3 de julio, de Defensa de la Competencia*).

[46] BMWi, press release of 19 August 2019, Altmaier: Ministererlaubnis im Verfahren Miba/Zollern, available (in German language) at: https://www.bmwk.de/Redaktion/DE/Pressemitteilungen/201 9/20190819-altmaier-ministererlaubnis-im-verfahren-miba-zollern.html.   Accessed 28 October 2022.

(1) of the Media Act,[47] which provides: "The [GVH] shall obtain the opinion of the Media Council relevant to the notification of concentration of enterprises under Section 24 of the [Competition Act], such enterprises or the affiliates of two groups of companies as defined in Section 15 of the Competition Act bearing editorial responsibility and the primary objective of which is to distribute media content to the general public via an electronic communications network or a printed press product."

The Media Act further sets out the rules of this administrative procedure; among others, it makes it clear that the GVH may prohibit a merger that was approved by the Media Council in respect of the public interest set out in Section 171 of the Media Act. The Media Council may also provide an opinion prior to the notification of the merger so that the notifying parties can assess whether (or under which circumstances) the notification could be approved by the Media Council.

If sustainability and environmental protection become an institutionalised policy of the Hungarian State, i.e. a professional administrative body is set up to carry out tasks related to furthering sustainability aims and policies and/or providing supervision over the proper implementation thereof, a public interest test related to sustainability could be introduced based on the template of the media plurality test described above. In this case, the regulation could make the approval of the merger dependent on the prior approval of the administrative/regulatory body responsible for sustainability. This would allow for the sustainability aspect of the merger to be examined by professionals in the field, as well as taking over the burden from the GVH regarding follow-up investigations and/or other obligation(s) undertaken for the merger to be compatible with sustainability goals and policies.

8.4.2 Competition Analysis

This section will focus on the analysis of public interest considerations in merger control. First, we focus on the use of sustainability as a sword, i.e. when the theory of harm is based on sustainability-related harm. Second, we consider the use of sustainability as a shield, i.e. when the otherwise anti-competitive merger is exempted based on sustainability-related benefits.

To preface, we must note that the analysis of complex mergers is in and of itself a complicated task, which—especially for Phase II mergers—requires economic expert support. Accordingly, regarding sustainability cases, it is likely that additional, more specialised, expertise may be required in order to carry out the necessary assessment.

---

[47] Act CLXXXV of 2010 on Media Services and on the Mass Media.

8.4.2.1 Use of Sustainability as Sword

As a preliminary note, it is to be underlined that sustainability as a sword in merger control is currently more of a theoretical concept than an enforcement reality. There are two very apparent scenarios that we can envisage here.

8.4.2.1.1 First Scenario: Effects on Sustainability Pointing in the Same Direction as Effects on Prices

In the first scenario, the merger, which adversely affects prices on the relevant market, also has a harmful effect on sustainability. In such a case, sustainability operates as a sword, but not as a stand-alone theory of harm, since it points in the same direction as price increases.

To date, there is very limited case law to refer to in this scenario. The GVH, for instance, has thus far not conducted such an assessment, where a clearly articulated theory of harm in a merger case was based on sustainability considerations. Nevertheless, such investigations are not without precedent in the EU. The Dutch Competition Authority (ACM) opened an in-depth probe into a deal in the calf-purchasing sector where one of the theories of harm that the ACM investigated was based on sustainability and animal welfare. President Snoep said that investigators were concerned the deal could lead to "purchasing power and monopsony pricing" and wanted to know "whether these possible lower prices will lead to a degradation of animal welfare and less investment in sustainable dairy farming in the Netherlands".[48] The ACM was concerned that lower prices can, particularly in the long term, lead to fewer investments in sustainability and animal welfare, for example.[49] Eventually, the merger was cleared as the ACM determined that it was not likely that the acquisition would lead to lower purchase prices of calves or the foreclosure of competitors.[50]

8.4.2.1.2 Second Scenario: Effects on Sustainability Pointing in a Different Direction Than Effects on Prices

The second scenario arises if the merger does not adversely affect the prices on the market but instead has harmful effects on sustainability. In this case, sustainability operates as a stand-alone theory of harm.

Since the effects on sustainability and prices do not point in the same direction, competition authorities need to undergo a complex balancing exercise to establish which consideration should eventually prevail.

The questions arise as to how one should measure the effects of the merger on sustainability and what the applicable test is to balance these effects towards each other. It is a well-established concept that a decrease in quality, choice or innovation

---

[48] Claf-purchasing deal a 'practical example' of sustainability concerns in M&A reviews, Dutch regulator says, MLex, 19 April 2021.

[49] *Concentration in veal sector: ACM clears acquisition of Van Dam by Van Drie*, ACM press release, 19 August 2021.

[50] Idem.

can be as harmful to competition as an increase in price. Therefore, it could be a viable option to consider sustainability as a non-price dimension of competition (i.e. dynamic effect).[51]

Measuring the non-price effects of mergers is a difficult but not an entirely new exercise as competition authorities, particularly the EC,[52] recently put special emphasis on their analysis, especially in regard to the effects of mergers on innovation.[53] This is also reflected in paragraph 38 of the Horizontal Merger Guidelines,[54] which underlines that "in markets where innovation is an important competitive force [...] effective competition may be significantly impeded by a merger between two important innovators, for instance between two companies with 'pipeline' products related to a specific product market".

Innovation is a good analogue not only because it can be assessed as a non-price effect but also because it directly relates to sustainability. For instance, in the *Dow/DuPont* merger, the EC underlined that the innovation in crop protection is not only of the utmost importance to farmers or consumers but must be evaluated due to its given impact on food, environmental safety and human health.[55]

As for the GVH, we cannot mention such cases where sustainability-related harm was established as a stand-alone theory of harm. However, the above-mentioned approach to focus on non-price effects is not unprecedented in the GVH's practice. Recent investigations into mergers in the digital markets[56] indicate that the GVH recognised that in certain markets, the assessment of the effects on prices is highly difficult and will therefore likely not lead to a verifiable theory of harm. Consequently, the GVH also determined that more emphasis should be put on the assessment of the non-price effects, namely the merger's effect on quality, variety and innovation.[57] Thus, it would seem that the GVH has already developed a way to deal with such non-price effects in the event sustainability-related cases emerge in the near future.

8.4.2.2 Use of Sustainability as a Shield

Sustainability as a "shield" encompasses situations where anti-competitive effects are found but, taking into consideration the possible positive effects of the merger on sustainability, the merger is nonetheless likely to be cleared. This situation differs

---

[51] A. C. Volpin, Sustainability as a quality dimension of competition: protecting our future (selves), *CPI Antitrust Chronicle*, 2020(1), pp. 9-18., p. 14.

[52] European Union's contribution to the OECD, Non-price effects of Mergers 2018, p. 6.

[53] . C. Volpin, Sustainability as a quality dimension of competition: protecting our future (selves), *CPI Antitrust Chronicle*, 2020(1), pp. 9-18., p. 15.

[54] Guidelines on the assessment of horizontal mergers under the Council Regulation on the control of concentrations between undertakings, OJ 2004 C 31, p. 5-18.

[55] M.7932 *Dow/DuPont* case (27 March 2017), paragraph 1980.

[56] VJ-12/2019 *Netrisk / Biztosítás.hu*; VJ-34/2018 *Media Markt / Tesco part of undertakings*; VJ-14/2019 *eMAG / Extreme Digital*; VJ-16/2019 *Szallas.hu / PK Travel*.

[57] J. Buránszki, Experiences with merger investigations in digital markets in recent cases of GVH, *Versenytükör*, 2020(2), pp. 5–25, p. 16.

8 Sustainability and Competition Law in Hungary

from the balancing exercise described above as competition agencies reach this determination long after establishing SIEC, not before.

One of the already existing instruments that can be mentioned here is the efficiency test, which could be applicable to mergers involving environmental benefits. In paragraphs 76–88 of the Horizontal Merger Guidelines, the EC sets out the three cumulative conditions that efficiency claims must satisfy if they are to lead to a merger being cleared. They have to (i) benefit consumers, (ii) be merger specific and (iii) be verifiable. The relevant guidance document[58] of the GVH largely echoes the EC's approach to accepting efficiency claims. Some argue[59] that sustainability claims are easy to fit into these requirements as environmental benefits are clearly consumer benefits. However, with regard to its verifiability, given that many environmental benefits may take some time to materialise and can thus be correspondingly difficult to quantify, it would be advisable to avoid taking an overly narrow financial approach,[60] which would make the application impossible.

Although we find it reasonable to accept efficiency claims based on sustainability reasons, it has to be noted that such claims have not yet been accepted by the EC or the GVH in their enforcement practices relating to mergers. Moreover, the above-mentioned concerns regarding verifiability raise practical concerns as competition agencies seem to be reluctant[61] to accept such claims even in normal cases, let alone in such sustainability cases where the possible benefits cannot be predicted with high certainty at the time of the merger.

Finally, as a specific application, we can mention the role of sustainability-related remedies as shields. Such remedies are not shields in the sense that they cannot be used as good arguments to mitigate the harmful effects of the SIEC. However, under merger control, remedies are a way of avoiding the prohibition of mergers due to SIEC, and therefore, from a certain perspective, sustainable remedies can be regarded as shields. As an example, the merger inquiry of *South East Water Ltd/Mid Kent Water Ltd* can be put forward where the behavioural remedy accepted by the UK Competition Commission was designed to preserve the water resource benefits arising from the merger.[62] A practical difficulty here lies in the fact that competition agencies normally have a strong preference for structural remedies, which are more

---

[58] "Az összefonódások elemzésének módszertani megközelítése", available (in the Hungarian language) at: https://www.gvh.hu/pfile/file?path=/gvh/elemzesek/vitaanyagok/vitaanyagok/elemzesek_vitaanyagok_Fuzios_altalanos_elvek_GYIK.pdf&inline=true. Accessed 28 October 2022.

[59] S. Holmes, Climate change, sustainability, and competition law, *Journal of Antitrust Enforcement*, 2020(8), pp. 354–405, p. 396.

[60] *Ibid*, p. 393.

[61] D. Cardwell, The Role of the Efficiency Defence in EU Merger Control Proceedings Following UPS/TNT, FedEx/TNT and UPS v Commission, *Journal of European Competition Law & Practice*, Volume 8, Issue 9, November 2017, pp. 551–560, available at https://doi.org/10.1093/jeclap/lpx049. Accessed 28 October 2022.

[62] S. Holmes, Climate change, sustainability, and competition law, *Journal of Antitrust Enforcement*, 2020(8), pp. 354–405, p. 394.

148   A. M. Horváth

difficult to be shaped in an environment-friendly way, as opposed to the less desirable behavioural remedies.

8.5   Greenwashing

In this section, we examine the role of sustainability in the assessment of unfair commercial practices.

In Hungary, as in some other Member States (e.g. Italy), the GVH also has the competence to deal with unfair commercial practices that may affect (distort) free competition in the Hungarian markets. The GVH mainly deals with B2C advertising – if the advertisement in question is broadcasted to a sufficient number of people to influence the course of free competition – based on the UCPD[63] and its national equivalent, the UCP Act.[64] Sometimes, however, the GVH also uses its powers to step up against unfair B2B practices to maintain the healthy competition process in Hungary and even has competence to investigate and, if necessary, ban comparative advertising.[65]

According to the established legal theory (specifically mentioned in the official reasoning by the Ministry of Justice of the various amendments to the Competition Act) and case law, the GVH was provided this competence not long after Hungary's

---

[63] Directive 2005/29/EC of the European Parliament and of the Council of 11 May 2005 concerning unfair business-to-consumer commercial practices in the internal market and amending Council Directive 84/450/EEC, Directives 97/7/EC, 98/27/EC and 2002/65/EC of the European Parliament and of the Council and Regulation (EC) No 2006/2004 of the European Parliament and of the Council, OJ 2005 L 149, p. 22-39.

[64] Act XLVII of 2008 on the Prohibition of Unfair Business-to-Consumer Commercial Practices.

[65] Section 6/A of the Competition Act: "Any form of communication, information or the making of a representation in any form with the aim or having the effect of promoting the sale or any other manner of placing on the market of goods or services, including natural resources that can be utilized as capital goods, covering real estate properties, securities, financial instruments or rights (hereinafter referred to collectively as "goods"), or in connection with this objective, the representation of the name, the trade mark or the activities of a business entity that directly or indirectly identifies that business entity's competitor, or the goods manufactured, sold or introduced by such other business entity for the same or similar purpose as those featured in the advertising."
Section 10 of the Competition Act:
"Comparative advertising shall be permitted only, if:

(a) it compares goods meeting the same needs or intended for the same purpose;
(b) it objectively compares one or more material, relevant, verifiable and representative features of those goods, which may include price if the price is also part of the comparison;
(c) for products with designation of origin, it relates in each case to products with the same designation."

accession to the EU, so that its legal toolkit would be enhanced to safeguard free competition within Hungarian markets. Misleading advertising may distort competition: when an undertaking is allowed to use false claims to manipulate the transactional decisions of consumers and consequently gain additional market share, it can have a similarly adverse effect on competition as (both horizontal and vertical) restrictive practices and the abuse of a dominant position.

Against this background, the authors of this report are of the view that when discussing the role of sustainability and environmental aspects in antitrust enforcement, the description of the Hungarian landscape could not be complete without mentioning the more and more frequent phenomenon of greenwashing and the competition authority's approach towards it.

By stepping up against greenwashing—as sustainability becomes a priority for more and more consumers when making their choices—the GVH can ensure free and healthy competition: undertakings introducing actual green technologies or solutions may win additional consumers in the competitive process, whereas false claims must be acted against so that they do not award their users unfair advantages in competition.

The constantly growing importance of green claims in the consumers' decision-making process has been supported by a Factsheet published by the EC.[66] The document highlights:

– Eighty per cent of webshops, webpages and advertisements contain information about the environmental impact of products.
– Fifty-six per cent of EU consumers said they had encountered misleading green claims.

8.5.1 Soft Legal Framework: Strict Approach

The GVH first addressed the issue of greenwashing in 2020 when it warned consumers of more frequent misleading green claims[67] and then published its so-called Green Marketing Notice.[68] However, in 2020, the GVH seemed to put on the top of its agenda the curbing of the spread of unlawful greenwashing activities and green claims; there has not been much development regarding this matter since then (see Sect. 8.5.2 below).

---

[66] Factsheet: Empowering Consumers for the Green Transition: https://ec.europa.eu/commission/presscorner/api/files/attachment/872173/Factsheet%20Empowering%20Consumers.pdf.pdf (06.06.2022.).

[67] Available (in Hungarian language) at https://gvh.hu/fogyasztoknak/gondolja_vegig_higgadtan/zold_hirdetesek (03.06.2022.).

[68] Green marketing – Guidance for undertakings from the Hungarian Competition Authority. Available at: https://gvh.hu/pfile/file?path=/en/for_professional_users/guidance-documents/szakmai_felhaszn_tajekoztatok_zold-iranymutatas_2020_a&inline=true. Accessed 6 June 2022.

In its warning, the GVH attempted to draw consumers' attention to possibly misleading green claims of businesses and urged consumers to check the accuracy of such claims before making transactional decisions. The authority has set out that untrue environmentally friendly statements, as well as true statements in misleading contexts, are to be considered unfair commercial practices.

According to GVH's Green Marketing Notice, "[g]reenwashing means the marketing or PR strategy of an undertaking with the intention of giving the impression that the undertaking in question is environmentally friendly and responsible for environmental protection while in its actual operation no substantive steps can be identified to achieve these goals".[69]

In the Green Marketing Notice, the GVH stated that it aims to assist undertakings in developing appropriate advertising practices regarding the environmentally friendly and sustainable nature of their products and services and, by this way, to help undertakings avoid infringing the law. The Green Marketing Notice specifies that it provides possible self-evaluation factors only with regard to the assessment of misleading commercial practices under the UCP Act and comparative advertising under the Competition Act.[70]

On the basis of the above, it is clear that GVH not only considers untrue statements about the product/service in question, e.g. "green", "organic", or "recyclable", to fall under the term greenwashing, but also includes the so-called priority claims, e.g. "the greenest" or "most eco-friendly" etc. This means that a green commercial communication may be found infringing the UCP Act and the Competition Act at the same time, e.g. "the most environmentally friendly product on the market", as a misleading commercial practice and unlawful comparative advertising. In our view, the latter is only applicable when—by the use of such a claim—one or few competitors are clearly identifiable: in highly competitive markets with numerous market players, such priority claim does not identify all the competitors (we refer to the definition of comparative advertising laid down in Section 6/A of the Competition Act), whereas in oligopolistic markets, the term "greenest" may refer to the fact that the identifiable competitors are less green, and as a result, their products/services are less desirable.

The guide summarises the criteria for undertakings to be kept in mind when designing eco-related advertisements/green claims:

(i) it provides general recommendations with examples, such as express the claim clearly and concretely, in clear and understandable language, in a realistic and accurate manner, without exaggeration, and in a verifiable and substantiated manner;[71]

---

[69] Green Marketing Notice, page 2.

[70] Green Marketing Notice, Chapter I.

[71] Green Marketing Notice, Chapter III.

8 Sustainability and Competition Law in Hungary

(ii) it discusses the requirements for providing proof in relation to certain typical green claims concerning ingredients/material, production process and future commitments;[72]

(iii) it discusses the requirements for substantiating comparative and priority claims;[73] and

(iv) it describes the legal framework of using certification marks.[74]

Based on the Green Marketing Notice, the GVH intends to apply a strict approach when assessing green claims as it requires a high standard for the appropriate substantiation of such claims, as well as it requires their regular review (e.g. whether the claim that was true at the beginning of a marketing campaign is still applicable later):

> Green claims must be based on solid, independent, verifiable and well-supported evidence, which takes into account the latest scientific findings and methods. Such evidence may include independent professional research, relevant test results or other credible data. Regarding justification, it is always important to keep in mind the specificity of the claim in question.

> The relevant evidence must already be available to the undertaking when it publishes its green claims for the first time. (...) In addition, it is recommended to consider the constant development of science and technology, as well as any changes to the relevant results; any green claims made based on these may be required to be reviewed from time to time.[75]

It is worth mentioning that according to Section 14 of the UCP Act, "the undertaking shall provide proof to verify the authenticity of any fact comprising a part of commercial practices. In the event of the business entity's failure to comply, the fact in question shall deemed to be untrue". Regarding green claims, this provision also means that a company using such a claim must be able to appropriately substantiate it when requested by the GVH: if it fails to do so, it might expect a hefty fine[76] from the GVH.

---

[72] Green Marketing Notice, Chapter IV Sections A)-B).

[73] Green Marketing Notice, Chapter IV Section D).

[74] Green Marketing Notice, Chapter V.

[75] Green Marketing Notice, Chapter VI.

[76] Even though not in cases involving "greenwashing", the GVH seems more and more courageous to impose heavy fines on companies breaching the UCP Act: Facebook was fined over EUR 3 million for claiming it can be used by users free of charge (while according to the GVH, they had to "pay" with their personal data); Telenor (a large Hungarian telecommunication service provider) was fined over EUR 5.4 million for advertising a free-of-charge phone, whereas its price was included in the monthly tariff fee charged by the company; and Booking.com was fined for a record of EUR 6.4 million for urging its consumers to make their online hotel bookings sooner than they would normally do.

8.5.2 Case Law: Less Focus on Enforcement

While the GVH has previously expressed that greenwashing might be a priority on its agenda, presumably due to the outburst of the pandemic, its focus shifted to misleading commercial practices concerning COVID-related communications.[77] At the time of completing this report (15 July 2022), we are aware of only one case[78] where an advertisement of an "organic" feature of a product was investigated by the GVH.

The GVH found that FOX CONSULTING Ltd., which operates the bio tanning salon franchise "KiwiSun", had breached the prohibition of unfair commercial practices with its advertisements by using statements such as—among others—"green bio" and "with the power of nature". In this case, the claims about the greenness of the product were considered "health claims" (statements on the positive impact on health) and not green claims (statements on the positive impact on the environment). The GVH found that FOX CONSULTING failed to substantiate and explain the meaning of its statements "green bio", "bio" and "with the power of nature", which in themselves, but especially when put together with other health claims, constituted unsubstantiated health claims.

The GVH imposed a fine of approx. EUR 21,200 and prohibited the undertaking from continuing the unlawful practice. Although the GVH assessed the compliance of classic "green claims" in this case, it did not implement a clear test on how to use such green claims as the authority found that the claims under investigation were rather "health claims" than "greenwashing".

8.5.3 Expectations for the Near Future Connected to Greenwashing

It would not be surprising if the GVH—together with other national competition authorities in other Member States—would step in the footsteps of the EC when setting "greenwashing" as a new priority (not only in terms of soft law anymore but also through actual enforcement) in its UCP enforcement agenda.

The EC published its New Consumer Agenda[79] in late 2020, where it states the following important indications: "Consumers need to be better protected against

---

[77] The GVH warned companies early on that it would protect consumers from false health claims in connection with the coronavirus pandemic and then used its extended powers – which were given to the GVH by the Hungarian Government in the course of its special "emergency legislative procedure" – to conduct "fast track" sector inquiries on the markets of rapid COVID tests in Hungary.

[78] VJ-4/2019 *Tanning Salon* case (21 February 2021). Press release available at https://www.gvh.hu/en/press_room/press_releases/press-releases-2020/the-gvh-imposed-a-fine-because-of-bio-solarium. Accessed 3 June 2022.

[79] Communication From The Commission To The European Parliament And The Council [COM (2020) 696 final] https://eur-lex.europa.eu/legal-content/EN/TXT/HTML/?uri=CELEX:52020DC0696&from=EN. Accessed 6 June 2022.

8 Sustainability and Competition Law in Hungary

information that is not true or presented in a confusing or misleading way to give the inaccurate impression that a product or enterprise is more environmentally sound, called 'greenwashing'. Actions to that effect are also being developed in the area of sustainable finance. Furthermore, the Commission will propose that companies substantiate their environmental claims using Product and Organisation Environmental Footprint methods to provide consumers with reliable environmental information."[80] Based on this policy paper, the EC already issued Recommendation (EU) 2021/2279 on the use of the Environmental Footprint methods to measure and communicate the life-cycle environmental performance of products and organisations. The logical next step for enforcement agencies would be to start assessing whether advertisers comply with the recommendations when advertising their products/services with green claims.

According to the press release of the EC, further legislative proposals are on the agenda of the European Union in the field of greenwashing. On 30 March 2022, the EC announced that it proposed amendments to the UCPD in relation to green claims as well as the early obsolescence of products.[81] The proposal[82] supplements the list of main product characteristics (Article 6 (1) of the UCPD) with the environmental and social impact/reparability/durability of the products, about which the trader must not mislead the consumers. Furthermore, the proposal includes new practices that are considered misleading after a case-by-case assessment (Article 6 (2) of the UCPD), such as making an environmental claim related to future environmental performance without clear, objective and verifiable commitments and targets and without an independent monitoring system. The proposal introduces new items on the so-called Blacklist (Annex I of the UCPD) as well. These practices shall in all circumstances be regarded as unfair should the proposal be adopted. According to the communication of the EC, these new Blacklist commercial practices would be the following:

- Making generic, vague environmental claims where the excellent environmental performance of the product or trader cannot be demonstrated
- Making an environmental claim about the entire product when it really concerns only a certain aspect of the product
- Displaying a voluntary sustainability label that was not based on a third-party verification scheme nor established by public authorities

The proposed changes to the UCPD are fully in line with the expectations set forth by the Green Marketing Notice of the GVH. Should the proposal be adopted by the competent EU bodies and later by the national legislators, the enforcement

---

[80] New Consumer Agenda, Section 3.1 (p. 8).

[81] Circular Economy: Commission proposes new consumer rights and a ban on greenwashing: https://ec.europa.eu/commission/presscorner/detail/en/ip_22_2098. Accessed 6 June 2022.

[82] Available at: https://eur-lex.europa.eu/legal-content/EN/TXT/?uri=CELEX%3A52022PC0143&qid=1649327162410. Accessed 6 June 2022.

actions to detect, substantiate and punish greenwashing practices would be easier and more effective by the application of the legal provisions.

In our view, the GVH will primarily focus on B2C commercial practices as consumers are becoming more conscious about environment-friendly technologies. On the other hand, it is also possible that green claims will become increasingly important in B2B aspects as well: eco-friendly and ethical business models are to be followed, especially by larger publicly listed companies, and consequently, their green claims will have to be substantiated in B2B relations as well.

8.6 Private Enforcement and Sustainability

In this section, we examine the role of sustainability in the assessment of the private enforcement of competition law.

8.6.1 Public and Private Enforcement of Competition Law

If (the public enforcement of) competition law will take into consideration sustainability benefits, this will automatically affect the private enforcement of competition law: on the one hand, the substantiation of damage claims could become more difficult if defendants have an additional form of defence; on the other hand, there could be more damage claims if plaintiffs can rely on violations of sustainability as the basis for their claim.

8.6.2 Sustainability, Competition Law and Private Enforcement: Setting the Framework for Interpretation in Hungary

8.6.2.1 Underlying Principles Serving as Potential Cornerstones of Interpretation

Under Article 28 of the Fundamental Law of Hungary, "[i]n the course of the application of law, the courts shall in principle interpret the laws in accordance with their objective and with the Fundamental Law. The objectives of a law shall in principle be determined relying on its preamble, and/or on the explanatory memorandums of the relevant legislative or amendment proposal. When interpreting the Fundamental Law or any other law, it shall be presumed that they are reasonable and of benefit to the public, serving virtuous and economical ends."

All goals and aims codified in the Fundamental Law (see Sect. 8.1 above) are to be considered by the courts when applying the relevant laws, including competition law. Therefore, it may be held that when Hungarian courts are applying competition law, they may indeed consider the aims and goals of environmental protection and sustainability as an overarching interpretative framework.

8.6.2.2 The Prohibition of Agreements Restricting Competition and the Prohibition of the Abuse of Dominance as the Basis for Claiming Damages

8.6.2.2.1 The Prohibition of Agreements Restricting Competition

The prohibition of anti-competitive agreements has been serving as a basis for damage claims for a number of years, and there have indeed been many actions in the EU Member States that have been filed before national courts.[83]

There have, however, not yet been any judicial actions in Hungary in which a plaintiff based its claim on an anti-competitive agreement that had as its subject a sustainability- and/or environment-related issue.

Nevertheless, an illustrative example of bringing damage actions before courts based on a decision of the EC is the truck cartel (AT.39824 *Trucks*), which has been the basis of claims not only in Hungary but also in Spain, Germany, the Netherlands, and the United Kingdom. In Hungary, these private actions all concern damages that allegedly resulted from the cartel practices of truck producers (MAN, Volvo/ Renault, Daimler, Iveco and DAF), whereby they coordinated their prices at the gross list level for certain types of trucks in the European economic area. However, the truck manufacturers have also engaged in a cartel relating to (i) the timing for the introduction of emission technologies to comply with the European emission standards and (ii) the passing on to customers of the costs for the emission technologies required to comply with the European emission standards.

We know certain large Hungarian transportation companies, which had purchased trucks in the time period affected by the cartel, have considered the possibility of launching damage actions on the basis of the truck producers' market practices that relate to the above-mentioned environmental issues. The damages that the transportation companies have suffered because of the cartel at stake may have been higher fuel prices (due to the less efficient methods of fuel consumption of the trucks), shorter life cycle of the products and higher maintenance costs. However, we are not aware of any private actions in Hungary that have been based on the aforementioned environment-related competition law infringements.

8.6.2.2.2 The Prohibition of Abuse of Dominance

As to the prohibition of the abuse of dominance, there is currently an emerging line of academic works contemplating the possibility of applying the prohibition as a means of limiting the pollution of certain undertakings and promoting social equality.[84] According to the aforementioned school of thought, competition law should and can act as a public policy tool to further environmental and social sustainability

---

[83] As to the Hungarian examples, we refer to P. Szilágyi, Antitrust Practice of Courts in Hungary in Follow-on Private Actions 2020, *Global Competition Litigation Review*, 2021(4), pp. 158-161.

[84] M. Iacovides and C. Vrettos, Falling Through the Cracks no More? Article 102 TFEU and Sustainability – the Relation Between Dominance, Environmental Degradation, and Social Injustice (September 25, 2020). *10:1 Journal of Antitrust Enforcement*, pp. 32-62.

goals. In Hungary, there have not yet been any cases in which private parties claimed damages in connection with the abuse-of-dominance scenarios that were in connection with sustainability- and/or environment-related issues.

8.6.3 The Material Criteria for Awarding Damages on the Basis of Competition Law Violations in Hungary

The below criteria must be demonstrated by the plaintiff to obtain damages on the basis of a competition law infringement.

8.6.3.1 Infringement of Competition Law

According to Section 88/B(9) of the Competition Act, the plaintiff must prove that the conduct of the defendant was unlawful, while the defendant must prove the conditions of exemption (*de minimis*, block exemption or individual exemption). Thus, a sustainability defence in damage litigation should be available, at least in the case of stand-alone claims. Notably, the standard for individual exemption based on sustainability benefits is likely to be at least as stringent as in the case of public enforcement (the burden of proof is higher before the courts).

8.6.3.2 Damages Suffered

For the claim to succeed, the plaintiff must prove that it has suffered damages (direct loss, loss of profit and costs of claiming damages[85]). Sustainability considerations are relevant from this aspect because the plaintiff would need to show that non-compliance with sustainability standards lead to actual damages (public interest harm caused by unsustainable practices are unlikely to qualify for damages).

8.6.3.3 Causal Link

The plaintiff must demonstrate that there is a causal link between the competition law infringement and the damages that the plaintiff has suffered. Sustainability and environmental aspects may come into play here as well. However, as has been elaborated above, establishing a close enough causal link between the environmentally damaging behaviour of an undertaking that violates competition law seems difficult to prove.

8.6.4 Procedural Considerations of Damages Actions

8.6.4.1 Follow-on and Stand-Alone Damage Claims

As has been set out above, a damage claim in Hungary may be filed before courts either as a follow-on action (subsequent to the decision of the GVH/EC) or as a

---

[85] Section 6:522 (2) a)-c), Civil Code.

8 Sustainability and Competition Law in Hungary

stand-alone action (which does not follow on from a previous finding of a competition law infringement by the GVH/EC).

8.6.4.2 Action in the Public Interest

It should also be mentioned that where an infringement falling within the competence of the GVH caused damages to a large number of consumers, the GVH may bring a civil law claim against the undertaking concerned on behalf of the consumers. The condition of such a claim is that an investigation has already been launched by the GVH regarding the infringement (the GVH may request the court to suspend litigation until the end of the investigation). Such an action does not affect the right of consumers to commence proceedings against the defendant in their own right.[86]

In this respect, it should be highlighted that environmental damages are, by their very nature, capable of affecting a large number of consumers; therefore, it may be expected that the GVH will identify cases that concern sustainability and environmental issues and that are thus capable of being the subject of an action in the public interest, as set out above.[87]

8.6.4.3 Collective Redress (Class Action)

The rules governing collective redress have recently been codified by the Civil Procedure Code.[88] The Civil Procedure Code regulates both opt-in collective actions and representative actions. Notably, opt-in collective actions are limited to certain cases. Interestingly, competition law infringements are not included in this list (although it could be debated that competition law infringements fall under the category of "claims arising in connection with consumer contracts"), while claims stemming from health impairment resulting directly from unforeseeable environmental pollution caused by human activities or arising in consequence of negligence are included (although it could be debated that simple unsustainable practices do not fall under this category).[89]

---

[86] Section 85/A of the Competition Act.

[87] We refer to paragraph 130 of AG Kokott's Opinion in Case C-435/18 (Otis) according to which "[t]here may be cases in which it is difficult to identify the specific legal entity to which harm sustained by the 'general public' should be repaid. However, in such cases, it is possible to consider having a representative of the public interest demand compensation for the harm sustained and making the injuring party pay the compensation into a fund that benefits the general public. A similar construct underpins, for example, 'parens patriae antitrust actions' under US law, which enable the state attorneys general to demand collective compensation on behalf of their citizens for scattered damage incurred by the public from inflated prices for consumer goods as the result of a cartel and to appropriate the proceeds from such actions for public welfare projects; see, in this regard, Farmer, S. B., 'More lessons from the laboratories: Cy pres distributions in parens patriae antitrust actions brought by state Attorneys General', Fordham Law Review, Vol. 68, 1999, p. 361 et seq."

[88] Act CXXX of 2016 on the Code of Civil Procedure.

[89] Section 583 (2) c) of Civil Procedure Code.

8.7 Agency Perspective

Sustainability considerations in competition law enforcement are regarded by the GVH as an important emerging topic. It is also regarded as a significant challenge. Both importance and challenge are manifested in the recent activities of the GVH in the field and the current situation, described below.[90]

8.7.1 Activity in the Area

8.7.1.1 Enforcement

The GVH has a modest case experience involving sustainability considerations. Parties have put forward sustainability considerations only in a few cases, which the GVH qualified as unfounded attempts at individual exemption. Those cases are discussed elsewhere in this report.[91]

So far, the GVH has not encountered genuine sustainability initiatives/agreements (either lawful or unlawful under the Competition Act and/or the TFEU). Similarly, significant sustainability considerations have not emerged in merger cases of the GVH or its abuse of dominant position investigations either.

8.7.1.2 Non-enforcement

The GVH took part in several activities aimed at exploring the topic. In its special project as the host of the ICN Annual Conference in 2021, it conducted a global survey among competition agencies and non-governmental advisors of the ICN, stocktaking existing experience and expectations, primarily, but not exclusively, about anti-competitive agreements.[92]

Results of the survey showed, *inter alia*, that there is very little enforcement experience and that the presence of the issue is not temporary. Competition agencies need to bridge expertise gaps, and there is a vocal demand from practitioners for more and better guidance. However, competition policy research to solve difficult analytical problems and measurement issues is a precondition for moving forward. These and other findings and conclusions have been disseminated broadly in Hungary as well as internationally.[93]

---

[90] The importance of the topic is also shown by several speeches and interviews given by the president of the GVH in recent years.

[91] See Sect. 8.2 above.

[92] GVH special project for the 2021 ICN annual conference: Sustainable development and competition law. Available at: https://www.gvh.hu/en/gvh/Conference/icn-2021-annual-conference/special-project-for-the-2021-icn-annual-conference-sustainable-development-and-competition-law. Accessed 28 October 2022.

[93] E.g.: Cs. Kovács and A. Nagy: Special project on sustainable development and competition law: The beginning of an even more beautiful friendship? In: The International Competition Network at twenty: Origins, accomplishments and aspirations. Eds: P. Lugard, D. Anderson. Intersentia,

8 Sustainability and Competition Law in Hungary

The survey was complemented with a high-level panel discussion at the conference with the participation of the president of the GVH.[94] Another high-level panel discussion at the next ICN Annual Conference in 2022 in Berlin, moderated by the president of the GVH, can be regarded as a follow on of the 2021 initiative.[95]

The GVH also took part in an ECN-level effort to deal with sustainability in the context of anti-competitive horizontal agreements, which in turn provided input for the Draft Revised Horizontal Guidelines.

On top of that, the GVH is closely following international developments, including the efforts of leading competition agencies and international organisations to develop and identify good practices.[96]

In summary, the GVH has little case experience but keeps an eye on international developments and evolving good practices. Experience gained from all these activities suggests that there is no inherent trade-off between sustainability and competition. Competition is, in fact, consistent with sustainability as a default. Whether there is a trade-off depends on the facts of the case. International good practices have not much evolved yet, and while the general analytical framework seems to be applicable, many details are yet to be explored. The degree of progress across agencies is likely explained by differences in agency parameters (such as size and the general level of development) and by differences in the overall societal, political and business environment (including the quantity and quality of sustainability-oriented private initiatives).

8.7.2 State of Affairs and Related Considerations

The GVH is open to incorporating relevant sustainability considerations into its competition enforcement. In the case of anti-competitive agreements, this is backed by a statutory provision that explicitly identifies positive effects on the environment as a benefit to be considered in an Article 101 (3) TFEU type of analysis.[97]

However, this does not mean that the GVH is naïve when receiving sustainability claims from parties. Cautionary tales are cases where such claims were put forward without any serious substantiation in the context of hard-core cartels. The GVH refused such greenwashing attempts and learned the lesson that greenwashing is a

---

Concurrences, 2022, pp. 261-270. Available at: https://www.concurrences.com/en/livre/the-international-competition-network-at-twenty-1112#id=77519. Accessed 28 October 2022.

[94] Host's Special Plenary Session: Sustainable Development and Competition Law. ICN Annual Conference 13 October 2021, Budapest (on-line).

[95] Special Break-out Session on Sustainability. ICN Annual Conference 5 May 2022, Berlin.

[96] E.g.: Roundtable discussion of the OECD Competition Committee on environmental considerations in competition enforcement, December 2021. Available at: https://www.oecd.org/daf/competition/environmental-considerations-in-competition-enforcement.htm.   Accessed 28 October 2022.

[97] Section 17 (a) of the Competition Act.

real danger when it comes to alleged sustainability benefits.[98] Also, the burden of proof is on the parties when it comes to Article 101 (3) TFEU and its Hungarian equivalent.

As far as using sustainability as a sword is concerned, the GVH is certainly open to the theory of harm that was applied by the EC in its *Car emissions* case. It is worth noting that in the theory of harm in question, there was an element of restriction of competition, albeit not in terms of pricing. In other words, it was about a sustainability-related competitive concern.

Also, the emergence of sustainability considerations (either as a shield or as a sword) is a factor that makes launching a formal procedure by the GVH more likely. This means a *ceteris paribus* higher probability of case launch, instead of automatically triggering a formal investigation. When it comes to case selection, sustainability is only one factor, and parties' sustainability claims are not taken at face value by the GVH. This overall approach reflects the interest of the GVH in the topic due to its importance and novelty.

In the absence of proper case experience and solid international technical metrics, the GVH has not issued any own guidance documents concerning sustainability in competition enforcement, and currently, no such guidance is under preparation. Nevertheless, the Draft Revised Horizontal Guidelines are available for parties and practitioners and provide an appropriate level of guidance for the time being.[99]

The lack of GVH guidance documents does not reflect any lack of recognition of the merit of useful guidance or a lack of intention to provide guidance when possible. Indeed, the GVH issued guidance on how to make green claims in the context of consumer protection at the end of 2020, a somewhat related topic, based on international best practices.[100] International and/or domestic developments may lead to the issuance of a GVH guidance document about sustainability in competition enforcement in the future.

The GVH is not aware of any ongoing or planned legislative actions with regard to sustainability in competition enforcement. In addition to the reasons discussed already in the context of guidance documents, sustainability is already referred to in the Competition Act, as mentioned above.

So far, the GVH has not engaged in "hard" capacity-building projects, i.e. those entailing a structural change of the organisation or its personnel, such as hiring specialists or setting up a dedicated department to deal with cases involving sustainability considerations. At the same time, "soft" capacity building, such as getting knowledgeable about the topic and following international developments, is ongoing, as described above.

---

[98] See Section 3.2 of Appendix B of the survey report of the GVH special project for the 2021 ICN annual conference.

[99] This is not a remote relevance, given the proximity of EU and Hungarian competition laws and the ability of Hungarian businesses to engage in activity that also affects trade between EU Member States.

[100] See Section 8.5.1 above.

The GVH regards its current level of engagement as adequate, given the low number and the type of sustainability-related cases up to now. Being aware of international developments and evolving practices will be useful and can serve as a basis for a more informed adaptation in the future if and when genuine sustainability cases emerge in Hungary and parties put forward better-founded sustainability claims and analyses.

Open Access This chapter is licensed under the terms of the Creative Commons Attribution 4.0 International License (http://creativecommons.org/licenses/by/4.0/), which permits use, sharing, adaptation, distribution and reproduction in any medium or format, as long as you give appropriate credit to the original author(s) and the source, provide a link to the Creative Commons license and indicate if changes were made.

The images or other third party material in this chapter are included in the chapter's Creative Commons license, unless indicated otherwise in a credit line to the material. If material is not included in the chapter's Creative Commons license and your intended use is not permitted by statutory regulation or exceeds the permitted use, you will need to obtain permission directly from the copyright holder.

Sustainability and Competition Law in Italy

9

Elisa Teti

9.1 Introduction

Before outlining the sustainability initiatives with a competitive impact that have been undertaken at the Italian level by the various competent authorities, it is necessary to briefly clarify some preliminary notions concerning the concept of sustainability and ecological transitions, the role that these issues have in the framework of competition law, and the role played by the Italian Antitrust Authority (IAA) in the context of ecological transition and in explaining how in concrete terms sustainability issues intersect with antitrust issues. Finally, some specific cases related to sustainability issues investigated by the IAA will be mentioned.

9.1.1 The Concept of Sustainability and Ecological Transition

The concept of sustainability is inextricably linked to the concept of ecological transition, which is one of the fundamental pillars and objectives of policies at the European and national levels underpinning the new Italian and European development models.

For this reason, to fully understand the meaning of this expression, reference should be made to the Recovery and Resilience Facility of 12 February 2021 (Regulation 2021/241) establishing the Pandemic Recovery and Resilience apparatus, as well as to the *Piano Nazionale di Ripresa e Resilienza – National Recovery and Resilience Plan* (PNRR) of April 2021, which constitutes its implementation at the Italian level.

E. Teti (✉)
Rucellai&Raffaelli, Milano, Italy
e-mail: e.teti@rucellaieraffaelli.it

© The Author(s) 2024

P. Këllezi et al. (eds.), *Sustainability Objectives in Competition and Intellectual Property Law*, LIDC Contributions on Antitrust Law, Intellectual Property and Unfair Competition, https://doi.org/10.1007/978-3-031-44869-0_9

From this perspective, the concept of ecological transition and thus sustainability constitutes an economic model aimed at increasing the competitiveness of the production system of goods and services, stimulating new entrepreneurial activities, and encouraging the creation of stable employment.[1]

9.1.2 The Role of Sustainability in Competition Law

The current economic context is characterised by the transformation of production activities under the impetus of environmental sustainability policies. Business choices towards sustainable innovation of production processes are also conditioned by the increasing environmental awareness of consumers.

It is in this sense that sustainability is an element on which to develop competition on the merits, i.e. a competitive situation from which consumers profit through lower prices, better quality, and a wider choice of new or more efficient goods and services.[2]

The protection of competition can be seen as complementary to public interest in environmental safeguard and sustainability insofar as antitrust and consumer protection rules are instrumental in enhancing the environmental sustainability of economic activities.

The Authority itself noted that competition can contribute to sustainable development by complementing existing instruments, such as regulation and taxation, to foster the transition process towards an environmentally sustainable growth model. Competitive pressure encourages companies to make the best use of available resources by producing at the lowest cost; it favours a better allocation among companies of production factors, including natural resources; it allows the most virtuous companies that adopt more energy-efficient technologies to remain on the market. In addition, competition stimulates firms to innovate by improving production processes and creating new products that limit CO2 emissions and use energy from renewable sources.[3]

In this context, the task of competition law to correctly balance the need to ensure dynamic and competitive markets with the promotion of investments by companies in terms of environmental sustainability remains in any case. The Authority has declared its readiness to apply competition law in evolutionary terms and to assess,

---

[1] See PNRR, p. 14. The PNRR, approved on 13 July 2021 by the European Council, is the document that the Italian government prepared to illustrate to the European Commission how Italy intends to invest the funds allocated at the European level under the Next Generation Eu programme. The PNRR is divided into 6 Missions subdivided into 16 functional components to achieve the economic and social objectives that are part of the Government's strategy.

[2] On the notion of competition on the merits with reference to an abuse, see most recently CJEU, C-377/2020, *Servizio Elettrico Nazionale SpA, ENEL SpA, Enel Energia SpA/Autorità Garante della Concorrenza e del Mercato*, ECLI:EU:C:2022:379, pt. 85-86.

[3] IAA Annual Report of 31 March 2021, pp. 15 ff.

in coordination with the European Commission and the other authorities of the Member States, the possible expansion of the instruments available to support development that is both sustainable and competitive at the same time.[4]

9.1.3   The Role of the IAA in the Ecological Transition

Concerning the role played by the Authority in terms of stimulation and impulse for the development of policies on sustainability and competition, special mention should be made of the advocacy activity carried out by the IAA. The Italian antitrust law provides that the Authority's powers include the power to report to the government, Parliament, regions, and local authorities regulatory and administrative measures already in force, or in the process of being drafted, that introduce restrictions on competition.

In report S4143 of March 2021 pursuant to Articles 21 and 22 of Law 287/90, concerning a series of competition reform proposals for the Annual Market and Competition Law, there is a section on some reform proposals to facilitate the achievement of sustainability objectives.[5] In this context, the IAA highlighted how competitive pressure can contribute to the promotion of sustainable development and focused its proposals on the infrastructural nodes of sustainability, namely:

(a) infrastructures for recharging electric cars to ensure that the necessary conditions are in place to ensure that no distortions are created in the development phase of the sector that could jeopardise the efficient functioning of competition in the future (see paragraph 3);
(b) differentiated waste management in the sense of introducing regulatory changes to eliminate unjustified discrimination between public and private operators in municipal waste management and prevent the notion of integrated waste management from being misused (see paragraph 3);
(c) incineration/waste-to-energy plants, proposing to amend Legislative Decree no. 152 of 3 April 2006, to introduce appropriate measures to further streamline the bureaucracy of the authorisation processes, providing for greater recourse to forms of self-certification at each authorisation stage and the certainty of the timeframe for the conclusion of the procedures also through the activation of substitutive powers in the event of inertia on the part of the public administrations concerned; as well as providing, through legislation, appropriate incentives and/or compensation for the populations and local authorities affected by waste-to-energy plant developments, without introducing improper extensions of price regulation to activities that can be guaranteed by the market.

---

[4] IAA Annual Report of 31 March 2021, pp. 15 ff.

[5] Section V, Competition at the Service of Environmental Sustainability, S4143, pp. 65 ff.

(d) a proposal to eliminate the improper weight of system charges on the electricity sector, gradually bringing the financing of the various items, including those incentivising renewable energy sources, which could selectively burden the consumption of fossil fuels in heating and transport, back into the general taxation sphere. In this sense, the Authority has proposed a reform of the financing for renewable energies that would exclude system charges from the electricity bill and instead envisage forms of taxation consistent with the pursuit of environmental principles, thus providing for these charges to be selectively borne on the consumption of fossil fuels in heating and transport and resorting to appropriate gradualness in the implementation of the reform.

More recently, in its report AS1824 of 31 March 2022, the IAA addressed, in particular, some issues relating to the electricity sector: the development and upgrading of networks, both nationally (energy transmission) and locally (energy distribution), are fundamental preconditions for the development of renewable energies; the development of second-generation smart metres, in this perspective, will allow sellers to formulate dynamic price offers (differentiated by time bands), facilitating virtuous demand behaviour, in favour of rationalisation and reduction of consumption (so-called demand response); the final transition to the free market of domestic consumers could contribute to the proliferation of the free market in the electricity sector, which could lead to a reduction in the number of consumers, and to the development of the electricity market in the future demand response); the definitive transition to the free market for domestic consumers could contribute to the proliferation of differentiated offers to the end consumer based on consumption profiles, stimulating both energy saving and efficiency objectives and a more competitive retail market structure. In this regard, it has been observed that the sustainability of energy markets is contributing to the development of innovative services that could not only complement the pure energy commodity offered by sellers but also allow the development of new markets and the evolution of existing ones.

9.1.4 The Intersection Between Sustainability and Competition Law

At the legislative level, there are numerous initiatives taken by the Italian legislator that are a part of the environmental policies defined at the European and national levels and that impact competition law (see Sect. 9.2.3).

However, on a preliminary note, concerning the intersection between sustainability and competition law, it should be pointed out that almost all of the cases dealt with by the competition authorities on the subject of sustainability concern cases in which environmental sustainability is used as an expedient to implement conduct that is not quite virtuous from a competitive point of view (so-called greenwashing).

Greenwashing is a transversal issue as it may concern, with different modalities and forms, both conduct relevant to antitrust rules in breach of Articles 101 and 102 TFEU (for an analysis of the cases, see paragraph 3) and conduct relevant to consumer protection (matter that is in charge of the same IAA; for some examples of IAA measures, see paragraph 3). Greenwashing practices not only harm competition but also have environmental effects as they delay the transition to more sustainable products or make it less convenient.

In particular, as far as restrictive agreements of competition are concerned, the expression "greenwashing" refers to forms of collaboration between companies, realised through agreements falsely presented as aimed at and necessary to achieve significant environmental benefits (e.g. in terms of saving raw materials or energy resources, lower greenhouse gas emissions, etc.) to the benefit of users and/or the community. About abuses of dominant position, the expression greenwashing is used when the dominant company instrumentally uses sustainability objectives to pursue policies of market monopolisation and the exclusion of competitors.

Furthermore, the practice of suggesting, in the context of a commercial, marketing or advertising communication, in a completely generic, unverifiable or misleading manner, that a product or service is sustainable in terms of, for example, its production, packaging, distribution or disposal methods (so-called green claim) is considered an unfair commercial practice, in breach of Articles 18 and following of the Italian Consumer Code.

Further on green claims and consumer protection, it should be noted that at the national level, there is an additional form of protection provided by Article 12 of the self-regulatory code, which provides for a specific discipline on environmental protection: "commercial communication claiming or evoking environmental or ecological benefits must be based on truthful, relevant and scientifically verifiable data. Such communication must make it clear to which aspect of the advertised product or activity the claimed benefits refer".[6]

Finally, it should be noted that a significant role in the intersection of sustainability and the protection of competition is also played by international ISO standards and public and private product certifications.

In this regard, it should be observed that the proposal for a Directive of the European Parliament and the Council amending Directives 2005/29/EC and 2011/83/EU, empowering consumers for the green transition through better protection against unfair practices and better information of 30 March 2022, will give binding force to a series of general principles, not only already established by ISO standards but which have been elaborated and applied over the years by the decisions of the IAA and IAP. In particular, reference is made to the following general principles: the

---

[6]The self-regulatory code is an instrument prepared by the Institute of Advertising Self-Discipline (IAP), a recognised association that has set the parameters for "honest, truthful, and fair" commercial communication at a private level to protect consumers and fair competition between companies. The IAP is an institute of a private nature, but to which the main operators in the sector at the national level adhere. The multiplicity of its members means that the Institute has broad control over the entire market.

prohibition of generic, misleading, and unverifiable assertions; the evaluation and provision of data necessary for the verification of assertions; the use of evaluation methods leading to reliable and reproducible results; when selecting a method for the evaluation and/or verification of an assertion, give preference to ISO standards, followed by other internationally recognised standards and methods used in industry and commerce that have been peer reviewed; the provision and preservation of supporting documentation based on the life of the product.

9.2 Sustainability Initiatives Undertaken at the Italian Level by the Relevant Authorities

9.2.1 Working Documents and Guidelines Provided by Administrative and Political Authorities

Among the working documents and guidelines provided by administrative and political authorities on the subject of sustainability, including for the implementation of the National Recovery and Resilience Plan (PNRR) (see paragraph 2.3), the following documents should be mentioned, without claiming to be exhaustive.

Circular No. 32 of 30 December 2021 issued by the Ministry of Economy and Finance (MEF) is an operational guide for compliance with the principle of not causing significant damage to the environment. This guide was issued taking into account that it is expressly established in Article 18 of EU Regulation 241/2021 that all measures of the National Recovery and Resilience Plans must comply with this principle. The guide provides guidance on taxonomic requirements, the corresponding legislation, and useful elements for documenting compliance with these requirements.[7]

Among the activities carried out by the Regulatory Authority for Energy Networks and the Environment (ARERA) are the following documents through which ARERA has expressed:

- Its comments on the Draft Law on the Annual Market and Competition Act 2021 (see paragraph 2.3), in particular concerning natural gas distribution concessions, large hydroelectric derivation concessions, the provision requiring the government to adopt a legislative decree to reorganise matters concerning local public services, as well as the provision on waste management services (Article 12)[8]

---

[7] The guide is available at the following link: https://www.rgs.mef.gov.it/VERSIONE-I/circolari/2021/circolare_n_32_2021/. Accessed 16 October 2022.

[8] See Doc. 82/2022/I/COM, 4 March 2022, available at the following link https://www.arera.it/allegati/docs/22/082-22.pdf. Accessed 19 November 2022.

- Its guidelines on operational modalities for the first urgent application of tariff concessions to natural gas-intensive companies referred to in the Decree of the Minister of Ecological Transition No. 541/2021. The aforementioned decree defined an aid scheme under Article 44 of Regulation (EU) No 651/2014 through the redetermination, as of 1 April 2022, of the fees covering general gas system charges applied to natural gas-intensive companies, linked to the financing of measures aimed at achieving common decarbonisation targets[9]
- Its contribution to the parliamentary examination of the PNRR.[10]

9.2.2 IAA Activities on Sustainability and Advisable Actions

Instead, with particular reference to the activities carried out by the IAA, it should be noted that there are currently no specific interventions by the Authority aimed at clarifying the application of antitrust regulations on the subject of sustainability through, for example, guidelines, working papers, or individual guidance. Such an intervention on the part of the IAA would be desirable since the Authority plays a fundamental role in guiding companies through the path of ecological transition. Indeed, in the current context where sustainability issues are constantly evolving and where companies are called upon to invest and innovate in this respect, the latter needs to have a clear frame of reference for antitrust compliance purposes.

In this sense, with particular reference to business-to-business cooperation, it would be desirable, following the example of the Dutch Competition Authority, to draw up guidelines on sustainability agreements. On this point, it should be noted in particular that sustainability agreements are defined as agreements *aimed at identifying, preventing, limiting or mitigating the negative impact of economic activities on people (including working conditions), animals, the environment or nature.*[11] However, while the concept of sustainability has been finalised, the list of initiatives that could fall into this category remains open. A definition of the objectives and initiatives likely to be assessed as "sustainable" under antitrust law would therefore be useful to guide companies wishing to support sustainability investments.

On the subject of sustainability agreements, some have suggested, without prejudice to the competence of the IAA in the final assessment of sustainable agreements, the advisability of requesting a non-binding opinion from the Ministry of Ecological Transition established in Italy by Article 2, Decree-Law no. 22 of

---

[9] See Doc. 59/2022/R/GAS, 15 February 2022, available at the following link https://www.arera.it/allegati/docs/22/059-22.pdf. Accessed 19 November 2022.

[10] See Doc. XXVII, No. 18, 2 March 2021, available at the following link https://www.arera.it/allegati/docs/21/086-21.pdf. Accessed 19 November 2022.

[11] The Dutch Authority for Consumers & Markets (ACM), Second draft version: Guidelines on Sustainability Agreements – Opportunities within competition law, Section 3, par. 7, available at the following link https://www.acm.nl/en/publications/second-draft-version-guidelines-sustainability-agreements-opportunities-within-competition-law.

1 March 2021, converted into Law no. 55 of 22 April 2021. The prior involvement of the governmental authority, which would give a non-binding opinion on the social merits of the agreement, could favour a faster and more conscious approval of a regulatory measure aimed at validating what has developed in a regulatory vacuum.[12]

Another tool through which the national authority could help incentivise companies to invest in sustainability is through the so-called comfort letter instrument, which would allow companies to invest while minimising the risk of antitrust violation.

Also at the level of consumer protection and, in particular, concerning green claims, taking into account the high number of cases in which the Italian Authority has sanctioned professionals for ecological statements contrary to professional correctness (see paragraph 3) and the fact that the current average consumer is an increasingly attentive and sensitive consumer to the issue of sustainability, it is hoped that the Italian Authority will define guidelines on green claims in the wake of the example of the Dutch Authority and the United Kingdom.[13]

In this context, it should be noted that, at the same time, since the risk of a speculative use of the environmental objective to pursue anti-competitive aims is high, another task of the Authority should be to counter the phenomenon of greenwashing through more intensive use of its sanctioning power, making the most of the instrument of aggravating circumstances if it is ascertained that the environmental objective has been used instrumentally to implement illegal practices.[14]

9.2.3 Legislative Initiatives on Sustainability at the National Level

In the context of national legislative initiatives concerning environmental sustainability, the recent approval of Constitutional Law No. 1 of 11 February

---

[12] A. Moliterni, Antitrust e ambiente ai tempi del Green Deal: il caso dei "sustainability agreements", in Giornale di diritto amministrativo 2021 (3), p. 363.

[13] The Dutch Authority for Consumers & Markets (ACM) published Guidelines on Sustainability Claims in January 2021, highlighting five key principles: make clear what sustainability benefit the product offers; substantiate your sustainability claims with facts, and keep them up-to-date You will have to be able to prove that your sustainability claims are true; comparisons with other products, services, or companies must be fair; be honest and specific about your company's efforts concerning sustainability; make sure that visual claims and labels are useful to consumers, not confusing. See https://www.acm.nl/sites/default/files/documents/guidelines-suistainability-claims.pdf ; the UK Competition & Markets Authority also published Guidelines on Green Claims (Green Claims Code) in September 2021, available at https://www.gov.uk/government/publications/green-claims-code-making-environmental-claims/environmental-claims-on-goods-and-services, Accessed 19 November 2022.

[14] A. Moliterni, Antitrust e ambiente ai tempi del Green Deal: il caso dei "sustainability agreements", in Giornale di diritto amministrativo 2021 (3), p. 362.

2022, which includes environmental protection among the fundamental principles of the Constitution, is of particular significance.

Articles 9 and 41 of the Constitution were amended by Constitutional Law No. 1 of 2022.

In particular, a new paragraph was introduced to Article 9 of the Constitution to recognise within the fundamental principles set out in the Constitution the principle of protecting the environment, biodiversity, and ecosystems, also in the interest of future generations. In this regard, it is interesting to emphasise how the concern for future generations is a hitherto unheard-of expression in the Italian constitutional text.

At the same time, Article 41 of the Constitution on the exercise of the economic initiative was amended. The amendment added, among the limits to private economic initiative, a reference to the environment, supplementing the only provision of the Constitution that the Italian antitrust law (Law No. 287/90) in Article 1 expressly refers to as a constitutional foundation. In particular, it was established that private economic initiative cannot be carried out to the detriment of health and the environment, adding these two limits to those already in force, namely security, freedom, and human dignity. Article 41 of the Constitution was also amended by reserving to the law the possibility of directing and coordinating economic activity, both public and private, for not only social but also environmental purposes.

The purpose of the amendment is, first of all, to give articulation to the principle of environmental protection, which goes beyond the mention of "protection of the environment, the ecosystem and the cultural heritage" provided for in Article 117, second paragraph, of the Constitution, introduced with the 2001 reform, in the part where it enumerates the matters over which the State has exclusive legislative competence.

Concerning national sustainability policies, mention should first be made of the Integrated National Energy and Climate Plan 2030 (PNIEC), which is the fundamental instrument at the national level of the energy and environmental policy for implementing European climate policy. The PNIEC is structured in five lines of action: from decarbonisation to energy efficiency and security via the development of the internal energy market, research, innovation, and competitiveness. On this point, it should be noted that, following the update of the European targets, a commission of experts was appointed in September 2021 to update the PNIEC targets as well as to ensure their consistency with Mission 2 of the PNRR.

The PNRR, the document that the Italian government has prepared to illustrate to the European Commission how Italy intends to invest the funds allocated at the European level under the Next Generation Eu programme, includes a package of investments and reforms divided into six missions. The projects envisaged in the mission "Green Revolution and Ecological Transition" (Mission 2) are aimed at fostering the country's green transition by focusing on energy produced from renewable sources, increasing resilience to climate change, and supporting investments in research, and innovation, and incentivising sustainable public transport. Mission 2 is divided into four components for which specific objectives are envisaged: sustainable agriculture and circular economy; renewable energy, hydrogen, grid, and sustainable mobility; energy efficiency and building upgrading; and land and water resource protection.

Taking into account the interest of the IAA in this topic, it is necessary to highlight in particular the investments destined for the development of electric recharging infrastructures. To reach the European decarbonisation objectives, a fleet of around 6 million electric vehicles is expected by 2030, for which 31,500 public fast-charging points are estimated to be necessary. The measure, therefore, aims to build the enabling infrastructure to promote the development of sustainable mobility and accelerate the transition from the traditional model of fuel-based refuelling stations to electric vehicle refuelling points. To enable the realisation of these objectives, the measure aims at the development of 7500 fast-charging points on motorways and 13,755 in urban centres, as well as 100 experimental charging stations with energy storage technologies.

In particular, concerning the issue of public infrastructures for electric recharging on national territory, mention should be made of (i) Article 57 of Decree-Law No. 76 of 16 July 2020, converted with amendments by Law No. 120 of 11 September 2020; the aforementioned article prescribes the simplification of the rules for the construction of electric vehicle charging points and stations: transparent and non-discriminatory criteria will have to be provided for the allocation of spaces and/or the selection of operators for the installation of electric car charging stations, and the regulatory obstacles that, especially from a tariff point of view, still stand in the way of the free performance of the activity of supplying electric energy for vehicle charging will also have to be overcome, and (ii) Article 1, paragraph 697, of the Budget Law 2021 (Law No. 178 of 30 December 2020), which provided for the obligation for motorway concessionaires to equip their concession networks with an adequate number of high-power recharging points within 180 days of the law coming into force.

With respect to the aforementioned provisions, the IAA, in the context of its advocacy activity (see paragraph 1.3), has proposed to amend paragraph 6 of Article 57 of Decree-Law No. 76 of 16 July 2020 and paragraph 697 of Article 1 of Law No. 178 of 30 December 2020 in order to provide, with respect to the activities of public administrations and public concessionaires, for the adoption of transparent and non-discriminatory procedures for the allocation of public spaces and/or the selection of operators for the installation of public spaces in order to provide, with regard to the activities of public administrations and public concessionaires, for the adoption of transparent and non-discriminatory procedures for the allocation of public spaces and/or the selection of operators for the installation of charging stations, identifying the price of the charging services offered as an evaluation parameter and ensuring the technological neutrality and interoperability of the installations, as well as to repeal paragraph 12 of Article 57 of Decree-Law No. 76 with the intention of ensuring a reasonable level of tariffs for the supply of electricity for recharging through the presence of a plurality of operators and genuinely competitive dynamics.

Among the legislative activities that are a part of the implementation of the PNRR are the following interventions by the Ministry of Agricultural, Food, and Forestry

Policies (Mipaaf): on 13 June, a decree was issued establishing the agri-food logistics contract instrument to strengthen agri-food logistics and storage systems, reduce environmental and economic costs, and support innovation in production processes.[15] In addition, on 28 June, the Ministerial Decree of 25 March 2022 was published in the Official Journal, providing the necessary directives for the launch of the "Agrisolar Park" measure, to which EUR 1.5 billion of PNRR funds are dedicated.[16]

Lastly, among the most recent legislative interventions at the national level, reference should be made to Draft Law No. 2469, Annual Market and Competition Law 2021, approved by the Senate on 30 May 2022 and currently under consideration by the committee. The Draft Law, which is linked to the public finance manoeuvre, consists of 36 articles, divided into eight chapters, including a chapter dedicated to competition, energy, and environmental sustainability (Chapter IV, Articles 11 and 12).

Article 11 amends Article 1, paragraph 697, of Law No. 178 of 2020 (Financial Plan Law 2021), concerning the provision of the motorway network with fast electric recharging points, stipulating that motorway concessionaires must select the operator for the installation of the recharging columns through competitive, transparent, and non-discriminatory procedures.

Article 12 introduces some amendments to the Environmental Code (Legislative Decree No. 152/2006) concerning the choice by non-household users producing waste assimilated to urban waste to use the public service operator or resort to the market (paragraph 1), the tasks of ARERA (paragraph 2), as well as the exclusion, from the list of subjects involved in the CONAI programme agreement on packaging waste, of the operators of sorting platforms (paragraph 3).

More specifically, paragraph 1 of Article 12 amends paragraph 10 of Article 238 of the Environmental Code (Legislative Decree 152/2006), which provides that the choice of using the public service manager or resorting to the market must be made for a period of no less than 5 years, reducing this minimum period to only 2 years, accepting the observation formulated by the IAA in its report no. 4143 (see paragraph 1.3). Paragraph 2 of Article 12 integrates the text of Article 202 of the Environmental Code by providing for new tasks for the Regulatory Authority for Energy, Networks and the Environment (ARERA). Paragraph 3 of Article 12 amends paragraph 5 of Article 224 of the Environmental Code, where it provides for the conclusion of a programme agreement on a national basis between CONAI and autonomous systems and all the operators in the reference sector with the National Association of Italian Municipalities (ANCI), the Union of Italian Provinces (UPI), or the management entities of the Optimal Territorial Ambit, accepting also, in this case, the observation made by the IAA in its report no. 4143.

---

[15] See https://www.politicheagricole.it/misura_logistica_pnrr. Accessed 19 November 2022.

[16] See   https://www.politicheagricole.it/flex/cm/pages/ServeBLOB.php/L/IT/IDPagina/18319. Accessed 19 November 2022.

9.3 Specific Cases Related to Sustainability Issues Investigated by IAA

In the experience of the Italian Antitrust Authority, there are currently no cases in which the issue of sustainability has been used by the Authority as a sword, to protect competition in general and competition with positive environmental repercussions in particular, or by companies as a shield, to protect and justify operations that are likely to produce competitive damage.

Nevertheless, in the context of the antitrust enforcement activity carried out in recent years, the Authority has shown itself to be particularly sensitive to sustainability issues: this is due to an awareness of the complementary function that competition policies can play concerning environmental protection. As mentioned above, the existence of such synergy was noted by the Authority itself in its Annual Report on its activities in 2020, where it was observed that *"competition, while not having the primary purpose of promoting sustainable development, can contribute, by complementing existing instruments such as regulation and taxation, to fostering the process of transition towards an environmentally sustainable growth model", acting as an incentive "to use the scarce resources of our planet efficiently"*[17] (see par 1.2).

The aforementioned antitrust enforcement activity "with an environmental background" has, in particular, addressed certain unilateral or concerted conduct, such as the hindering of the introduction of innovations that could favour a greater circularity of production processes or the implementation of sustainable mobility infrastructures or services, which could harm competition and, at the same time, the environment.

On this point, it is necessary, first of all, to recall the interventions of the Authority which have concerned the waste management market and which have resulted in support for the already mentioned circularity of the production processes in the relative sector through the maintenance of a fair competitive pressure in the economic relations between the operators concerned or through the opening to competition of previously monopolistic markets.[18] These mentioned cases both involved abusive exclusion strategies implemented by consortia active in the waste collection and recycling sector to the detriment of minor operators wishing to enter the market by proposing innovative solutions.

The first of the cited cases ended with the proposal, by the Conai consortium, of commitments whose acceptance allowed new autonomous waste management systems to enter the market, thus ensuring the emergence of unprecedented recovery and recycling methods for special plastic waste and thus contributing to the achievement of the sustainability objective represented by a broader development of the circular economy. In the second case cited, on the other hand, the Corepla

---

[17] IAA Annual Report of 31 March 2021, pp. 15 ff.

[18] Case A476 – Conai plastic waste management, decision no. 25609, 3 September 2015 and Case A531 - Recycling primary packaging/abusive Corepla wrapping, decision no. 28430, 27 October 2020.

consortium was fined more than EUR 25 million for obstructing the market access of Coripet, a consortium that had introduced an alternative and innovative plastic recycling system, which would have led to an increase in the collection and recycling of differentiated waste, especially in geographical areas with lower environmental performance.[19]

A further example of the interconnection between antitrust and environmental objectives is a proceeding concerning an alleged cartel involving different levels of the spent lead-acid accumulator recovery chain. This agreement, aimed at guaranteeing the historical members of Cobat and Cobat Ripa, active in the spent vehicle lead-acid accumulator recovery chain, would have a continuous flow of waste at controlled prices and exclude competing collection systems from the market.[20]

The Authority concluded the proceedings by finding that the remedies proposed by the companies involved were adequate to address the competition concerns expressed in the initiating order. Particularly from an environmental point of view, it is necessary to highlight the important structural measure presented by the parties concerning the definitive divestment of the shares held by the smelters in Cobat and the exercise by them of their right to withdraw from Cobat Ripa, thus making Cobat a producers-only system, as is the case for all other national operators of spent accumulator collection. More generally, the proposed remedies were considered capable of creating the conditions for the realisation of competitive dynamics in the recovery chain of these highly polluting materials that are also more efficient from an environmental point of view.

In the same line as the above-mentioned case is a case in which the Authority accused the Erion Wee consortium (a collective waste management system for electrical and electronic equipment) of exclusionary abuse on the supply side in the markets for compliance services provided to producers of electrical and electronic equipment and on the demand side in the markets for waste treatment services for electrical and electronic equipment supplied by recovery facilities to collective systems.[21] The alleged abusive conduct engaged in by the consortium was described by the IAA as an impediment to effective competition in the supply of services necessary for producers to be able to comply with their environmental obligations; the procedure was concluded with the acceptance of the commitments of the consortium.

The IAA's activity also concerned proceedings with a potential impact on the decarbonisation process. In particular, regarding this issue, the Authority assessed the environmental implications of potentially harmful conduct by companies active

---

[19] The decision was confirmed by Lazio Regional Administrative Court, decision no. 11997 of 22 November 2021.

[20] Case I838 - Restrictions in the purchase of exhausted lead accumulators, decision no. 29718, 15 June 2021.

[21] Case A544 - Erion WEEE, decision no. 30130, 27 April 2022.

in the electric mobility sector. In a case the IAA investigated an alleged abuse perpetrated by Google against Enel. This abuse consisted in the refusal, motivated by the intention to protect Google Maps' business model, to give access to the Android Auto platform to the search and navigation app JuicePass (formerly Enel X Recharge), implemented by Enel for location and booking services of electric car charging stations. In the context of the measure, issued after the investigation, with which the Authority imposed a fine of over 100 million euro on Google, the Authority made some environmental considerations. In particular, it was noted that the contested conduct consisted in an exclusionary abuse *whose effects affect consumer welfare and market structure and may hinder innovation in services related to electric mobility provided through apps*, and it was then found that the conduct *could influence the development of an adjacent sector, i.e. that of electric mobility, in a crucial phase of the latter's start-up, about the development of a network of infrastructures for recharging electric cars adequate to the phase of growth and evolution of the demand for recharging services, with repercussions also on a more rapid diffusion of electric vehicles and the transition towards a more environmentally sustainable mobility.*[22]

Concerning the assessment of mergers between companies, the IAA has not yet had to deal with the most critical cases, i.e. those that require the Authority to perform a difficult trade-off between the benefits to the environment on the one hand and the harm caused to competition on the other. Nonetheless, the IAA has also had to deal with sustainability issues when assessing certain merger operations.

In this regard, two recent operations authorised by the Authority are particularly noteworthy.

The first, which concerned the electric mobility sector, concerns the establishment of a joint venture between the company Enel X and the car manufacturer Volkswagen, aimed at installing and managing, in the domestic market, 3000 ultra-fast public charging points for recharging the batteries of electric cars (so-called high-power charge (HPC)).[23]

In particular, the initiating investigative decision hypothesised that the transaction was likely to constitute or strengthen a dominant position in the market in which JVC will be active and/or in two vertically related markets, that of the provision of eMobility services (EMP) in Italy and that of the production and marketing of automobiles (OEM), with particular reference to the production and marketing of electric cars (BEV) in Italy. The transaction at issue is relevant insofar as it is a part of a legislative and regulatory context marked, both at the Euro-Union and national levels, by the promotion of electromobility as an appropriate measure to contribute to the pursuit of the decarbonisation objectives assumed at the European level and mitigate the environmental impact in the transport sector. The Authority carried out

---

[22] Case A529 - Google/compatibility of Enel app XItaly with android car system, decision no. 29645, 27 April 2021.

[23] Case C12404 - Enel X-Volkswagen finance Luxembourg/JVC, decision no. 29945, 9 December 2021.

an in-depth investigation of the markets involved and in particular of the market for the construction and management of charging infrastructures for public or private HPC electric cars with public access (so-called CPO market) since the latter is a market in an embryonic state susceptible to strong expansion in the coming years.

In conclusion, the Authority has cleared the transaction in question, recognising the concrete interest in the entry of other major players into the HPC market, which will also be able to benefit from the public incentives provided by the PNRR for the development of HPCs and from sectoral regulation that guarantees non-discriminatory access to electricity distribution networks.

A second merger was carried out between Bolton Manitoba, a company belonging to a multinational group active in the production and marketing of consumer products, including products for household cleaning and personal care, and Madel, an Italian company of which the former acquired exclusive control, which markets a wide range of similar products with a low environmental impact, which can be defined as "green products".[24] With particular reference to the latter case, within the measure by which the Authority authorised the merger, it questioned, for the first time, the possibility/necessity of identifying a separate market for "green" products. The doubt raised by the IAA, which has been resolved in the negative in the present case, is, however, emblematic of how the issue of sustainability can, for example, condition the very definition of the relevant market when taking into account the impact of environmental issues on consumer preferences. From this point of view, if a given consumer good was the result of sustainable production methods and at the same time there was a low degree of substitutability between traditional and "green" products, it would be conceivable to identify different markets for the same type of good.

Lastly, it is worth mentioning several cases in which the IAA, in the application of consumer protection law, has sanctioned certain companies that have adopted *greenwashing* strategies, i.e., put in place unfair commercial practices to convey a deceptively positive self-image to the consumer in terms of environmental impact. On this point, it is interesting to note the transversal nature of this practice, where, reviewing the decisions taken on the subject by the Authority, it can be seen how companies operating in the most diverse sectors, such as the food sector[25] or transport and mobility[26] or even personal hygiene,[27] have resorted to it.

---

[24] Case C12416 - Bolton group/Madel, decision no. 30050, 1 March 2022.

[25] Case PS6302 - Acqua Sant'Anna bio bottle, decision no. 24046, 14 November 2012.

[26] Case PS10211 - Volkswagen pollutant emissions of diesel vehicles, decision no. 26137, 4 August 2016.

[27] Case PS10389 Olive Italia-pannolini nappynat, decision no. 26298, 15 December 2016. On this issue, see also Case PS4026 - Acqua San Benedetto-la scelta naturale, decision no. 20559, 10 December 2019; Case PS7235 - Ferrarelle-impatto zero, decision no. 23278, 8 February 2012; Case PS8438 - Wellness innovation project-pannolini naturaè, decision no. 24438, 3 July 2013; Case PS11400 - Eni diesel+-pubblicità ingannevole, decision no. 28060, 20 December 2019; PS11848 - Dolomiti energia/offerte commerciali, decision no. 29774, 13 July 2021.

Among the interventions of the IAA carried out through the instrument of moral suasion, it has to be mentioned the case in which the Authority invited EasyJet to remove the profiles of possible unfairness related to the fact that it advertised, as a characteristic of its typical activity, any environmental initiative unrelated to it and presented its activity with a neutral environmental footprint or characterised by an absence/exiguity of emissions. Following the aforementioned solicitation, the company disclosed that it had already voluntarily changed the text of some pages on its website in which it describes its commitment to the environment.[28]

Open Access This chapter is licensed under the terms of the Creative Commons Attribution 4.0 International License (http://creativecommons.org/licenses/by/4.0/), which permits use, sharing, adaptation, distribution and reproduction in any medium or format, as long as you give appropriate credit to the original author(s) and the source, provide a link to the Creative Commons license and indicate if changes were made.

The images or other third party material in this chapter are included in the chapter's Creative Commons license, unless indicated otherwise in a credit line to the material. If material is not included in the chapter's Creative Commons license and your intended use is not permitted by statutory regulation or exceeds the permitted use, you will need to obtain permission directly from the copyright holder.

---

[28] MS-PS11598 – Easyjet-emissioni zero di CO2, notices posted on 30 November 2021.

Sustainability and Competition Law in Malta

10

Clement Mifsud-Bonnici

10.1 The Concept of Sustainability

The concept of "sustainability" is not defined under Maltese law, but there are a number of acts of parliaments and subsidiary legislation for different sectors that use the term "sustainability".

The closest definition in the law is found in the Sustainable Development Act, which defines "sustainable development" as "development that meets the needs of the present without compromising the ability of future generations to meet their own needs".[1] This definition has been lifted from the *Report on the World Commission on Environment and Development* of 1987.[2]

The Constitution of Malta does provide for a positive obligation on the Government of Malta to "protect and conserve the environment and its resources for the benefit of the present and future generations and shall take measures to address any form of environmental degradation in Malta, including that of air, water and land, and any sort of pollution problem and to promote, nurture and support the right of action in favour of the environment".[3] Unfortunately, this "positive obligation" is not judicially enforceable, but it remains "fundamental to the governance of the country and it shall be the aim of the State to apply these principles in making laws".[4]

---

[1] (Chapter 521 of the Laws of Malta), Article 3(1).

[2] Available at https://digitallibrary.un.org/record/139811?ln=en. Accessed 16 August 2022.

[3] Constitution of Malta, Article 9(2).

[4] Constitution of Malta, Article 21.

C. Mifsud-Bonnici (✉)
Ganado Advocates, Valletta, Malta
e-mail: cmifsudb@ganado.com

© The Author(s) 2024

P. Këllezi et al. (eds.), *Sustainability Objectives in Competition and Intellectual Property Law*, LIDC Contributions on Antitrust Law, Intellectual Property and Unfair Competition, https://doi.org/10.1007/978-3-031-44869-0_10

This provision was only introduced in 2018,[5] and prior to that, the Constitution of Malta had no provision on environmental protection or sustainable development; however, a similarly worded obligation of the State has already been imposed by an act of parliament since 1991.[6]

Further, sustainable development is specifically mentioned in the Treaty of the European Union as an objective of the European Union,[7] and it is also mentioned in the Treaty on the Functioning of the European Union (TFEU)[8] and the Charter.[9] Malta, as a Member State of the EU, has a duty of sincere cooperation in the pursuit of Union objectives.[10]

There is a general consensus that "sustainability" is relevant to all sectors and industries, and more recently, there is much talk of "ESG" factors of sustainability being the (i) environmental, (ii) social, and (iii) governance factors—the concept of ESG has been expressly endorsed by the Government of Malta in its policy[11] but not in its law.

10.2 Competition Law and Sustainability

10.2.1 Complementary Objectives

The Competition Act does not make any express reference to "sustainability" or "sustainable development".[12] Neither does any other law in Malta.

It is submitted that sustainable development, environment protection and the protection of competition in the markets are not concepts and objectives that are necessarily incompatible with each other. Rather, there are objectives pursued by sustainable development and competition law that are complementary to each other.[13]

In practical terms, it is submitted that sustainability can be a part of the assessment to be made by the Office for Competition within the Malta Competition and

---

[5] Act XXII of 2018. There is no mention of "competition law" or "markets" in the parliamentary debates leading up to the adoption of this amendment to the Constitution.

[6] Act V of 1991, entitled *An act to protect the environment*, promulgated on 26 February 1991. Interestingly, and at this time, there was no mention that such an obligation imposed on the State was not judicially enforceable. However, later, iterations of the "Environment Protection Act" qualified that such an obligation on the State was not judicially enforceable.

[7] Treaty on the European Union, Article 3(3).

[8] Article 11 TFUE.

[9] Charter of Fundamental Rights of the European Union, Article 37.

[10] Treaty on the European Union, Article 4(3).

[11] See https://sustainabledevelopment.gov.mt/esg-reports-2020/. Accessed 16 August 2022.

[12] Chapter 379 of the Laws of Malta.

[13] For example, the breach of environmental legislation by dominant undertakings can be classified as an abuse within Article 9 of the Competition Act and/or Article 101 of the TFEU.

Consumer Affairs Authority (hereinafter the "Malta NCA"),[14] and by the Civil Court (Commerce Section) and other courts and tribunals where applicable (hereinafter the "Malta courts"), when assessing claims of breaches of competition law and proposed concentrations—as illustrated below. For the readers' benefit, the public enforcement of competition law is vested in Malta NCA, as the prosecutor, and in Malta courts, specifically the Civil Court (Commerce Section), which will decide on whether there is a competition law infringement, as well as on the extent of any fines.[15]

However, there may be instances where the conduct of undertakings and even laws or government policies are anti-competitive, and their effect is capable of undermining the objectives pursued by competition law.

10.2.2 Sustainability and the Enforcement and Advocacy of Competition Law

Firstly, and as a matter of principle, coordinated conduct that pursues a sustainability objective has the potential to fall within the scope of Article 101(1) TFEU and its national counterpart, Article 5(1) of the Competition Act. However, sustainability can be invoked by undertakings as a defence against an allegation or claim of coordinated conduct as an efficiency under subparagraph 3 of both Article 101 TFEU and Article 5 of the Competition Act.

This is the line that the position the European Commission took in the *Guidelines on vertical restraints*[16] and the one which it appears to be taking with the draft *Guidelines on the applicability of Article 101 of the Treaty on the Functioning of the European Union to horizontal co-operation agreements.*[17] Once the latter guidelines are implemented, it is envisaged that they would apply to Malta NCA and the Malta courts. The Competition Act also imposes a duty as such by requiring that either Malta NCA or the Malta courts, as the case may be, "shall have recourse [. . .] to relevant decisions and statements of the European Commission including interpretative notices on the relevant provisions of the TFEU and secondary legislation relative to competition [. . .]".[18]

Secondly, sustainability may also be invoked as a defence against claims of abuse of dominance conduct under Article 102 TFEU and its national counterpart, Article 9 of the Competition Act, by way of an "objective justification". Elsewhere, the case

---

[14] See https://mccaa.org.mt/Section/index?sectionId=1060. Accessed 16 August 2022.

[15] The decision of the Civil Court (Commerce Section) is subject to a second and last stage of review before the Court of Appeal.

[16] European Commission, Commission Notice Guidelines on vertical restraints, JO2022, C 248, p.1), para 9.

[17] Para 555. Available at https://eur-lex.europa.eu/legal-content/EN/TXT/?uri=CELEX%3A52022 XC0419%2803%29. Accessed 16 August 2022.

[18] Competition Act, Articles 12A (7) and 13(7).

was made for Article 102 TFEU to be used as an enforcement tool where a dominant undertaking engages in unsustainable business practices[19]—it is submitted that the same case can be made for Article 9 of the Competition Act to the extent that they can be characterised as "abuses".

Thirdly, sustainability may also be invoked as an efficiency gain in merger control cases where the Malta NCA is reviewing a notifiable concentration in terms of the law.[20]

Fourthly, Malta NCA may also promote a competitive environment where sustainability and ESG measures are proposed and implemented by engaging in its advocacy role[21] with the State and public authorities in the exercise of its public powers or functions or otherwise with persons who are not undertakings. Although Malta NCA can only issue non-binding recommendations through its advocacy role, as a matter of practice, public authorities heed Malta NCA's advice and recommendations. In any case, in one reported case, the Malta courts have directed public authorities to give considerable weight to such recommendations made by Malta NCA by way of advocacy.[22]

There are no reported cases of competition law infringements or advocacy or merger control involving sustainability in Malta. Although the Office for Consumer Affairs, within the same Malta Competition and Consumer Affairs Authority, has reported in the Authority's annual report for 2020 that they carried out checks on 344 claims on misleading sustainability claims,[23] and this on the basis of consumer protection legislation, not competition law.

However, it is submitted that sustainability should not be relegated to a mere "sword" or "shield" in competition law enforcement. There is a greater likelihood that the 17 sustainable development goals are achieved through engagement by undertakings, in particular, where there is a market failure that has not been addressed by regulation and law. Undertakings should be given precise written

---

[19] M. Iacovides and C. Vrettos, 'Falling through the cracks no more? Article 102 TFEU and sustainability: the relation between dominance, environmental degradation and social injustice', Journal of Antitrust Enforcement 2022, 10, 32-63. Available at https://academic.oup.com/antitrust/article/10/1/32/6352604. Accessed 16 August 2022.

[20] Control of Concentrations Regulations (Subsidiary Legislation 379.08), Regulation 4(4): "Concentrations that bring about or are likely to bring about gains in efficiency that will be greater than and will offset the effects of any prevention or lessening of competition resulting from or likely to result from the concentration shall not be prohibited if the undertakings concerned prove that such efficiency gains cannot otherwise be attained, are verifiable and likely to be passed on to consumers in the form of lower prices, or greater innovation, choice or quality of products or services."

[21] Malta Competition and Consumer Affairs Authority Act (Chapter 510 of the Laws of Malta), Article 14(1) (g) (h) and (i).

[22] Vivian Corporation Limited vs Central Procurement and Supplies Unit et, Court of Appeal decided on 17 March 2021 (Ref. 12/2021/1) para 24.

[23] MCCAA Annual report 2020. Available at https://mccaa.org.mt/media/6963/2020-annual-report-mccaa_one-page.pdf. Accessed 16 August 2022.

guidance on whether proposed cooperation, standards, or conduct is likely to be justified or fall foul of competition rules.

Malta NCA did not follow suit with other national competition authorities, and no guidance has been issued on sustainability and competition law. This issue of such guidance should be considered to provide for a degree of legal certainty, in particular, within the context of the publicly documented increase of initiatives of horizontal cooperation taken by the private sector in Malta on the basis of "sustainability" and "ESG" objectives.

Malta NCA has not been active in exploring the relationship between sustainability and competition law, but it is understood that Malta NCA is closely following, through the European Competition Network, the workings of other national competition authorities, and those of the European Commission.

The Malta NCA must probably consider ad hoc block exemption regulations or re-introducing negative clearance, at least, where conduct does not affect the internal market and is to be regulated purely by national counterparts of Articles 101 and 102 TFEU. The latter was revoked from the Competition Act upon Malta's entry into the European Union in 2004.

10.2.3 Competence and Cooperation

Having said all this, it is submitted that it is not Malta NCA's function to verify whether Malta is meeting the sustainable development goals.[24]

In fact, the Sustainable Development Directorate within the Ministry for the Environment, Energy and Enterprise was specifically designated[25] as the competent authority, in terms of the Sustainable Development Act, "to ensure the development and implementation of Malta's sustainable development strategy".[26] However, there are other regulators and public authorities that are specifically vested with the competence to monitor and enforce specific sectorial legislation that may have sustainability objectives—such as the Environment Resources Authority.[27]

This is for good reason. The resources, knowledge and proper delegated authority lie with the Ministry and other specific regulators and public authorities and not with Malta NCA.[28] However, Malta NCA must retain its competence to decide on competition law infringement or to advocate where there are distortions on the market as a result of sustainability-driven measures and conduct.

---

[24] No reference to "sustainability" or "sustainable development" can be found in the operative allocating functions and responsibilities to the Malta NCA. Malta Competition and Consumer Affairs Authority Act (Chapter 510 of the Laws of Malta), Article 14.

[25] Government Notice 472 of 4 June 2013 published in the 19,095th edition of *The Malta Government Gazette*.

[26] Sustainable Development Act (Chapter 521 of the Laws of Malta), Article 5(a).

[27] Environment Protection Act (Chapter 549 of the Laws of Malta).

[28] See Malta Competition and Consumer Affairs Authority Act (Chapter 510 of the Laws of Malta), Article 13.

Malta NCA can pursue cooperation with other regulators and public authorities through its advocacy role, as explained above, and generally through good relationships maintained in public administration.

10.2.4 Laws, Guidance and Policies

At the time of writing, there are no laws, guidance or policies that expressly address the relationship between sustainability and competition law.

However, it is submitted that the promulgation of laws, the establishment of standards and regulatory rules by public authorities and the issuance of policy documents in the name of sustainability have the potential to impact market dynamics and increase or restrict competition. Moreover, such laws and policies have the potential to be used as a point of reference in competition law enforcement.

An example is *Malta's Sustainable Development Vision for 2050*, which is a policy document published by the Ministry for the Environment, Energy and Enterprise in September 2018 pursuant to the Sustainable Development Act. This document espouses the Government of Malta's vision and strategy to promote sustainable development in Malta's economy, and in various instances, the document encourages the private sector to engage with the public sector to achieve this vision—specific proposals for the establishment of cooperatives and public-private partnerships are made. It is submitted that, within the context of potential competition law enforcement, any claims or arguments made on sustainable development may be corroborated with this document to verify whether such is aligned with this vision for sustainable development.

Outside the realm of Malta NCA and competition law, the Government of Malta has promulgated a number of laws and even created agencies in the name of sustainability that may impact competition on the market.

A noteworthy example is the following: in 2018, Circular Economy Malta (previously known as the Resource Recovery and Recycling Agency) was established. The primary function of this agency is "to foster a transition towards, and implement measures, for the growth and development of the circular economy", but it has also been tasked to issue to private operators a licence to run a scheme for a beverage container refund system under the Beverage Containers Recycling Regulations.[29] These Regulations are essentially a transposition of Directive 94/62/EC on packaging and packaging waste, and essentially, the scheme is an application of the extended producer responsibility principle. The European Commission has classified such "producer responsibility organisations" as undertakings.[30]

---

[29] Subsidiary Legislation 549.134.

[30] DG Competition Paper Concerning Issues of Competition in Waste Management Systems, para 6-7. Available at https://ec.europa.eu/competition/sectors/energy/waste_management.pdf. Accessed 16 August 2022.

Earlier in 2022, Circular Economy Malta issued a 10-year licence to BCRS Malta Ltd—a limited liability company established and owned by the majority of producers and importers of single-use beverage containers in Malta.[31] It is not clear whether this licence confers exclusivity, but the licensed operator is the only one running such a scheme in Malta, and Circular Economy opted for a direct grant of this licence rather than through a competitive bid.[32] The Regulations impose that participation in the scheme is permitted on a fair, reasonable and non-discriminatory (FRAND) basis.[33] Within this context, it is submitted that competition on the market has and will continue to be taken into account in any interventions made by the State to address perceived market failure, even when such interventions are legislative and where such laws encourage or even require coordination between undertakings.

10.3 Conclusion

It is obvious that the relationship between sustainability and competition law in Malta has not been investigated and that competition law is not featured in the discussion in Malta on sustainable development. It is submitted that this is the case because competition law is not necessarily seen as an obstacle to the pursuit of agreements, measures, policies and legislation inspired by sustainability. However, competition law has the potential to deter, rightly so, such agreements, measures, policies and legislation in view of perceived risks of competition law infringements and concerns of harm to markets.

The reality is that competition law must be taken into account by both undertakings and public authorities when pursuing sustainability measures, and therefore, guidance on what is permitted and what is not might encourage that such measures, which are not harmful to competition, are adopted. It is submitted that the guidance issued and to be issued by the European Commission will go a long way in providing clarity on this, and perhaps Malta NCA should look into providing ad hoc and specific guidance with respect to the provisions of the Competition Act. This guidance has the potential to provide a degree of certainty that can allow competition law and sustainable development to coexist and complement each other in the pursuit of their objectives. It is further submitted that even though Malta NCA is not competent *ex lege* to verify or ensure the achievement of sustainable development goals in Malta, Malta NCA should, in compliance with the State's "positive obligation" in the Constitution of Malta, explore whether there is a need for such ad hoc and specific guidance.

---

[31] See https://bcrsmalta.mt/about-us/. Accessed 16 August 2022.

[32] Subsidiary Legislation 549.134, Regulation 4(2) and (3).

[33] Ibid., Regulation 11(1).

Open Access This chapter is licensed under the terms of the Creative Commons Attribution 4.0 International License (http://creativecommons.org/licenses/by/4.0/), which permits use, sharing, adaptation, distribution and reproduction in any medium or format, as long as you give appropriate credit to the original author(s) and the source, provide a link to the Creative Commons license and indicate if changes were made.

The images or other third party material in this chapter are included in the chapter's Creative Commons license, unless indicated otherwise in a credit line to the material. If material is not included in the chapter's Creative Commons license and your intended use is not permitted by statutory regulation or exceeds the permitted use, you will need to obtain permission directly from the copyright holder.

Sustainability and Competition Law in Switzerland

11

Johana Cau and Alexandra Telychko

11.1 Introduction

While environmental protection and sustainability more generally are topics of increasing importance in public policy, the question arises as to the extent to which they may interact with the Swiss competition law framework. For example, sustainability initiatives may stem from behaviour affecting competition or such behaviour may be justified on environmental protection grounds.

This contribution therefore discusses the role of sustainability goals in competition policy and enforcement. It will present how the Swiss competition authorities, particularly the Competition Commission ('Comco') and its Secretariat, address sustainability within the existing legal framework.

We will first provide a brief description of the relevant legal framework on sustainability in general in Switzerland, as well as the interaction between sustainability and competition law, focusing mainly on the role of sustainability in the application of Article 5 para 2 of the Swiss Cartel Act [1] ('CartA') (Sect. 11.2). We will then briefly illustrate this interaction with cases dealing with sustainability (Sect. 11.3). Finally, we will give an overview of recent policy developments in terms of sustainability by the Swiss authorities (Sect. 11.4).

---

[1] Federal Act on Cartels and other Restraints of Competition. RS 251, available at https://www.fedlex.admin.ch/eli/cc/1996/546_546_546/en. Accessed 8 November 2022.

J. Cau
Lenz & Staehelin, Geneva, Switzerland
e-mail: Johana.cau@slaughterandmay.com

A. Telychko (✉)
Faculty of Law, University of Geneva, Geneva, Switzerland
e-mail: oleksandra.telychko@unige.ch

© The Author(s) 2024

P. Këllezi et al. (eds.), *Sustainability Objectives in Competition and Intellectual Property Law*, LIDC Contributions on Antitrust Law, Intellectual Property and Unfair Competition, https://doi.org/10.1007/978-3-031-44869-0_11

11.2 General Legal Framework

11.2.1 Introductory Remarks on the Concept of Sustainability in Swiss Law

Switzerland bases itself on the definition provided by the World Commission on Environment and Development (the Brundtland Commission) in its report 1987 report 'Our Common Future', namely that 'sustainable development is a development that meets the needs of the present without compromising the ability of future generations to meet their own needs'.[2]

Sustainable development is a principle to which the Swiss Confederation and the cantons are bound. Article 2 (Aims) of the Federal Constitution[3] establishes sustainable development as a national goal,[4] and Article 73 (Sustainability) calls on the Confederation and the cantons to strive for '[...] a balanced relationship between nature and its ability to renew itself, on the one hand, and the demands placed on it by the human race, on the other'.[5] To date, the Federal Council has implemented these constitutional requirements through sustainable development strategies.

The Federal Council understands sustainable development as follows: Sustainable development enables the basic needs of all human beings to be met and ensures a good quality of life throughout the world, now and in the future. It encompasses the three dimensions of environmental responsibility, social solidarity and economic efficiency in an equal, balanced and integrated manner, while considering the tolerance limits of the planet's ecosystems.[6] The 2030 Agenda for Sustainable

---

[2] World Commission on Environment and Development (1987), Our Common Future. Report to the World Commission on Environment and Development, United Nations General Assembly document A/42/427, available at https://www.are.admin.ch/dam/are/en/dokumente/nachhaltige_entwicklung/dokumente/bericht/our_common_futurebrundtlandreport1987.pdf.download.pdf/our_common_futurebrundtlandreport1987.pdf. Accessed 8 November 2022.

[3] Federal Constitution of the Swiss Confederation ('Federal Constitution'), RS 101. Available at https://www.fedlex.admin.ch/eli/cc/1999/404/en. Accessed 8 November 2022.

[4] Article 2 of the Federal Constitution (Aims) (highlighted in italics by the authors): '[1] The Swiss Confederation shall protect the liberty and rights of the people and safeguard the independence and security of the country. [2] It shall promote the *common welfare, sustainable development, internal cohesion and cultural diversity of the country.* [3] It shall ensure the greatest possible equality of opportunity among its citizens. [4] It is committed to the *long-term preservation of natural resources and to a just and peaceful international order'.*

[5] In addition, see the following excerpt from the preamble of the Federal Constitution (highlighted in italics by the authors): 'The Swiss People and the Cantons, mindful of their *responsibility towards creation*, resolved to renew their alliance so as to strengthen liberty, democracy, independence and peace in a *spirit of solidarity and openness towards the world*, determined to live together with *mutual consideration* and respect for their diversity, conscious of their common achievements and their *responsibility towards future generations...'.*

[6] Swiss Confederation Federal Council (2022) Sustainable Development Strategy 2030 ('SDS 2030'), p. 6, available at https://www.are.admin.ch/dam/are/en/dokumente/nachhaltige_entwicklung/publikationen/sne2030.pdf.download.pdf/sne2030.pdf. Accessed 8 November 2022.

Development ('Agenda 2030') provides the framework[7] with its underlying principles and 17 global sustainable development goals.

Some cantons have also adopted laws on public action for sustainable development, which govern the respective cantonal policies on sustainable development.[8]

11.2.2 The Role of Sustainability in Swiss Competition Law

There is no specific mention or obligation of sustainability in the CartA. Thus, Articles 2 and 73 of the Constitution serve as guiding principles for the actions of the federal authorities, including the Comco.

11.2.3 The Intersection Between Sustainable Development and Competition Law

The sustainable use of available resources, a component of sustainable development, can play a certain role in Swiss competition law. Indeed, Article 5 par. 2 lit. a CartA mentions the more rational use of resources as a reason for economic efficiency[9] that can justify[10] significant restrictions on competition.[11] This is particularly the case for

---

[7] In its SDS 2030, the Federal Council presents the priorities it intends to set for the implementation of the 2030 Agenda over the next ten years. The SDS 2030 and the associated Action Plan 2021–2023, adopted on 23 June 2001 (in German) are available at https://www.are.admin.ch/dam/are/de/dokumente/nachhaltige_entwicklung/publikationen/aktionsplan2021-2023.pdf.download.pdf/Aktionsplan%202021-2023%20zur%20Strategie%20Nachhaltige%20Entwicklung%202030.pdfm. Accessed 8 November 2022.

[8] See, for example, Geneva law on public action for sustainable development of 23 March 2001 (Agenda 21), RS/GE A 2 60; Neuchâtel law on public action for sustainable development of 31 October 2006 (Agenda 21), RS/NE 805.7.

[9] A restriction of competition is considered economically efficient when the result is more efficient or as efficient with the agreement as without the agreement.

[10] Horizontal price fixing, quantity-restricting or geographic market-sharing agreements cannot in principle be justified on this ground (Article 5 par. 3 CartA).

[11] The parliamentary work behind the CartA shows a clear objective, according to which competition must encourage, or even oblige, companies to optimise the use of resources, Message on the Federal Act on Cartels and other Restraints of Competition of 23 November 1994 (Feuille fédérale 1995 I 472, p. 515) ('CartA Message'). In addition to the rational use of resources, the CartA Message also mentions the rational use of public goods: 'Agreements that improve the efficiency of the company's resources will be considered as increasing economic efficiency. This possibility of justification will allow companies to achieve, in particular, general objectives (environmental protection, rational use of energy, health, education, safety) in the sense of self-regulation based on cooperation, while increasing economic efficiency' (CartA Message, p. 556). However, as the reference to the public good was finally dropped from the text of the CartA, a narrower view of the rational exploitation of resources than that contained in CartA Message must be adopted (I. Chabloz, L'autorisation exceptionnelle en droit de la concurrence, Étude de droit suisse et comparé, Schulthess Verlag 2002, p. 189).

environmental protection agreements which at the same time bring economic benefits to consumers.[12]

The term 'resources' includes (i) entrepreneurial resources, such as money, (ii) public resources, (iii) natural resources and (iv) knowledge.[13]

An agreement that aims at the efficient use of resources or public goods is by its very nature positive for the consumer and therefore for the community, even if it is not obvious that companies always derive an immediate benefit that can be passed on to the consumer through the market.[14] The CartA Message cites the case of confectionery manufacturers who choose to use a more environmentally friendly (in terms of energy consumption and waste disposal) packaging material for their products instead of aluminium as an example of a possible efficiency justification.[15]

The 'polluter-pays' principle enshrined in the Environmental Protection Act ('EPA') requires external costs to be internalised.[16] For environmental protection aspects to be considered under Article 5 par. 2 lit. a CartA, the alleged environmental benefits or resource savings must be sufficiently closely linked to the operations of the companies involved in the agreement or to the product in question.[17]

The criterion of rational exploitation of resources contained in Article 5 para 2 CartA must, in other words, be interpreted in accordance with the productive function of competition.[18] That is, the agreement in question is only justified on grounds of economic efficiency when it is necessary to rationalise the use of the undertaking's resources in the production of a good or service. However, this does not prevent the agreement from being environmentally friendly at the same time.

On the other hand, if an agreement protects the environment, but does not reduce the production costs of the companies concerned, then an exceptional authorisation from the Federal Council is required on the basis of overriding public interests[19] under Article 8 CartA.

---

[12] Decision of the Comco 'SWICO/Sens', LPC 2005/2, 266, para 92.

[13] Decision of the Swiss Federal Court 129 II 18 para 10.3.3.

[14] CartA Message, p. 557.

[15] CartA Message, p. 555.

[16] Article 2 of the Federal Act on the Protection of the Environment, RS 814.01. Available at https://www.fedlex.admin.ch/eli/cc/1984/1122_1122_1122/en. Accessed 8 November 2022.

[17] R. Zäch, Schweizerisches Kartellrecht, Stämpfli Verlag AG 2005, N 415; Decision of the Comco 'SWICO/Sens', LPC 2005/2, 266, para 92; Decision of the Comco 'Klimarappen', LPC 2005/1, 242, paras 28 ff.

[18] I. Chabloz, L'autorisation exceptionnelle en droit de la concurrence, Étude de droit suisse et comparé, Schultess Verlag 2002, p. 189.

[19] The notion of public interest is an indeterminate legal concept (U. Häfelin/G. Müller, Grundriss des allgemeinen Verwaltungsrechts, 3ème éd., Stämpfli Verlag 1998, n. 452). However, a public interest can be defined as 'a considerable interest which affects a large number of citizens and which they are unable or unwilling to satisfy by their own means' (A. Grisel, Traité de droit administratif, Éditions Ides et Calendes 1984, vol. I, p. 339). This definition gives broad discretion to the competent authority, in this case the Federal Council. The content of the public interest can vary considerably according to place, time and ideological conceptions (P. Moor, Droit administratif, Les fondements généraux, 2e éd., vol. I, Stämpfli Verlag 1994, p. 388).

The justification of a more efficient use of resources can in turn be considered for both horizontal[20] and vertical[21] competition agreements. Such agreements are not regarded as horizontal agreements.[22]

In conclusion, the motive of more rational exploitation of resources in Art. 5 para 2 CartA must not serve to achieve a general objective, such as environmental protection.[23]

11.3 Specific Interaction/Cases

In this section, we examine in detail specific cases relating to sustainability and competition law. We will discuss the lack of cases in Switzerland where sustainability could have played a role in competition law enforcement in the form of a 'sword' (Sect. 11.3.1), cases where sustainability played a role in competition law enforcement in the form of a 'shield' (Sect. 11.3.2), and illegal anti-competitive conduct that occurred in the context of sustainability initiatives (Sect. 11.3.3).

11.3.1 Absence of Cases Where Sustainability Has Played a Role in Competition Law Enforcement in the Form of a 'Sword'

As mentioned above (Sect. 11.2.2), there is no specific mention or obligation of sustainability in the CartA. Therefore, there is no case law in Switzerland pertaining to situations where sustainability has played a role in the enforcement of competition law in the form of a 'sword'. The Comco has not been active in using competition rules protecting competition for the benefit of sustainability goals (e.g. protecting competition in industries crucial to sustainability).

In fact, Swiss competition law pursues only purely economic objectives. If confronted with unsustainable business practices, the Comco would not consider, according to the authors, sustainability or environmental arguments, but only abuses of monopoly power if these abuses affect the economy.

---

[20] Decision of the Comco '*SWICO/Sens*', LPC 2005/2, 266, paras 90–96.

[21] Decision of the Comco '*Hors-Liste Medikamente*', LPC 2010/4, 688, para 298; Decision of the Comco '*Feldschlösschen/Coca Cola*', LPC 2005/1, 124 ff., para 116.

[22] P. Krauskopf/O. Schaller, in: M. Amstutz/M. Reinert (ed.), Basler Kommentar, Kartellgesetz, 2. Auflage, Helbing Lichtenhahn Verlag 2022, Art. 5 N 331.

[23] I. Chabloz, L'autorisation exceptionnelle en droit de la concurrence, Étude de droit suisse et comparé, Schulthess Verlag 2002, p. 189; C. J. Meier-Schatz, Unzulässige Wettbewerbsbreschränkungen, arts. 5–8, in: R. Zäch (ed.), Das neue schweizerische Kartellgesetz, Schulthess Verlag 1996, pp. 26 ff.; E. Homburger/B. Schmidhauser/F. Hoffet/P. Ducrey, Kommentar zum schweizerischen Kartellgesetz vom 6. Oktober 1995 und zu den dazugehörenden Verordnungen, Schulthess 1996/1997, Art. 5 n. 104.

11.3.2 Cases Where Sustainability Has Played a Role in the Enforcement of Competition Law in the Form of a 'Shield'

There have been very few decisions illustrating cases where sustainability has been used as a defensive argument in the enforcement of competition law, mainly in the context of the grounds for justifying a non-competitive agreement under Article 5 para 2 CartA (Sect. 11.3.2.1). In such cases, any environmental protection arguments considered in the application of Article 5 para 2 CartA must be sufficiently closely related to the production process of the product that is the subject of the agreement.

The Comco also issued opinions on the compliance of several regional initiatives with the Internal Market Act[24] ('IMA'), rejecting the argument that these initiatives could be justified on grounds of public environmental interest (Sect. 11.3.2.2).

In contrast, in merger control, the Comco only analyses the purely competitive aspects affecting the merger. It does not take into account environmental policy aspects.[25] For example, in the merger control proceedings for the creation of the joint venture 'Swiss H2 Generation' by Groupe E and ENGIE, Comco mentioned that the creation of the joint venture was aimed at meeting the growing demand for hydrogen and related services for hydrogen-powered vehicles in the context of the global energy transition. However, the merger considerations were only about market shares and the criterion of creating or strengthening a dominant position, not about potential environmental benefits. Therefore, we do not address merger control in this section.

11.3.2.1 Cases Where Competition Law Has Not Been Infringed

In 1997, the Environmental Convention of the Swiss Business Association for Information, Communication and Organisation Technology ('SWICO') on the disposal of electronic waste allegedly constituted a (horizontal) price agreement within the meaning of Article 5 para 3 CartA. The agreement regulated the uniform return and disposal of discarded equipment, the uniform introduction of advance disposal fees and their amount (between competitors). Competition in the organisation of disposal was thus hindered, as the members of the SWICO Environmental Convention took on this task. Nevertheless, the Comco Secretariat carried out a preliminary investigation pursuant to Article 26 CartA, which was subsequently closed without opening an investigation in the final report.[26] In the same year, the Comco also issued an opinion indicating that, as a rule, there are more competitive solutions than the introduction of a uniform, industry-wide advanced disposal fee. Such a fee

---

[24] Federal Act on the Internal Market, RS 943.02. Available (in French) at https://www.fedlex. admin.ch/eli/cc/1996/1738_1738_1738/fr. Accessed 8 November 2022.

[25] Comco (2019) Press release '*Gateway Basel Nord*', p. 2. Available at https://www.newsd.admin. ch/newsd/message/attachments/57304.pdf. Accessed 8 November 2022.

[26] Decision of the Comco '*SWICO Recycling*', LPC 1997/2, 142 ff.

11 Sustainability and Competition Law in Switzerland

should only be introduced in exceptional cases, as it would restrict customers' freedom of choice in terms of disposal services. In the Comco's view, it is sufficient to designate the parties responsible for waste disposal and to leave the subsequent disposal to the market.[27]

In another decision concerning an advance recycling fee for the benefit of SWICO and Sens, the Comco further indicated that it believed the term resources also include public goods and natural resources, so that ecological parameters can also be taken into account when assessing the efficiency criterion. Environmental protection arguments can be considered in the application of Article 5 para 2 CartA, but they must be sufficiently closely related to the operation (production process) of the undertakings involved in the agreement or the product in question. However, in this case, other grounds for economic efficiency related to production costs were analysed by the Comco, which did not accept this argument.[28]

11.3.2.2 Cases Where the Comco Weighed the Environmental Benefits Against the Harm to Competition

In cases concerning a label for products from the Geneva region and a draft new law project giving priority to regional products for collective gastronomy in the Fribourg region, the Comco ruled on the conformity of these sustainability initiatives with the provisions of the IMA.[29]

Particularly, under Article 3 para 1 lit. a to c IMA, the cantons may restrict market access by means of conditions or requirements, provided that these are indispensable for the protection of overriding public interests and respect the principle of proportionality. Environmental protection may constitute a public interest that may justify a restriction of market access.

In both cases, the cantons of Geneva and Fribourg argued that local procurement and short supply routes can contribute to sustainability in the catering industry.

In the first case, the Comco stated that, in relation to the public interest in protecting agriculture and the environment, from the point of view of sustainability and ecology, the environmental footprint of the transport of the grain and flour was negligible compared to the production of the grain itself.[30]

In the second case, the Comco considered it conceivable that a non-regional product delivered has a better environmental footprint than a regional product. The canton was silent on the question of the effects of transport in Switzerland on the ecological footprint of products in relation to their production and processing. Therefore, the new law obliging public sector restaurants to systematically give

---

[27] Activity report of the Comco, LPC 1997/4, p. 422.

[28] Decision of the Comco 'SWICO/Sens', LPC 2005/2, 266, para 92.

[29] The Act guarantees free and non-discriminatory access to the market for all persons having their registered office or place of business in Switzerland, so that they can carry out a lucrative activity on the entire Swiss territory.

[30] Statement of the Comco of 26 June 2017 in the proceedings 2C_261/2017, LPC 2018/1, pp. 171 ff., fn. 17.

preference to regional products in their purchases, thus restricting market access for non-regional suppliers, could not be justified on environmental grounds.[31]

It is interesting to note that before the revision of the CartA in 1995, such an argument was accepted to justify hard-core cartels: the old law allowed even very drastic restrictions of competition to be justified from the point of view of economic and social policy by using the balance sheet method.[32] For example, territorial agreements between competitors lead to a reduction in road transport and can therefore be assessed positively in terms of environmental protection. Based on these considerations, a former Swiss cement cartel was approved under the old CartA: these benefits outweighed the harmful effects on competition.[33]

The cement cartel had set up a combined rail/road cement transport system which, as the former Cartel Commission (now Comco) recognised, had beneficial effects from an environmental point of view. Thus, the Cartel Commission opined that agreements which artificially increased the cost of road transport to partly finance rail transport were justified on the basis of overriding public interest grounds within the meaning of Article 29 para 3 CartA (in its version of 20 December 1985).

However, the current CartA now pursues different objectives and these considerations are no longer considered apart from the use of resources under Article 5 para 2 CartA as explained above. Thus, the described scheme could only be justified under Article 5 para 2 CartA if it reduced transport costs (which was not

---

[31] Recommendation of the Comco of 4 April 2016 to the Canton of Fribourg concerning the bill on collective public catering, LPC 2016/2, 578, para 20.

[32] Article 29 CartA (in its version of 20 December 1985) reads as follows (emphasis added):

1. The Commission shall examine, at the request of the Federal Department of Economic Affairs or on its own initiative, whether a cartel or similar organisation has harmful economic or social consequences.
2. In examining whether there are harmful economic or social consequences, the Commission shall weigh up the useful and harmful effects. If it finds that there are significant restrictions or distortions of competition, it shall weigh up the useful and harmful effects. In doing so, it shall take into account the effects on freedom of competition and on the extent of competition. In addition, it considers all other relevant effects such as those on the quality of supply, on the structure of the economic sector, on the regional economy, on the ability of Swiss companies to compete at home and abroad, and on the interests of the employees and consumers concerned.
3. The economic and social consequences are detrimental if, as a result of this examination, the detrimental effects prevail. Harmfulness shall, however, be established in all cases where effective competition in the market for certain goods or services is prevented, unless the examination shows that there are overriding reasons of public interest which make such prevention indispensable.

---

[33] Comco (1993), Cement Market Investigation, PCCPr 1993/5 pp. 27–184; French summary, pp. 13–23; M. Baldi/F. Schraner, Gaba-Urteil des Bundesverwaltungsgerichts als wettbewerbspolitischer Markstein, SJZ 2014, p. 502.

the case here) and if this advantage outweighed the disadvantages from a competition point of view.[34]

11.3.3 Illegal Anti-Competitive Behaviour in the Context of Sustainability Initiatives

Illegal anti-competitive behaviour can occur in the context of sustainability initiatives, for example in the case of misleading advertising (Sect. 11.3.3.1) or in the form of an agreement affecting competition (Sect. 11.3.3.2).

Under Swiss law, green washing in advertising can be invoked as a misleading commercial practice, over which the Comco has no jurisdiction. Cases of green washing have been brought before other authorities in connection with violations of the Unfair Competition Act ('UCA').[35]

There was one case where the Comco issued an opinion on a potentially illegal anti-competitive agreement within the meaning of Article 5 CartA, developed in the context of a sustainability initiative, namely the implementation of the CO2 Act.[36]

11.3.3.1 Green Washing

The Comco has no jurisdiction in the area of unfair competition and deceptive marketing practices. Therefore, the Comco has not dealt with cases where companies give the (false) impression of pursuing a sustainability initiative when in fact it serves as a cover for anti-competitive behaviour (green washing) and a violation of the CartA.

However, the Swiss Commission for Fairness ('SCF'), a neutral and independent institution aiming to guarantee the self-control of advertising in the communication sector, monitors compliance with the rules of fairness in commercial communication (all forms of advertising, marketing, sponsoring, sales promotion and public relations). The SCF is the executive body of the Foundation for Fairness in Commercial Communication, which brings together the most important organisations in the Swiss advertising industry.

Anyone has the right to complain to the SCF about a commercial communication that he or she considers unfair. In assessing a complaint, the SCF bases itself on the UCA and on the Consolidated Code of Advertising and Commercial Communication Practice of the International Chamber of Commerce. Particularly, under Article 3 para 1 lit. b UCA, advertising statements must not be false or misleading.

---

[34] In other words, it must be established whether the loss of allocative efficiency is outweighed by the beneficial effect of the agreement on innovative efficiency.

[35] Federal Unfair Competition Act, RS 241. Available (in French) at https://www.fedlex.admin.ch/eli/cc/1988/223_223_223/fr. Accessed 8 November 2022.

[36] Federal Act on the Reduction of CO2 Emissions ("CO2 Act"), RS 641.71. Available at https://www.fedlex.admin.ch/eli/cc/2012/855/en. Accessed 8 November 2022.

The SCF is often confronted to complaints about green marketing. Although SCF decisions are not legally binding, they are accepted by advertisers in most cases.

In recent decisions, the SCF decided that various statements made by a company in its advertising campaign for the sustainability of Swiss dairy production (poster 'xxxxxxxx and biodiversity: a true love story') did not intentionally mislead consumers and were therefore not unfair.[37] On the other hand, advertising statements in a commercial and on the company's website that gave the false impression that feeding exclusively with feed from the farm is characteristic of Swiss meat and that farms that fatten on pasture without purchased fodder are the exception, were partially unclear and could therefore be misleading if not corrected and clarified.[38]

There are therefore cases of green washing, which consists of misleading consumers by giving them the false impression of pursuing a sustainable approach. Such behaviour in Switzerland is, however, apprehended by the UCA and not by the CartA, thus not by the Comco.

11.3.3.2 Genuine Sustainability Initiatives that Served as a Springboard for Other Anti-Competitive Behaviour

In 2004, the Comco issued an opinion[39] to the Swiss Agency for the Environment, Forests and Landscape (formerly the Swiss Federal Office for the Environment) on the climate cent initiative. This initiative was developed by the Climate Cent Foundation, a voluntary system set up by the Swiss business community with the aim of ensuring effective climate protection and investing its funds in greenhouse gas reduction projects abroad. The Climate Cent Foundation was financed by a tax of 1.5 cents per litre levied on gasoline and diesel imports in the years 2006 to 2012. The emission reduction certificates generated by these projects were handed over to the Swiss Confederation free of charge.

The Comco considered that the planned levy of about one cent per litre of gasoline or diesel undoubtedly represented a conscious and deliberate cooperation of the companies concerned and thus an agreement between oil importers. Interestingly, however, Comco did not qualify the agreement as a direct or indirect price-fixing agreement pursuant to Article 5 para 3 CartA, since the companies concerned did not agree on the final price but only on a very small cost element, too small to have a price harmonising effect. The Comco therefore assumed that an indirect price agreement was not intended by the parties. Furthermore, the primary objective of the agreement was to reduce CO_2 emissions to meet the requirements of the CO2 Act: if the reduction target had not been achieved by voluntary measures of the industry alone, the Confederation would have been entitled to levy an incentive tax on fossil fuels. The importers could therefore also be considered as fulfilling the mandate of the law.[40]

---

[37] SFC Decisions No. 106/22 of 4 May 2022 and No. 220/21 of 19 January 2022.

[38] SFC Decision No. 234/21 of 13 March 2022.

[39] Opinion of the Comco '*Klimarappen*', LPC 2005/1, 239 ff.

[40] Opinion of the Comco '*Klimarappen*', LPC 2005/1, 240 ff., paras 11 ff.

Nevertheless, the Comco analysed the qualitative and quantitative factors of the agreement to determine whether it could significantly affect competition. First, the climate cent concerns a cost element, which represents 0.7% to 1.5% of the final price of gasoline. Secondly, if one deducts the share of taxes and duties, the share of the climate cent amounts to 2% to 4%. According to the Comco, such a cost element is not negligible and must be considered significant. Furthermore, for the successful implementation of the climate cent project, as many oil importers as possible should participate in the competition agreement, which implies an increase in market shares. Overall, the Comco found that the climate cent project significantly affected competition.[41]

On possible grounds for justification based on Article 5 para 2 CartA and the objective of more rational use of resources, since the revenues generated by the agreement would be used for national measures to reduce CO2 emissions and the purchase of foreign certificates, and thus environmental benefits, the Comco analysed whether there was a sufficient link between the agreement and the production process (gasoline and diesel, not fuel oil).[42] To be justified under competition law, it had to be ensured that the climate cent would lead to the internalisation of the environmental costs of gasoline and diesel, which was not the case. It was only assumed that the uniform climate cent charge for gasoline and diesel would lead to a reduction in environmental pollution and thus in the environmental costs for petroleum products (not only for gasoline and diesel). This could nevertheless be achieved either by a corresponding reduction in the costs of consumption in Switzerland or, according to the objective of the CO2 Act, by a corresponding reduction of global consumption.

Due to the current state of the project when it was submitted to the Comco, the latter found it difficult to assess to what extent the reduction target of the CO2 Act would be achieved. Therefore, the Comco considered that the agreement could not be justified on grounds of economic efficiency in accordance with Article 5 para 2 CartA.

11.4 The Activity of Agencies, Legislator and Specific Commissions in the Field of Sustainability

The Comco Secretariat has set up a 'Core Group Sustainability', which monitors developments in Switzerland and abroad.[43] Comco has not undertaken any other specific initiatives in the field of sustainability.

However, we would like to mention the notable developments in the field of public procurement (Sect. 11.4.1) and sustainable finance (Sect. 11.4.2).

---

[41] Opinion of the Comco 'Klimarappen', LPC 2005/1, 241, paras 18 ff.

[42] Opinion of the Comco 'Klimarappen', LPC 2005/1, 241 ff., paras 25 ff.

[43] A. Heinemann, Nachhaltigkeitsvereinbarungen, sic! 5 | 2021, pp. 213 ff., p. 226.

11.4.1 Public Procurement

The new and completely revised Public Procurement Act[44] came into force on 1 January 2021. The new law marks a paradigm shift towards sustainable public procurement, on the one hand, by introducing the three dimensions of sustainable development among the objectives of the law and, on the other hand, by creating a place for the integration of a sustainability-related criteria at the different stages of the procurement process.

The adoption of the new law has led to numerous works and initiatives: (i) On 29 November 2019 the Federal Procurement Conference adopted the 'Guiding Principles for Sustainable Public Procurement (of Goods and Services)';[45] (ii) the Federal Office for the Environment offers a Guide to Green Procurement;[46] (iii) the report of the expert group 'Corporate Social Responsibility: Der Bund als Beschaffer', published in 2019, provides information on the implementation status of these measures;[47] (iv) the Federal Council's report 'Sanctions at the place of performance: Ensuring compliance with minimum social requirements in public procurement' details how sustainability (including social sustainability) is and must be taken into account in public procurement;[48] (v) gradually, sustainability is

---

[44] Federal Act on Public Procurement ('Public Procurement Act'), RS 172.056.1. Available at https://www.fedlex.admin.ch/eli/cc/2020/126/en. Accessed 8 November 2022.

[45] Federal Procurement Conference (2019), Guiding Principles for Sustainable Public Procurement (of Goods and Services). Available (in French) at https://www.bkb.admin.ch/dam/bkb/fr/dokumente/Oeffentliches_Beschaffungswesen/BKB_Leitsaetze_fr.pdf.download.pdf/BKB_Leitsaetze_fr.pdf. Accessed 8 November 2022.

[46] Swiss Confederation Federal Office for the Environment (2021) Guide to Green Public Procurement. Available (in French, German and Italian) at https://www.bafu.admin.ch/bafu/fr/home/themes/economie-consommation/info-specialistes/marches-publics-ecologiques/recommandations-pour-les-achats-publics-ecologiques.html. Accessed 8 November 2022. See also U. Bolz/M. Mettler (2019) Nachhaltiges Beschaffungswesen: Beschaffung von Innovationen – innovative Beschaffung – Grundlagen – ein Diskussionsbeitrag im Auftrag des Bundesamts für Umwelt (BAFU). Available at http://www.nachhaltige-beschaffung.ch/pdf/Innovative_Beschaffungen_Grundlagen_20190523.pdf. Accessed 10 November 2022.

[47] U. Bolz/P. Lüthi/B. Eicher/P. Müller (2018) Corporate Social Responsibility (CSR): Der Bund als Beschaffer. Available at https://www.are.admin.ch/dam/are/fr/dokumente/nachhaltige_entwicklung/publikationen/corporate-social-responsibility-csr-der-bund-als-beschaffer.pdf.download.pdf/corporate-social-responsibility-csr-der-bund-als-beschaffer.pdf. Accessed 8 November 2022. The expert report in German (with a summary in French and Italian) offers an analysis of the Confederation's performance in the area of corporate social responsibility (CSR) in its role as a purchaser.

[48] Swiss Confederation Federal Council (2022) Sanctions au lieu d'exécution des travaux : garantie du respect des exigences sociales minimales dans les marchés publics. Available at https://www.efd.admin.ch/dam/efd/fr/das-efd/gesetzgebung/berichte/bericht-bbl-bundesratsitzung-17082022.pdf.download.pdf/beilage-bericht-fr.pdf. Accessed 10 November 2022.

11 Sustainability and Competition Law in Switzerland

influencing all procurement, including the choice of automobiles for federal employees.[49]

11.4.2 Sustainable Finance

In June 2020, the Swiss government published a report on sustainable finance in Switzerland,[50] taking a position on a series of ongoing European initiatives.

In December 2020, the Swiss government then outlined the next steps in its strategy to make the Swiss financial centre more sustainable by adopting a package of measures.[51]

Since then, the Federal Council has launched the consultation on an ordinance on climate reporting by large companies.[52]

In May 2021, the Swiss prudential regulator — the Financial Market Supervisory Authority ('FINMA') — introduced reporting requirements for supervised financial institutions in accordance with the Task Force on Climate-related Financial Disclosures ('TCFD') by amending its circulars 'Disclosure – Banks'[53] and 'Disclosure – Insurers',[54] which came into force on 1 July 2021. Initially, only large banks and insurance companies will be subject to the transparency requirements.

The Federal Council intends to make Switzerland a world leader in sustainable investment through climate transparency. To this end, the Federal Council adopted various measures at its meeting on 17 November 2021. Particularly, it recommends that financial market participants create transparency in all financial products ad

---

[49] See in particular Art. 23 para 3 of the Federal Ordinance on Federal Motor Vehicles and their Drivers, RS 514.31. Available (in French) at https://www.fedlex.admin.ch/eli/cc/2005/163/fr. Accessed 8 November 2022.

[50] Swiss Federation Federal Council (2020) Sustainability in Switzerland's financial sector: Situation analysis and positioning with a focus on environmental aspects. Available at https://www.sif. admin.ch/dam/sif/en/dokumente/dossier/int_finanz-waehrungsfragen/int_ waehrungszusammenarbeit/bericht_sustainable_finance.pdf.download.pdf/24062020- Nachhaltigkeit%20Bericht%20Executive%20Summary-EN.pdf. Accessed 8 November 2022.

[51] Press release available at https://www.sif.admin.ch/sif/en/home/documentation/press-releases/ medienmitteilungen.msg-id-81571.html. Accessed 8 November 2022.

[52] Press release and related documents available at https://www.sif.admin.ch/sif/en/home/ documentation/press-releases/medienmitteilungen.msg-id-87790.html. Accessed 8 November 2022.

[53] FINMA (2015) Circulaire 2016/01 Publication – Banques : Exigences prudentielles de publication. Available at https://www.finma.ch/~/media/finma/dokumente/dokumentencenter/myfinma/ rundschreiben/finma-rs-2016-01.pdf?sc_lang=fr&hash=CCB6D96A5A8340CE86E63822DBE3 8086. Accessed 8 November 2022.

[54] FINMA (2015) Circulaire 2016/02 Publication – Assureurs (public disclosure) : Bases du rapport sur la situation financière. Available at https://www.finma.ch/~/media/finma/dokumente/ dokumentencenter/myfinma/rundschreiben/finma-rs-2016-02-20210506.pdf?sc_lang=fr&hash= 5B43C223D06BB05B4703BA18601C2701. Accessed 8 November 2022.

client portfolios by means of comparable and meaningful climate compatibility indicators.

The Federal Council also invites the financial sector to join international 'net zero' alliances and intends to promote the conclusion of sectoral agreements to this end. In addition, the Federal Council has instructed the Federal Department of Finance ('FDF'), in cooperation with the Federal Department of the Environment, Transport, Energy and Communications ('DETEC'), to report back to the Federal Council by the end of 2022 on the implementation of the above-mentioned recommendations by the financial sector and, if necessary, to submit proposals for measures.

Finally, the Federal Council has instructed the FDF, in cooperation with DETEC and FINMA, to submit proposals by the end of 2022 on how financial market law could be adapted, particularly with regard to transparency, to prevent greenwashing.

11.5 Conclusion

Although sustainability does not play a significant role in Swiss competition law, whose objectives are primarily economic in nature, it is worth noting that there is scope for justifying agreements that significantly affect competition on environmental grounds by explicitly mentioning a more rational use of resources.

In terms of interactions and specific cases, Comco has dealt with very few cases impacted by sustainability considerations. These few cases nevertheless show that it is necessary to carefully examine in each case whether there is a competition agreement and whether environmental aspects need to be considered. Particularly, when sustainability is used as a defensive argument, it is necessary to check in each case whether the agreement leads to the expected improvements. Furthermore, there must be no milder means to achieve the environmental objectives set and the restriction of competition and the environmental benefits must be proportional to each other.[55]

Finally, Swiss authorities have developed and implemented several initiatives in related areas, particularly public procurement law and sustainable finance. However, the authors are not aware of any initiatives or activities planned by the Comco or its Secretariat to promote sustainability in Swiss competition law.

---

[55] A. Heinemann, Nachhaltigkeitsvereinbarungen, sic! 5 I 2021, pp. 213 ff., p. 224.

Open Access This chapter is licensed under the terms of the Creative Commons Attribution 4.0 International License (http://creativecommons.org/licenses/by/4.0/), which permits use, sharing, adaptation, distribution and reproduction in any medium or format, as long as you give appropriate credit to the original author(s) and the source, provide a link to the Creative Commons license and indicate if changes were made.

The images or other third party material in this chapter are included in the chapter's Creative Commons license, unless indicated otherwise in a credit line to the material. If material is not included in the chapter's Creative Commons license and your intended use is not permitted by statutory regulation or exceeds the permitted use, you will need to obtain permission directly from the copyright holder.

Sustainability and Competition Law in the United Kingdom

12

Simon Holmes, Nicole Kar, and Lucinda Cunningham

12.1 Introduction

The interaction of competition law and sustainability has been a live question for the UK government and the UK's Competition and Markets Authority ('CMA') in recent times. With the hosting of COP26 in Glasgow in 2021, the fight against climate change concerns has benefitted from renewed political will.

Momentum across a wide range of environmental social and governance ('ESG') issues has continued over the last year. The pandemic spurred environmental and social concern prompting, for example, the UK government's strategy in 2021 to 'build back greener'. This strategy reflected a move more broadly across Europe. These initiatives demonstrate the continued—and growing—sentiment that ESG

Simon Holmes is a visiting Professor in law at the University of Oxford, a member of the UK's Competition Appeal Tribunal and has published widely on sustainability and competition policy (for completeness see the bibliography at Appendix 1 below). Nicole Kar is a Partner and Global Head of Antitrust and Foreign Investment at Linklaters LLP. She acknowledges the kind assistance of William Langridge and Aditi Aggarwal of Linklaters LLP in relation to this contribution. Lucinda Cunningham is a Barrister at Matrix Chambers and was formerly a Référendaire at the UK's Competition and Appeal Tribunal. The opinions and views expressed in this contribution are those of the authors.

S. Holmes · N. Kar (✉) · L. Cunningham
University of Oxford and Competition Appeal Tribunal, Oxford, UK

Linklaters, London, UK

Matrix Chambers, London, UK
e-mail: nicole.kar@linklaters.com; lucindacunningham@matrixlaw.co.uk

© The Author(s) 2024
P. Këllezi et al. (eds.), *Sustainability Objectives in Competition and Intellectual Property Law*, LIDC Contributions on Antitrust Law, Intellectual Property and Unfair Competition, https://doi.org/10.1007/978-3-031-44869-0_12

concerns are no longer an afterthought, but rather seen as a pre-condition (in Europe at least) to economic growth.

Increasing sustainability concerns and public policies are also placing growing pressure on businesses and consumers to make investment, innovation and purchasing decisions that promote, or at least are not harmful to, the environment. While competition policy is far from the primary driver of these social and economic changes, competition policy should, nevertheless, recognise this social and political shift. This includes, e.g. consideration of environmental benefits as potentially a key aspect of competition between firms, as well as a source of benefits to consumers, and avoiding chilling business interest in pro-environmental initiatives. A respect and concern for sustainability measures must, however, distinguish between genuine environmental benefit and 'greenwashing' that acts as a fig leaf for collusive or anticompetitive behaviour.

By way of context, the UK competition regime is strongly rooted in the European approach, but has a number of distinctive quirks. Reflecting the UK's former EU membership, much of the UK competition regime currently mirrors the EU competition law rules contained in Article 101/102 TFEU, and the underlying principles. While in the EU, the UK was obliged to follow EU case law, generating much relevant precedent at EU level. However, even during its EU membership, UK merger control had a notably independent streak and the scope and use of its market's regime had no analogue in the EU. Following the UK's departure from the EU and the end of the transition period on 31 December 2020, EU competition law no longer applies in the UK. UK competition authorities and courts must still ensure there is no inconsistency with principles laid down by the EC and the European courts before 31 December 2020,[1] but—significantly—UK regulators and courts now have the freedom to diverge.

The CMA is increasingly active in exploring sustainability issues both in its publications and in its wider ambitions for global thought leadership. Citing sustainability as a 'strategic priority' in its recent annual reports,[2] a key theme for the CMA is supporting the transition to low carbon growth, including through developing healthy and competitive markets in sustainable products and services. The CMA further notes it will continue to prioritise cases that could impede the UK's Net Zero goals for a low-carbon economy.[3]

The CMA has consulted on and issued guidance on how it considers sustainability-related issues across the antitrust, mergers and markets regimes. Particularly, it: (i) issued an information sheet on sustainability agreements in January 2021; (ii) conducted a market study into electric vehicle ('EV') charging

---

[1] Section 60A of the Competition Act 1998.

[2] Competition and Markets Authority Annual Report and Accounts 2022 to 2023, published on 24 March 2022, available at https://assets.publishing.service.gov.uk/government/uploads/system/uploads/attachment_data/file/1062414/Final_Annual_Plan_for_2022_23.pdf.   Accessed 19 November 2022.

[3] Ibid., para 2.51.

points and, subsequently, undertook an antitrust investigation and accepted commitments in relation to long-term exclusivity agreements for EV charge points at motorway service areas; (iii) provided advice to the UK Government on how the current competition and consumer law frameworks facilitate or hinder sustainability and Net Zero objectives; (iv) established a Sustainability Task Force to spearhead further engagement with these issues; and (v) under its consumer powers, has published a 'Green Claims Code' to help businesses and consumers avoid 'greenwashing' claims.

The CMA has generally focused upon environmental sustainability (and, in its advice to government, it specifically considered the linkage with the UK's commitment to Net Zero by 2050). The CMA has not indicated interest in a wider formulation of sustainability which would incorporate, for example, social and ethical issues (in line with broader corporate ESG priorities and the UN Sustainable Development Goals ('SDGs')), as has been the case for example in Germany and more broadly, the EU.[4]

However, to date, and despite the extensive discussion around consideration of sustainability issues, sustainability considerations have so far only played a limited role in actual enforcement. The CMA has (except for its market study into the EV charging market) yet to apply sustainability considerations in competition law (either antitrust enforcement or merger control) as either a sword or accepted its use as a shield (likewise, nor have sustainability considerations been significant in private competition proceedings).

This contribution is split into three parts which consider the interaction of competition law and sustainability by assessing the role and impact of the UK's relevant legal instruments. First, it considers the impact of the Chapter I prohibition in the Competition Act 1998 ('CA98') (i.e. the prohibition on anti-competitive agreements) on cooperation between competitors. Second, how the CA98 Chapter II prohibition (i.e. the prohibition on abuse of a dominant position) can be used to help advance sustainability aims. Third, how the merger control and markets regimes can and should be used to support sustainability goals in the UK.

---

[4] The German Federal Cartel Office assessed three ESG initiatives in early 2022: (i) Living Wages in the Banana Sector; (ii) Animal Welfare Initiative or Initiative Tierwohl; and (iii) in the milk sector. See also, EU Commission, *Draft Guidelines on the applicability of Article 101 of the Treaty on the Functioning of the European Union to horizontal co-operation agreements* (March 2022) (the '*Draft Horizontal Guidelines*') at para 543 which defines sustainable development in terms of '*environmental and social (including labour and human rights) development*', available at: https://ec.europa.eu/competition-policy/public-consultations/2022-hbers_en. Accessed 19 November 2022.

12.2 Sustainability Agreements and the Chapter I Prohibition

This section will focus on the role that the Chapter I prohibition (and its EU equivalent, Article 101 TFEU), contained in Section 2 of the CA98, plays, (and more importantly) could play, in supporting action towards a more sustainable future. In short, it will be argued that the Chapter I prohibition has the capacity to be applied in a way which furthers sustainability goals. While it is clear competition law does not prohibit sustainability agreements outright, the ambiguity caused by a lack of clear guidance as to the application of the provisions to such agreements and fear of enforcement (and significant penalties) prevents businesses from taking bolder action in pursuit of sustainability goals. Under the existing legal framework in the UK, this contribution will explore the various ways in which cooperation agreements directed at achieving sustainable goals may be 'shielded' from the Chapter I prohibition and how the shield may be strengthened.

This section will refer primarily to the position and the 'state of play' in the UK under Chapter I CA98. Nonetheless, given the similarity between the UK and EU regimes on cooperation agreements, reference will be made to EU law and practice.

There are currently no reported cases in the UK, nor examples of antitrust enforcement by the CMA, which involve a consideration of the application of the Chapter I prohibition (or the exemption criteria) to sustainability agreements. Accordingly, the impact of the existing framework to such agreements will be considered, highlighting any areas where a degree of flexibility and clarity of approach would be welcomed.

12.2.1 What Is a 'Sustainability Agreement'?

Before diving into the detail of the Chapter I prohibition, it is worth briefly addressing what a sustainability agreement is and the importance of cooperation—to put the role competition law has to play into context. Simply put, a sustainability agreement is any agreement between competitors which pursues one or more sustainability objectives.[5] The term 'sustainability agreement' was recently defined by the EU Commission in its Draft Horizontal Guidelines as: 'any type of horizontal

---

[5] 'Sustainability' or 'sustainable development' is a broad concept which lacks a universally agreed definition. The Brundtland Commission in 1987 defined sustainable development as: 'development that meets the need of the present without compromising the ability of future generations to meet their own needs'. Others adopt the definition used by the UN in Resolution, 'The future we want' (adopted by the General Assembly on 27 July 2021) 66/288 at para 1: 'an economically, socially and environmentally sustainable future for our planet and for the future present and future generations', available at: https://www.un.org/en/development/desa/population/migration/generalassembly/docs/globalcompact/A_RES_66_288.pdf. Accessed 19 November 2022.

12 Sustainability and Competition Law in the United Kingdom

cooperation agreement that genuinely pursues one or more sustainability objectives, irrespective of the form of cooperation'.[6]

The CMA provided the following example of such agreements in its 'information sheet' to businesses: 'For example, businesses may decide to combine expertise to make their products more energy efficient or agree to use packaging material that meets certain standards in order to facilitate package recycling and reduce waste'.[7]

There are various types of cooperation agreements that can contribute to sustainability goals (the following list is non-exhaustive):

(i) Standard setting agreements or codes of conduct promoting practices—'by which businesses, often through trade associations or standardisation organisations, set standards on the environmental performance of products, production processes, or the resources used in production'[8]—including joint standards and certification labels (i.e. about the use of raw materials or production methods);

(ii) Binding commitments and agreements that incentivise participants to contribute to sustainability objectives;

(iii) Joint development of new products, technologies or markets, where pooling of resources or expertise is required to achieve sufficient scale—for example, the development of innovative carbon capture technologies;

(iv) Agreements to phase out, improve or replace unsustainable products with more sustainable ones (i.e. reducing or phasing out packing materials or technologies where greener alternatives are available);

(v) Agreements to combine resources—for example, coordinating stock, warehouse and transportation, or agreeing to make joint purchasing decisions;

(vi) Agreements to reduce environmental damage; and[9]

(vii) Agreements by which participants, their suppliers and/or their distributors, agree to adhere to, or respect, particular sustainability laws including labour laws (e.g. on child labour and minimum wages), environmental rules (i.e. banning illegal logging practices), and fair-trade rules.

---

[6] EU Commission, Draft Guidelines on the applicability of Article 101 of the Treaty on the Functioning of the European Union to horizontal co-operation agreements, March 2022, para 547.

[7] CMA, Guidance 'Environmental sustainability agreements and competition law' (27 January 2021), available at, https://www.gov.uk/government/publications/environmental-sustainability-agreements-and-competition-law/sustainability-agreements-and-competition-law. Accessed 19 November 2022.

[8] The CMA, in its information sheet, dedicated the majority of its attention to 'standard-setting agreements'. However, it is important to note that standard setting agreements, although important, are only one type of sustainability agreement. See CMA, Guidance 'Environmental sustainability agreements and competition law' (27 January 2021), available at, https://www.gov.uk/government/publications/environmental-sustainability-agreements-and-competition-law/sustainability-agreements-and-competition-law. Accessed 19 November 2022.

[9] For example, see ACM, 'Guidelines on Sustainability agreements – Opportunities within competition law' (26 January 2021), at para 8.

12.2.2 Why Is Cooperation Between Businesses Important?

Cooperation is key to tackling climate change and supporting the transition towards a more sustainable future. Although action by individual companies can take the agenda a long way, the problem of climate change will not be solved without strong and collaborative partnerships between businesses. Cooperation between businesses has the potential, not only to achieve a much greater impact (through economies of scale), but to decrease risk and drive innovation—particularly in relation to new and enhanced technologies. Through collaboration, businesses are able to: '[...] tap into expertise, skills, time, or other assets that they may not have access to internally... multisector collaborations enable each contributor to bring their strengths to the table and tackle a challenge that would not be possible to solve individually'.[10]

In addition, the UN SDGs specifically call for cooperation and highlight the need for businesses to work together to solve common challenges.[11]

12.2.3 Recent Developments in the UK

The CMA made 'supporting the transition to a low carbon economy' one of its strategic objectives in both its annual plan for 2021–22 and 2022–23.[12] It is likely that this will remain a key objective in the forthcoming years. Thus far, the CMA has sought to address this objective principally through two different lenses, namely a competition lens and a consumer protection lens. For our purposes, we focus on the competition lens however, it is worth noting that that the CMA is also doing work in the consumer space (for example, on 20 September 2021, the CMA published a 'Green Claims Code' aimed at protecting consumers from misleading environmental claims amidst concerns over 'greenwashing'[13] and there is further work in the pipeline).

The CMA first published guidance in relation to sustainability and competition law on 27 January 2021 in the form of an 'information sheet' aimed at businesses and trade associations setting out some key points to be considered when deciding to

---

[10]The Sustainability Institute '*Leveraging the Power of Collaborations*', (December 2020), p. 22, accessible at: https://www.sustainability.com/globalassets/sustainability.com/thinking/pdfs/report-leveraging-the-power-of-collaborations.pdf. Accessed 19 November 2022.

[11]SDG 17 entitled 'Advancing Partnerships' seeks to meet the goals of the UN 2030 Agenda through partnerships, whether between national governments, the international community, civil society, the private sector, or any other actor.

[12]CMA, Annual plan for 2021 to 2022 of 23 March 2021, accessible at: https://www.gov.uk/government/publications/competition-and-markets-authority-annual-plan-2021-to-2022. Accessed 19 November 2022.

[13]CMA, Guidance on environmental claims on goods and services (20 September 2021), accessible at: https://www.gov.uk/government/publications/green-claims-code-making-environmental-claims. Accessed 19 November 2022.

enter into sustainability agreements.[14] The guidance is very high-level and does not go beyond stating general principles.

Further, on 19 July 2021, Kwasi Kwarteng, the Secretary of State for Business, Energy and Industrial Strategy ('BEIS'), wrote to the CMA asking it to provide advice to the UK Government on how competition and consumer law frameworks could be enhanced to better support Net Zero and sustainability goals, including preparing for climate change.[15]

Following a public consultation, on 14 March 2022, the CMA published its advice to the UK Government, on how competition and consumer laws can better support the UK's Net Zero and sustainability goals (the 'Advice to Government').[16] In its Advice to Government, the CMA concludes that the current competition laws in the UK are not themselves an obstacle to sustainability initiatives, and therefore do not require fundamental change or reform. The CMA considers that many initiatives aimed at achieving sustainability goals can take place under the existing competition regime. The CMA does however acknowledge that stakeholders emphasised the need for more clarity around how competition law will be applied in an environmental sustainability context, including in relation to the exemption criteria.

The CMA has established a 'Sustainability Taskforce' dedicated to sustainability issues to lead the development of additional guidance (in relation to Chapter II of the CA98) and to further the debate with government and industry stakeholders.

12.2.4 Sustainability as a 'Shield' and the Chapter I Prohibition

There are various ways in which sustainability agreements might 'escape' the prohibition on anti-competitive agreements[17] and in this sense sustainability considerations act like a 'shield'. The following four routes, have been the subject of extensive debate in this context:[18]

---

[14] CMA, Guidance 'Environmental sustainability agreements and competition law' (27 January 2021), available at, https://www.gov.uk/government/publications/environmental-sustainability-agreements-and-competition-law/sustainability-agreements-and-competition-law.   Accessed 19 November 2022.

[15] Letter from Kwasi Kwarteng, Secretary of State for BEIS, to Andrea Coscelli for the CMA on sustainability (2021), accessible at: https://assets.publishing.service.gov.uk/government/uploads/system/uploads/attachment_data/file/1004016/sos-letter-to-andrea-coscelli-on-sustainability.pdf. Accessed 19 November 2022.

[16] CMA, Environmental sustainability and the UK competition and consumer regimes: CMA advice to the Government (14 March 2022), available at https://www.gov.uk/government/publications/environmental-sustainability-and-the-uk-competition-and-consumer-regimes-cma-advice-to-the-government/environmental-sustainability-and-the-uk-competition-and-consumer-regimes-cma-advice-to-the-government. Accessed 19 November 2022.

[17] S. Holmes, Climate change, sustainability and competition law, Journal of Antitrust Enforcement 2020, Vol. 8, Issue 2, pp. 368.

[18] Particularly, see the series of essays in Concurrences, 'Competition Law, Climate Change & Environmental Sustainability' (March 2021), edited by S. Holmes, D. Middelschulte and M. Snoep.

First, not all agreements between competitors which pursue sustainability goals are caught by the Chapter I Prohibition, i.e. where they do not affect key parameters of competition (such as price, choice, quality, innovation etc) or are entered into by competitors with low market share (i.e. the anticipated effect on competition would be minimal).

Second, by satisfying the exemption criteria in Section 9 of the CA98 (and its EU equivalent Article 101(3) TFEU).

Third, on the basis of the reasoning in Case C-67/96 *Albany International BV v Stichting Bedrijfspensioenfonds Textielindustrie* [1999] ECR I-5751 (commonly known as the 'Albany route').[19]

Fourth, by applying the 'ancillary restraints' or 'objective necessity' doctrines to sustainability agreements.

This contribution focuses on the first two 'routes'. Although the *Albany, Wouters and Meca-Medina* line of jurisprudence (routes three and four) potentially provide an escape for sustainability agreements, arguably the preferred approach from a participating company's point of view is through the application of the exemption criteria.[20]

12.2.4.1 Agreements Not Caught by the Chapter I Prohibition

As to the first 'route', the CMA recognises that 'many forms of collaboration between businesses for the achievement of sustainability goals are unlikely to raise any competition law issues' but does not elaborate.[21] The CMA intends to provide further guidance on 'when sustainability agreements will not restrict competition',[22] which is welcomed. The fact that an agreement genuinely pursues a sustainability objective should be considered in determining whether the restriction in question is a restriction by object or a restriction by effect within the meaning of the Chapter I prohibition—as has been suggested in the EU Commission's Draft Horizontal Guidelines. Accordingly, it can be considered that the following types of agreement

---

[19] S. Holmes, Climate change, sustainability and competition law Journal of Antitrust Enforcement 2020, Vol 8, Issue 2, pp. 370.

[20] That said, there are strong parallels to be drawn between sustainability and the objectives protected in these cases (i.e. in *Meca-Medina*, rules to safeguard 'equal chances, [. . .] the integrity and objectivity of competitive sport and ethical values in sport'). The concept of 'ethical values in sport' could be said to be analogous to the values inherent in sustainability objectives. We therefore encourage the CMA to at least indicate that it will consider if genuine sustainability considerations exclude the application of Chapter I on a case-by-case basis.

[21] CMA, Guidance 'Environmental sustainability agreements and competition law' (27 January 2021), available at, https://www.gov.uk/government/publications/environmental-sustainability-agreements-and-competition-law/sustainability-agreements-and-competition-law. Accessed 19 November 2022.

[22] CMA, Environmental sustainability and the UK competition and consumer regimes: CMA advice to the Government (14 March 2022), available at https://www.gov.uk/government/publications/environmental-sustainability-and-the-uk-competition-and-consumer-regimes-cma-advice-to-the-government/environmental-sustainability-and-the-uk-competition-and-consumer-regimes-cma-advice-to-the-government. Accessed 19 November 2022.

are, in principle, likely to be 'shielded' (or fall outside the Chapter I prohibition) by reason of a lack of appreciable effect on competition: (i) agreements imposing loose environmental commitments with no specific commitments or obligations contained therein, (ii) commitments as to general outcomes or objectives absent any precise/specific mechanics for achieving those, and (iii) commitments as to specific actions on matters that do not affect key parameters of competition. However, while these forms of cooperation may be beneficial, since there is unlikely to be any appreciable effect, they will also be inherently limited in their impact and effectiveness in achieving sustainability goals.[23]

12.2.4.2 Exclusion from UK Competition Law

The ultimate form of 'shield' would be to grant sustainability agreements (or at least those aimed at reaching Net Zero goals) an exclusion order under the CA98, such that they are excluded from the Chapter I prohibition altogether (either generally or in specified circumstances).[24] While these powers should be used sparingly, there can be no doubt that the climate crisis provides 'exceptional and compelling reasons of public policy' to exclude agreements from the ambit of the Chapter I prohibition in certain circumstances.

12.2.5 The Exemption Criteria and Sustainability Agreements

The ambit of the exemption criteria, contained in Section 9 of the CA98 (and its EU equivalent, Article 101(3) TFEU), is highly relevant to ongoing efforts to address climate change and progress sustainable development objectives. The CMA has stated that sustainability agreements can be individually exempt on a case-by-case basis but has expressed no view as to the establishment of any block exemptions that cover sustainability initiatives.

In short, a sustainability agreement (like any other agreement)[25] caught by the Chapter I prohibition may be exempted provided the parties to the agreement can show that the following four cumulative conditions in Section 9 of the CA98 are satisfied:[26]

---

[23] E. Brink and J. Ellison, 'Article 101(3) TFEU: the Roadmap for Sustainable Cooperation' in Concurrences, Competition Law, Climate Change & Environmental Sustainability (March 2021), at pp. 42.

[24] Paragraph 7 of Schedule 3 to the CA98 provides that, an exclusion order can be made where the relevant minister is satisfied that there: 'are exceptional and compelling reasons of public policy why the Chapter I prohibition ought not to apply'. Exclusion orders were granted in relation to COVID-19 — and there are analogies that could and should be drawn with the climate crisis.

[25] There is nothing inherently different about sustainability benefits in the context of an exemption.

[26] The CMA stated it would have regard to the EU Commission's Guidelines on the application of Article [101(3)]: see CMA, 'Guidance on the functions of the CMA after the end of the Transition Period' (December 2020), para 4.21, accessible at https://www.gov.uk/government/publications/

Condition 1: the agreement must contribute to improving (i) the production or distribution of goods or (ii) to promoting technical or economic progress.

Condition 2: consumers must receive a fair share of the resulting benefit(s).

Condition 3: the restrictions must be indispensable to achieving these objectives.

Condition 4: the agreement must not give the parties any possibility of eliminating competition in respect of substantial elements of the products in question.

The CMA has not yet provided any particular guidance as to the application of these criteria to sustainability agreements, we discuss the application of each below.

12.2.5.1 Condition One: 'Benefits'

The types of 'benefit' set out in condition one in Section 9 of the CA98 are exhaustive but are framed in the alternative. Arguably all four types of benefit can and do encompass sustainability in the following ways:[27]

(i) 'improving production' such as by using fewer, or more sustainable, resources resulting in a better allocation of resources;
(ii) 'improving distribution' such as by sharing or pooling of logistics to reduce transport emissions;
(iii) 'promoting technical progress' such as through the development of new, greener or cleaner technologies; and
(iv) 'promoting economic progress' which could encompass 'anything that provides a higher standard of living'.[28]

Construing the 'benefits' in this way, is also consistent with the UK's (and the EU's) commitment to moving towards a more 'circular economy' whereby resources are used optimally, particularly by extracting the maximum use out of those resources already in circulation and creating new resources in a sustainable way.[29]

---

guidance-on-the-functions-of-the-cma-after-the-end-of-the-transition-period. Accessed 19 November 2022.

[27] EU Commission, 'Communication from the Commission to the Council and the European Parliament On Environmental Agreements' (1996) COM(96) 561 final, pp. 17: 'In analysing individual cases under Article [101(3)], [...] the protection of the environment might be considered as an element which contributes to improving the production or distribution of goods and to promoting technical and economic progress', accessible at https://eur-lex.europa.eu/legal-content/EN/TXT/PDF/?uri=CELEX:51996DC0561&from=DE. Accessed 19 November 2022. Further, the EU Commission's Draft Horizontal Guidelines also recognise at pp. 577–578 that '[...] cleaner technology, less pollution, measures to preserve or restore biodiversity, improved conditions of production and distribution, more resilient infrastructure or supply chains, better quality products' constitute relevant benefits.

[28] M. Dolmans, Sustainable Competition Policy and the 'Polluter Pays' Principle, in Concurrences, Competition Law, Climate Change & Environmental Sustainability, March 2021.

[29] See, for example, the UK Government's policy statement, accessible at: https://www.gov.uk/government/publications/circular-economy-package-policy-statement/circular-economy-package-policy-statement and the EU's Circular Economy Action Plan, accessible at: https://ec.europa.eu/

12 Sustainability and Competition Law in the United Kingdom

The transition to a circular economy specifically envisages businesses working together to create a framework for more sustainable products, and emphasises the need to make changes to existing supply chains and business and market models. It is expressly recognised by the UK government that 'the move to a more circular economy will bring the four UK nations environmental, financial and social benefits'.[30] It is also stated that moving towards a circular economy results in reductions in GHG emissions and the creation of new jobs. Accordingly, there is currently enough scope under the existing rules to allow benefits deriving from sustainability agreements to be recognised as 'economic progress' for the purposes of the exemption framework. In any event, one should resist the temptation to squeeze everything into the economic box given the additional three alternatives (see above).

This reading is also consistent with the CMA's merger assessment guidelines ('MAGs'), discussed further below, which expressly recognise 'environmental sustainability benefits' as relevant consumer benefits.[31] The same principle equally applies to benefits flowing from sustainability agreements but confirmation and guidance from the CMA on this point would be valuable. The CMA's Advice to Government notes the possibility of this definition being expanded in the future: '. . .it may become apparent that the concept of "relevant customer benefits" in the mergers and markets regimes should be expanded to explicitly include sustainability, or that a new "sustainability" public interest consideration should be added'.[32]

It would plainly be wrong (and unhelpful) to classify environmental or sustainability benefits as 'non-economic' or as requiring an unduly broad interpretation of the law.[33]

12.2.5.2 Condition Two: 'Fair Share'

The second condition requires that consumers receive a 'fair share of the resulting benefits' (i.e. the improvements generated by the agreement). The meaning of 'fair share' has been the subject of debate, particularly whether 'full compensation' is required. 'Fair share' does not, in our view, require consumers to be 'fully compensated' for the costs of any agreement as a matter of ordinary language or law. First, the wording of Section 9 CA98 (and Article 101(3) TFEU) makes no

---

environment/pdf/circular-economy/new_circular_economy_action_plan.pdf. Accessed 19 November 2022.

[30] UK Department for Environment, Food and Rural Affairs (Defra), the Department of Agriculture, Environment and Rural Affairs (DAERA), '*Circular Economy Package policy statement*' (30 July 2020) https://www.gov.uk/government/publications/circular-economy-package-policy-statement/circular-economy-package-policy-statement. Accessed on 19 November 2022.

[31] CMA, 'Merger assessment guidelines' (CMA129) updated 18 March 2021 (the 'MAGs'), para 8.21, accessible at <https://www.gov.uk/government/publications/merger-assessment-guidelines>. Accessed 19 November 2022. These are discussed further in Sect. 3 below.

[32] This reflects the suggestion made by Simon Holmes in 'Climate change, sustainability and competition law (2020)' Journal of Antitrust Enforcement, Volume 8, Issue 2, at pp. 391–392.

[33] Ibid., fn. 36 pp. 354–405 and pp. 372–373.

reference to 'full compensation'. Second, the EU jurisprudence does not support such an interpretation.

The CJEU's judgment in Case C-382/12 *MasterCard Inc. and others v European Commission* [2014] ECLI 2201 (*'Mastercard'*) is often misinterpreted as requiring 'full compensation'. However, *Mastercard* itself contains no reference to the concept of full compensation. The case concerned an assessment of fair share in the context of two-sided markets. The CJEU considered whether benefits which accrued to consumers on a different side of the market, could be taken into account for the purposes of the fair share condition—and if so, whether those benefits were alone sufficient to meet the fair share condition. It was held that:[34] (i) 'it is necessary to take into account [...] all the objective advantages flowing from the [agreement]' (para 237) to 'all consumers in the relevant markets', including on separate but connected markets (para 234), and (ii) benefits that accrue to consumers in connected markets may be counted towards the fair share requirement, provided those consumers negatively affected by the restriction also receive some of the benefits (para 242).[35] On the facts of *Mastercard*, there was no overlap between the two sets of consumers (namely, cardholders and merchants), and accordingly the CJEU upheld the General Court's finding that the advantages to cardholders were not of a character to compensate for the disadvantages to competition alone (paras 240–242). Accordingly, as a matter of law, *Mastercard* did not establish a requirement for full compensation and requires no more than 'appreciable objective advantages' for those affected by the restriction.

In any event, it is worth highlighting that *Mastercard* is confined to cases involving two-sided markets. Sustainability agreements are generally unlikely to face the same issues as those in *Mastercard*. Environmental or climate-related benefits are likely to have broad impacts on society and will naturally include those consumers who are 'negatively' affected by the restriction of competition in the agreement. In addition, requiring consumers to be fully compensated in this context would be inconsistent with the 'polluter pays' principle enshrined in EU law.[36]

Separately, in 2020 the UK Supreme Court issued a judgment in *Sainsbury's Supermarkets Ltd v Visa Europe* concerning the fair share requirement in two-sided markets. The Supreme Court relied on the Opinion of Advocate General Mengozzi in *Mastercard* in holding that:

---

[34] See the discussion at paras 230 to 248 of the *Mastercard* judgment.

[35] At para 248 of *Mastercard*, the CJEU also recognised that the extent of the 'share' of benefits received need not be the same and may even differ between different groups of relevant consumers — such a conclusion runs counter to any suggestion of full compensation.

[36] We consider that 'fairness', in this context, should reflect the 'polluter pays' principle under Article 191(2) TFEU. For a fuller discussion of this point see M. Dolmans, 'Sustainable Competition Policy and the "Polluter Pays" Principle' in Concurrences, Competition Law, Climate Change & Environmental Sustainability (March 2021).

The merchants are the consumers of the services which are subject to the restriction of competition, and are therefore the consumers which the second condition is presumably intended to protect. *If the merchants are not fully compensated for the harm inflicted on them by the restrictive measure, it is difficult to see how they can be said to receive a "fair" share of the resultant benefits. As the Advocate General indicated at point 158 of his Opinion, it is not the purpose of competition law to permit anti-competitive practices to harm consumers in one market for the sake of providing benefits to those in another.*[37] (Emphasis added)

The Supreme Court relies on the Opinion of AG Mengozzi, which was not binding on, nor was it followed by, the CJEU in *Mastercard* in relation to fair share. The Supreme Court finds that 'fair share' requires 'consumers' to be 'fully compensated' for the harm inflicted by the restriction of competition. We respectfully consider this part of the judgment to be confined to its own facts—and reflects the Supreme Court's assessment of what was 'fair' in the context of the two-sided market governing multilateral interchange fees. We do not consider that this establishes a wider requirement for full compensation in every case. For the reasons set out above, a full compensation requirement is inconsistent with the decision in *Mastercard*. This point appears to be recognised by the CMA in its Advice to Government:

> Our view is that, if a particular agreement or practice restricting competition leads to environmental benefits to a broader group of consumers than just those adversely affected by the restriction of competition, such benefits, in principle, can be taken into account in the "fair share" assessment under section 9 CA98. This is because the harmed consumers are also part of the broader group of consumers that receive the environmental benefits. However, based on existing case law, a share of those benefits must accrue to those consumers who suffer from the restriction of competition, and those consumers must also be fully compensated for the detriment they suffer. This suggests that the "fair share" assessment would not, in principle, permit a situation where benefits to one group of consumers are offset against net harm to an entirely different group of consumers (without sufficient benefit accruing to the harmed consumers). Notwithstanding this limitation under the existing [UK] case law, the CMA considers that the current framework gives scope to take into account environmental benefits as part of the assessment under section 9 CA98. Existing precedents show that the "fair share" criterion has been applied flexibly in practice in light of the specific circumstances of each case. Moreover, this is an area where the case law may evolve further in the future, including in light of the scope for the CMA and courts to depart from EU precedent in certain circumstances under section 60A CA98, which potentially provides additional flexibility.[38]

The CMA expressly refers to the limitation in the case law in relation to environmental benefits and indicates its intention to apply the 'fair share' condition

---

[37] *Sainsbury's Supermarkets Ltd and others v Visa Europe Services LLC and others* [2020] UKSC 24, at para 174.

[38] CMA, Environmental sustainability and the UK competition and consumer regimes: CMA advice to the Government (14 March 2022), available at https://www.gov.uk/government/publications/environmental-sustainability-and-the-uk-competition-and-consumer-regimes-cma-advice-to-the-government/environmental-sustainability-and-the-uk-competition-and-consumer-regimes-cma-advice-to-the-government. Accessed 05 January 2023.

'flexibly'. The CMA raises the possibility of using Section 60A of the CA98 as a means of departing from the EU jurisprudence (as applied by the Supreme Court),[39] which would arguably be a sensible approach to adopt. As it stands, the current approach of 'full compensation' is unduly narrow, inconsistent with the Section 9 exemption of the CA98 and Article 101(3) TFEU,[40] and runs contrary to the CMA's strategic objective of 'supporting the transition to a low carbon economy'.

The application of the 'fair share' condition in Section 9 CA98 is one of the most important areas in the context of competition law and sustainability agreements. Any suggestion that 'full compensation' is required (which we do not consider is required as a matter of law for the reasons set out above) acts as a real barrier to businesses entering into sustainability agreements in the first place. Therefore, there is a need for a shift away from concepts of 'full compensation' and towards a global approach to 'benefits', in recognition of the global nature of the fight against climate change.

12.2.5.3 Collective Benefits

Benefits arising from sustainability agreements on the whole tend to benefit consumers (and even society) collectively, rather than individually. The EU Commission, in its Draft Horizontal Guidelines, acknowledges 'collective benefits', albeit in a limited way. However, the CMA's position is less clear.

For sustainability agreements to have the greatest possible role/impact in the fight against climate change, competition authorities must begin to recognise 'appreciable collective benefits' and benefits to 'indirect users' (or society at large). This includes collective benefits outside of the relevant product market and geographic market. Benefits such as the prevention and reduction of GHG emissions, and protection or conservation of biodiversity are inherently collective (and global) in nature.

---

[39] CMA, Environmental sustainability and the UK competition and consumer regimes: CMA advice to the Government (14 March 2022) at footnote 11, available at https://www.gov.uk/government/publications/environmental-sustainability-and-the-uk-competition-and-consumer-regimes-cma-advice-to-the-government/environmental-sustainability-and-the-uk-competition-and-consumer-regimes-cma-advice-to-the-government. Accessed 19 November 2022.

[40] Respectfully, the CJEU in *Mastercard* did not go so far as to mandate 'full compensation' in every case. On our reading of *Mastercard*, which is consistent with that of the ACM, the CJEU requires 'appreciable objective advantages' for the affected consumers (at para 234) and left the door open to the possibility that this may be less than 'full compensation' in certain circumstances. Rather, we consider that the statement (at para 242) of the CJEU judgment, interpreted by the Supreme Court as disregarding 'out of market' benefits entirely from the assessment under fair share, is actually more nuanced. On our reading, it simply supports the proposition that 'out of market benefits cannot 'in themselves' compensate for in market disadvantages'. The key point is that there is evidence that such benefits are enjoyed by affected consumers. See, M. Dolmans, 'Personal comments in response to the CMA Call for Inputs on competition policy and the UK's net zero and environmental sustainability goals' (10 November 2021) pp. 13, fn. 1, available at https://assets.publishing.service.gov.uk/government/uploads/system/uploads/attachment_data/file/1061313/Maurits_Dolmans.pdf. Also see ACM, Legal Memo, 'What is meant by fair share for consumers in article 101(3) TFEU in a sustainability context' (27 September 2021), p. 3 available at https://www.acm.nl/sites/default/files/documents/acm-fair-share-for-consumers-in-a-sustainability-context.pdf. Accessed 19 November 2022.

12 Sustainability and Competition Law in the United Kingdom

Additionally, the EU Commission deals with this exact point at paras 601–608 of the Draft Horizontal Guidelines, recognising the role 'collective benefits' have to play in internalising negative externalities associated with non-sustainable consumption decisions of consumers.[41]

12.2.5.4 Qualitative or Quantitative?

Both qualitative and quantitative data can be adduced to demonstrate that the agreement will lead to the benefits claimed. This is accepted by the CMA in its Advice to Government: '[the fair share condition] does not necessarily require a quantitative assessment; as the EU's guidelines on Article 101(3) exemption acknowledge, the assessment can factor in qualitative efficiencies which require a value judgment (EU exemption guidelines Sect. 3.4.3)'.

In terms of quantifying environmental benefits from agreements, there exists a number of economic tools to effectively 'convert' environmental effects into monetary terms.[42]

Once quantified, they need to be balanced against the potential reduction in competition. Particularly, two key areas which would benefit from further guidance from the CMA are:

Out of market benefits: Although the CMA has indicated its view of 'consumers' for the purposes of the 'fair share' requirement, it is not clear whether this extends to consumers outside the UK.[43] If competition law is to facilitate businesses' cooperation in pursuit of beneficial sustainability objectives—it would be advisable to allow out of market benefits. This could be done, for all sustainability agreements or a subset.[44]

Discounting: Economic literature offers various views on how to estimate the discount rate for environmental effects—it would be helpful if the CMA provided clarity on the discounting rate it would accept.[45]

---

[41] However, although the reference to collective benefits by the EU Commission is helpful and encouraging, we note the inherent inconsistency with the final sentence of paras 601 and 602, which, on one reading, seem to suggest collective benefits may only be counted where they do in fact accrue to all those consumers affected by the restriction of competition (i.e. they are in the same group) — which we consider to be an undesirable outcome. We prefer the expressions in paras 603 and 606© — which refer to the in-market consumers as substantially overlapping or being 'part of' the larger group of beneficiaries — which is much more flexible and realistic.

[42] An extremely helpful overview of such methods can be found in the study commissioned by the Dutch ACM and the Hellenic Competition Commission, 'Technical Report on Sustainability and Competition' (26 January 2021), available at <https://www.acm.nl/en/publications/technical-report-sustainability-and-competition>. Accessed 19 November 2022

[43] CMA, 'Merger assessment guidelines' (CMA129) updated 18 March 2021, para 8.20, accessible at <https://www.gov.uk/government/publications/merger-assessment-guidelines>. Accessed 19 November 2022.

[44] For example, ACM permits wider benefits to be taken into account when considering environmental-damage agreements.

[45] For a fuller discussion of this point and quantification generally, see Linklaters and Oxera, 'Response to CMA Call for Inputs on environmental sustainability and the competition and

12.2.5.5 Conditions Three and Four – 'Indispensability' and 'Elimination of Competition'

Conditions three and four are not at the centre of the 'sustainability' debate (and are largely uncontroversial) but it is important to remember that they are vital (and ever present) safeguards when considering the correct approach to the scope of conditions one and two. For example, the 'indispensability condition' is considered to be an important part of the architecture of Section 9 CA98 (and Article 101(3) TFEU). It acts as a vital safeguard, ensuring that agreements (whether sustainability agreements or otherwise) are not exempted when it is not necessary to do so. However, it is precisely because we have these powerful safeguards that we should not be afraid to interpret and apply conditions one and two of Section 9 CA98 in a manner which is consistent with the wording of the statute, the transition to a low carbon economy, and the pressing need to combat climate change.

It should also be borne in mind that necessity has the potential to act a barrier to firms going beyond existing national law, regulations or industry standards. However, such cooperation may be precisely what is needed to achieve the goal more quickly (particularly where regulations are outdated, and the legislative process is slow) or to go further than the regulations require.

12.2.6 Using Competition Enforcement as a 'Sword' to Promote Sustainability

Although the EU Commission has made clear that, as a starting point, if the agreement pursues a genuine sustainability objective, it will not amount to a restriction 'by object' but its effects on competition will need to be assessed, the CMA has been less clear.[46] What the CMA is very clear on is that 'sustainability agreements must not be used as a cover for a business cartel or other illegal anti-competitive behaviour'. In this sense, the provisions may be used as a 'sword' to prohibit cartels hiding behind sustainability initiatives. In March 2022, the CMA and EU Commission announced that they are investigating anti-competitive behaviour involving a number of vehicle manufacturers relating to arrangements for recycling old or written-off vehicles. The CMA stated its investigation 'reflects the CMA's commitment—outlined in its draft Annual Plan 2022 to 2023—to prioritise promoting environmental sustainability through effective competitive markets'.[47] It seems

---

consumer law regimes' (19 November 2021) at para 19, available at: <https://assets.publishing.service.gov.uk/government/uploads/system/uploads/attachment_data/file/1061310/Linklaters_and_Oxera.pdf>. Accessed 19 November 2022.

[46] Thus far, the CMA has been silent as to whether agreements which pursue sustainability objectives could be considered to be an object restriction but makes clear that 'by object' restrictions must be avoided in any case.

[47] CMA, Press release, 'CMA launches investigation into recycling of cars and vans' (15 March 2022), available at: <https://www.gov.uk/government/news/cma-launches-investigation-into-recycling-of-cars-and-vans>. Accessed 19 November 2022.

that this could be a sign of competition enforcement increasingly being used as a 'sword' to promote sustainability.

The same point applies equally to information exchanges. Parties to genuine sustainability agreements ought to take care not to exchange any competitively sensitive information unless strictly necessary — and sustainability should not be used as a cover to share such information.

12.2.7 Conclusion

Overall, what is most important is that businesses have certainty as to the approach which will be taken by competition authorities to sustainability agreements so that businesses are able to properly self-assess and are not deterred from entering into these agreements in the first place. If the UK, or elsewhere, intends to support the transition to a more sustainable future, there needs to be a degree of flexibility in respect of the application of the Chapter I prohibition and exemptions to sustainability agreements to enable the positive effects of such private arrangements to be realised.

This is particularly so in relation to the exemption criteria, given the requirement to self-assess. In the absence of detailed, context-specific, and example-driven guidance, it is extremely difficult for firms to anticipate how the CMA will treat their particular venture under the CA98, with the result being that many firms will often decide, on balance, not to proceed with the agreement.

12.3 Using Abuse of Dominance Laws to Help Take Action for a Sustainable Future

This section of the contribution will focus on the role that the Chapter II prohibition (and its EU equivalent, Article 102 TFEU), contained in Section 18 of the CA98, play, and (more importantly) could play, in supporting action for a more sustainable future. It will be argued there is more scope to use these 'abuse of dominance' provisions than is often appreciated (and less of a tension between competition law and sustainability goals than some may suggest).

This section will refer primarily to the position and the 'state of play' in the UK under Chapter II of the CA98. Nonetheless, given the similarity between the UK and EU regimes on abuse of dominance, to aid in the discussion of Chapter II, significant reference will also be made to EU law and practice.[48] On this point we note that the CMA in its Advice to the UK government on 'Environmental sustainability and the

---

[48] It should be noted that the national competition laws of many countries (not only in Europe but around the world) are modelled on EU law — with many containing identical wording. Furthermore, even laws which use different words (such as Section 2 of the US Sherman Act) are generally trying to tackle the same fundamental problem-the control of market power and negative effects on the economy, society (and, ultimately, the planet).

UK Competition and consumer regimes' of 14 March 2022[49] did not refer to Abuse of Dominance and there was only a brief reference to it in its 29 September 2021 'Call for Inputs' on this.[50]

Before diving into the nooks and crannies of the Chapter II prohibition and Article 102 TFEU, it is important to place the analysis in its proper context. Particularly, the following five points are worth emphasising.

Whatever one's views on the extent to which *'dominant'* companies (or companies with wider market power) are particularly responsible for climate change and unsustainable business practices, it should however be accepted that, other things being equal, big companies are likely to have a bigger impact on the market/planet than small ones and are more likely to have a *'dominant position'* — or at least market power.[51] Regardless, it should be recognised that, if we have the tools to mitigate the impact of such companies on climate change (or other unsustainable practices) we have a duty to use them. Not only is this a moral duty but it makes sense from a basic efficiency of time and resources perspective given the urgency of the task. In this context it is also worth recalling that there is no requirement in either UK or EU law to show a causal relationship between the dominant position and the abuse.[52]

If the concept of a dominant position captured circumstances where there was no real market power of concern, it would instinctively (and rightly) merit taking a cautious approach to the concept of abuse. However, the opposite is also true. If, as the evidence suggests, there is widespread (harmful) market power out there,[53] which is not necessarily caught by the narrower concept of a dominant position, then arguably a more robust approach to the concept of an abuse can, and should, be

---

[49] CMA, Environmental sustainability and the UK competition and consumer regimes: CMA advice to the Government (14 March 2022), available at https://www.gov.uk/government/publications/environmental-sustainability-and-the-uk-competition-and-consumer-regimes-cma-advice-to-the-government/environmental-sustainability-and-the-uk-competition-and-consumer-regimes-cma-advice-to-the-government. Accessed 19 November 2022.

[50] CMA, 'Environmental Sustainability Advice to Government: Call for Inputs', available at: https://www.gov.uk/government/consultations/environmental-sustainability-advice-to-govern ment-call-for-inputs. Accessed 19 November 2022.

[51] It has been estimated that just 100 companies are responsible for over 70% of global industrial greenhouse gas emissions since 1988-and these companies are likely to be big ones (P. Griffin, 'The Carbon Majors Database-CDP Carbon Majors Report 2017' (CDP, July 2017)).

[52] See, for example, CJEU, T-321/05, *Astra Zeneca v EU Commission* [2010 5 CMLR 28]. This approach was confirmed in the UK by the Privy Council in *Carter Holt Harvey Building Products Group Limited v The Commerce Commission* [2004] UKPC 37 at para 49.

[53] See the IMF paper, 'Rising Corporate Market Power: Emerging Policy Issues', SDN/21/01, particularly at pages 5, 7 and 16. This is also shown clearly in the work of scholars like De Loecker and Eeckhout. They show that corporate mark ups; profit rates and the valuation of companies (relative to sales) have all gone up dramatically in the last 40 years or so. See further for example, J. De Loecker, & J. Eeckout, (2018), Global market power (No w24768). US National Bureau of Economic Research; 'The Profit Paradox' by J. Eeckout Princeton University Press, 2021 and T. Philippon's 2019 book 'The Great Reversal: how America gave up on free markets'.

taken.[54] To be clear, this approach does not suggest stepping outside the ambit of the legal prohibition in Chapter II or Article 102, only not to be unduly timid in interpreting it.

Those with the greatest power have the greatest responsibility to use it properly. This sentiment has been echoed by the EU courts when analysing abuse of dominance, where they have often made it clear that the 'special responsibility' of a company with a dominant position depends on the 'degree of dominance' held by that company.[55] Such statements emphasise how in the face of extreme market power, there is no need to take a restrictive approach to its abuse.

To determine the correct approach to abuse, it is important to shed light on what the Chapter II Prohibition and Article 102 were supposed to (and can) achieve. How can we decide whether something is an 'abuse' if we have lost sight of the purpose of the prohibition? As Iacovides and Vrettos argue,[56] competition lawyers and economists have got so trapped in a narrow (so-called) 'more economic approach' or narrow so-called 'consumer welfare' standard that they 'are unable to think outside its narrow market confines'.[57] As they rightly argue, by 'accepting that unsustainable business practices can be abuses of a dominant position...we focus on what we as a society and Article 102 TFEU care about'. As Iacovides and Vrettos conclude 'our approach is about more competition, just not the toxic kind. It is a call for refocusing competition policy and reconnecting concepts such as "abuse" with the general goal of the system of EU competition law. Our proposals are activist, but they are certainly not radical'. This approach rightly emphasises that, none of this is as radical as it might seem at first (superficial) sight.

None of this should suggest that the concept of abuse is a static one. Quite the contrary: it needs to be considered in the light of current economic, social and environmental priorities.[58] Right now the number one priority for the UK and the

---

[54] In this context we note the IMF's conclusion in its paper (at page 24) on 'Rising Corporate Market Power': 'the effects of corporate power can be partly mitigated by enforcing restrictions on the abuse of a dominant position more actively' (emphasis added). [Ibid., fn. 56].

[55] See, for example CJEU, Case T-201/04, *Microsoft* [2007] ECR II-3601, paragraph 775. It is also clear that many cases of abuse of dominance have concerned so-called 'super dominance' with market share of 70, 80 or even 90%. See, for example, the Intel Case, ECJ Case C-413/14P *Intel Corp v Commission* ECLI:EU:C:2017:632

[56] 'Radical for whom? Unsustainable Business Practices as abuses of dominance' in 'Competition Law, Climate Change & Environmental Sustainability' (March 2021), edited by Simon Holmes, Dirk Middelschulte and Martijn Snoep

[57] Iacovides and Vrettos ask a very pertinent question: 'a market logic may work, but do we really want everything to be filtered through that logic if that is only possible because we contort concepts (e.g. consumer welfare) and tests that were devised in a different time and on the basis of discredited assumptions and failed ideologies?'.

[58] What may have been the top priorities in 1957 -the date of the Treaty of Rome that first set out what is now Article 102 TFEU- (or even 1979 when the Hoffman La Roche case set out the classic definition of an abuse) such as the establishment of a 'Common Market' and the potential exclusion of competitors (especially from other member states) are not necessarily the top priorities in 2022 (or at least not the only ones). Of course, these still include a system of healthy competition etc. but

EU (and, indeed, the world) is the fight against climate change. This is reflected in the CMA's recognition of [fighting] 'climate change and supporting the transition to a low carbon economy' as one of its top 'strategic priorities'[59] and from an EU perspective, the EU's Green Deal and numerous statements by EU institutions. While this section must base its analysis in what the Chapter II Prohibition and Article 102 say, the wider political, economic and, indeed, existential imperative cannot be ignored.

It is against that background that this section will now consider the two ways in which abuse of dominance provisions such as the UK Chapter II Prohibition and Article 102 TFEU are most relevant to climate change and unsustainable business practices:

First, using them as a 'sword' to attack unsustainable practices.[60]

Second, recognising sustainability as a potential 'shield' against accusations that genuine practices to mitigate climate change or increase sustainability are an 'abuse' of a dominant position.

12.3.1 The Chapter II Prohibition as a 'Sword' to Attack Unsustainable Practices

Before diving into the detail of specific unsustainable practices that may infringe the Chapter II Prohibition (and the examples given within those provisions themselves) it is again important to understand the context and the general purpose and meaning of the prohibition. Ten points should be particularly borne in mind:

   (i) The classic definition of an 'abuse' is that given by the CJEU in the *Hoffman La Roche* case: it is conduct 'through recourse to methods different from those which condition <u>normal</u> competition in products or services'[61] (emphasis

---

this is not inconsistent with the overall priority of combatting climate change with all available tools.

[59] CMA, Annual plan for 2021 to 2022 (23 March 2021), accessible at: https://www.gov.uk/government/publications/competition-and-markets-authority-annual-plan-2021-to-2022. Accessed 12 May 2022.

[60] Abuse of dominance provisions can also be used as a 'sword' to attack steps taken (or purportedly taken) in the name of sustainability if they are anti-competitive-either in the sense of 'green washing' or because, on analysis they fall foul of Chapter II/Article 102. A sustainability motive, or simply being in the environmental sector, is no defence. An example of the latter from EU case law is the so-called 'Green Dot' case [ECJ Case C-385/07 P [2009] ECR 1-6155]. On this see S. Kingston, Greening EU Competition Law, Cambridge University Press 2011, at pp. 312–318 and C. Thomas 'Exploring the Sustainability of Article 102' in Competition Law, Climate Change & Environmental Sustainability, 2021.

[61] ECJ, Case 85/76 *Hoffmann – La Roche*, ECR 1979 461, pt 91.

12 Sustainability and Competition Law in the United Kingdom 223

added). Exactly what this means is far from clear,[62] but two guiding principles may be suggested at this stage. First, what is 'normal' may change over time and should reflect society's values at the time the assessment of potential abuse is made (e.g. disposing of chemicals in a river may have been 'normal' and acceptable in the 1960s but is not now). Second, abuses should be as consistent as possible with what an ordinary citizen would consider to be an abuse: it is odd (to say the least) that loyalty rebates which *reduce* prices and which are widely given by companies regardless of their size are (generally) condemned as an 'abuse' if given by a dominant company, but charging exorbitant prices for a product, or paying a supplier so little that (s)he cannot feed a family, is something which many in the competition law bubble have difficulty seeing as an 'abuse'. Anyone outside that bubble would probably come to the opposite conclusion and, therefore, in the competition sphere, one should not be afraid to call out abuses which fit with one's innate sense of what an abuse of power is.[63]

(ii) Probably the most obvious and illogical disconnect, between the competition bubble and both the person in the street and the original meaning of the Chapter II Prohibition and Article 102, is the former's focus on 'exclusionary' abuses (such as loyalty rebates) and the paucity of cases brought against 'exploitative' abuses.[64] Not only are three quarters of the examples of abuses given in both the Chapter II Prohibition and Article 102 themselves, exploit-ative, but exploitative abuses fit more easily with one's innate sense of what is 'fair' and what an 'abuse' of power really is[65]. This is important in the current context as most instances of unsustainable activities will be exploitative, rather than exclusionary, in nature. There are some tentative signs of a renewed interest in exploitative abuses (and indeed beyond) both in the area

---

[62] The argument is often somewhat circular. Having decided that a practice is an abuse it is held not to be 'normal competition'. Nor does the term 'competition on the merits' really take the analysis any further for the same reason.

[63] The point here is not to challenge the loyalty rebate cases like Intel (CJEU, C-413/14P *Intel Corp v Commission* ECLI:EU:C:2017:632), or to argue for more intervention generally against high prices — although that may be warranted — but simply to make the point not to lose touch with the basic idea of what an 'abuse' is. There may be instances where an abuse is complex/technical and not easily understood by non-experts (loyalty rebates and self-preferencing are probably examples of this) but these should be the exceptional cases-and certainly not blind us to the more obvious abuses.

[64] It is however understandable in the context of a prevailing so-called 'free market' ideology premised on the (somewhat naïve) idea that as long as there is sufficient competitive pressure, and competitors are not excluded from participating in the market, there will be no opportunity for abuse of dominance or exploitation because dominance, if ever held, will be fleeting. Hence, the focus on exclusion on the assumption that this forestalls a need to target exploitation directly.

[65] This is consistent with the 'travaux preparatoires' of the Treaty of Rome (the predecessor to the TFEU and which first set out what is now Article 102 TFEU) which indicated that the intention behind Article 102 was primarily to sanction exploitative abuses. (P. Akman, 'Searching for the Long Lost-Soul of Article 82 EC': (2009) 29(2) Oxford Journal of Legal Studies 267, 271.)

224 S. Holmes et al.

of big tech—particularly in relation to platform and ecosystem power[66]—and in the number of excessive pricing cases brought by the CMA and other national competition authorities across Europe in recent years.[67]

(iii) Many of the reasons for a reluctance on the part of competition authorities to make full use of abuse of dominance provisions' potential are not relevant to the power and sustainability concerns which this contribution focuses on. For example, while it can be argued (not always correctly) that the market will correct in the case of excessive prices,[68] the same cannot be said for unfairly low purchase prices. Furthermore, potential concerns over the risk of harm through intervention when no underlying harm exists ('false positives') are vastly outweighed by the risk of not intervening when harm is being done ('false negatives') in the face of climate change—particularly given the uncertainties of tipping points and recognition that much of the damage being done is irreversible.

(iv) In principle, it should not matter how an abuse is classified: something either is, or is not, an abuse; the examples in the Chapter II Prohibition and Article 102 are just that—examples; and both the UK and European courts have consistently held that the categories of abuse are not closed.[69] In practice, it is often easier to convince a conservative competition establishment that an unsustainable practice is an abuse if it falls within a well-established category ('box ticked'), but equally there is no need to try to squeeze an unsustainable practice into a particular box into which it does not fit easily but which is clearly abusive. Not only is this not required as a matter of law, but it helps ensure that abuse of dominance provisions continue to evolve and prove themselves capable of dealing with the most pressing issues of our time. If it does not, they will seem increasingly arcane and risk becoming increasingly irrelevant.

(v) On the face of it, the Chapter II Prohibition and Article 102 would seem inherently well suited to attack unsustainable and exploitative actions by dominant (and often 'super-dominant') companies. Furthermore, there is nothing in the jurisprudence of the UK or European courts to suggest to the contrary.[70] The question therefore is not so much, is it possible to use the

---

[66] See I. Lianos and B. Carballa, Economic Power and New Business Models in Competition Law and Economics: Ontology and New Metrics, CLES Research Paper Series 3/2021, March 2021.

[67] See, for example Pfizer Flynn in the UK (CMA, 18, December 2015 — and subsequent appeals.)

[68] This was a widespread view of US antitrust prior to its recent boost under President Biden and FTC chair, Lina Khan.

[69] See for example, CJEU, Case T-321/05, *Astra Zeneca v EU Commission* [2010 5 CMLR 28] or, in the UK courts, *Purple Parking v Heathrow Airport* [2011] EWHC 987 (Ch) paras 75–108.

[70] If, contrary to this view, the CJEU were to take a different view, it would be open to the CMA and UK courts (which, post Brexit, are not constrained by the CJEU's (future) judgments) to take a different and more progressive approach to the UK's Chapter II prohibition and tackle unsustainable practices by dominant companies. This would be a rare positive example of the UK putting into practice the political slogan of 'taking back control'.

12 Sustainability and Competition Law in the United Kingdom

Chapter II Prohibition to attack these practices, but is there the will to do so? This means a willingness by civil society and injured parties to bring cases to the attention of the competition authorities (or courts), and a willingness by the latter to take the cases on and look at the UK abuse of dominance laws with a fresh pair of eyes.

(vi) There is a strong legal case for factoring in environmental and other sustainability factors when considering whether conduct does, or does not, amount to an abuse, when the Chapter II Prohibition is read in light of other UK legislation including the UK's Climate Change Act 2008, the Paris Agreement of 2015, and the Human Rights Act of 1998, which all provide persuasive support for this.[71] The position is similar under Article 102 TFEU, when read (as it must be) in the light of the 'constitutional' provisions of the treaties.[72] Particularly, (i) The goals in Article 3 of the TFEU of a 'high level of protection and improvement of the environment' and 'the sustainable development of the earth'; and (ii) The clear requirement in Article 11 TFEU that 'environmental protection requirements *must* be integrated into [*all EU*] . . .policies and activities' (emphasis added).

(vii) It is interesting that neither the Chapter II Prohibition nor Article 102 contain a general requirement that the abuse must have an adverse effect on competitors,[73] which makes the general focus on exclusionary, rather than exploitative, abuses all the more odd. In fact, neither provision explicitly requires there to be an effect (let alone an adverse effect) on competition— only that it 'may affect trade' (within the UK or between Member States as the case may be). This should not, however, be read into too much, given that these provisions should be read in context and they are to be found within the UK CA98 and in the chapter of the TFEU headed 'Rules on Competition'. While this almost certainly means an unsustainable practice which has absolutely nothing to do with competition cannot amount to an abuse, it does suggest there is more scope for finding an unsustainable practice to be an abuse so long as there is some reasonable nexus to the competitive structure of the market or competitive process.

(viii) There is also no general requirement in the Chapter II Prohibition or Article 102 that the practice must prejudice consumers for it to amount to an abuse.[74]

---

[71] S. Holmes, Climate change, sustainability and competition law in the UK, 41 ECLR 2020, Issue 8, pp. 387–390.

[72] On which see S. Holmes, Climate change, sustainability and competition law (2020) Journal of Antitrust Enforcement, Vol. 8, Issue 2, Part IV.

[73] There is a requirement that 'trading partners' be placed at a 'competitive disadvantage' in the example of an abuse set out in Article 102(c) and Section 18 (2) (c) CA98 dealing with discrimination (a point confirmed by the CJEU in Case C-525/16, *Meo-Servicios de Communioes e multimedia,* ECLI:EU:C:2018:270). This does not, however, apply to Article 102 or the Chapter II Prohibition as a whole.

[74] Although this is a requirement in the example of an abuse given in Section 18 (2) (b) of the CA98 and in Article 102(b) concerning 'limiting production, markets or technical development'. See more

This suggests that injury to other stakeholders such as suppliers or employees (or perhaps the environment) is sufficient for a practice to amount to an abuse.[75] At the very least it suggests that these provisions are not just concerned with consumers' direct or short-term interests but that the abuse may consist of damage to consumers' longer term interests—whether through the weakening of the structure of competition or, arguably, in terms of the impact of the practices on the planet and the environment in which those consumers live and breathe.

(ix) Notwithstanding the points made above, it is clear that many or most unsustainable practices with which competition authorities are likely to be concerned will affect competition and affect competitors and/or prejudice consumers. Indeed, most such practices will have most of the characteristics of other well-established categories of abuse and competition law violations. For example, those dominant companies which avoid paying the true price for inputs or off-load costs onto third parties and society, whether by paying unfair prices to suppliers, dumping waste in rivers, avoiding tax liabilities, or polluting the atmosphere, or which delay introducing more sustainable products or fail to open up their product ecosystems to more sustainable alternatives or components, or refuse to licence new green technologies on fair terms, are: (i) gaining an unfair competitive advantage over rivals who are not doing so; (ii) raising barriers to entry and excluding sustainable competitors (and these competitors may be just 'as efficient' in financial terms—and perhaps more efficient in natural resource and planetary terms—as the dominant company but simply not be engaging in the same unsustainable practices);[76] (iii) potentially reducing incentives to innovate (as companies may question: why bother innovating to reduce costs if you can cheat the system?); and (iv) not engaging in 'normal competition' or 'competition on the merits'.

(x) The approach advocated in this section is in no way inconsistent with the so-called 'more economic' approach, (even if there are some doubts about that approach generally). On the contrary, it is an approach which is far more in

---

generally, P. Colomo, Anticompetitive Effects in EU Competition Law, Journal of Competition Law & Economics 2021, Vol. 17, Issue 2.

[75] It also means that there is nothing equivalent in the Chapter II Prohibition or Article 102 themselves to the requirement in each of Section 9 of the CA98 and Article 101(3) TFEU; namely that consumers must get a fair share of the benefits of an agreement if it is to be exempt from the prohibition in Chapter I on anti-competitive agreements.

[76] As Iacovides and Vrettos point out, in assessing whether the competitors are 'as efficient' as the dominant company it would not seem appropriate to make the comparison based upon the dominant company's costs as these are artificially suppressed by the very abuse complained of — consistent with the recognition that prices that are a result of market power cannot be a proper baseline for the conduct of the SNNIP test-to avoid the so-called 'cellophane fallacy'. See, 'Radical for whom? Unsustainable Business Practices as abuses of dominance' in 'Competition Law, Climate Change & Environmental Sustainability' (March 2021), edited by Simon Holmes, Dirk Middelschulte and Martijn Snoep.

tune with the original (and better) meaning of 'economics'.[77] When competition authorities strengthened their economic capabilities in the early 2000s, it was largely in response to criticisms of their (relative) lack of these capabilities in some earlier cases. Nothing in that made (or makes) it inevitable that the approach to competition law and economics should focus on a narrow 'Chicago' version of the consumer welfare standard or neoclassical price theory—and certainly does not require or permit an unduly narrow approach to the Chapter II prohibition. The approach to competition policy and sustainability in fact, provides more, not less, scope for the intelligent use of economics—especially environmental economics.[78]

This contribution will now consider some examples of abuses contained within the Chapter II prohibition and Article 102 TFEU themselves.

12.3.2 Unfair Prices and Conditions [Section 18 (2)(a) CA98 and Article 102(a) TFEU]

Section 18(2)(a) CA98 and Article 102(a) give as examples of potential abuses, 'unfair purchase or selling prices or other unfair trading conditions' of a dominant company. This is potentially broad ranging and there is no reason, in principle, why this could not be used more widely to condemn unsustainable practices which are unfair from an economic, political, environmental or climate change point of view.

12.3.2.1 Unfair Purchase Prices
An example is the incredibly low prices paid by many retailers and intermediaries to farmers for their produce (e.g. bananas, coffee and cocoa)—prices which do not enable those farmers to feed their families; do not cover the true costs of production; can lead to an excessive use of scarce resources (land, water etc); and often discourage the development of more sustainable methods of production. To those brought up in the competition bubble this might seem radical, but looking at the wording and purpose of Section 18(2)(a) CA98 and Article 102(a) afresh, with a clear focus on fairness and on all aspects of prices and trading terms, it can quickly be seen that it is not. If it is possible to challenge unfair *selling* prices for being too *high* ('excessive pricing') or for being too *low* ('predatory pricing'), it begs the question: why not challenge depressingly *low purchase* prices?—and, as already

---

[77] For an account of how the original and more holistic approach to economics has changed, see J. Aldred, License to be Bad-How Economics Corrupted Us, Allen Lane, 2019, e.g. at Chapter 1.

[78] See also S. Holmes, Climate change, sustainability and competition law Journal of Antitrust Enforcement, 2020 Vol. 8, Issue 2, at pp. 9–11 and point (vii) on pp. 45–46.

pointed out, markets are hardly likely to 'self-correct' in the case of *low* (as opposed to *high*) prices.[79]

12.3.2.2 Predatory Pricing

Often selling prices are unsustainably low because they do not reflect the true costs of production.[80] Obvious examples are where some of the costs of production have been off loaded onto society (e.g. in the form of carbon emissions not captured or effluent not treated and dumped on land or in rivers)—the so-called 'negative externalities'. Another example is where the prices paid for inputs (whether raw materials like cocoa or exploitative labour) are unsustainably low (e.g. because they do not reflect the true costs of purchasing those inputs).

In these instances, the prices might be shown to be predatory once the 'true' costs of production are properly taken into account (but otherwise applying the usual tests for predation as set out by the courts in cases like *Akzo*).[81] While this may not be straightforward, (but nor have historic predatory pricing cases been), it does merit further consideration—especially by environmental economists.

12.3.2.3 Other Unfair Trading Conditions

Wherever a dominant company imposes unsustainable practices on a customer or supplier, there is no reason these could not be condemned as an abuse. One example (outside the area of pricing) might be requiring a supplier to produce a product in an environmentally damaging way.

Although, in the past, exploitative cases have tended to relate to pricing practices, there is nothing in the Chapter II prohibition or Article 102 TFEU to suggest that that should be the case, quite the contrary: it explicitly refers to 'other' unfair trading terms. Furthermore, some of the reasons for the low level of enforcement action against exploitative pricing practices do not apply (or are less relevant) in the case of non-pricing practices; for example, the enforcer does not risk becoming a price regulator and the market is less likely to self-correct in relation to abuses which do not concern excessive prices. Indeed, many of the criteria identified in the cases when assessing 'unfairness' in Article 102(a) can be readily seen in the case of the sort of unsustainable practices with which we are concerned: they are often 'unnecessary'; 'disproportionate' 'unilaterally imposed' etc.

---

[79] For a fuller discussion of this see Holmes, Climate change, sustainability and competition law (2020) Journal of Antitrust Enforcement, Volume 8, Issue 2. Another example is the unsustainable production of meat discussed by Iacovides and Vrettos in 'Radical for whom? Unsustainable Business Practices as abuses of dominance' in 'Competition Law, Climate Change & Environmental Sustainability' (March 2021), edited by Simon Holmes, Dirk Middelschulte and Martijn Snoep.

[80] For an excellent discussion of 'true costs' and 'true pricing' see True Price Foundation, 'A Roadmap for True Pricing. Vision Paper — Consultation draft' (2019), <https://trueprice.org/a-roadmap-for-true-pricing/> accessed 19 January 2020. This paper article includes some helpful ideas on how to determine a 'true price' in terms of which external costs should be taken into account, how negative externalities should be quantified and how to 'monetise' them.

[81] CJEU, Case C-550/07P, *Akzo Nobel v European Commission*, ECLI:EU:C:2010:512.

12.3.3 Limiting the Production of Products or Services [Section 18(2)(b) CA 98 and Article 102(b) TFEU]

Section 18(2)(b) CA98 and Article 102(b) TFEU give as examples of potential abuses: 'limiting production, markets or technical development to the prejudice of consumers'. Suzanne Kingston provides some colour on examples of practices by a dominant company which are damaging from a sustainability perspective that might constitute such abuses:[82]

First, limiting the ability of third parties to develop environmentally friendlier production method or products.[83]

Second, failing to satisfy a clear demand for an environmental service [or product].[84]

Third, be extremely inefficient in refusing to use an environmentally friendly technology thus increasing environmental costs.[85]

Such examples may seem novel to some but that is not a problem either as a matter of law or policy. The courts have consistently made it clear that the categories of abuse are not closed[86] and the climate crisis demands thinking afresh, re-interpreting old ideas, and using all the tools available.

12.3.4 Using Sustainability as a 'Shield'

The second way in which sustainability and monopoly power interact is the potential for sustainability considerations to act like a 'shield' against accusations that genuine efforts to fight climate change or prevent unsustainable practices amount to an abuse under abuse of dominance provisions.

Environmental considerations are sometimes seen as a 'defence', or as an 'objective justification', for conduct by a dominant company that might otherwise be considered to amount to an abuse.[87]

---

[82] See S. Kingston, Greening EU Competition Law, Cambridge University Press 2011, pp. 325.

[83] By analogy to cases like Suiker Unie, (CJEU, Case 40/73, *Coöperatieve Vereniging 'Suiker Unie' UA and others v Commission*, ECR 1663).

[84] By analogy to cases like P&I Clubs IGA [OJ 1999 L 125/12].

[85] By analogy to cases like Port of Genoa [OJ 1997 L 301/27].

[86] CJEU, T-321/05, *Astra Zeneca v EU Commission* [2010 5 CMLR 28].

[87] On this approach sustainability would be an 'objective justification' for conduct which is prima facie abusive where a dominant company (or exceptionally companies which are collectively dominant) engage in proportionate behaviour to tackle environmental or climate change issues (and where there is no way of achieving these objectives in a way that is less restrictive of competition). See, for example, the excellent discussion of this by S. Kingston, Greening EU Competition Law, Cambridge University Press 2011 at pp. 304–312). She identifies three categories of 'objective justification': (1) where a dominant company takes 'reasonable steps' to protect its commercial interests; (2) if the efficiencies justify the conduct such that there is 'no net harm to consumers'; and (3) legitimate public interest grounds.

This approach is reflected in the CMA's 'Call for Inputs' referred to above[88] where (at para 24) it suggested that: 'If a business with market power conducts or takes part in a sustainability initiative that might otherwise be considered an abuse of dominance, this conduct may fall outside of the Chapter II CA98 prohibition if it can demonstrate that the conduct in question is objectively justified and proportionate. This would need to be considered on a case-by-case basis, and evaluating whether the conduct is 'proportionate' may, in practice, present challenges similar to those mentioned above in relation to section 9 CA98.

However, arguably it should not be necessary to mount a 'defence' in this way: something either is, or is not, an abuse and unlike the Chapter I prohibition (or its EU equivalent, Article 101 TFEU)[89] there is no two part test set out in either the Chapter II prohibition or Article 102 TFEU.

As noted at above, there is a strong legal case for factoring in environmental and other sustainability factors when considering whether conduct does, or does not, amount to an abuse when the Chapter II prohibition is read in light of the UK's Climate Change Act 2008, the Paris Agreement of 2015, and the Human Rights Act of 1998 (with a similar analysis applying when Article 102 is read (as it must be) in the light of the 'constitutional' provisions of the treaties as referred to above).[90]

There is an equally strong moral and logical case. If conduct is genuinely intended to combat climate change, reduce environmental damage or otherwise contribute to sustainable development, it is difficult to see how, as a matter of common sense and language, that it can be seen as an 'abuse' that the law intended to prohibit. There may be exceptions to this, but these will be rare. Indeed, in these cases it is likely that the conduct was not genuinely intended to reduce environmental harm or it was done in an anticompetitive manner.[91]

There is a further practical reason why it is better to take account of sustainability factors in the initial assessment of a potential abuse rather than as an 'objective justification' or as a 'defence'. In the former case, it is for the competition authority to establish the abuse; in the latter case, there is (an evidential) burden of proof on the dominant company to establish the objective justification. Since there is no two-part test in the Chapter II prohibition or Article 102 (unlike Article 101 and the Chapter I prohibition), the competition authorities continue to bear the burden of proof

---

[88] CMA, Environmental Sustainability Advice to Government: Call for Inputs, available at https://www.gov.uk/government/consultations/environmental-sustainability-advice-to-government-call-for-inputs. Accessed 19 November 2022.

[89] Article 101 (1) TFEU prohibits anti-competitive agreements etc and Article 101(3) sets out the conditions under which they may be exempt from that prohibition.

[90] On which see S. Holmes, Climate change, sustainability and competition law (2020) Journal of Antitrust Enforcement, Volume 8, Issue 2, Part IV.

[91] See, for example the 'Green Dot' case discussed in footnote [97].

12 Sustainability and Competition Law in the United Kingdom

throughout and therefore it makes sense for the assessment to form part of the analysis of the alleged harm.[92]

Although there are few decided cases of direct relevance, the following are examples of instances where environmental (or other sustainability) considerations could act as a 'shield' against, or alter, a conclusion/finding that such conduct is potentially abusive:

(i) charging a higher price to cover environmental costs or reinvest in environmental protection[93] to counter allegations of 'excessive pricing';
(ii) charging different customers different prices according to the use to which the product is put — such as how environmentally friendly it is (e.g. whether products are recycled or the energy efficiency of the downstream production process) to counter allegations of 'discriminatory pricing':
(iii) making the purchase of one product from the dominant company conditional on the purchase of another environmentally friendly product (e.g. sale of a printer conditional on the purchase of recyclable toner cartridges)[94] to counter an allegation of 'tying';
(iv) offering exceptionally low prices to generate trial of a new environmentally friendly product to counter an allegation of 'predatory pricing';[95]
(v) refusing to grant access to an essential facility to a user who intends to use the facility for environmentally unfriendly purposes (e.g. denying access to diesel vehicles — provided this was done on a non-discriminatory basis) to counter an allegation of 'refusal to supply'.

The CMA in its 'Call for Inputs' referred to above suggests that 'where a business (or businesses) with market power enters into sustainability initiatives or agreements, this *may* give rise to a Chapter II infringement *risk* if the initiative or

---

[92] In practice, the difference may not be enormous as it will always be important for the dominant company to convince the authority of the objective facts relevant to the sustainability benefits of its conduct and (in reality) of its genuine motives.

[93] This approach would be consistent, not only with the 'polluter pays' principle, but also the approach suggested above in relation to challenging abusively low prices for failing to properly reflect environmental costs (see above on the Chapter II Prohibition as a 'sword').

[94] Although it would be necessary to show that there was no less restrictive solution. For example, this might mean requiring that the environmentally friendly product was bought but not necessarily from the dominant company.

[95] Suzanne Kingston suggests two ways in which environmental considerations may be relevant to accusations of predatory pricing. 'In the first place, evidence that the intention of the dominant undertaking pricing above AVC but below ATC was genuinely pursuing environmental protection aims in so doing should mean that the conduct is not considered abusive. This follows from the Akzo test itself without needing to consider the effects of Article 11 TFEU. In the second place, evidence that a dominant undertaking pricing below AVC was genuinely pursuing legitimate environmental protection aims, and that there is no less restrictive way of achieving these aims, should rebut the AKZO presumption of abuse'. See for example, S. Kingston, Greening EU Competition Law, Cambridge University Press 2011, pp. 322–323).

agreement involves an abuse of this power'. (emphasis added). It goes on to give various examples of where this 'risk may arise'.[96] While it is agreed that there 'may' be a 'risk' of infringement, these are also good examples of where this conduct *may not* amount to an abuse — either on a proper analysis of the conduct it cannot reasonably be considered to be an 'abuse' or (as the CMA suggests in the passage quoted above) it may be 'objectively justified' and 'proportionate'.

While sustainability considerations may mean that something which might otherwise look like an abuse is not an abuse on closer analysis, sustainability is no 'get out of jail free' card. Competition law applies just as much where sustainability issues are at stake. It is no defence that a company or companies operate in an environment related sector. [97]

12.3.5 Some Concluding Remarks on the 'Shield'

It is well established that a dominant company can take 'reasonable steps' to protect its own commercial interests if they are attacked by rivals so long as the actual purpose is not to 'strengthen this dominant position and abuse it' and its actions are proportionate to the threat which it faces.[98] If this is the case in relation to commercial interests, then, *a fortiori*, a dominant company should be able to take such steps where its motives are not those of commercial gain at all.

---

[96] 'For example, Chapter II infringement risk may arise where: (a) A business with market power changes its pricing policies in connection with a sustainability initiative to incentivise customers to purchase more sustainable products and/or to use the relevant products or services in a sustainable way. (b) A business with market power seeks to ensure it can recoup the cost of significant environmental investments through increasing prices or entering into long-term exclusive arrangements. (c) A business with market power changes the terms on which it sells its products or services in connection with a sustainability initiative, for example making the purchase of one product or service conditional on the purchase of another sustainable product. (d) A business with market power refuses to deal with suppliers or customers who do not meet sustainability criteria that the business with market power has set and which are not required by law'.

[97] This is illustrated by the so-called 'Green Dot' decision of the European Commission (and subsequent judgements of both the General Court and the ECJ). In this case DSD was found by the Commission to have abused its dominant position on the market for the organisation of the take cash back and recovery from consumers of used sales packaging in Germany. DSD owned the famous 'Green Dot' trademark and under a network of agreements it required companies to pay them a fee for all sales packaging distributed within Germany which bore the Green Dot trademark regardless of whether or not those manufacturers were actually using DSD's services (i.e. they were disposing of the packaging themselves or using a competing system). This meant that such companies had to pay twice for the disposal of packaging. The Commission found it to be exploitative of customers to require the payment of a fee for a service that was not actually provided. It was also exclusionary in that it made it harder for competing systems to be set up. It is noteworthy in two respects: that the case concerned re-cycling was no reason to treat the case any differently from any other; and the concerns which the Commission had here were classic competition concerns: anti-competitive exploitation and foreclosure. ECJ Case C-385/07 P [2009] ECR I-6155].

[98] For example, CJEU Case T-30/89, *United Brands* [1976] ECR 207, para 189.

12 Sustainability and Competition Law in the United Kingdom

It is sometimes objected that it is not for private companies to take action in the public interest and that such matters should be left to the public authorities.[99] Such arguments now appear outdated in the face of a growing climate emergency (and the lack of political appetite to commit unilaterally without the aid of private enterprises, evident at COP26). Rather, all the resources of the public and private sector must be engaged to combat climate change—as is well recognised by the UK government (with its commitment to 'deliver a Net Zero and more environmentally sustainable economy'); by the CMA in its Call for Inputs;[100] and by the EC—particularly in relation to the Green Deal).[101] Regulation is indeed necessary, but it is often limited in scope or geographic reach and is often slow in coming or simply not ambitious enough to fight climate change (and combat unsustainable practices) on a sufficient scale or at the necessary speed (not least because some companies and trade associations lobby against it).[102]

As a matter of law, it is clear that the concept of 'abuse' is an objective concept and does not depend on the motives of the dominant company. In practice, whether sustainability initiatives are genuine or not is very relevant. It will be very rare that enforcement action will be taken against a company where the evidence (e.g. internal documents) shows clearly that that company was genuinely doing something for sustainability reasons.[103]

---

[99] See cases like CJEU, Case C-53/92 P, *Hilti,* ECR II-1439, 1994, para 118. However, in such cases one has the suspicion that the public interest arguments (re safety) were added on once its anti-competitive tying conduct was attacked. See also the Commission's 2009 Guidance on Article 82 (now Article 102 TFEU) [OJ 2009 C 45/7].

[100] 'The CMA believes that regulation and government policy are the primary means to achieve the UK's Net Zero and sustainability goals. However, the CMA also believes that other public bodies and businesses can play an important role through a wide range of initiatives (including cooperation agreements and unilateral initiatives), translating into more sustainable supply chains and more environmentally-friendly products and services for consumers'. (CMA, 'Environmental Sustainability Advice to Government: Call for Inputs', available at: https://www.gov.uk/government/consultations/environmental-sustainability-advice-to-government-call-for-inputs. Accessed 19 November 2022.)

[101] See, for example, the EC's policy brief of 10 September 2021, 'Competition Policy in Support of Europe's Green Ambition' (the 'Green Deal Policy Brief'): '*In order to reach the goals set out in the European Green Deal, everyone, private and public, must play their part. This includes competition enforcers*' (emphasis added).

[102] See 'How a powerful US lobby group helps big oil to block climate action', Chris McGreal, The Guardian, 19 July 2021, reporting that: 'Critics accuse Shell and other major oil firms of using API [the American Petroleum Institute] as cover for the industry. While companies run publicity campaigns claiming to take the climate emergency seriously, the trade group works behind the scenes in Congress to stall or weaken environmental legislation'.

[103] This may be contrasted with cases where the public interest considerations appear to have been added on after the event or where the actions amount to fairly orthodox anti-competitive conduct (as per the Green Dot case — discussed in footnote [97]. Note also the Commission's acknowledgment at para 559 of its Draft Horizontal Guidelines of 1 March, 2022 that 'the fact that an agreement genuinely pursues a sustainability objective may be taken into account in determining whether the restriction in question is a restriction by object or a restriction by effect within the meaning of Article 101(1)' — see EU Commission, Draft Guidelines on the applicability of Article 101 of the

Dominant companies should not be discouraged from 'doing the right thing' or trying to make a contribution to combat climate change for fear of the competition law consequences. This is important as dominant companies are often (not always) large multinationals which have the economic clout and the potential to make a real difference. While it is correct to be cognisant of the possibility of some companies 'green washing'[104] there are companies (and certainly many individuals within companies) who are genuinely trying 'to make a difference'. Competition law should not make it more difficult to put these good intentions into practice. Allowing abuse of dominance provisions to act like a 'shield' may, in some circumstances assist with this.

12.3.6 Conclusion on Abuse of Dominance and Sustainability

This section of the contribution has sought to argue that there is more scope to use abuse of dominance provisions such as the UK Chapter II prohibition than is often appreciated (and there is less of a tension between competition law and sustainability goals than is sometimes suggested).

Particularly, competition lawyers and authorities alike should not be afraid to re-visit the fundamental purpose of these provisions. When this is done, it will often be clear that unsustainable conduct by dominant companies is an 'abuse' such that the Chapter II prohibition can be used as a 'sword' to attack them.

Conversely, there is scope to recognise sustainability as a shield against (false) accusations that genuine practices to mitigate climate change or to pursue sustainability goals are an abuse of a dominant position.

In this way, the laws governing abuse of dominance can indeed help take action for a sustainable future—and play an active part in ensuring that future (rather than being an obstacle to it).

12.4 UK Merger Control and Sustainability in the UK[105]

Although the focus of the sustainability debate has, for good reason, often focused on potential obstacles to competition cooperation, merger control regimes neverthe-less have a role to play in both advancing and protecting the sustainability agenda. This could arise in two different ways, for example, a competition authority

---

Treaty on the Functioning of the European Union to horizontal co-operation agreements (March 2022).

[104] This concept is explored further in the CMA's new Green Claims Code, accessible at: <https://www.gov.uk/government/publications/green-claims-code-making-environmental-claims>. Accessed 19 November 2022.

[105] For further discussion of how sustainability considerations can be factored into merger control see articles by Kar, Cochrane and Spring and by Burnside, De Backer and Strohl in Holmes, Climate change, sustainability and competition law (2020) Journal of Antitrust Enforcement,

prohibiting or requiring remedies in relation to a merger on the basis of sustainability concerns, i.e. because it may result in environmental harm (akin to using sustainability considerations as a 'sword'); or alternatively, approving a merger on the basis that resulting environmental benefits can justify/offset any harm that may stem from the loss of competition (akin to the 'shield' approach).

The CMA has not, to date, considered sustainability issues in-depth in any merger control review, but we anticipate there will be an increasing number of transactions being reviewed by the CMA which have a sustainability angle for the following reasons: (i) the CMA is likely to proactively seek out such deals in line with its 'strategic priority' to encourage sustainability goals; (ii) businesses operating in sustainability-related sectors are expanding incredibly fast, which is likely to increasingly drive M&A and investment in these sectors; and (iii) even in the wider economy, ESG considerations are driving commercial decision-making and strategies, leading to sustainability-related investments.

The CMA's ability to scrutinise transactions involving a sustainability-element is facilitated by its relatively broad and flexible jurisdiction. It has jurisdiction to review transactions involving either: (i) a Target that has UK turnover of GBP 70 million or greater; or (ii) a share of supply of goods or services in the UK of 25% or more. This latter threshold is far more flexible and 'elastic' than a market share test and provides the CMA with very significant discretion in determining whether the parties overlap. It is, therefore, well positioned to 'call in' transactions with sustainability implications.

12.4.1 Sustainability as a 'Sword': Assessing Sustainability as a Parameter of Competition

There are two principal routes for the CMA to account of sustainability considerations during the merger control review process—where sustainability is a key parameter of competition between the parties, or where the parties anticipate environmental efficiencies or customer benefits arising from the merger, which can be used to 'offset' any harm to consumers from the loss of competition.

First, it can consider sustainability as part of the initial competition assessment. As part of its merger assessment, the CMA will identify the competitive parameters that are most important to the process of rivalry. The nature of competition will influence the theories of harm considered, and it will consider both price and non-price aspects of competition as relevant in the circumstances.[106] If one of the parameters for competition between the merging parties is sustainability-related

---

Volume 8, Issue 2, Part VII Part VII; and S. Holmes and M. Meagher, 'A Sustainable Future: how can control of monopoly power play a part', SSRN May 3, 2022.

[106] CMA, 'Merger assessment guidelines' (CMA129) updated 18 March 2021 (the 'MAGs'), para 2.3–2.4, accessible at <https://www.gov.uk/government/publications/merger-assessment-guidelines>. Accessed 19 November 2022.

(e.g. the parties compete on energy efficiency, 'green' credentials, or sustainability innovation) the CMA can account for that in its competitive assessment.

The CMA has, in principle, taken important steps towards accounting for sustainability considerations during merger control reviews. In 2021, the CMA updated its MAGs and explicitly recognised that the sustainability of a product or service can constitute a non-price aspect of competition that can be an important part (or even the primary focus) of the competitive process,[107] i.e. where firms compete on factors relevant to environmental sustainability. Applying this principle to a specific transaction could, therefore, enable the CMA to challenge a merger on the basis that it would result in a substantial lessening of competition ('SLC') on sustainability parameters—and either prohibit the merger or require remedies to prevent the loss of competition (and resulting environmental harm).

This is particularly relevant in the context of 'sustainability-killer acquisitions'. As in the wider pharmaceuticals and technology context where this concept was initially popularised, this could involve an established firm acquiring a new and innovative firm to pre-empt future competition. This could be a polluting firm acquiring a 'green' competitor that has innovated a more sustainable approach. The consequence of this transaction may be that the innovating business is shut down ('killed') or that its innovative technology remains on the market as a niche product but is not expanded across the market.

In this scenario, where competition is on sustainability-related terms, prohibiting the transaction may be unnecessary as cleaner rivals may be able to persuade environmentally conscious customers to switch to them and exercise an ongoing competitive constraint. However, if there would, nevertheless, be a negative environmental impact, this may require remedies or even prohibition—particularly if consumers may not accurately assess future costs (hyperbolic discounting), have status quo biases which make them reluctant to try new products, or may believe that choices made at an individual level cannot make a difference.

12.4.2 Sustainability as a 'Shield': Offsetting SLCs with Sustainability Benefits

The second way in which the CMA can account for sustainability is by recognising environmental benefits that stem from the transaction—and considering whether these environmental benefits (whether independently or in conjunction with other efficiencies) are sufficient countervailing factors that they prevent or mitigate any SLC arising from a merger by offsetting anticompetitive effects.[108]

---

[107] CMA, 'Merger assessment guidelines' (CMA129) updated 18 March 2021 (the 'MAGs'), para 2.5, accessible at <https://www.gov.uk/government/publications/merger-assessment-guidelines>. Accessed 19 November 2022.

[108] Ibid., para 8.1.

12 Sustainability and Competition Law in the United Kingdom

Merger efficiencies fall into two categories: (i) rivalry-enhancing efficiencies (which must not only be timely, specific and sufficient to prevent an SLC from arising, but also merger-specific and benefit customers in the UK); and (ii) relevant customer benefits ('RCBs') (i.e. benefits to UK customers, other than through improved competition in the relevant market).

The updated MAGs recognise that environmental benefits can constitute merger efficiencies. Efficiencies can include sustainable elements to the extent they are rivalry-enhancing, and may include enabling more efficient/sustainable production processes, better innovation or more environmentally-friendly R&D. RCBs can be broader, and capture environmental benefits such as reduced carbon emissions and supporting the transition to a low carbon-economy (to the extent that firms do not normally compete on sustainability).[109] The CMA also notes that a merger may lead to lower energy costs and some benefits that customers may value (such as a lower carbon footprint of the firm's products).

Such RCBs may act as countervailing factors that prevent or mitigate any SLC arising from the merger.[110] A significant advantage of the CMA accounting for environmental RCBs is that, unlike rivalry-enhancing efficiencies, they can be realised in markets other than the relevant market.[111]

12.4.3 Consideration of Sustainability Issues in CMA Merger Reviews to Date

Although the CMA has now laid the conceptual framework for addressing sustainability in merger cases, there has been very limited analysis of such issues in merger control cases to date—although, as noted above, this is likely to change as sustainability issues become increasingly important for businesses, both as a parameter of competition (in response to shifting consumer preferences) and as a driver of M&A activity.

The CMA recognised the role of sustainability as a non-price parameter of competition between firms in *Cargotec Corporation/Konecranes Plc* (2021).[112] Particularly, it considered that the parties were, in part, close competitors for the supply of 'reach stackers' ('RS') because both parties were taking active steps to

---

[109] Ibid., para 8.3(b), 8.21.

[110] Although the CMA notes that it is rare for a merger to be cleared on the basis of rivalry enhancing efficiencies or relevant customer benefits alone (for example, as of March 2021, the CMA had only cleared three cases on the basis of RCBs), CMA, 'Merger assessment guidelines' (CMA129) updated 18 March 2021, para 8.27, accessible at <https://www.gov.uk/government/publications/merger-assessment-guidelines>. Accessed 19 November 2022.).

[111] However, this is perhaps limited by the concept of 'relevant customers' which only includes the parties' direct consumers under Section 30(4) of the Enterprise Act 2002.

[112] Cargotec Corporation / Konecranes Plc Merger Inquiry, Final report dated 29 March 2022, available at: https://assets.publishing.service.gov.uk/media/62458e408fa8f52773d76abf/310322_Cargotec_Conecranes_Final_Report.pdf Accessed 12 May 2022.

develop electrified mobile equipment and considered themselves as being among the few suppliers that could offer a full range of RS (including value, premium and eco-friendly).[113] However, the CMA accounted for a wide variety of evidence across numerous products in reaching its decision that the transaction would result in a SLC, and did not specifically find or indicate that the SLCs would result in environmental harm (e.g. that eco-friendly RS may become more expensive or no longer be available).

The CMA also considered potential sustainability efficiencies in *Cargotec/ Konecranes*. The parties argued that 'an important part of the deal rationale is to ensure that the parties would be better placed to address sustainability challenges in the industry, by providing a platform for innovation'. However, the CMA decision states that the parties did not provide further evidence to support this argument. Given that the CMA applies a high evidentiary standard for accepting rivalry-enhancing efficiencies arising from a merger (irrespective of whether sustainability-related or not), it—unsurprisingly—did not consider it had sufficient evidence to either assess or give credit for any sustainability-related efficiencies that would offset the SLC.[114]

The CMA also considered environmental benefits arising from a merger in *Pennon Group plc/Bristol Water*.[115] This merger involved regulated water companies and, as such, was assessed under a distinct regime which does not apply the usual SLC test. Nevertheless, as in the standard process, the parties were able to argue that the prejudice arising from the merger could be offset by merger-specific relevant customer benefits (RCBs), and therefore still has comparative value. In this case, the parties argued that the RCBs would include, among other benefits, 'environmental benefits'.[116] However, the CMA noted that it was unclear whether the environmental benefits would be merger-specific and whether the proposed benefits would be cost effective. The CMA also noted that the acquirer was a 'poor performer on environmental outcomes'.[117] It concluded, that there was insufficient evidence that the potential RCBs would outweigh the merger's adverse impacts.

As such, the CMA has not accepted proposed environmental benefits resulting from a transaction as a basis on which to offset the harm from loss of competition. This emphasises the importance of the CMA being prepared to put sufficient weight on sustainability benefits as offsetting potential competitive harm. However, in at least one, if not both, of the recent cases where the parties mooted environmental

---

[113] Ibid., para 201.

[114] Ibid., para 58.

[115] Case ME/6946/21, Pennon Group plc / Bristol Water Holdings UK Limited merger enquiry, full text decision available at: https://assets.publishing.service.gov.uk/media/622a2efcd3 bf7f5a86be8f8a/Pennon_Bristol_fulltext.pdf. Accessed 12 May 2022.

[116] Ibid., para 57.

[117] Ibid.

benefits, it does not appear that the parties provided sufficiently detailed evidence to enable the CMA to rely on those benefits.[118]

12.4.4 Incorporating Sustainability into Merger Control Going Forward

It is clear, therefore, that if the CMA is to act on its 'strategic priority' by using merger control to support the sustainability goals, this requires action on behalf of both regulator and dealmakers.

The CMA, for its part, has laid out a high-level framework for incorporating sustainability considerations into its merger analysis through its updated MAGs. More, however, must be done. Particularly, the CMA must have a clearer, publicised process for quantifying such benefits and demonstrating how they can be applied to offset anticompetitive harms. The CMA has said that it is considering whether it can provide practical guidance on how it weighs sustainability-related efficiencies and RCBs against competition concerns.[119] Without guidance on these steps, dealmakers and their advisers will struggle to assess, justify and present (both for internal approval and to the CMA) even genuinely 'green' transactions.

Nevertheless, if the CMA is to accept and give credit to sustainability benefits, much of the onus (initially at least) is on merging parties. Dealmakers (and their advisers), need to be willing to present and effectively defend mergers motivated by sustainability objectives or that will generate sustainability benefits. Business rationales need to be clearly articulated and the efficiency analyses well-evidenced. Absent such evidence, even genuine sustainability benefits risk being perceived as greenwashing and competition authorities will not have a credible basis to give meaningful and sufficient credit to such benefits to outweigh any competitive harm. The following three points are particularly important for merging parties to be able to demonstrate:

Merger-specificity: sustainability benefits must be 'reliant on the merger in question' and cannot also be brought about by other means. The CMA will investigate, and the parties will need to explain, what barriers there are against merging firms achieving the same improvements absent the merger. Particularly, the CMA may be sceptical whether the sustainability benefits could also be achieved by the

---

[118] CMA, 'Merger assessment guidelines' (CMA129) updated 18 March 2021, para 8.27, accessible at <https://www.gov.uk/government/publications/merger-assessment-guidelines>. Accessed 19 November 2022.

[119] CMA, Environmental sustainability and the UK competition and consumer regimes: CMA advice to the Government (14 March 2022), available at https://www.gov.uk/government/publications/environmental-sustainability-and-the-uk-competition-and-consumer-regimes-cma-advice-to-the-government/environmental-sustainability-and-the-uk-competition-and-consumer-regimes-cma-advice-to-the-government. Accessed 19 November 2022.

firms acting independently or if consumers could access the sustainability benefits without the merger.[120]

Verifiability: merging parties will need to provide sufficiently detailed evidence in order for the CMA to verify efficiency claims. The CMA notes in its Guidelines that many efficiency claims by merger firms are not accepted by the CMA because the evidence supporting those claims is difficul to verify and substantiate.[121] Yet in most cases those efficiencies are usually cost and revenue synergies and more easily balanced against, for example, potential price rises, than less quantifiable sustainability benefits.

Internal documents: although these are not a specific part of the CMA's criteria, internal business documents are likely to be reviewed by the CMA and will be key in supporting (or undermining) the deal rationale and potential efficiencies (i.e. they must be consistent with the purported environmental goals, and not instead indicate an anticompetitive rationale). A company's wider ESG record may also be considered for evidence as to the credibility of the parties' aims (as in *Pennon/Bristol Water* where the CMA noted that Pennon was already a poor performer on environmental outcomes).

12.4.5 Public Interest Regime

Beyond what is currently permitted under the legislative framework, one way of strengthening the role of sustainability considerations within the merger control regime would be to enable the UK government to intervene in merger cases on sustainability grounds, by adding sustainability as a 'public interest consideration' under Section 58 Enterprise Act 2002 ('EA02'). This would allow the government to intervene to refer a merger for an in-depth investigation, or block or clear a transaction on sustainability grounds (e.g. clearing it where the merger gives rise to competition concerns and it considers that, in the interests of public policy, the merger's environmental benefits outweigh any potential restriction of competition). Indeed, in its Advice to the Government, the CMA noted that the Government may wish to add a new 'sustainability' public interest consideration, or legislate to specifically include sustainability benefits as RCBs.

---

[120]CMA, 'Merger assessment guidelines' (CMA129) updated 18 March 2021, para 8.16–8.19, accessible at <https://www.gov.uk/government/publications/merger-assessment-guidelines>. Accessed 19 November 2022.

[121] Ibid., para 8.6.

12.4.6 Sustainability and the UK's Market Inquiry Regime[122]

A particularly important tool for the CMA (as well as other sectoral regulators in the UK which also have competition powers) in advancing the sustainability agenda is the power to undertake market inquiries. These generally involve a 12-month market study, and may be followed by a subsequent 18–24-month in-depth market investigation. In contrast to investigations under the Enterprise Act for a merger control review, market inquiries can be launched by the CMA to examine whether, and if so why, markets may not be working well and if there are competition or consumer problems within the market. Particularly, it will take an overview of regulatory or other economic drivers as well as patterns of consumer and business behaviour.[123]

In contrast to its competition enforcement capacity, there is no need for any market participant to be suspected of infringing competition law in order for the CMA to commence an investigation, conclude that the market is not working effectively, or to recommend/require remedies. Indeed, market inquiries can lead to a range of outcomes, including: (i) a clean bill of health; (ii) encouraging businesses to self-regulate; (iii) making recommendations to government to change regulation or public policy; (iv) taking competition or consumer enforcement actions; (v) behavioural remedies, such as market-opening measures and informational remedies; and (vi) structural remedies, such as divestiture or IP remedies.

Market inquiries are, therefore, a particularly effective means for the CMA to ensure and shape effective competition in emerging markets relevant to sustainability and Net Zero goals before competition issues develop or become acute. Moreover, the CMA considers that it can, where appropriate, consider environmental sustainability to the extent that it leads to economic harm to consumers.[124]

In 2021, the CMA completed a market study into electric vehicle charging, an emerging market of central importance to the green transition and Net Zero ambitions. The market study identified a number of challenges for the sector as it develops and set out a number of remedies. These remedies included recommendations to government to increase the pace and scale of charging roll-out across the UK. The CMA also used the market study as the basis to launch a Chapter I investigation into long-term exclusivity arrangements for charging along motorways. It subsequently accepted commitments by the relevant firms in March 2022. The CMA noted in its decision that the transition from petrol and diesel cars to

---

[122] See, also the discussion in S. Holmes, Climate change, sustainability and competition law (2020) Journal of Antitrust Enforcement, Volume 8, Issue 2 at pp. 392–393.

[123] Market studies and investigations – guidance on the CMA approach: CMA3, published on 10 January 2014, para 1.5 available at: <https://www.gov.uk/government/publications/guidance-on-the-cmas-investigation-procedures-in-competition-act-1998-cases/guidance-on-the-cmas-investigation-procedures-in-competition-act-1998-cases#issuing-the-cmas-provisional-findings%2D%2Dthe-statement-of-objections-and-draft-penalty-statement.>. Accessed 19 November 2022.

[124] CMA, 'Environmental Sustainability Advice to Government: Call for Inputs', para 79, available at: https://www.gov.uk/government/consultations/environmental-sustainability-advice-to-government-call-for-inputs. Accessed 19 November 2022.

EVs is key to the UK's Net Zero commitments by 2050. As such, the CMA considered it was 'essential that there is a comprehensive and competitive EV charging network in place'.[125] In an open letter to the motorway service area operators and EV charge point operators, the CMA noted its decision will help develop a competitive charging network to support the government's ban on new petrol and diesel car sales by 2030.

Reflecting the importance of the sector and the importance of the policy prerogative in shaping its market monitoring and enforcement work, the CMA has publicly committed to launching at least one market study in a Net Zero-related market in the next financial year.

Moreover, as with mergers, sustainability could be included as a public interest consideration under Section 153 of EA02, enabling the UK government to intervene in markets cases on sustainability grounds.

12.4.7 Conclusion

This contribution has sought to consider the ways in which sustainability considerations can be used as (i) a 'sword' when applying competition law by finding agreements/practices/mergers that are unsustainable or harmful to society as anticompetitive, and (ii) a 'shield' by protecting agreements/practices/mergers genuinely aimed at achieving sustainability goals from the application of competition law.

Although UK competition law is alive to the issues in this context, it can and must go further than it currently has to facilitate businesses in achieving sustainable goals in the fight against climate change and the move towards a more sustainable economy.

Appendix 1: Simon Holmes, Selected Papers on Climate Change, Sustainability and Competition Policy

Simon Holmes, Visiting Professor of Law, Oxford University, Judge, UK Competition Appeal Tribunal.

1. A Sustainable Future: how can control of monopoly power play a part? Simon Holmes and Michelle Meagher, 3 May, 2022. Available at: <https://papers.ssrn. com/sol3/papers.cfm?abstract_id=4099796>

---

[125] CMA, Case number 51050, 'Decision to accept binding commitments in relation to certain exclusive arrangements for the supply of electric vehicle charging points' (8 March 2022), para 1, available at: https://assets.publishing.service.gov.uk/media/622634d28fa8f5490d52ef91/EV_decision_to_accept_commitments_V2_070322.pdf. Accessed on 19 November 2022.

2. Revised draft horizontal guideless on sustainability agreements. ClientEarth and Simon Holmes' contribution to the European Commission's public consultation, April, 2022. Available at: https://www.clientearth.org/latest/documents/horizon tal-agreements-between-companies-revision-of-eu-competition-rules/
3. JAE Paper 'Climate Change, Sustainability and Competition Law', accessible at: https://academic.oup.com/antitrust/article/8/2/354/5819564? login=true
4. 'Climate Change, Sustainability and Competition Law in the UK' (published in the ECLR, July, 2020); available at: https://www.law.ox.ac.uk/sites/files/oxlaw/ cclp_working_paper_cclpl51.pdf
5. Climate Change and Competition Law – Simon Holmes note for OECD Roundtable on sustainability and competition law: available at https://one.oecd.org/ document/DAF/COMP/WD(2020)94/en/pdf
6. Horizontal agreements between companies-revision of EU competition rules. ClientEarth and Simon Holmes contribution to the Commission's consultation, October, 2021. Available at: <https://www.clientearth.org/media/3c4lsaex/ clientearth-and-s-holmes-contribution-horizontal-agreements-04-10-2021.pdf>
7. 'Competition Policy supporting the Green Deal'. [Concurrences No 1-2021]. Available at https://www.concurrences.com/en/review/issues/no-1-2021/ conferences/sustainable-development-what-role-for-competition-policy-new-frontiers-of
8. 'Climate Change is an existential Threat: competition law must be part of the solution and not part of the problem' [CPI Antitrust Chronical, July 2020]. Available at: <https://www.competitionpolicyinternational.com/climate-change-is-an-existential-threat-competition-law-must-be-part-of-the-solution-and-not-part-of-the-problem/>
9. 'Consumer welfare, sustainability and competition law goals' [Concurrence No 2- 2020]. Available at: <https://www.concurrences.com/en/review/issues/no-2-2020/foreword/consumer-welfare-sustainability-and-competition-law-goals-93496-en>
10. Covid 19 Lessons for climate change blog, April 2020: Available at: <http:// competitionlawblog.kluwercompetitionlaw.com/2020/04/23/climate-change-sustainability-and-competition-law-lessons-from-covid-19/>
11. 'Competition Law, Climate Change & Environmental Sustainability', edited by Simon Holmes, Dirk Middelschulte and Martijn Snoep, [Concurrence]. Available at: <https://www.concurrences.com/en/all-books/competition-law-climate-change-environmental-sustainability>

Open Access This chapter is licensed under the terms of the Creative Commons Attribution 4.0 International License (http://creativecommons.org/licenses/by/4.0/), which permits use, sharing, adaptation, distribution and reproduction in any medium or format, as long as you give appropriate credit to the original author(s) and the source, provide a link to the Creative Commons license and indicate if changes were made.

The images or other third party material in this chapter are included in the chapter's Creative Commons license, unless indicated otherwise in a credit line to the material. If material is not included in the chapter's Creative Commons license and your intended use is not permitted by statutory regulation or exceeds the permitted use, you will need to obtain permission directly from the copyright holder.

Part II

Sustainability Objectives in Intellectual Property Law

Sustainability and Intellectual Property in Austria

13

Georg Kresbach

13.1 Introduction

Intangible assets such as patents, trademarks, designs and data are becoming increasingly important in today's knowledge economy. According to the EU Commission's Action Plan on Intellectual Property to strengthen EU's economic resilience and recovery ('IP Action Plan')[1] published on 25 November 2020, IP-intensive industries account for 45% of GDP and 93% of EU exports, while the added value of IP is growing across most European industrial ecosystems. Globally, IP filings are on the rise, as intangible assets become crucial in the global race for technological leadership.[2] It goes without saying that the success of sustainable economies depends to a large extent on new and innovative solutions. Intellectual property rights, therefore, undoubtedly play an increasingly significant role in sustainable and green businesses.

Although the term sustainability is used as a buzzword for more and more economic sectors, there is still no uniform legally binding definition of it. Nevertheless, in some cases, sustainable economic behaviour is already impacting companies' strategies. For instance, in the real estate industry, compliance with ESG

---

[1] Communication from the Commission to the European Parliament, the Council, the European Economic and Social Committee and the Committee of the Region - Making the most of the EU's innovative potential. An intellectual property action plan to support the EU's recovery and resilience of 25 November 2020, COM(2020) 760 final, CELEX number 52020DC0760.

[2] https://ec.europa.eu/commission/presscorner/detail/en/ip_20_2187. Accessed 19 November 2022.

G. Kresbach (✉)
Wolf Theiss Rechtsanwälte GmbH & Co KG, Vienna, Austria

Wolf Theiss Rechtsanwälte GmbH & CoKG, Wolf Theiss Rechtsanwälte GmbH & CoKG, Vienna, Austria
e-mail: georg.kresbach@wolftheiss.com

© The Author(s) 2024
P. Këllezi et al. (eds.), *Sustainability Objectives in Competition and Intellectual Property Law*, LIDC Contributions on Antitrust Law, Intellectual Property and Unfair Competition, https://doi.org/10.1007/978-3-031-44869-0_13

('Environmental, Social & Governance') criteria is considered to be a desirable guiding principle, and in the financial services sector, ecologically sustainable investments are to be promoted through the so-called EU Taxonomy Regulation.[3] For the purpose of this report, the term sustainability is used as a common paraphrase of this word and is understood to mean a principle of action for the use of resources, according to which no more should be consumed than can be regrown, regenerated or provided again in the future. Sustainability thus aims to achieve the most resource-conserving and long-lasting effect possible.[4] In other words, sustainability means that current thinking and actions should improve the living situation of the present generation without worsening the future of the next generation.[5]

13.2 Current Status of Sustainability in Austrian IP Law

13.2.1 Overview

In Austrian industrial property law (patent and utility model law, trademark law and design law) and copyright law, there are no special legal provisions that explicitly deal with the concept of sustainability.

However, this does not mean that industrial property protection has no practical relevance for 'sustainability' in Austria. For example, in May 2022, a search in the national trademark register for the terms 'Nachhaltigkeit' or 'sustainability' or 'Green' yielded more than 200 results, such as:

- 'Carbon View SUSTAINABILITY SOLUTIONS',[6]
- 'CLOUD SUSTAINABILITY',[7] or
- 'GREENROAD'.[8]

These results show quite clearly that the topic of *sustainability* has long since found its way into—at least trademark—law.[9]

---

[3] Council Regulation (EU) 2020/852 of 18 June 2020 on the establishment of a framework to facilitate sustainable investment, and amending Regulation (EU) 2019/2088, OJ 2020, L 198, p. 13.

[4] B. Müller, D. Widl, E. Sonnleithner In: Zahradnik, Richter-Schöller (eds), Handbuch Nachhaltigkeitsrecht, Manz 2021, p. 32.

[5] See http://webarchiv.bundestag.de/archive/2008/0506/wissen/analysen/2004/2004_04_06.pdf. Accessed 5 January 2023.

[6] Austrian trademark AT 1179764, protected – inter alia – for *Computer software; computer programs (downloadable software)* in class 9.

[7] Austrian trademark AT 12244703, protected – inter alia – for *material processing; recycling, recirculation, processing, sorting, treatment and disposal of waste* in class 40.

[8] Austrian trademark AT 267537, protected – inter alia – for *engines, other than for land vehicles* in class 7 and *accumulators* in class 9.

[9] The situation is even more pronounced at the European level: as per May 2022, the search term 'sustainability' yielded around 300 hits and 'green' over 9000.

13 Sustainability and Intellectual Property in Austria

Apart from these examples, intellectual property law can have both a positive and a negative impact on the issue of sustainability. This can probably be traced back to the fact that the shift to sustainable economic performance requires a transformation of long-established economic systems, and frictions between the old system and the new evolving system are unavoidable. Traditional production processes that no longer meet the specifications of sustainability ('climate neutrality', 'resource conservation', 'circular economy') must be replaced by innovative production methods. All these changes will have an impact on the established system of industrial property rights, which are tailored to traditional manufacturing methods.[10]

Consequently, the existing legal framework of intellectual property law will have to be adapted to meet the needs for adequate legal protection for sustainable innovations. In addition, the owners of intellectual property rights will also have to adapt to changed framework conditions to better meet business objectives related to sustainability.

13.2.2 Patent Law

In patent law, the prospect of a patent being granted can naturally be an additional motivating factor to develop environmentally friendly and sustainable solutions. One example of such an invention is the process for the production of green hydrogen by means of electrolysis for the generation of green oxyhydrogen gas, which can be mixed with natural gas resulting in a product with a lower carbon content.[11]

In the US, the number of patents granted in the field of clean energy (so-called 'green patents') increased dramatically in the first decade of the 21st century and continue to show increasing growth rates. Between 1998 and 2007, an increase of more than 135% was recorded.[12] Furthermore, given the fact that sustainable ideas are very popular, patented inventions that represent climate-friendly or climate-neutral alternatives can generate corresponding licensing income for the patent holder through their commercialisation.

However, in addition to the positive aspects of intellectual property rights, which support the sustainable economic output described above, there are a number of downsides.

---

[10] E. Epplinger. P. Vimalnath, A. Jain, E. Kushnir, A. Gurtoo and F. Tietze: Sustainability Transition in Manufacturing: The Role of New Entrants and how they use IPR. In: Science and innovation – an uneasy relationship? Rethinking the roles and relations of STI policies. In: European Forum for Studies of Policies for Research and Innovation. Oslo: 2021, pp. 1–18e; https://euspri-forum.eu/eu-spri-annual-conference-oslo-june-9-11-2021/. Accessed 1 June 2022.

[11] See www.key-energy.eu. Accessed 21 April 2022. Whether patent protection has actually been granted for this process is beyond the knowledge of the author of this article.

[12] E. Koester, Green Entrepreneur Handbook: The Guide to Building and Growing a Green and Clean Business, CRC Press 2011, pp. 103–114.

A patent can always be blocked for use by third parties by its holder for a certain period of time. Thus, inventions that could help humanity achieve a much-needed more sustainable way of life, in most cases, become difficult to access for the masses. While patent protection is valid, third parties can use inventions only if they enter into often very expensive licence agreements. The only way out of this are compulsory licences, which the owner of a patent is obliged to grant. However, the requirements for this are very strict under both the Austrian Patent Act[13] and the TRIPS Agreement[14] and often do not represent a realistic alternative. If there are no licence agreements—of whatever kind—third parties can only freely access the protected inventions 20 years after the filing date. In the fight against climate change, every day counts and inventions that will only be made available to the masses decades later are thus blocked or difficult to access for a period of time that is immensely important.

Ultimately, however, it also depends on how holders of intellectual property rights (especially patents) manage them. There are certainly owners who see 'the greater cause' behind their invention and are very well aware that their inventions represent important contributions in the fight against climate change and thefore pass on their protected technologies at low licence prices to third parties. At the same time, there will certainly also be entrepreneurs who see the lucrative business behind sustainable inventions and will offer licensing agreements only at very high fees.

13.2.3 Trademark Law

Trademark law and design law (design patent law) already provide valuable contributions to the promotion of sustainable products and economic output. The consumer behaviour of Austrians who tend to opt for environmentally friendly products, for example, reflects the growing trend towards a sustainable lifestyle. The ever-increasing demand for sustainable products can be measured in figures. There are already more than 800 registered 'organic' brands in Austria, and based on the trademarks granted according to the Austrian Trademark Register, the number of 'organic' brands has increased significantly from 2018/2019. The share of organic food in the retail sector has also more than tripled within two years from 2.7% in 2019 to 11.3% in 2021.[15] Brands and product designs that convey that products were produced in a climate-neutral way or are otherwise good for the environment

---

[13] Section 36 Austrian Patent Act, Federal Law Gazette No. 137/1971 as amended by Federal Law Gazette I. No. 61/2022.

[14] Articles 30 and 31 TRIPS Agreement.

[15] Statista Research Department (2022): Organic share of sales of various food products in Austria 2021. See https://de.statista.com/statistik/daten/studie/427038/umfrage/anteil-ausgewaehlter-bio-lebensmittel-im-lebensmitteleinzelhandel-in-oesterreich/#:~:text=Insgesamt%20hat%20sich%20innerhalb%20von,Euro%20f%C3%BCr%20Bio%2DLebensmittel%20aus. Accessed 19 November 2020.

immediately catch the eye of buyers and certainly contribute to such products becoming a bestseller.

In this context, quality labels and seals of approval are of particular importance in Austria. They denote products and services that meet specified certification requirements as organic, climate-neutral or climate-friendly or similar and can be protected as guarantee marks. A rough search in May 2022 shows that there are well over 100 organic quality labels in Austria.[16]

An example of such a quality label is the Austrian Ecolabel,[17] which is licensed by the Austrian Federal Ministry for Climate Protection, Environment, Energy, Mobility, Innovation and Technology in compliance with strict environmental criteria that can probably also be subsumed under the term 'sustainability'.

Another example is the AMA Quality Seal[18] that confirms the quality of products (foodstuffs) has been tested. The seal is even protected as a trademark in Austria.

The AMA Quality Seal is the quality mark administered and controlled by Agrarmarkt Austria, a company under public law.

The seal is awarded to foodstuffs that exceed the legal requirements in terms of quality and whose origin can be traced. The guidelines for awarding the AMA Quality Seal must be approved by the Federal Ministry for Sustainability and Tourism. The products' compliance with the standards is ensured by independent, state-accredited inspection bodies and laboratories. The seal is the best-known quality mark for food in Austria and is recognised by the majority of people living in the country.[19]

Even in trademark and design law, sustainable intellectual property rights do not always have only positive effects. Consider, for example, 'greenwashing', i.e. cases in which trademarks are registered for goods and services to present them to consumers as sustainable, climate-friendly or the like, or at least as good for the environment, when in reality they are not at all—or at least not to the extent that is made credible to consumers.[20] This issue is exacerbated by the fact that the certification criteria for granting licences for certification marks are not always transparent and largely unregulated.[21] Here, the ground for refusal of registration for deceptive signs according to Sec. 4 para 1 no. 8 of the Austrian Trademark Act and/or the prohibition of misleading advertising (Sec. 2 Austrian Unfair Competition Act) and/or aggressive business practices (Sec. 1a Austrian Unfair Competition Act)

---

[16] https://www.bewusstkaufen.at/ratgeber/fleischersatzprodukte/. Accessed 27 May 2022.

[17] https://www.umweltzeichen.at/de/home/start. Accessed 19 November 2022.

[18] AMA-Gütesiegel: AMA (amainfo.at). Accessed 19 November 2022.

[19] https://www.amainfo.at/konsumenten/siegel. Accessed 19 November 2022.

[20] More on this topic e.g. Gütezeichen auf dem Prüfstand – Kaufhilfe oder Greenwashing? https://konsum.greenpeace.at/guetezeichen/. Accessed 19 November 2022.

[21] J. Luksan, Zur Intransparenz von Gütesiegeln – Wo Nachhaltigkeit endet und Greenwashing beginnt, NR 2021, pp. 428–435.

could come into play as corrective measures.[22] For example, in Austrian case law, the designation 'made of biological building materials' for bricks was considered misleading within the meaning of Sec. 2 Austrian Unfair Competition Act, because consumers understand 'biological' to mean free of chemicals, but actually a petrochemical product was used in the production of these bricks.[23]

Another corrective measure for combatting unfair business practices against inaccurate claims of supposedly environmentally friendly properties in the presentation and advertising of products or services will be the Directive of the European Parliament and of the Council proposed on 30 March 2022. It aims to empower consumers for the green transition through better protection against unfair practices and better information.[24] One of the key objectives of this directive is to enable consumers to make informed purchasing decisions and therefore to contribute to more sustainable consumption. It also targets unfair commercial practices that mislead consumers away from sustainable consumption choices. [25]

13.2.4 Recent M&A Transactions in the Field of Sustainability

In recent years, there has been a noticeable increase in corporate transactions (mergers, acquisitions, investments, as well as the formation of joint ventures) and licensing in Austria the main value-creating factor of which was intellectual property rights. These include, for example, the acquisition of an Austrian biotech company focused on the research and development of vaccines against SARS viruses by an international pharmaceutical company, whereby patented processes for the production of vaccines constituted a significant share of the target company value.[26]

Another trend in Austria is an increase in transactions or licensing related to goods and services that have the provision of sustainable economic output as their objective or that are otherwise related to sustainable business practices. As examples

---

[22] According to the case law of the German Federal Supreme Court, the award of a quality mark can be prohibited if the associated (over-)examination processes do not show sufficient objectivity; see German Federal Supreme Court 4 July 2019, I ZR 161/18 – IVD-quality seal. See also A. Anderl, A. Ciarnau. In: Zahradnik, Richter-Schöller (eds), Handbuch Nachhaltigkeitsrecht, Manz 2021, pp. 75–77.

[23] See Austrian Supreme Court 20 September 1994, 4 Ob 90/94.

[24] Proposal for a Directive of the European Parliament and of the Council amending Directives 2005/29/EC and 2021/83/EU as regards empowering consumers for the green transition through better protection against unfair practices and better information of 30 March 2022, COM(2022) 143 final, 2022/0092 (COD).

[25] See reasons for and objectives of the Proposal for a Directive of the European Parliament and of the Council amending Directives 2005/29/EC and 2021/83/EU as regards empowering consumers for the green transition through better protection against unfair practices and better information of 30 March 2022, COM(2022) 143 final, 2022/0092 (COD), p. 1.

[26] See https://www.wolftheiss.com/press/press-release/wolf-theiss-advises-msd-on-the-acquisition-of-the-austrian-biotech-company-themis-bioscience/. Accessed 19 November 2022.

13 Sustainability and Intellectual Property in Austria

from the recent past, the following transactions can be cited, particularly in the automotive and e-mobility sectors:

- A joint venture with an Austrian company that produces battery systems for mobile applications;[27]
- Participation in a company that specialises in, among other things, charging point management for e-cars;[28]
- The acquisition of a company for the rebalancing of rental cars based on data sets for the optimisation of car-sharing models.

However, a more comprehensive analysis of comparable transactions in Austria with a focus on sustainable economic performance is unfortunately not possible due to the lack of availability of the relevant data.

13.3 The Role IP Should Play in Sustainability

13.3.1 Overview

Austria is one of the most modern industrialised nations in the world with a high number of highly-specialised domestic players and a well-developed start-up scene. In this high-tech and innovation-driven environment, intellectual property will play an important role in the development of sustainable solutions. IP should foster innovation by securing the rights of inventors on the one hand and facilitating knowledge sharing on the other. Another important role will be to facilitate investment in local businesses by ensuring IPRs through a strong and clear legal framework and efficient enforcement.

In efforts to overcome climate and environmental challenges and for a more careful use of natural resources, all available possibilities should be used to help mankind achieve a climate-neutral and resource-saving way of life as quickly as possible. For Austria and the EU to achieve the goal of being climate-neutral by 2040 or 2050, as set by the *Green Deal*,[29] we need clever minds with inventive skills and the courage to think in new directions. In terms of the 'reward and incentive theory', the prospect of or granting of patent protection can undoubtedly be a strong motivation factor for making an effort and accepting the potential financial risk of failure. From this perspective, intellectual property rights, especially patents, can make an important contribution to promoting sustainability.

---

[27] See https://www.wolftheiss.com/press/press-release/wolf-theiss-advises-miba-on-the-expansion-of-itseMobility-business/. Accessed 19 November 2022.

[28] https://www.enio-management.com. Accessed 19 November 2022.

[29] Communication from the Commission to the European Parliament, the European Council, the Council, the European Economic and Social Committee and the Committee of the Regions – The European Green Deal of 11 December 2019, COM(2019) 640 final, CELEX number 52019DC0640.

The role of intellectual property rights should not be reduced to no longer promoting less sustainable inventions; rather, the focus should be on making the development of climate-friendly ideas more attractive to the masses. A competition between 'traditional' inventions and 'sustainable' inventions appears to be less than beneficial and one should not exclude the other. The goal should be to ensure that ideas that can possibly contribute to stopping or at least slowing down climate change and which support the careful use of natural resources can be realised as quickly as possible, regardless of the financial means or previous experience of their inventor. Especially for start-ups that carry out research and development in the field of sustainable solutions, an adequate protection of their intellectual efforts is key to the successful marketing of their ideas and services either to investors or for selling licences.

13.3.2 Educational Challenges to Foster Sustainable Innovations

Although Austria has a well-developed legal system for the protection of intellectual property rights and can rely on efficient law enforcement by courts and the Austrian Patent Office, little has been done so far to specifically address the issues of sustainability. In intellectual property legal practice, there are no special provisions or other measures in Austria, either in the legal framework or in the law enforcement practice, that could contribute to the promotion of sustainable economic performance.

The Austrian Patent Office has reviewed over 7000 Austrian start-ups and found that only 6% of them apply for a trademark and only 2% for a patent.[30] The number is therefore surprisingly low and shows that Austria still has a lot of catching-up to do to promote awareness among young companies of the essential role that intellectual property rights can play in their success. In addition to the dissemination of appropriate information and educational work, financial support for SMEs lacking considerable expertise in intellectual property rights will play an essential role in facilitating access to the individual IP rights.

Nevertheless, initial efforts to promote sustainability are already evident in industrial property protection. Every two years, the Climate Protection Ministry jointly with the Austrian Patent Office awards the 'State Prize Patent' ('Staatspreis Patent')[31] for inventions and trademarks. The awards are intended to honour Austrian inventions and companies that have implemented in practice particularly original, distinctive and complex ideas. Increasingly, patents and trademarks that

---

[30] Austrian Patent Office (2021): Austrian Patent Office takes stock: More inventions despite Corona crisis. https://www.patentamt.at/alle-news/news-detail/artikel/oesterreichisches-patentamt-zieht-bi/#:~:text=Neue%20Studie%3A%20Das%20Patentamt%20hat,Das%20ist%20 schockierend%20wenig. Accessed 19 November 2022.

[31] See https://www.patentamt.at/staatspreis-patent-2022/. Accessed 19 November 2022.

13 Sustainability and Intellectual Property in Austria

represent significant contributions to combating climate change are recognised by this prize.[32]

Moreover, the Austrian Patent Office has introduced a number of *educational initiatives*, although they are not specifically geared towards sustainable inventions. It offers free advisory services (in person and online) on basic questions in connection with obtaining patent and trademark protection,[33] organises targeted events and has established the so-called 'IP Hub', an online service providing an overview of various topics, including how to obtain IP rights, to interested parties.[34] It also allows online searches (partly for a fee) in the patent[35] and trademark[36] register.

Apart from this, initial educational work could start with broadening the basic understanding of the importance of intellectual property rights and their types among the target audiences, who generally do not have a sound legal education. In counselling practice, it is quite common that companies are not familiar even with the basic distinction between patent law and trademark law and their different protective purposes and requirements for protection.

For example, in trademark law the concepts are often confused. The requirement of distinctiveness of a sign for a company does not necessarily have to be equated with creativity or creative activity (although both characteristics can increase the chances for trademark protection).

More information about the different IP protection systems available for the results of research and development work in Austria and internationally, as well as about the availability of financial resources to inventors and founders, can help companies to become more inventive. Furthermore, advice could be offered to inventors in SMEs and start-ups to support them in developing their ideas. For example, they would benefit from assistance with determining whether third parties already have IP rights for a particular research result, since such searches can be time-consuming and cost-intensive.

13.3.3 Easier IP Application Procedures

Considering the current state of the applicable regulation and the established grant requirements, sustainability still plays a subordinate role in Austrian intellectual property law. The searches for the term *sustainability* on the page of the Austrian Patent Office delivers 20 results, all of which, however, are only indirectly linked to

---

[32] See DER STANDARD (2021): How sustainability can also be beautiful; https://www.derstandard.at/story/2000131803300/wie-nachhaltigkeit-auch-schoen-sein-kann. Accessed 19 November 2022.

[33] https://www.patentamt.at/kontakt/.

[34] https://www.patentamt.at/ip-hub/.

[35] https://www.patentamt.at/en/patents/services-searches/searches-and-expert-opinions-section-57a-of-the-austrian-patent-act/. markenaehnlichkeitsrecherche/. Accessed 19 November 2022.

[36] https://www.patentamt.at/markenaehnlichkeitsrecherche/. Accessed 19 November 2022.

the topic of sustainability. Although there are support programmes for companies (such as the 'Patent.Scheck' of the Austrian Development Corporation),[37] there are no support measures specifically targeting sustainable companies or sustainable economic performance.

Another facilitation measure should be the introduction of *fast-track procedures* for registering sustainable inventions in Austria. EUIPO[38] and WIPO[39] already have such processes in place. The idea behind this is simple: by speeding up the patent registration, inventions are protected more quickly and rights holders get a clear idea of whether their creations can be protected or not. With these shorter registration procedures, companies can start developing appropriate licensing models and licensing their environmentally friendly inventions more quickly.

Another area in which improvements appear to be necessary to increase the use of industrial property rights for a sustainable economy concerns the *financial support* of SMEs. The goal should be to use benefits and subsidies in a targeted way to make sustainable start-ups and projects even more attractive. For example, financial assistance can be provided for patent protection, which is associated with initial and ongoing costs.[40] Although Austria already has projects, such as 'Patent.Scheck', that provide relief from application fees, these are available to all types of entrepreneurs, regardless of whether they are sustainable or not.[41] Similarly, there are subsidies for young entrepreneurs in Austria, but these can also be claimed by all types of entrepreneurs and thus they do not form an additional source of motivation to make companies more sustainable.

Furthermore, tax relief for sustainability-related inventions should be considered. For example, such relief can be an additional temporary or permanent reduction or even abolition of the taxability of income from licensing,[42] as well as the elimination of turnover tax for exploitation proceeds.

---

[37] FFG (2022): *Patent.Scheck – Subsidies, Conditions;* see https://www.ffg.at/programm/patentscheck. Accessed 19 November 2022.

[38] https://euipo.europa.eu/ohimportal/de/fast-track-conditions. Accessed 19 November 2022.

[39] See https://www.wipo.int/wipo_magazine/en/2013/03/article_0002.html. Accessed 19 November 2022.

[40] The minimum costs of an application in Austria depend on the number of claims; the application fees amount to at least EUR 322, with a fee of EUR 104 being charged for each additional claim from the 11th claim onwards; if the patent is granted, the publication fees of the patent specification amount to at least EUR 208. EUR 208, with an additional fee of EUR 135 for each 15 pages from the 16th page onwards; from the sixth year of registration onwards, the patent proprietor must pay an annual fee; the costs increase from EUR 104 in the first year of payment to EUR 1755 in the twentieth year, Austrian Patent Office https://www.patentamt.at/. Accessed 19 November 2022.

[41] FFG (2022): Patent.Scheck – Subsidies, Condition, https://www.ffg.at/programm/patentscheck. Accessed 9 May 2022; see also the funding programme provided by Austria Wirtschaftsservice (AWS) Energie & Klima: AWS: https://www.aws.at/aws-energie-klima/. Accessed 19 November 2022.

[42] Under Section 38 para 1 of the Austrian Income Tax Act, if the income includes income from the exploitation of patent-protected inventions by other persons, the tax rate for the inventor shall be reduced to half of the average tax rate applicable to the total income.

13.3.4 Easier Access to Obtain Patent Protection

Generally, the absolute novelty requirement for inventions creates hurdles for obtaining patent protection for inventions, especially in relation to novel technical challenges and novel technologies which will also include inventions in the field of sustainability. Inventors often either lack experience or rush to make use of their findings, while the publication of inventions prior to filing a patent application makes obtaining the patent impossible. Therefore, it is important to promote the understanding that newly created know-how, which can be the basis for a possible invention, can be protected from the outset by appropriate measures (non-disclosure agreements, no prior publications detrimental to novelty).

At the same time, it would be beneficial to introduce a *statutory grace period* for certain inventions, which would allow the inventor to publicly present or test their invention without jeopardising the protection requirement of novelty to better assess the future success of the invention at an early stage. The narrow exceptions for harmless prior publications currently provided for in Sec. 3 (4) of the Austrian Patent Act and Art. 55 (1) of the EPC would therefore have to be extended or supplemented accordingly. The provisions on the protection of trade secrets under the Know-How Directive (EU) 2016/943[43] and Sections 26a–26j of the Austrian Unfair Competition Act[44] are unlikely to provide sufficient compensation for the need to prematurely publish inventions without putting the (absolute) novelty requirement at risk and endangering the prospects of obtaining patent protection.[45]

Another type of inventions, for which the granting of patent protection should be facilitated, are sustainability-related inventions that fulfil the requirement of novelty and inventive step, but whose *industrial applicability* is not yet clearly recognisable or provable at the time of the patent application. In this respect, the discussion about possible patent protection for sustainable inventions is similar to the discussion in connection with patent protection for inventions in biotechnology and genetic engineering that took place in the 1970s and 1980s. In the early years, when the industrial applicability of research results in these sectors was not yet clear, the patenting of many valuable results did not appear to be secure. It was therefore proposed that so-called '*application-related*' research results could be submitted for patent protection and thus invention protection could be granted for research results even if the applicant could only submit and prove the possibility of their industrial applicability in the course of the patent granting procedure.[46] This approach deserves to be examined more closely for its application to the area of sustainability.

---

[43] Directive 2016/943 of the European Parliament and of the Council of 8 June 2016 on the protection of undisclosed know-how and business information (trade secrets) against their unlawful acquisition, use and disclosure, OJ 2016, L 157, p. 1.

[44] Federal Law Gazette No. 448/1984 as amended by Federal Law Gazette I. No. 110/2022.

[45] See on the principle of absolute novelty M. Horkel, W. Poth. In: M. Stadler/, A. Koller (eds), Patent Act, Linde 2019, pp. 61–62.

[46] F.-K. Beier, Zukunftsprobleme des Patentrechts, GRUR 1972, pp. 214–225; F.-K. Beier, J. Straus, Der Schutz wissenschaftlicher Forschungsergebnisse. Zugleich eine Würdigung des

13.3.5 Easier Access to Innovative Processes Through Knowledge Sharing

Another significant factor in fostering sustainable economic performance is knowledge and experience sharing and cross-border cooperation. As the COVID-19 pandemic clearly demonstrated, international cooperation between science and industry was a key driver for the creation of effective vaccines within an extremely short period of time. The discussion on patent exemption for developing countries, triggered by the Corona pandemic and which resulted in a decision by the WTO Ministerial Conference on 17 June 2022[47] to restrict the rights of patent holders to COVID-19 vaccines for the benefit of countries in need (developing and least-developed countries), could also be pursued in the area of sustainability. The so-called 'TRIPS waiver' covers the free access to patent-protected technologies, insofar these are necessary for the development and maintenance of health systems in specified countries. The discussion of this approach will have to be conducted in the same way with regard to the protection for sustainable inventions, especially those related to global climate change.

Easier access to innovative processes and goods will certainly be another important factor, especially if these are innovations that can be classified as *standard essential patents* ('SEPs'). While under the current legal framework, access to such SEPs is a time-consuming and costly process, the creation of clear legal guidelines could make this process easier and more affordable. One of the focal points of the IP Action Plan of the European Commission[48] is therefore rightly the creation of such an EU-wide uniform legal framework for licensing of SEPs.

To sum up, no measures facilitating IP sharing, including licensing, or otherwise promoting an increase in cross-company or cross-sectoral cooperation in sustainability-related initiatives exist in Austria. Such measures could comprise financial incentive schemes, but compulsory licensing should be avoided. The solution will continue to be the contractual design of efficient licensing models enabling the commercialisation of new ideas and creative solutions and their use by third parties.

13.3.6 Easier Access Through Digitalisation

Another type of innovation where the promotion of protection will be required is *software* and *algorithms*. Here, first and foremost, copyright protection for software

---

Genfer Vertrages über die internationale Eintragung wissenschaftlicher Entdeckungen, Verlag Chemie 1982; G. Kresbach, Patentschutz in der Gentechnologie, Springer-Verlag 1994, pp. 88–94.

[47] WTO Draft Ministerial Decision on the TRIPS Agreement of 17 June 2022, WT/MIN(22)/W/15/Rev.2.

[48] See Sect. 13.1 above.

used for the collection and evaluation of data obtained through the use of green technologies should be considered.

It is not wrong to assume that *digitalisation* is expected to contribute a significant share of inventions in the field of sustainability. The increased use of artificial intelligence (AI) and machine learning (ML) should make it possible to achieve commercially applicable results faster and in a more targeted manner. This in turn can lead to a more sparing use of natural resources (energy, water, waste avoidance). To promote the use of AI and ML, it would be helpful to obtain clarity and legal certainty about whether and which outputs produced using AI and ML can be protected by intellectual property rights. It remains to be seen how the legal plans for regulating the use of artificial intelligence will develop at EU level, namely under the proposal for a regulation of the European Parliament and of the Council laying down harmonised rules on artificial intelligence (Artificial Intelligence Act).[49]

13.3.7 Drafting New IP Laws for Sustaibable Innovations?

At the same time, it does not seem very expedient to generally change or loosen the substantive requirements for obtaining protection under intellectual property law (patent: novelty, inventive step and industrial applicability; trademark: distinctiveness) for innovations that have the promotion of sustainability as their objective, or to introduce new intellectual property rights for such inventions at all. Such measures would, first of all, require a consensus on what constitutes a *sustainable invention* and according to which criteria *sustainability* is to be examined.

Even if a consensus is reached, it is obvious that such measures will lead to numerous, not easily solvable delimitation issues. It would also ultimately result in a dilution of property rights, because there would then be at least two different levels of rights. Since in most cases patent protection is not only sought in Austria, but inventors file their inventions centrally with the EPO, the consensus on what constitutes a *sustainable* invention, for which there should be facilitated protection requirements, would have to be reached not only at the national level, but at least at the European level (by the members of the EPC). All this seems to have little prospect of materialising in the near future.

In addition, it is reasonable to expect that the number of applications claiming to be related to sustainable products and services will rise sharply for the sake of getting the registration through quickly. An application under the sustainable inventions category may be made purely to qualify for special treatment with more relaxed requirements for protection. This could then lead to a devaluation of the entire system of protective rights. In trademark law, the situation is not much different

---

[49] Proposal for a Regulation of the European Parliament and of the Council laying down harmonized rules on artificial intelligence (Artificial Intelligence Act) and amending certain Union legislative acts of 21 April 2021, COM(2021) 206 final, 2021/0106 (COD), CELEX number 52021PC0206.

and the protection requirement of distinctiveness of trademarks should not be softened.

Summarising, the IP protection measures strengthening sustainable economic performance should focus less on the creation of new legal provisions under the individual substantive IP laws[50] or on a new legal framework for the protection of sustainable ideas and inventions. They should rather support companies contributing to sustainable economic performance through financial subsidies, and the deferral or complete waiver of official fees for registration of trademarks, designs and patents.

It is also important to differentiate between large, multinational companies on the one hand and small to medium-sized enterprises (SMEs) on the other. While the former can be assumed to have substantial experience in dealing with new ideas and their implementation in practice, the latter are likely to have very limited experience or none at all. In addition, obtaining and maintaining intellectual property rights can be very expensive (especially if protection is sought not only in Austria, but in several countries or 'worldwide') for SMEs and start-ups. They may not have sufficient capital yet to adequately protect their assets such as innovative ideas, concepts and services. Any kind of support should therefore primarily benefit smaller or young companies with little market experience.

13.4 Improving the Success of IP's Role in Sustainability

As already mentioned in Sect. 13.3.2 above, in the short term, it seems most appropriate to support and promote innovations and ideas related to sustainability through appropriate professional educational and information-sharing measures as well as through financial relief for the registration and maintenance of formal property rights (patents, trademarks and designs).

Adapting existing national IP laws and international treaties (Paris Convention, TRIPS Agreement) is feasible in view of the global dimension of climate change. This step will be necessary, but it seems a long way off and has little prospect of being implemented within a reasonable timeframe.

A much more promising approach to promoting the protection of intellectual innovations and creative ideas is the *IP Action Plan* of the European Commission published on 25 November 2020.[51] In short, in this action plan, the European Commission set themselves the goal to help companies, especially SMEs, to make the most of their inventions and creations and to ensure that the EU economy and society benefit from these. Not least against the backdrop of the Corona crisis, the Action Plan aims at enabling Europe's creative and innovative sectors to remain

---

[50] Patent law, Trademark law, Utility Model law, Copyright law.

[51] Communication from the Commission to the European Parliament, the Council, the European Economic and Social Committee and the Committee of the Region - Making the most of the EU's innovative potential. An intellectual property action plan to support the EU's recovery and resilience of 25 November 2020, COM(2020) 760 final, CELEX number 52020DC0760.

world leaders and to accelerate Europe's environmental and digital transformation. The Action Plan sets out important steps to better protect intellectual property rights, promote the use of intellectual property by SMEs, facilitate the sharing of intellectual property to optimise the dissemination of technologies, fight trademark and product piracy, improve the enforcement of intellectual property rights, and ensure a level playing field worldwide.[52]

One concrete measure proposed by the EU IP Action Plan in relation to modernising design law has already been acted upon. In April 2021, the European Commission launched a consultation process to adapt the Design Protection Directive[53] to the requirements of sustainability and digitalisation, among other things. One of the focal points is the facilitation of design protection for 3D printing technologies, as these technologies are considered to have sustainability advantages over conventional manufacturing processes. The European Commission planned to publish a proposal based on the results of the consultation in the 2nd quarter of 2022. However, neither the consultation results, nor a directive proposal have been published so far.[54]

The application of less strict legal requirements for the protection of sustainable inventions through (national) patents and trademarks seems to be less effective. This would lead to numerous delimitation issues, thereby increasing legal uncertainty and ultimately diluting legal protection.

Whatever measures should be planned at the legislative level or in terms of financial support, it will be important that such initiatives are implemented not only on a national level, but also internationally. A great example of such a cross-border initiative is the 'WIPO GREEN' project. 'WIPO GREEN' is a platform where inventors and buyers of inventions can meet and work together on sustainable ideas.[55] The portal connecting suppliers and buyers creates a stronger link between entrepreneurs who can support each other in further developing ideas and projects.

13.5 Conclusion

Intellectual property rights play an important role in achieving sustainable economic performance, which to a large extent relies on new and innovative solutions. The importance of such solutions will continue to increase as the challenges ahead multiply. Therefore, the creation of appropriate incentives for research and development to ensure an efficient and easy access to the protection of intellectual property

---

[52] https://ec.europa.eu/commission/presscorner/detail/en/ip_20_2187. Accessed 19 November 2022.

[53] Directive 98/71/EC of the European Parliament and of the Council of 13 October 1998 on the legal protection of designs, OJ 1998, L 289, p. 28.

[54] For the status of the consultation process see https://ec.europa.eu/info/law/better-regulation/have-your-say/initiatives/12609-Intellectual-property-review-of-EU-rules-on-industrial-design-Design-Directive-en. Accessed 19 November 2022.

[55] See https://www3.wipo.int/wipogreen/en/. Accessed 19 November 2022.

assets will be crucial. This can be achieved through information dissemination and educational measures and the re-shaping of the existing legal framework. Incentives for increased cross-company and cross-sectoral cooperation will also be a decisive factor in boosting the creation of sustainable solutions. The challenges for carrying out sustainable economic activities and the careful use of natural resources, especially against the backdrop of accelerating climate change, are by their very nature not limited to individual countries, but require global efforts. The improvement measures should be taken at least at EU level to create uniform standards across the economic block. Cross-country collaboration in intellectual property law will be essential to accelerate progress and facilitate individual access to sustainable solutions. Purely national measures will not be sufficient.

Open Access This chapter is licensed under the terms of the Creative Commons Attribution 4.0 International License (http://creativecommons.org/licenses/by/4.0/), which permits use, sharing, adaptation, distribution and reproduction in any medium or format, as long as you give appropriate credit to the original author(s) and the source, provide a link to the Creative Commons license and indicate if changes were made.

The images or other third party material in this chapter are included in the chapter's Creative Commons license, unless indicated otherwise in a credit line to the material. If material is not included in the chapter's Creative Commons license and your intended use is not permitted by statutory regulation or exceeds the permitted use, you will need to obtain permission directly from the copyright holder.

Sustainability and Intellectual Property in Brazil

14

Lucas Bernardo Antoniazzi

14.1 The Current Role IP Play in Sustainability in Brazil

Intellectual property rights, in short, are a set of rules that aim to protect certain types of intangible assets, with specific and—more traditional—objective of fostering innovation.

However, as a legal rule, considering the role of the right to public policies, Intellectual Property can act as a tool for the development and application of a variety of public policies, such as the development of national industry, foreign trade aspects and competition.[1]

In this context, sustainability—or rather, the promotion of sustainability—is just another public policy that can benefit from the tension between protection and access provided by IP rules.

As an example, greater IP enforcement aimed at "green" technologies can be an important instrument for fostering innovation in this sector (exercising the traditional IP function of fostering innovation).

The rapporteur would like to thank his colleagues from the Study Committee in Competition from the Brazilian Intellectual Property Association—Ana Luiza Pinheiro, Felipe Oquendo and Juliana Libman—for the inestimable aid with research and responding the questions.

[1] M. Castro and M.-T. Leopardi Mello, Uma abordagem jurídica de análise de políticas públicas, Revista de Estudos Empíricos em Direito, 2017, pp. 12–13.

L. B. Antoniazzi (✉)
Di Blasi, Parente & Associados, Rio de Janeiro, Brazil
e-mail: lucas.antoniazzi@diblasi.com.br

© The Author(s) 2024
P. Këllezi et al. (eds.), *Sustainability Objectives in Competition and Intellectual Property Law*, LIDC Contributions on Antitrust Law, Intellectual Property and Unfair Competition, https://doi.org/10.1007/978-3-031-44869-0_14

On the other hand, depending on the objectives of a particular public policy, instead of fostering innovation, it may be much more important to guarantee public, general, and unrestricted access to technologies that are protected by IP. In this scenario, the way IP could contribute to sustainable development is to have its application exempted.

The role of IP, therefore, must be to serve as one (of several) tools for public policies seeking sustainable development.

It will be up to the public agents and the stakeholders to define public policies that promote sustainable development, indicating how the application of intellectual property rules can contribute to such public policies.

Specifically, in Brazil, the relationship between intellectual property and sustainability, in the Federal Constitution and in the Brazilian special legislation, is not clear at first, as it is not direct.

The Brazilian Federal Constitution mentions intellectual property rights in its Article 5, items XXIX (copyrights) and XXIX (industrial property rights), providing them not only with constitutional protection but also with a finalistic feature in relation to the objectives of the Brazilian Republic.

Indeed, at least for industrial property rights and industrial secrets, the Constitution determines that the protection ensured to industrial creations, brands, business names, and other distinctive signs is granted *"with a view to the social interest and the technological and economic development of the Country."*

Additionally, every property right, whether material or immaterial, is subject to a social function under item XXIII of the same constitutional Article.

However, in the Brazilian laws that govern intellectual property rights, there are no provisions that explicitly relate them to the achievement of sustainability, understood in the context of the Environment, Social and Governance (ESG) triad.

Although there are undoubtedly provisions in such laws that echo the interest and social function of intellectual property, such as, for example, the possibility of compulsory licensing of patents due to a state of emergency and lack or insufficiency of exploitation, as well as the prohibition of registration, as a trademark, of expressions that go against morals and good customs, and although it is defensible that the protection granted to holders of intellectual property rights, business secrets, and the repression of unfair competition has as a reflex objective the protection of consumers and the social order, the fact is that the current Industrial Property Law still seems to reflect the more traditional concerns on the matter, established since the first Brazilian document on the subject, the Rule of 1809, promulgated by the then Prince Regent D. João VI, of Portugal, Brazil and Algarves, when the Portuguese court was exiled in Rio de Janeiro, driven away by persecutions. Napoleonic ions. The aforementioned Rule had the objective of promoting the progress of industries and commerce, and the only sign of any social concern comes from the proclaimed final objective of the act, which would be the *"public happiness of the faithful vassals."*

Undoubtedly, the encouragement of commerce and inventors, on the one hand, and the repression of bad practices in commerce, on the other, lead to the achievement of social objectives. To give one example, we highlight the patenting of new

14 Sustainability and Intellectual Property in Brazil

substances for medical use, which can cure, prevent, or at least mitigate the effects of diseases, as well as the patenting of substances and equipment used to increase agricultural productivity and improve conditions population health. Although it can be argued that it is not the granting of the patent, but the invention and its exploitation itself, that achieve sustainability goals, it can be objected to this argument, equally, that without the reward system that is the patent system, there would never be an economic incentive for the realization and dissemination of such inventions.

Nevertheless, Brazilian rules contain more explicit provisions on sustainability.

One of the most obvious examples is the existence of one exclusivity program focused on "green patents," established by the Brazilian Patent and Trademark Office (BPTO) almost six years ago, through its Resolution n° 175/2016. This PPH, although attracting criticism in the sense that it would be violating the CUP by giving priority treatment to certain inventions based on discrimination of type of industry or invention, has accelerated the examination and granting of several patents related to inventions that reduce, prevent, or mitigate the environmental impact, stimulating research and development in this regard.

Other examples are certification marks and geographical indications, which, however, act in different ways in relation to sustainability.

The certification marks are those used to certify the conformity of a product or service with certain norms or technical specifications, notably in terms of quality, nature, material, and methodology used (Art. 123, II of the Brazilian IP Law— BIPL). Evidently, in addition to the quality and intrinsic safety of products and services, such brands may also be aimed at certifying good environmental, social, and governance practices, stimulating, with the exclusivity and prestige that their registration guarantees, the performance of their holders and helping in building public trust in certification and in certified products and services.

Geographical indications, divided into indications of origin and appellations of origin, are the geographical name of a country, city, region, or locality in its territory, which has become known as a center for the extraction, production, or manufacture of a particular product or provision of a given service (indication of origin) or designating a product or service whose qualities or characteristics are exclusively or essentially due to the geographical environment, including natural and human factors (designation of origin) (Art. 177 of the BIPL).[2]

By ensuring the origin and/or natural and human factors related to certain products and services, geographical indications can not only have a similar effect to good practice certification marks, by designating products and services that adopt "green" processes or that meet certain positive parameters of respect for social rights, but also tend to promote the enrichment of those involved in the production and supply of products and services identified by geographical indications, many of which belong to traditional and impoverished communities, by increasing the

---

[2]The legal regime for the protection of geographical indications was the subject of the Brazilian report published in Kobel et. al, *Antitrust in Pharmaceutical Markets & Geographical Rules of Origin*, Springer, 2018.

perception of value by the consuming public. In this way, it is not uncommon for geographical indications to end up promoting the improvement of conditions in an entire community that is historically poor and poorly served by government policies.

Aside from the rights directly established by the laws that govern intellectual property in Brazil, it is necessary to mention the legal protection granted, in our country, to traditional knowledge associated with biodiversity. This protection is currently made through the requirement of authorization for access to Associated Traditional Knowledge (in Portuguese CTA) for purposes of scientific research, bioprospecting, or technological development, which are very closely related to technological creations subject to patent protection and, eventually, for industrial secrets. Authorization is provided after prior consent of the traditional peoples who hold the knowledge whose access is sought and the Genetic Heritage Management Council (in Portuguese CGEN).[3] The fees involved, as well as the revenue arising from contracts for the economic exploitation of the product or process originated from the samples, are intended, among others, to provide economic support and improve the living conditions of these populations.

As tools, intellectual property rights are also subject to misuse, either by omission in their exploitation, or by abuse of the rights conferred. Such misuse can certainly have negative impacts on the environment and society, however, currently the legislation does not provide for punishments specifically related to damage to the environment.

The Brazilian Industrial Property Law provides for the penalty of compulsory licensing of patents not only in the event of a national emergency and in the absence or insufficiency of the exploitation of the object of the patent, but also as a punishment for the abuse of the rights conferred by the patent, particularly when they have anti-competitive effects (Article 68 of BIPL). However, to date, there is no known precedent for the application of the compulsory patent license for this reason.

The behavior of patent holders in the exercise of their exclusivity rights has raised the concern of the Brazilian judiciary, in some cases, especially for the maintenance of the monopoly prices that patents allow.

One of the peaks of this concern—whose legitimacy is questioned by most intellectual property lawyers—was the decision handed down by the Federal Supreme Court in the records of the Declaratory Action of Unconstitutionality n°. 5529, which deemed the only mechanism provided by law in Brazil to compensate the holder for excessive delays by the BPTO in the examination of patents unconstitutional (Article 40, sole paragraph of the BIPL). The provision, which was revoked by Law No. 14,195/2021, guaranteed a minimum term of ten years for invention patents and seven years for utility model patents, counting from the grant date. One of the main grounds for the Federal Supreme Court's decision was the alleged abuse of depositors who would use various expedients to delay the examination and,

---

[3] CGEN, in turn, is part of the Instituto do Patrimônio Histórico e Artístico Nacional (IPHAN), a federal agency linked to the Ministry of Culture.

14 Sustainability and Intellectual Property in Brazil

therefore, obtain patents with a longer term, thus delaying the availability of the object of the patent to the public domain.

14.2 How This Role Is Pursued

In Brazil, specifically for sustainable technologies, there is the "Green Patents" program (BPTO Resolution No. 175/2016),[4] which aims to accelerate the examination of patent applications for projects that support the sustainable economy. Its focus are projects that aim to promote less environmental impact, since the BPTO understood that it would be a measure to combat the advance of climate problems, which has been increasing in recent decades.

More broadly, as an instrument to foster innovation in several fields, there is also the so-called "Law of the Good" (Federal Law n° 11.196/05), which grants tax incentives for companies that invest in innovation.

It is worth mentioning that those examples of speeding up the examination or encouraging the filing of patents or other types of IP can be considered, in general, as rules that promote the fostering of innovation.

In relation to IP rules that seek to guarantee access to innovations, except for compulsory licensing, mentioned in a previous answer, there is no other rule in this sense, which could be potentially positive for the promotion of sustainability.

After all, it is not enough for new technologies to be developed (in terms of sustainability or others), but for these technologies to be more accessible to society.

This can be done, for example, as a reduction in the protection of intellectual property rights over sustainable technologies or, what seems like a less extreme solution, through the reduction or exemption of taxes on the use of these sustainable technologies. In Brazil, although there is a debate about this topic, there is still no rule that guarantees this tax benefit.

In fact, the relation of IP and the fostering of sustainability raises a debate which is not restricted to sustainable technologies, but to all intellectual property: the tension between the enforcement (protection) of technology and access to technology, that is, which one promotes greater social well-being.

Apparently, except for pharmaceutical patents, there is still not enough evidence to conclude that greater enforcement promotes the promotion of more technology,[5] but, on the other hand, there is also no evidence that greater access will bring greater social well-being.

Ideally, public policy that seeks to promote sustainability through IP should plan how to best use the tension between protection and access to technologies, balancing both, and establish metrics for evaluating the corresponding results.

---

[4]Patentes Verdes (2015), BPTO https://www.gov.br/inpi/pt-br/servicos/patentes/tramite-prioritario/projetos-piloto/Patentes_verdes. Accessed 19 November 2022.

[5]M. Lemley, Faith-Based Intellectual Property, UCLA Law Review, 2015.

14.3 How to Improve It

Considering the fundamental role of a tool for public policies, we understand that intellectual property rights in our country do fulfill their role of promoting the development of sustainability.

Indeed, we mentioned practical cases, in our country, of policies which encourage innovation and "green patents" through IP rules.

It is worth noting, however, that IP rules alone are not sufficient to achieve sustainable development and must be applied in the context of broader public policies that have this objective.

Generally, it can be said that the IP registering authority, the BPTO, provides information about registered IP rights that enable assessment of how IP is supporting sustainability in your country, provides a specific procedure and fees, and offers an incentivization scheme for promoting sustainable development.

As explained above, the "green patents" program is a living example of a practice of encouraging technological innovation focused on the environment, widely spread in the country and the data of such program has been disclosed by the BPTO.

Nonetheless, despite the BPTO efforts, other similar programs or policies are welcomed.

The IP enforcement system in Brazil, although it still needs improvement, is quite strong, with a system of judicial protection that guarantees the defense of IP holders, as well as with practices by Customs and by the local police seeking to repress IP infringements and counterfeits.

In any case, we refer to our previous comment on Sect. 14.2 above that there is no evidence that IP enforcement necessarily guarantees the promotion and development of innovations (sustainable or not) to comment that there is no way to confirm whether this enforcement system in Brazil, which is strong and effective, ends up fostering the development of sustainable practices and policies.

Considering the indirect role of IP in relation to the promotion of sustainability—that is, it is not property rights themselves that aim to promote sustainability, but public policies that use IP rights as an instrument to do so—we understand that there is no need to improve intellectual property rights in the country, but rather to improve the debate between government, society, and all stakeholders about the development of sustainable public policies.

In this sense, any changes in patentability conditions/requirements specifically for technologies or other intangible assets that promote sustainability, in principle, could imply discrimination against other technological fields, in violation of Article 27 of the Paris Union Convention.

Furthermore, it seems to us that any changes in the conditions/requirements for patentability are just a way to promote greater access to sustainable technology or foster its protection, which, as seen, can be achieved in other less impactful ways than the alteration of legal requirements already consolidated and that are practically uniform in different legislations.

14 Sustainability and Intellectual Property in Brazil

Likewise, creating new PI types does not seem to be necessary. On the one hand, it seems that the current IP types cover all forms of intangible assets that can contribute to the development of public policies on sustainability.

On the other hand, the effect of creating new types of intellectual property would only constitute another way of promoting greater protection for sustainable technologies, which, as seen, can be achieved by other means, without generating any potential negative impacts resulting from the creation of a new type of IP (such as debates about the scope of protection, protection requirements, overlap with other types of IP, etc.).

Regardless, although we understand that the current IP system does not need to be modified (but the public policies that use it as a tool to promote sustainability), if any local changes do not conflict with provisions of international treaties, such as amendments to conditions/requirements for granting a patent, it is, indeed, possible for such a change to be carried out at the local level.

Again, as we have the opinion that the IP system, locally or internationally, does not need an improvement to promote the sustainable development, there would be no need to modify any international treaty.

It is worth to mention, however, that if the makers of public policies—locally or internationally—intend to provide to green patents a special treatment, in prejudice of patents in other technological fields, it is recommended to have Article 27 of the Paris Union Convention to be amended in this way.

14.4 Conclusion

Intellectual property rights have the power to function as an efficient and important instrument for promoting sustainability, through incentives for innovation and/or access to technologies, thus contributing to public policies that seek to advance new sustainable technologies or that seek to spread the use of existing technologies.

In Brazil, without prejudice to the need for improvements, this is already happening, either through the very effects of granting an IP right, as with certification marks and geographical indications, but, mainly, through some specific policies of tax benefit to encourage innovation and an accelerated examination program for "green patent" applications.

For this reason, it is believed that the current system of protection of IP rights is already sufficient to contribute to the promotion of sustainable development, making new legislative changes or the creation of new types of IP unnecessary.

As the IP is just a tool, the important thing at this point does not seem to be to change the tool, but what is done with it.

This improvement in the use of IP, in our opinion, will take place through better public policies to promote sustainability, remembering that, as it is an agenda that is important to society, not only public agents and public policymakers, but the whole society, including IP holders, are stakeholders who must contribute to the promotion of sustainable development.

Open Access This chapter is licensed under the terms of the Creative Commons Attribution 4.0 International License (http://creativecommons.org/licenses/by/4.0/), which permits use, sharing, adaptation, distribution and reproduction in any medium or format, as long as you give appropriate credit to the original author(s) and the source, provide a link to the Creative Commons license and indicate if changes were made.

The images or other third party material in this chapter are included in the chapter's Creative Commons license, unless indicated otherwise in a credit line to the material. If material is not included in the chapter's Creative Commons license and your intended use is not permitted by statutory regulation or exceeds the permitted use, you will need to obtain permission directly from the copyright holder.

Sustainability and Intellectual Property in Germany

15

Thomas Hoeren, Tabea Ansorge, and Oliver Lampe

15.1 Introduction

We are encountering the term 'sustainability' increasingly often in today's age. According to many people, they are striving for a more sustainable lifestyle and sustainable production is increasingly being advertised on consumer goods.

A more climate-friendly way of life is closely related to research and innovation in a wide variety of areas of life which make this lifestyle possible in our modern society. Resources should be used sparingly by perfecting recycling processes. Travel should become more environmentally friendly and thus more sustainable, for example through electric mobility. Energy is to be obtained sustainably. Behind all these developments are concepts and ideas, some of which can be protected by intellectual property rights such as patents, trade secrets or even trademarks. The question therefore arises what role intellectual property rights play in sustainability. Intellectual property rights are generally granted irrespective of the sustainability of a product. They are value neutral. However, this does not mean that they cannot have a positive or possibly also a negative effect.

T. Hoeren (✉) · T. Ansorge · O. Lampe
Westfälische Wilhelms-Universität, Münster, Germany
e-mail: hoeren@uni-muenster.de; tabea.ansorge@uni-muenster.de; oliver.lampe@uni-muenster.de

© The Author(s) 2024
P. Këllezi et al. (eds.), *Sustainability Objectives in Competition and Intellectual Property Law*, LIDC Contributions on Antitrust Law, Intellectual Property and Unfair Competition, https://doi.org/10.1007/978-3-031-44869-0_15

15.2 Definition of Sustainability

There is no generally accepted definition of the term sustainability.[1] It can be interpreted in socio-political, economic and legal terms. When the term 'sustainability' is used colloquially, it is usually understood as an ecological concept in the sense of a resource-conserving and thus environmentally and climate-friendly behaviour.[2] However, to work out the various effects of individual intellectual property rights on 'sustainability', this report will be based on a uniform definition.

15.2.1 Legal Understanding of the Term

From a legal perspective, a distinction can be made between a narrow and a broad understanding of the concept of sustainability. Narrowly understood, ecological sustainability aims at respecting the carrying capacity of nature in all state decisions directed towards the future.[3] These decisions are intended to create the basis of life for future generations within the framework of resource management. At the same time, ecological sustainability means maintaining or restoring economic performance in the long term.[4]

Social sustainability also means guaranteeing the conditions for sustainable coverage of resource-independent basic human needs through functioning social security systems and services of general interest.[5]

According to the broad understanding of the concept of sustainability, the three-dimensional integrative concept, sustainability aims at a balance of economic development, social security and preservation of natural resources for future generations.[6]

---

[1] M. Vogt, Nachhaltigkeit, In: Görres Gesellschaft (ed), Görres-Gesellschaft-Staatslexikon, 8th ed, Vol. 4, Herder 2020.

[2] M. Vogt, Nachhaltigkeit, In: Görres Gesellschaft (ed), Görres-Gesellschaft-Staatslexikon, 8th ed, Vol. 4, Herder 2020.

[3] W. Kahl, Nachhaltigkeit, In: Görres Gesellschaft (ed), Görres-Gesellschaft-Staatslexikon, 8th ed, Vol. 4, Herder 2020; V. Oschmann, In: C. Theobald and J. Kühling (eds), Energierecht, 114th Ed, Vol. 1, C.H. Beck January 2022, EEG § 1, para 22.

[4] W. Kahl, Nachhaltigkeit, In: Görres Gesellschaft (ed), Görres-Gesellschaft-Staatslexikon, 8th ed, Vol. 4, Herder 2020; United Nations (ed) (1987), Report of the World Commission on Environment and Development: Our Common Future, Chapter 2, http://www.un-documents.net/wced-ocf.htm, accessed 13 July 2022.

[5] W. Kahl, Nachhaltigkeit, In: Görres Gesellschaft (ed), Görres-Gesellschaft-Staatslexikon, 8th ed, Vol. 4, Herder 2020; K. Bosselmann, The Principle of Sustainability, 1st ed, Routledge 2008, pp. 826 ff.

[6] W. Kahl, Nachhaltigkeit, In: Görres Gesellschaft (ed), Görres-Gesellschaft-Staatslexikon, 8th ed, Vol. 4, Herder 2020.

15.2.2 Social-Ethical Understanding of the Term

In the 18th century, the term sustainability was first used primarily as a forestry conservation rule to describe an ethical attitude according to which the power of nature should be unfolded in a future-oriented manner.[7] The intention is to generate only as much yield as is necessary to maintain earning power. Furthermore, from a political perspective, the concept of sustainability in relation to the structure of society describes not only the preservation but also the development of technical and social innovations.[8]

In terms of justice theory, sustainability means the basis for global and intergenerational justice.[9] From this perspective, sustainability thus means creating equal life chances for future generations as well as equal rights for everyone regarding resources.[10]

15.2.3 Political Science Understanding of the Term

The political significance of sustainability is primarily reflected in the global model of sustainable development designed by the UN.[11] According to this concept, development is sustainable if it meets the needs of current generations without depriving future generations of opportunities.[12] To achieve intra- and intergenerational justice through universal participation, the preservation of natural systems is essential.[13]

---

[7] M. Vogt, Nachhaltigkeit, In: Görres Gesellschaft (ed), Görres-Gesellschaft-Staatslexikon, 8th ed, Vol. 4, Herder 2020.

[8] M. Vogt, Nachhaltigkeit, In: Görres Gesellschaft (ed), Görres-Gesellschaft-Staatslexikon, 8th ed, Vol. 4, Herder 2020.

[9] BVerfGE 157, 30 ("Klimabeschluss des BVerfG") of 24 March 2021; F. Ekardt (2010) Klimawandel und soziale Gerechtigkeit, p. 31, available at https://www.kas.de/c/document_library/get_file?uuid=5f333536-2589-7f03-c07d-1ceea90afc48&groupId=252038. Accessed 15 July 2022.

[10] M. Vogt, Nachhaltigkeit, In: Görres Gesellschaft (ed), Görres-Gesellschaft-Staatslexikon, 8th ed, Vol. 4, Herder 2020.

[11] K.-W. Brand, Nachhaltigkeit, In: Görres Gesellschaft (ed), Görres-Gesellschaft-Staatslexikon, 8th ed, Vol. 4, Herder 2020.

[12] UN (ed) (1987), Our Common Future: The Brundtland Report of the World Commission on Environment and Development, available at http://www.un-documents.net/wced-ocf.htm. Accessed 13 July 2022; W. Schön, "Sustainability" in Corporate Reporting, ZfPW 2022, p 207.208.

[13] K.-W. Brand, Nachhaltigkeit, In: Görres Gesellschaft (ed), Görres-Gesellschaft-Staatslexikon, 8th ed, Vol. 4, Herder 2020.

15.2.4 Economic Understanding of the Term

From an economic perspective, sustainability is understood as a course of action that does not aim at one-sided and short-term economic advantages and does not place a heavy burden on resources.[14] Furthermore, a distinction between different capital stocks is made: physical capital, human capital and natural capital. These should be used in such way that future and present generations are not restricted.[15] According to the concept of weak sustainability, different forms of capital are infinitely substitutable.[16] This is seen differently in context of the strong sustainability concept, primarily because of the dependence of economic and social systems on natural systems.[17] Based on the premise of strong sustainability, rules for dealing with natural resources have therefore been developed.[18] Particularly, exhaustible resources such as fossil energy sources should only be used if they can be replaced by renewable resources of equal value.

15.2.5 Essential Features of 'Sustainability'

Depending on the definition, sustainability therefore implies not only the protection of the environment, but also the establishment of ecological, economic and social justice. Future generations should continue to receive their natural basis of life.[19] The core of the term, as it is interpreted here, is consequently the resource-conserving handling of the economy and society, which can be achieved by innovation and saving resources.

---

[14] B. Hansjürgens, Nachhaltigkeit, In: Görres Gesellschaft (ed), Görres-Gesellschaft-Staatslexikon, 8th ed, Vol. 4, Herder 2020.

[15] W. Schön, "Nachhaltigkeit" in der Unternehmensberichterstattung, ZfPW 2022, pp. 207–208.

[16] R. Döring (2004) Wie stark ist schwache, wie schwach starke Nachhaltigkeit?, p. 4, available at https://www.econstor.eu/dspace/bitstream/10419/22095/1/08_2004.pdf. Accessed at 14 July 2022.

[17] B. Hansjürgens, Nachhaltigkeit, In: Görres Gesellschaft (ed), Görres-Gesellschaft-Staatslexikon, 8th ed, Vol. 4, Herder 2020.

[18] Aachener Stiftung Kathy Beys (2015) Lexikon der Nachhaltigkeit, Starke und schwache Nachhaltigkeit, available at https://www.nachhaltigkeit.info/artikel/schwache_vs_starke_nachhaltigkeit_1687.htm. Accessed 23 June 2022.

[19] W. Schön, "Nachhaltigkeit" in der Unternehmensberichterstattung, ZfPW 2022, pp. 207–208; similar to the commonly used definition of sustainability by Brundtland Reports of the United Nations: UN General Assembly (1987) Report of the World Commission on Environment and Development - Our Common Future, available at https://digitallibrary.un.org/record/139811. Accessed 14 July 2022.

15.3 Current Status: Intellectual Property Rights and Sustainability

The existence of various intellectual property rights is often justified with the argument that they are a driving force for innovation.[20] For this reason, the following section will examine the extent to which intellectual property rights play a role in achieving the goal of a sustainable world.

15.3.1 Patent Law

Under Section 1 Patent Act (PatG),[21] patents are granted to protect new inventions. This may include both devices and processes (Sec. 9 PatG). The aim of patent law is to strike a balance between the inventor's interest in using his invention exclusively for as long as possible and the general public's interest in further development of research.[22] By granting a patent, the inventor receives the possibility to exclude unauthorised third parties from using or marketing the invention for a certain period (Sec. 9 PatG). Thus, within the limits of the permitted acts under Sec. 11 PatG, the patent proprietor is free to decide which person is permitted to use the technology. As a result, the patent proprietor has a position that can be described as a monopoly and is only obliged to grant a compulsory licence to a third party in the cases of Sec. 24 (1) PatG respectively Sec. 102 TFEU.[23] At the same time, every patent proprietor is required to disclose his invention in such way that a person skilled in the field could carry it out (Sec. 34 (4) PatG). This is intended to compensate for the monopolistic protection and to provide specialist groups with the opportunity to take up the invention and develop it further, leading to continuous innovation competition.[24] Generally, patent protection is possible for 20 years (Sec. 16 PatG).

---

[20] C. Kilchemann (2005) Die Wirkung des Patentschutzes auf Innovation und Wachstum, p. 2, available at https://wwz.unibas.ch/fileadmin/user_upload/wwz/99_WWZ_Forum/Forschungsberichte/15_05.pdf. Accessed 7 July 2022; H. Zech (2021) Brauchen wir ein Patentrecht? available at https://www.ifo.de/publikationen/2021/aufsatz-zeitschrift/patentschutz-impulsgeber-fuer-innovationen-oder-behinderung. Accessed at 7 July 2022; World Intellectual Property Organization (2022) Innovation and Intellectual Property, available at https://www.wipo.int/ip-outreach/en/ipday/2017/innovation_and_intellectual_property.html. Accessed 7 July 2022; critical: F. Machlup, Die wirtschaftlichen Grundlagen des Patentrechts - 3. Teil, GRUR 1961, p. 524; W. Landes and R. Posner, The Economic Structure of Intellectual Property Law, 1st ed, Harvard University Press 2003, pp. 326–327; R. Mazzoleni and R. Nelson, The benefits and costs of strong patent protection: a contribution to the current debate, Vol. 27, Elsevier 1998, pp. 273–284.

[21] Patent Act, published on 16 December 1980 (Bundesgesetzblatt 1981 p. 1), last amended on 30 August 2021 (Bundesgesetzblatt I p. 4074).

[22] R. Rogge and K.-J. Melullis, Einleitung Patentgesetz, In: Benkard (ed), Patentgesetz, 11th ed, C.H. Beck 2015, para 1.

[23] R. Rogge, In: Benkard (ed), Patentgesetz, 11th ed, C.H. Beck 2015, § 24 PatG, paras 3, 18.

[24] A. Schäfers, In: Bernkard (ed), Patentgesetz, 11th ed, C.H. Beck 2015, § 34 PatG, paras 14a–34.

Goods that can generally be protected by patent law are the so-called 'green' technologies. Those are defined as technologies that contribute to environmental protection by reducing the impact on the environment. They achieve this by using resources more sustainably, recycling more waste and disposing the remaining waste in an environmentally friendly manner. The comparative benchmark for this is the technology that would be used instead of the 'green' technology. However, the extent to which a technology is ultimately 'green' can usually only be assessed in retrospect. For example, the computer was expected to eliminate use of paper almost completely, which has not yet come true.[25]

15.3.1.1 Possible Positive Effect of Patent Protection

Patents can now create an incentive to invest in sustainable technology and to disseminate it profitably through the exclusive right. However, there is no 'green' patent law in this sense, as the granting of patents is neutral in terms of value,[26] meaning that not only sustainable, but all inventions can be protected.

Nevertheless, at least 'green' technologies can *also* be protected by patents and then be marketed through sales and licensing agreements. The profits generated by this can in turn be reinvested in new innovations.[27] In addition, licensing can lead to cooperation with other companies that can use the technology to create further new products. Furthermore, licensing can also help to bridge social barriers, e.g. by issuing targeted licences to companies that produce in poorer regions, thus creating new jobs there.[28] Especially with regard to 'green' technology, licensing offers an opportunity to promote its dissemination. Start-ups are considered to be very innovative in the field of sustainable technology, but they often lack the financial resources and capacities to make the technologies they develop a success.[29] Therefore, cooperation with established larger companies by way of licensing is a good way to promote the dissemination of the technology and achieve greater sales.[30]

---

[25] F. Klein (2020) GREEN IP - A look at how sustainability influences IP and how IP can help in achieving sustainability, available at https://www.ashurst.com/en/news-and-insights/legal-updates/a-look-at-how-sustainability-influences-ip-and-how-ip-can-help-in-achieving-sustainability/. Accessed 12 July 2022.

[26] A. Krefft, Patente auf human-genomische Erfindungen, Carl Heymanns 2003, p. 96; I. Schneider, In: Metzger (ed) Methodenfragen des Patentrechts, Mohr Siebeck 2018, pp. 6–7.

[27] C.-W. Davies, T. Nener and N. Pereira (2021) Green IP: the role of intellectual property in sustainability, available at https://www.financierworldwide.com/green-ip-the-role-of-intellectual-property-in-sustainability. Accessed 12 July 2022.

[28] International Science Council (2022) Three things to know about how Intellectual Property can contribute to sustainability transitions, available at https://council.science/current/blog/how-intellectual-property-sustainability-transitions/. Accessed 5 July 2022.

[29] International Science Council (2022) Three things to know about how Intellectual Property can contribute to sustainability transitions, available at https://council.science/current/blog/how-intellectual-property-sustainability-transitions/. Accessed 5 July 2022.

[30] International Science Council (2022) Three things to know about how Intellectual Property can contribute to sustainability transitions, available at https://council.science/current/blog/how-intellectual-property-sustainability-transitions/. Accessed 5 July 2022.

15 Sustainability and Intellectual Property in Germany

Due to the increased risk of extreme weather conditions or natural disasters caused by climate change, the question of how we can live more sustainably has become increasingly important in recent years. With the Green Deal,[31] the EU has agreed to become climate neutral by 2050. As a result, the demand for 'green' technology is increasing. The market is following suit by investing more in the research and development of sustainable technologies to create the basis for such a development. Thus, patents currently benefit the research and development of 'green' technology in particular.[32]

15.3.1.2 Possible Negative Effects of Patent Protection and Alternatives

The possible incentive effects, which have a positive impact on the inventor, can also have negative effects at the same time.

First, filing and maintaining a patent can be relatively costly, for which some companies may lack the financial and human resources. The application fee is at least EUR 40 for an electronic application and EUR 60 for a paper application. The costs for the subsequent examination procedure amount to at least 350 EUR.[33] These costs alone will still be bearable for most companies. However, the maintenance of the patent becomes more expensive, increasing from year to year. For example, an annual fee of EUR 70 is payable from the third year onwards, rising to EUR 2030 by the twentieth year of the patent.[34] If a declaration of willingness to grant a licence is submitted under Sec. 23 (1) PatG, the annual amount is reduced by 50%, but still leads to considerable costs, although it should be noted that such declaration will not be desired by every patent proprietor. In addition to these costs, there are costs for the patent attorney commissioned to file the patent application, which can easily range between EUR 3000 and EUR 6000.[35] This can be a problem for small or medium-sized enterprises (SMEs) in particular. Although there are funding programmes, such as that of the Federal Ministry for Economic Affairs and Climate Action (BMWK),

---

[31] See: https://ec.europa.eu/info/strategy/priorities-2019-2024/european-green-deal_en. Accessed 15 July 2022.

[32] F. Klein (2020) GREEN IP - A look at how sustainability influences IP and how IP can help in achieving sustainability, available at https://www.ashurst.com/en/news-and-insights/legal-updates/a-look-at-how-sustainability-influences-ip-and-how-ip-can-help-in-achieving-sustainability/. Accessed 12 July 2022; C.-W. Davies, T. Nener and N. Pereira (2021) Green IP: the role of intellectual property in sustainability, available at https://www.financierworldwide.com/green-ip-the-role-of-intellectual-property-in-sustainability. Accessed 12 July 2022; N. Yu, Assessment of the Mechanism for Mining Technology Transfer in the Area: Loopholes in ISA Practice and Its Mining Code, In: Z. Koyuan and C. Yen-Chiang (ed), Preserving Community Interests in Ocean Governance towards Sustainability, MDPI 2022, p. 159.

[33] Annex to Sec. 2 (1) PatKostG (Schedule of fees), Patent Costs Law, publication of 13 December 2001 (Bundesgesetzblatt p. 3656), last amended on 13 December 2001 (Bundesgesetzblatt I p. 4074).

[34] Deutsches Patent- und Markenamt (2022) Kostenmerkblatt, available at https://www.dpma.de/docs/formulare/allgemein/a9510.pdf. Accessed 12 July 2022.

[35] Freie Universität Berlin, available at https://www.fu-berlin.de/forschung/service/patente-und-lizenzen/faq/kosten.html. Accessed 12 July 2022.

which aims to promote the transfer of technology and knowledge through patents in small and medium-sized enterprises with the WIPANO funding programme,[36] it cannot be assumed across the board that every SME will receive such funding. Larger companies usually have established structures and large amounts of human and financial resources. This is not necessarily the case with SMEs, which can lead to a considerable competitive disadvantage.[37] Especially in an area where many companies are looking for their spot and competition is fierce, the basic disadvantage caused by patent law for SMEs compared to larger companies can be multiplied. Negative aspects can also be associated with monopolistic protection for a limited period. Thus, innovative products and processes are protected for a few years, but at the same time this can also prevent faster global dissemination of such innovations.[38] While patent law creates positive incentives for innovation in the development of sustainable technologies, it also hinders their diffusion, especially to the global South.[39] Even the use of compulsory licensing is not able to remedy this situation, as possible protector states in Africa or Asia do not have the necessary means to produce climate-friendly technologies.[40] As a consequence, the technologies would first require manufacturing in another state to be exported to the protected state.[41] However, such a procedure is not covered by a compulsory licence (Art. 31 lit. f TRIPS, Section 24 PatG).[42]

This has serious consequences, as innovations usually build on each other and only show significant differences in terms of sustainability when they interact.[43] The pace of sustainable innovation can thus be slowed down as a result of the barriers created.[44] This phenomenon can be mitigated using voluntary licences, striking a

---

[36] Bundesministerium für Justiz und Verbraucherschutz (2020) Richtlinie zur Förderung des Technologie- und Wissenstransfers durch Patente, Normung und Standardisierung, available at https://www.mw-patent.de/media/files/WIPANO-F%C3%B6rderung-17.01.2020.pdf.   Accessed 12 July 2022.

[37] Study of the Fraunhofer IAO Stuttgart, "Professionelles Patentmanagement für kleine und mittlere Unternehmen in Baden-Württemberg", available at https://www.dpma.de/docs/service/kmu/fraunhoferinstitut_studie-patentmanagement-kmu.pdf. Accessed 14 July 2022, pp. 10 ff.

[38] R. Ballardini, J. Kaisto and J. Similä, Developing novel property concepts in private law to foster the circular economy, Journal of Cleaner Production 2021, pp. 279 ff.; E. Eppinger, A. Jain, P. Vimalnath, A. Gurtoo, F. Tietze and R. Hernandez, Sustainability transitions in manufacturing: the role of intellectual property, COSUST 2021(49), pp. 118–126.

[39] C. Heinze, Patentrecht und Klimawandel – eine Skizze, GRUR Newsletter 2020(1), p. 6.

[40] C. Heinze, Patentrecht und Klimawandel – eine Skizze, GRUR Newsletter 2020(1), p. 7.

[41] C. Heinze, Patentrecht und Klimawandel – eine Skizze, GRUR Newsletter 2020(1), p. 7.

[42] Agreement on Trade-Related Aspects of Intellectual Property Rights (TRIPS), Notice of 15 April 1994 (Bundesgesetzblatt II pp. 1438, 1730), last amended on 6 December 2005 (OJ 2007 L 311 p. 37); C. Heinze, Patentrecht und Klimawandel – eine Skizze, GRUR Newsletter 2020(1), p. 7.

[43] Cf. E. Eppinger, A. Jain, P. Vimalnath, A. Gurtoo, F. Tietze and R. Hernandez, Sustainability transitions in manufacturing: the role of intellectual property, COSUST 2021(49), pp. 118–126.

[44] E. Eppinger, A. Jain, P. Vimalnath, A. Gurtoo, F. Tietze and R. Hernandez, Sustainability transitions in manufacturing: the role of intellectual property, COSUST 2021(49), pp. 118–126.

15 Sustainability and Intellectual Property in Germany

good balance between the interests of the inventor (recouping investments) and society (driving research).[45]

Particular attention should be paid to ensuring that the acquisition of licences is realistically accessible to all countries. Currently, developing and emerging countries in particular are mainly responsible for rising energy consumption.[46] The aim of technology transfer should therefore be to introduce sustainable innovations in these regions to mitigate the consequences of climate change and counteract existing impacts.[47] Currently, German exports of environmental technologies are still strongly focused on other OECD countries.[48] However, the research report 'Environmental innovation made in Germany: What contribution can they make to achieving the SDGs in emerging and developing countries?'[49] comes to the conclusion that Germany could make an important contribution to achieving the SDGs in emerging and developing countries, especially in the technology fields of mobility, energy generation and efficiency, and (partly) recycling after comparing the German technology profile with the technological needs of emerging and developing countries. For example, Germany has a high level of innovation in the areas of environmentally friendly drives, wind power, combined heat and power (CHP) or heating systems; technologies that are considered priorities by developing and emerging countries.[50] In addition, however, for effective technology transfer it must also be taken into account that existing technologies may have to be adapted to regional conditions.[51] This is countered by the fact that most developing countries

---

[45] E. Eppinger, A. Jain, P. Vimalnath, A. Gurtoo, F. Tietze and R. Hernandez, Sustainability transitions in manufacturing: the role of intellectual property, COSUST 2021(49), pp. 118–126.

[46] W. Hoffmann, J. Lewerenz and T. Pellkoffer, Spannungsfeld von Technologietransfer und Schutz geistigen Eigentums, In: Forschungsverbund Erneuerbare Energien -FVEE- (ed), Themen 2009. Forschen für globale Märkte erneuerbarer Energien, Selbstverlag 2009, p. 127.

[47] W. Hoffmann, J. Lewerenz and T. Pellkoffer, Spannungsfeld von Technologietransfer und Schutz geistigen Eigentums, In: Forschungsverbund Erneuerbare Energien -FVEE- (ed), Themen 2009. Forschen für globale Märkte erneuerbarer Energien, Selbstverlag 2009, p. 128.

[48] C. Gandenberger and F. Marscheider-Weidemann, Umweltinnovationen made in Germany: Welchen Beitrag können sie zum Erreichen der SDG in Schwellen- und Entwicklungsländern leisten? Ein Beitrag zur Weiterentwicklung der deutschen Umweltinnovationspolitik, Umweltbundesamt 2021, p. 34.

[49] C. Gandenberger and F. Marscheider-Weidemann, Umweltinnovationen made in Germany: Welchen Beitrag können sie zum Erreichen der SDG in Schwellen- und Entwicklungsländern leisten? Ein Beitrag zur Weiterentwicklung der deutschen Umweltinnovationspolitik, Umweltbundesamt 2021.

[50] C. Gandenberger and F. Marscheider-Weidemann, Umweltinnovationen made in Germany: Welchen Beitrag können sie zum Erreichen der SDG in Schwellen- und Entwicklungsländern leisten? Ein Beitrag zur Weiterentwicklung der deutschen Umweltinnovationspolitik, Umweltbundesamt 2021, pp. 16–17.

[51] W. Hoffmann, J. Lewerenz and T. Pellkoffer, Spannungsfeld von Technologietransfer und Schutz geistigen Eigentums, In: Forschungsverbund Erneuerbare Energien -FVEE- (ed), Themen 2009. Forschen für globale Märkte erneuerbarer Energien, Selbstverlag 2009, p. 128; C. Gandenberger and F. Marscheider-Weidemann, Umweltinnovationen made in Germany: Welchen Beitrag können

lack the structural, material and financial resources for adaptation.[52] One approach to still fully exploit the potential of markets in emerging and developing countries is the development of 'frugal innovations' (i.e. affordable robust and functional innovations).[53] The aim of this is not to omit individual cost-intensive functions or features of the existing product, but on the contrary to develop radically new approaches to meet the requirements and conditions of the target market from the outset. In Germany, the idea of 'frugal design' has not yet arrived in the existing innovation processes or has not been strongly developed. In addition, small and medium-sized enterprises sometimes lack the capacities for research and development in this area.[54] This makes them all more dependent on the willingness of larger companies to license patented technologies under the fairest possible conditions.

Patents can also restrict third parties who do not have the corresponding licences from sustainably reprocessing (recycling).[55] Without a licence, they are basically barred from using and thus also from further developing and reprocessing. Nevertheless, the current legal situation also offers business models such as *refurbed,*[56] *Swappie*[57] or *BackMarket*[58] the opportunity to become active in the field of recycling and reselling patent-protected objects. Thus, the exhaustion principle covers both the maintenance without a licence and the restoration of the usability of the object in question as intended.[59] The limit of the exhaustion doctrine is only reached if the repair and reconditioning is equivalent to the production of a new object, for example when the destroyed smartphone is reassembled. Apart from this, these

---

sie zum Erreichen der SDG in Schwellen- und Entwicklungsländern leisten? Ein Beitrag zur Weiterentwicklung der deutschen Umweltinnovationspolitik, Umweltbundesamt 2021, p. 14.

[52] W. Hoffmann, J. Lewerenz and T. Pellkoffer, Spannungsfeld von Technologietransfer und Schutz geistigen Eigentums, In: Forschungsverbund Erneuerbare Energien -FVEE- (ed), Themen 2009. Forschen für globale Märkte erneuerbarer Energien, Selbstverlag 2009, p. 129.

[53] C. Gandenberger and F. Marscheider-Weidemann, Umweltinnovationen made in Germany: Welchen Beitrag können sie zum Erreichen der SDG in Schwellen- und Entwicklungsländern leisten? Ein Beitrag zur Weiterentwicklung der deutschen Umweltinnovationspolitik, Umweltbundesamt 2021, pp. 24–25.

[54] C. Gandenberger and F. Marscheider-Weidemann, Umweltinnovationen made in Germany: Welchen Beitrag können sie zum Erreichen der SDG in Schwellen- und Entwicklungsländern leisten? Ein Beitrag zur Weiterentwicklung der deutschen Umweltinnovationspolitik, Umweltbundesamt 2021, p 26.

[55] E. Eppinger, A. Jain, P. Vimalnath, A. Gurtoo, F. Tietze and R. Hernandez, Sustainability transitions in manufacturing: the role of intellectual property, COSUST 2021(49), p. 122.

[56] https://www.refurbed.de/unternehmen/. Accessed 14 July 2022.

[57] https://swappie.com/de/so-funktioniert-unser-service/. Accessed 14 July 2022.

[58] https://www.backmarket.de/de-de/l/smartphones/6c290010-c0c2-47a4-b68a-ac2ec2b64 dca. Accessed 14 July 2022.

[59] K. Haft, G. Baumgärtel, J. Dombrowski, B. Grzimek, B. Joachim and M. Loschelder, Die Erschöpfung von Rechten des Geistigen Eigentums in Fällen des Recyclings oder der Reparatur von Waren (Q 205), GRUR Int. 2008(11), p. 947.

sustainable business models are already possible without a licence.[60] This is a step in the right direction, as it at least prevents patent law from standing in the way of sustainable use of existing products. However, it would be desirable for patent law to not only not inhibit 'green' innovations, but to create incentives to drive them forward.

15.3.1.3 Efforts of the DPMA

In all this, the German Patent and Trade Mark Office (DPMA) was early aware of the importance of patent protection for the development of 'green' technologies.[61] Thus, as early as 2010, the President of the German Patent and Trade Mark Office spoke of the need to strike a balance between patent protection and the worldwide dissemination of 'green' technologies.[62] Although patent law serves to refinance the investments made in the technology, these technologies are mainly developed in industrialised countries.[63] Therefore, in addition to the indispensable protection of the high financial expenditure through patents, there must also be a worldwide dissemination of these technologies.[64]

In addition, the DPMA referred to the development of the number of patent applications in the field of renewable energies and its importance for combating climate change in various press releases.[65] The annual report of the DPMA also evaluated the patent applications in the field of climate-friendly technologies published by the DPMA and the EPO and classified their numerical development.[66]

---

[60] K. Haft, G. Baumgärtel, J. Dombrowski, B. Grzimek, B. Joachim and M. Loschelder, Die Erschöpfung von Rechten des Geistigen Eigentums in Fällen des Recyclings oder der Reparatur von Waren (Q 205), GRUR Int. 2008(11), p. 948.

[61] Press release DPMA, 22 July 2010 - Podiumsdiskussion über Umwelttechnologien im Deutschen Patent- und Markenamt, available at https://www.dpma.de/service/presse/pressemitteilungen/archiv/2010/20100722_1.html. Accessed 14 July 2022.

[62] Press release DPMA, 22 July 2010 - Podiumsdiskussion über Umwelttechnologien im Deutschen Patent- und Markenamt, available at https://www.dpma.de/service/presse/pressemitteilungen/archiv/2010/20100722_1.html. Accessed 14 July 2022.

[63] Press release DPMA, 22 July 2010 - Podiumsdiskussion über Umwelttechnologien im Deutschen Patent- und Markenamt, available at https://www.dpma.de/service/presse/pressemitteilungen/archiv/2010/20100722_1.html. Accessed 14 July 2022.

[64] Press release DPMA, 22 July 2010 - Podiumsdiskussion über Umwelttechnologien im Deutschen Patent- und Markenamt, available at https://www.dpma.de/service/presse/pressemitteilungen/archiv/2010/20100722_1.html. Accessed 14 July 2022.

[65] Press release DPMA, 29 March 2022 - Deutschland führend bei klimafreundlichen Innovationen, available at https://www.dpma.de/service/presse/pressemitteilungen/29032022/index.html. Accessed 13 June 2022; press release DPMA 24 April 2020 – Erfindungen für eine grüne Zukunft, available at https://www.dpma.de/service/presse/pressemitteilungen/20200424.html. Accessed 13 June 2022.

[66] DPMA - Annual Report 2021, available at https://www.dpma.de/digitaler_jahresbericht/2021/jb21_de/patente.html. Accessed 13 June 2022.

15.3.1.4 Interim Result

When it comes to the protection of 'green' technology, a conflict that patent law has been fighting ever since the beginning becomes relevant once again: It is about the balance of interests between inventor and society. While the inventor strives to amortise his development costs and usually to make a profit, the public wants access to innovations at low cost. This is particularly relevant in the case of 'green' technology, as man-made climate change is increasing the time pressure regarding climate-friendly technologies.

To address the conflict, the use of licensing can help, as the following example illustrates: The Coca-Cola company has licensed a patent for a plant bottle technology to selected companies that do not compete with Coca-Cola (e.g. Heinz tomato ketchup).[67] This led to a significant step towards sustainability, as such bottles feature a significantly lower carbon footprint than other bottles.[68] Ten years later, the patent was also made available to Coca-Cola's competitors.[69] However, the scenario presented only works if patent holders are willing to take such actions.

Consequently, the question that arises from all these considerations and will be addressed later within this contribution is: How can patent law provide further incentives to develop and share 'green' technologies without slowing down innovation in this field?

15.3.2 Trade Secrets Law

Trade secrets are company-related information that is not generally known and therefore has an economic value, that is protected by appropriate confidentiality measures and where there is a legitimate interest in keeping it secret (Sec. 2 No. 1 GeschGehG).[70] The protection of trade secrets is a less reliable way to protect 'green' technologies, as there is no registration procedure here.[71] The owner of the trade secret must have taken reasonable protective measures to keep the information

---

[67] Coca Cola Europe, WHY WE'RE SHARING OUR PLANTBOTTLE TECHNOLOGY WITH THE WORLD, available at https://www.coca-cola.eu/news/supporting-environment/why-were-sharing-our-plantbottle-technology-with-the-world. Accessed 13 July 2022.

[68] Coca Cola Europe, WHY WE'RE SHARING OUR PLANTBOTTLE TECHNOLOGY WITH THE WORLD, available at https://www.coca-cola.eu/news/supporting-environment/why-were-sharing-our-plantbottle-technology-with-the-world. Accessed 13 July 2022.

[69] Coca Cola Europe, WHY WE'RE SHARING OUR PLANTBOTTLE TECHNOLOGY WITH THE WORLD, available at https://www.coca-cola.eu/news/supporting-environment/why-were-sharing-our-plantbottle-technology-with-the-world. Accessed 13 July 2022.

[70] Trade Secret Directive, (Bundesgesetzblatt 2019 I p. 466).

[71] F. Klein, GREEN IP - A look at how sustainability influences IP and how IP can help in achieving sustainability, available at https://www.ashurst.com/en/news-and-insights/legal-updates/a-look-at-how-sustainability-influences-ip-and-how-ip-can-help-in-achieving-sustainability/. Accessed 13 July 2022.

secret (Sec. 2 No. 1, lit. b GeschGehG).[72] The concrete meaning of reasonable protective measures has been subject of much debate, especially since the Trade Secret Protection Act came into force.[73] Furthermore, protection is only granted against the unlawful acquisition, use or disclosure of the information, but not for its substance (cf. Sec. 4 GeschGehG).[74] If a person obtains the protected information without using illegal means, he or she may use the information.[75] Thus, the derivation of trade secrets from products put on the market ('reverse engineering') has also been liberalised considerably (Section 3 (2) No. 2 GeschGehG).[76] Compared to patent protection, the protection of trade secrets is therefore not as comprehensive and, due to the lack of registration (argument ex Sec. 4 No. 1 MarkenG,[77] Sec. 34 (1) PatentG) and the associated examination, more difficult to enforce and associated with greater legal uncertainty.[78] Nevertheless, a trade secret can be a useful complementary protection measure, especially for certain information that does not have to be disclosed in the patent application and therefore does not enjoy patent protection.[79]

---

[72] F. Klein, GREEN IP - A look at how sustainability influences IP and how IP can help in achieving sustainability, available at https://www.ashurst.com/en/news-and-insights/legal-updates/a-look-at-how-sustainability-influences-ip-and-how-ip-can-help-in-achieving-sustainability/. Accessed 13 July 2022.

[73] A. Ohly, Das neue Geschäftsgeheimnisgesetz im Überblick, GRUR 2019(5), pp. 441–443; T. Hohendorf, Know-How Schutz und Geistiges Eigentum, 1st ed, Mohr Siebeck 2020 p. 168.

[74] F. Klein, GREEN IP - A look at how sustainability influences IP and how IP can help in achieving sustainability, available at https://www.ashurst.com/en/news-and-insights/legal-updates/a-look-at-how-sustainability-influences-ip-and-how-ip-can-help-in-achieving-sustainability/. Accessed 13 July 2022.

[75] Recital 16, Directive 2016/943 of the European Parliament and of the Council; R. G. Bone, A New Look at Trade Secret Law: Doctrine in Search of Justification, California Law Review 1998(86), pp. 241–244; F. Klein, GREEN IP - A look at how sustainability influences IP and how IP can help in achieving sustainability, available at https://www.ashurst.com/en/news-and-insights/legal-updates/a-look-at-how-sustainability-influences-ip-and-how-ip-can-help-in-achieving-sustainability/. Accessed 13 July 2022.

[76] F. Klein, GREEN IP - A look at how sustainability influences IP and how IP can help in achieving sustainability, available at https://www.ashurst.com/en/news-and-insights/legal-updates/a-look-at-how-sustainability-influences-ip-and-how-ip-can-help-in-achieving-sustainability/. Accessed 13 July 2022.

[77] Act on the Protection of Trade Marks and other Signs, Bundesgesetzblatt 1994 I p. 3082, last amended on 10 August 2021 (Bundesgesetzblatt I p. 3490).

[78] F. Klein, GREEN IP - A look at how sustainability influences IP and how IP can help in achieving sustainability, available at https://www.ashurst.com/en/news-and-insights/legal-updates/a-look-at-how-sustainability-influences-ip-and-how-ip-can-help-in-achieving-sustainability/. Accessed 13 July 2022.

[79] F. Klein, GREEN IP - A look at how sustainability influences IP and how IP can help in achieving sustainability, available at https://www.ashurst.com/en/news-and-insights/legal-updates/a-look-at-how-sustainability-influences-ip-and-how-ip-can-help-in-achieving-sustainability/. Accessed 13 July 2022.

15.3.2.1 Possible Advantages and Disadvantages of Trade Secrets Protection

Trade secrets can, just like patents, also promote sustainability by creating incentives for innovation. In this context, trade secrets complement patents and act with them as complementary counterparts.[80] At the same time, trade secrets can also be a cheaper alternative to patents, with similar effects as innovation drivers.[81] This is promoted by the fact that the protection of trade secrets aims to promote the exchange of knowledge by protecting it.[82] In doing so, the law on the protection of secrets facilitates the exchange of knowledge, for example by eliminating the 'information paradox'.[83] This results from the fact that contracting parties will only be willing to pay money for knowledge if they also know it as a result of an exchange.[84] Companies that deal with sustainable technologies are more likely to enter into research cooperations if the protection afforded by the GeschGehG provides a protection concept that goes beyond contractual confidentiality agreements. This effect can therefore benefit 'green' technologies but is not specifically limited to this area. Therefore, this should rather be seen as a general effect of the law on the protection of secrets.

Like patents, trade secrets can also hinder the rapid dissemination of sustainable technologies since unlike patents, there is no disclosure within the meaning of Sec. 34 PatG, making further development even more difficult.[85] In the area of manufacturing technologies, trade secrets also frequently prevent the disclosure of data that is required for a comparison of the life cycle assessments of the various manufacturing technologies.[86] An example of this is steel production. Each company uses its own recipe, which also depends on how much energy is used in the production process. For this reason, the energy input is protected as trade secret and is not disclosed in the LCA. As a result, the comparability of the life cycle assessments of steel manufacturers is limited, which can inhibit the development of sustainable energies.

---

[80] A. Ohly, Der Geheimnisschutz im deutschen Recht: heutiger Stand und Perspektiven, GRUR 2014(1), pp. 1–3.

[81] C. Ann, Know-how – Stiefkind des Geistigen Eigentums? , GRUR 2007(1), pp. 39–40.

[82] Recital 3, Directive 2016/943 of the European Parliament and of the Council.

[83] K. J. Arrow, Economic Welfare and the Allocation of Resources for Invention, National Bureau of Economic Research, The Rate and Direction of Inventive Activity: Economic and Social Factors, Princeton Legacy Library 1962, p. 615; A. Ohly, Der Geheimnisschutz im deutschen Recht: heutiger Stand und Perspektiven, GRUR 2014(1), p. 3.

[84] K. J. Arrow, Economic Welfare and the Allocation of Resources for Invention, National Bureau of Economic Research, The Rate and Direction of Inventive Activity: Economic and Social Factors, Princeton Legacy Library 1962, p. 615; A. Ohly, Der Geheimnisschutz im deutschen Recht: heutiger Stand und Perspektiven, GRUR 2014(1), p. 3.

[85] S. Sandeen and D. Levine. In: Sarnoff (ed.), Research Handbook on Intellectual Property and Climate Change, Edward Elgar Publishing 2018, p. 355.

[86] E. Eppinger, A. Jain, P. Vimalnath, A. Gurtoo, F. Tietze and R. Hernandez, Sustainability transitions in manufacturing: the role of intellectual property, COSUST 2021(49), p. 118.

Trade secret protection can inhibit the dissemination of sustainable technologies to the extent that data is subject to secrecy protection. Almost any data, if it is subject of appropriate secrecy measures, can be protected by secrecy and thus be withheld from the public. The fact that it is irrelevant whether the data—e.g. machine-generated data from industry and science[87]—can already be used and exploited in practice appears to be problematic. Information for which there is currently no field of use or application can also be subject to secrecy protection.[88] Even purely electronic retrievability is sufficient for this.[89] Embodiment in physical form is not required. In this way, the protection of secrets opens the possibility for companies to exclude the public from the use of data that would be significant for the development of more sustainable technologies. Furthermore, companies are currently not forced to share 'green' information that could solve social and environmental problems. Rather, they are free to keep the information completely hidden from the public.[90]

15.3.2.2 Interim Result

Trade secrets face a similar dilemma to patents: on the one hand, they can provide an incentive for the development and discovery of information worthy of secrecy so that it can then be used in the best possible way. They can create cooperation networks between smaller and larger companies that promote innovation.[91] On the other hand, they are also attributed with innovation-inhibiting effects, especially in the case of new developments that rely on the interaction of several inventions/discoveries,[92] which also contributes to making progress towards a resource-saving, sustainable 'circular economy' more difficult.[93]

---

[87] S. Hessel and L. Leffer, Rechtlicher Schutz maschinengenerierter Daten, MMR 2020(10), p. 649.

[88] C. Alexander, In: Köhler, Bornkamm and Feddersen, Gesetz gegen den unlauteren Wettbewerb, 40th edition 2022, § 2 GeschGehG, para 28.

[89] C. Alexander In: Köhler, Bornkamm and Feddersen, Gesetz gegen den unlauteren Wettbewerb, 40th edition 2022, § 2 GeschGehG, para 25a.

[90] S. Sandeen and D. Levine. In: Sarnoff (ed.), Research Handbook on Intellectual Property and Climate Change, Edward Elgar Publishing 2018, p. 353.

[91] International Science Council, Three things to know about how Intellectual Property can contribute to sustainability transitions (2022), available at https://council.science/current/blog/how-intellectual-property-sustainability-transitions/. Accessed 13 July 2022; Directive 2016/943 of the European Parliament and of the Council, recital 1 and 2.

[92] E. Eppinger, A. Jain, P. Vimalnath, A. Gurtoo, F. Tietze and R. Hernandez, Sustainability transitions in manufacturing: the role of intellectual property, COSUST 2021(49), p. 122; cf. D. S. Karjala, Sustainability and intellectual property rights in traditional knowledge, Jurimetrics, 2012(53), p. 58.

[93] E. Eppinger, A. Jain, P. Vimalnath, A. Gurtoo, F. Tietze and R. Hernandez, Sustainability transitions in manufacturing: the role of intellectual property, COSUST 2021(49), p. 121.

15.3.3 Design Law

Design rights offer an easier and cheaper way to protect the aesthetic appearance of a product compared to other intellectual property rights like patents.[94] These rights are gaining importance especially in industries such as fashion, consumer goods, mechanical engineering, automotive and aerospace.[95] It is a so-called unexamined IP right, which arises by entry in the register at the DPMA. The application under Sec. 11 Design Act (DesignG)[96] must contain a request for registration, information allowing the identity of the applicant to be established and a representation of the design suitable for publication. Under Sec. 16 DesignG, the DPMA only examines whether the application fees under Sec. 5(1), first sentence, of the Patent Costs Law and the requirements for obtaining a date of filing under Sec. 11(2) are met and whether the application complies with the other filing requirements. The requirements for obtaining such protection are relatively low and the application is filed quickly since only a formal examination is undertaken and not all substantive requirements have to be elaborately examined first.[97]

For example, the specific designs of car or aircraft parts that reduce air resistance and thereby result in better fuel efficiency can be protected as 'green' products.[98] In the future, design rights may become even more relevant as 3D printing gains momentum as a potentially sustainable manufacturing process.[99] The sustainability potential of 3D printing lies in the fact that waste production is significantly reduced.[100] The reason for this is that unlike conventional production methods, the materials do not have to be moulded.[101]

---

[94] C. W. Davies, T. Nener, N. Pereira, Green IP, the role of intellectual property in sustainability (2021), Financier Worldwide, available at https://www.financierworldwide.com/green-ip-the-role-of-intellectual-property-in-sustainability. Accessed 13 July 2022.

[95] C. W. Davies, T. Nener, N. Pereira, Green IP, the role of intellectual property in sustainability (2021), Financier Worldwide, available at https://www.financierworldwide.com/green-ip-the-role-of-intellectual-property-in-sustainability. Accessed 13 July 2022.

[96] Act on the Legal Protection of Designs, Bundesgesetzblatt 2014 I p. 122, last amended on 10 August 2021 (Bundesgesetzblatt I p. 3490).

[97] F. Klein, GREEN IP - A look at how sustainability influences IP and how IP can help in achieving sustainability, available at https://www.ashurst.com/en/news-and-insights/legal-updates/a-look-at-how-sustainability-influences-ip-and-how-ip-can-help-in-achieving-sustainability/. Accessed 13 July 2022.

[98] F. Klein, GREEN IP - A look at how sustainability influences IP and how IP can help in achieving sustainability, available at https://www.ashurst.com/en/news-and-insights/legal-updates/a-look-at-how-sustainability-influences-ip-and-how-ip-can-help-in-achieving-sustainability/. Accessed 13 July 2022.

[99] C. W. Davies, T. Nener, N. Pereira, Green IP, the role of intellectual property in sustainability (2021), Financier Worldwide, available at https://www.financierworldwide.com/green-ip-the-role-of-intellectual-property-in-sustainability. Accessed 13 July 2022.

[100] C. W. Davies, T. Nener, N. Pereira, Green IP, the role of intellectual property in sustainability (2021), Financier Worldwide, available at https://www.financierworldwide.com/green-ip-the-role-of-intellectual-property-in-sustainability. Accessed 13 July 2022.

[101] C. W. Davies, T. Nener, N. Pereira, Green IP, the role of intellectual property in sustainability (2021), Financier Worldwide, available at https://www.financierworldwide.com/green-ip-the-role-of-intellectual-property-in-sustainability. Accessed 13 July 2022.

Design as such also plays a big role in sustainability. Products can be designed to be more streamlined and efficient or made from recycled materials so that they have less of a harmful impact on the climate. Modular mattresses can be mentioned here as an example, in which individual parts can be sustainably replaced.[102] Design law allows these innovative approaches to be fully protected, from marketing logos and packaging to the shape of furniture and vehicles to the user interfaces of computers and smartphones.[103] This is a great advantage for the designers of environmentally friendly products and services, as they can then fully benefit from the marketing of these products.[104]

However, as with the other types of intellectual property rights, this comprehensive protection is accompanied by the problem that third parties are excluded from using the design and only the designer can offer the specific product. Sustainable designs are thus restricted in their dissemination if the owners oppose licensing. Unlike the Patent Act, the Design Act does not provide any possibility of compulsory licensing.

15.3.4 Copyright

Copyright protects the author in his or her 'personal intellectual creation' (Sec. 2 (2) UrhG[105]).[106] It is therefore, at least from the earlier idea, not an industrial intellectual property right in the classical sense[107] but places the protection of the creative work of each individual in the foreground (Sec. 1 UrhG). On this basis, at first glance it seems far-fetched that copyright has any impact on the broad understanding of the concept of sustainability at all. However, there are several ways in which copyright can have an impact on sustainable action.

---

[102] WIPO (2020), World Intellectual Property Day 2020 - Innovate for a green future, Design rights and sustainability, available at https://www.wipo.int/ip-outreach/en/ipday/2020/articles/design_rights.html. Accessed 13 July 2022.

[103] WIPO (2020), World Intellectual Property Day 2020 - Innovate for a green future, Design rights and sustainability, available at https://www.wipo.int/ip-outreach/en/ipday/2020/articles/design_rights.html. Accessed 13 July 2022.

[104] WIPO (2020), World Intellectual Property Day 2020 - Innovate for a green future, Design rights and sustainability, available at https://www.wipo.int/ip-outreach/en/ipday/2020/articles/design_rights.html. Accessed 13 July 2022.

[105] Act on Copyright and Related Rights, Bundesgesetzblatt 1965 I p. 1273, last amended on 23 June 2021 (Bundesgesetzblatt I p. 1858).

[106] W. Bullinger, In: Wandtke and Bullinger (eds), Praxiskommentar Urheberrecht, ed 6, C.H.Beck 2022, § 1 UrhG para 1.

[107] U. Loewenheim, In: Schricker and Loewenheim, Urheberrecht, ed 6, C.H.Beck 2020, Einleitung zum UrhG para 10.

15.3.4.1 Sustainable Works

On the one hand, copyright can protect works such as films, books or plays that deal with the issue of sustainability and thus give creators an incentive to develop such works.[108] Admittedly, there also is a risk here that the works protected by copyright will only become known to a small part of the public. However, this problem is in no way comparable to that of patent law or the law on the protection of secrets, since in copyright law there is much greater pressure for the work to be offered, otherwise no profits can be generated. This is not the case with the law on the protection of secrets and patent law since inventors and/or bearers of secrets can also use the protected goods profitably for their own purposes. Moreover, in copyright law there is no danger of having a single solution that is then denied to all others by a protective right.

However, there may be exceptions to this in special case constellations. A possible problematic scenario might be if only one photographer has photographed a particular natural phenomenon.[109] This could have a particularly important function in drawing attention to the circumstances of climate change.[110] Under the protection of copyright law Sec. 2 (1) No. 5, Sec. 72 UrhG), a monopolistic position would then accrue to this unique photograph.[111] However, it should be emphasised that this covers rather rare exceptions, which are to be accepted in this way. Finally, it is also true here—and even more so—that a unique photograph only acquires value through its licensing.

15.3.4.2 Official Works

Another field in which copyright can become relevant is an exception to copyright protection.[112] Official works are excluded from copyright (Sec. 5 UrhG). According to this, *'laws, ordinances, official decrees and notices as well as decisions and officially drafted guidelines to decisions'* (Sec. 5 (1) UrhG) and *'other official works which have been published in the official interest for general knowledge'* (Sec. 5 (2) UrhG) are in the public domain.[113] This can become relevant in the case of

---

[108] E. Eppinger, A. Jain, P. Vimalnath, A. Gurtoo, F. Tietze and R. Hernandez, Sustainability transitions in manufacturing: the role of intellectual property, COSUST 2021(49), pp. 118–126.

[109] E. Derclaye (2012), The role of copyright in the protection of the environment and the fight against climate change: is the current copyright system adequate?, pp. 369 ff.

[110] E. Derclaye (2012), The role of copyright in the protection of the environment and the fight against climate change: is the current copyright system adequate?, pp. 369 ff.

[111] E. Derclaye (2012), The role of copyright in the protection of the environment and the fight against climate change: is the current copyright system adequate?, pp. 369 ff.

[112] E. Derclaye (2012), The role of copyright in the protection of the environment and the fight against climate change: is the current copyright system adequate?, p. 369.

[113] For a more detailed explanation of the eight different official works see O. Lampe, Informationen des Bundesamts für Sicherheit in der Informationstechnik: Urheberrechtlicher Schutz oder freie Nutzung?,ZdiW 2021(7), pp. 279–282.

15 Sustainability and Intellectual Property in Germany

contributions by the state that deal with the environment and sustainability.[114] However, by no means all contributions written with taxpayers' money are covered by this exception, so useful contributions are still protected by copyright.[115] In addition, case law on the subject shows that the state does take action against the use of its contributions,[116] so that there is legal uncertainty here when such contributions are used. This can make the exchange of knowledge and the dissemination of contributions financed with taxpayers' money more difficult.

15.3.4.3 Software, Algorithms and Data

Actual technical developments that achieve sustainable development can also be partially protected by copyright law. For example, algorithms can be used to optimise processes.[117] The scope of protection of copyright also extends to computer programs (Sec. 69a ff. UrhG), so that software and algorithms can enjoy copyright protection.[118] While the pure 'idea' does not enjoy copyright protection, the specific code is protected from use by third parties via copyright.[119] The evaluation of software and data can be crucial to improve existing technologies in an environmentally friendly way or to make entirely new ('green') technologies possible in the first place.[120] Software itself can thus contribute to a greener economy in various ways. Its possibilities range from recording and evaluating a

---

[114] E. Derclaye (2012), The role of copyright in the protection of the environment and the fight against climate change: is the current copyright system adequate?, p. 370.

[115] See among others: BGH, Judgement of 20 July 2006 - I ZR 185/03, GRUR 2007(2), p. 137; BGH, Judgement of 12 June 1981 – I ZR 95/79, GRUR 1982(1), p. 37.

[116] BGH, Judgement of 20 July 2006 - I ZR 185/03, GRUR 2007, pp. 137–138; BGH, Judgement of 12 June 1981 – I ZR 95/79, GRUR 1982, p. 37; OLG Stuttgart, Judgement of 14 July 2010 - 4 U 24/10, ZUM-RD 2011, pp. 20–22. See also the problems described in detail: O. Lampe, Informationen des Bundesamts für Sicherheit in der Informationstechnik: Urheberrechtlicher Schutz oder freie Nutzung?,ZdiW 2021(7), pp. 279–282.

[117] C. W. Davies, T. Nener and N. Pereira, Green IP, the role of intellectual property in sustainability (2021), Financier Worldwide, available at https://www.financierworldwide.com/green-ip-the-role-of-intellectual-property-in-sustainability. Accessed 13 July 2022; E. Eppinger, A. Jain, P. Vimalnath, A. Gurtoo, F. Tietze and R. Hernandez, Sustainability transitions in manufacturing: the role of intellectual property, COSUST 2021(49), pp. 118–126.

[118] Klein, Fabian (2020), GREEN IP - A look at how sustainability influences IP and how IP can help in achieving sustainability, available at https://www.ashurst.com/en/news-and-insights/legal-updates/a-look-at-how-sustainability-influences-ip-and-how-ip-can-help-in-achieving-sustainability/. Accessed 14 June 2022.

[119] Klein, Fabian (2020), GREEN IP - A look at how sustainability influences IP and how IP can help in achieving sustainability, available at https://www.ashurst.com/en/news-and-insights/legal-updates/a-look-at-how-sustainability-influences-ip-and-how-ip-can-help-in-achieving-sustainability/. Accessed 14 June 2022.

[120] Klein, Fabian (2020), GREEN IP - A look at how sustainability influences IP and how IP can help in achieving sustainability, available at https://www.ashurst.com/en/news-and-insights/legal-updates/a-look-at-how-sustainability-influences-ip-and-how-ip-can-help-in-achieving-sustainability/. Accessed 14 June 2022.

company's emissions[121] to streamlining and optimising supply and production lines.[122] Nevertheless, when it comes to copyright protection, it is necessary to bear in mind that only the exact source code can be protected but not the underlying and more profound idea.[123] This makes it possible to use modified source codes.

The data already mentioned can also be protected, at least in part, via copyright. However, this does not apply to 'raw climate data', as pure facts are excluded from protection.[124] If they only have any specific structure, protection will generally fail due to the requirements of Sec. 2 (2) UrhG.[125] If the required level of creation is nonetheless given, protection as a database work (Sec. 4 (2) UrhG) may be considered. This would entail the consequence that the duration of protection of 70 years after death would also apply (Sec. 64 UrhG). In turn this may result in (required) historical data also potentially becoming a liability risk over a long period of time as the data ages, making it increasingly difficult to trace the copyright holder (or his heirs) to acquire a right of use.[126] However, a database producer's right to the data collection or to the user interface to the database could still exist (Sec. 87a ff. UrhG). It is positive to note that there are exceptions to the rule for scientific research (Sec. 87c (1) No. 2 UrhG). However, these exceptions are quite limited.

15.3.4.4 Interim Result

It turns out that, contrary to first appearances, copyright law can indeed have an impact on the field of sustainable trade. However, the protection is incomplete and can only intervene in addition to other intellectual property rights. However, it also has its own key area through the protection of creative engagement with the topic of 'sustainability', which should not be underestimated. The problem of the monopolistic position of a creator is by far not as great as with patent law and secret protection law.

15.3.5 Trademark Law

The Act on the Protection of Trade Marks and other Signs (hereinafter MarkenG) protects trademarks, business designations and geographical indications in

---

[121] C. W. Davies, T. Nener and N. Pereira, Green IP, the role of intellectual property in sustainability (2021), Financier Worldwide, available at https://www.financierworldwide.com/green-ip-the-role-of-intellectual-property-in-sustainability. Accessed 13 July 2022.

[122] E. Eppinger, A. Jain, P. Vimalnath, A. Gurtoo, F. Tietze and R. Hernandez, Sustainability transitions in manufacturing: the role of intellectual property, COSUST 2021(49), pp. 118–126.

[123] G. Schulze, in: Dreier and Schulze (eds.), Urheberrechtsgesetz, 7th ed. 2022, § 2 para 127.

[124] M. Carroll, Intellectual Property and Related Rights in Climate Data, in: J. Sarnoff (ed.), Research Handbook on Intellectual Property and Climate Change, 2016, p. 386.

[125] M. Carroll, Intellectual Property and Related Rights in Climate Data, in: J. Sarnoff (ed.), Research Handbook on Intellectual Property and Climate Change, 2016, p. 387.

[126] M. Carroll, Intellectual Property and Related Rights in Climate Data, in: J. Sarnoff (ed.), Research Handbook on Intellectual Property and Climate Change, 2016, pp. 387–388.

15 Sustainability and Intellectual Property in Germany   291

accordance with Sec. 1 MarkenG. Under Sec. 3 MarkenG, all signs may be protected as trademarks which can distinguish the goods or services of one undertaking from those of other undertakings. Signs consisting exclusively of shapes or other characteristic features which are due to the nature of the goods themselves, which are necessary to achieve a technical effect or which confer substantial value on the goods are excluded. Section 4 of the Trade Mark Act distinguishes between registered trademarks (No. 1), utility marks (No. 2) and notoriety marks (No. 3).

The purpose of the use of certain brands is already not limited to a proof of origin, but also serves to communicate certain ideals and objectives of the using company as well as outstanding characteristics of the products.[127] Trademarks can have an impact on sustainability as consumers can see how sustainable a certain product is. They can communicate attitudes and principles of a company in relation to its efforts to design its products in an environmentally friendly way.[128] This results in transparency for consumers who can understand the higher prices of more sustainable products.[129] However, there are also certain terms and designations that cannot be protected as trademarks because they have a purely descriptive function, (Sec. 8 (2) No. 2 MarkenG). Such terms must be freely available to all companies so that they can also describe their goods accordingly. Nevertheless, trademark owners are not prevented from using phrases such as 'green' or 'eco' as part of the brand.[130]

In Sec. 106a MarkenG, the so-called certification mark[131] was implemented in the German Trademark Act. This mark guarantees certain characteristics of a product and thus focuses on the guaranteed function instead of the origin function.[132] The owner of a certification mark specifies corresponding quality features in a statute, which must at least meet the requirements of Sec. 106d (2) MarkenG, and grants licences only to those licence applicants who fulfil the criteria mentioned. Under Sec. 106b MarkenG, the trademark proprietor may not himself be active in the field for which he grants corresponding quality trademarks, which is intended to avoid a conflict of interests.[133] Classic seals such as the TÜV seal[134] can be considered as a

---

[127] G. Elskamp and S. Völker, Die neuen Markenfunktionen des EuGH, WRP 2010, p. 69.

[128] F. Klein, GREEN IP - A look at how sustainability influences IP and how IP can help in achieving sustainability, available at https://www.ashurst.com/en/news-and-insights/legal-updates/a-look-at-how-sustainability-influences-ip-and-how-ip-can-help-in-achieving-sustainability/. Accessed 13 July 2022.

[129] D. Seeliger and K. Gürer, Kartllrecht und Nachhaltigkeit: Neue Regeln für Umweltschutzvereinbarungen von Wettbewerbern? BB 2021, p. 2050.

[130] C. W. Davies, T. Nener and N. Pereira, Green IP, the role of intellectual property in sustainability (2021), Financier Worldwide, available at https://www.financierworldwide.com/green-ip-the-role-of-intellectual-property-in-sustainability. Accessed 13 July 2022.

[131] See also: https://www.dpma.de/marken/markenschutz/mamog/gewaehrleistungsmarke/index.html. Accessed 14 July 2022.

[132] M. Vohwinkel, In: Kur, v. Bomhard and Albrecht (ed.), BeckOK Markenrecht, 29th ed 2022, § 106a MarkenG.

[133] M. Vohwinkel, In: Kur, v. Bomhard and Albrecht (ed.), BeckOK Markenrecht, 29th ed 2022, § 106a MarkenG, para 3.

[134] https://register.dpma.de/DPMAregister/marke/registerHABM?AKZ=017277849&CURSOR=3. Accessed 14 July 2022.

certification mark. At the same time, trademark law with the certification mark also offers enormous potential to facilitate sustainable consumption decisions for consumers and can thus already make a decisive contribution to the promotion of sustainability in its current form.

For example, the EU Ecolabel,[135] the FSC label[136] or the Rainforest Alliance seal[137] are well-known seals for sustainable products. In all cases, these are certification marks. Consumers are thereby able to compare the sustainability of different products and subsequently make an informed and, in the best case, sustainable purchase decision without having to undertake more in-depth research. It turns out that consumers are more likely to choose the more environmentally friendly products, even if this means higher costs for them.[138]

Although 'green' appearing trademarks also harbour the danger of so-called 'greenwashing', in which trademarks that are not particularly environmentally friendly merely give the impression that consumers are making a particularly sustainable choice by a certain choice of trademark — for example, by using green colour. However, this is attempted to be counteracted by the prohibition of the registration of misleading trademarks in Sec. 8 (2) No. 4 MarkenG.[139]

In the case of the designated seal as a certification mark, the danger of 'greenwashing' is rather low due to the statutory requirement and corresponding inspection obligations of the trademark proprietor. Section 106e (2) MarkenG further strengthens the general prohibition of deception for certification marks.[140] As criticised by some environmental activists, the fact that such seals often only show a minimum standard and pretend to be more sustainable than the certified product actually offers could prove to be problematic.[141] Nevertheless, certification marks at least offer the possibility to point out certain features such as environmental balances and thus to distinguish more sustainable products from less sustainable products.

---

[135] https://register.dpma.de/DPMAregister/marke/registerHABM?AKZ=018055852&CURSOR=12. Accessed 14 July 2022.

[136] https://register.dpma.de/DPMAregister/marke/registerHABM?AKZ=018032783&CURSOR=0. Accessed 14 July 2022.

[137] https://register.dpma.de/DPMAregister/marke/registerHABM?AKZ=018323648&CURSOR=5. Accessed 14 July 2022.

[138] For example, EU Commission (2021): Commission's new consumer survey shows impact of COVID-19 and popularity of 'greener' choices consumer survey, available at: https://ec.europa.eu/commission/presscorner/detail/de/ip_21_1104. Accessed 14 July 2022.

[139] E. Rosati (2022), Green with ... IP, available at https://euipo.europa.eu/ohimportal/en/news?p_p_id=csnews_WAR_csnewsportlet&p_p_lifecycle=0&p_p_state=normal&p_p_mode=view&p_p_col_id=column-1&p_p_col_count=2&journalId=9135001&journalRelatedId=manual/. Accessed 4 July 2022.

[140] F. Albrecht, In: Kur, v. Bomhard and Albrecht (ed.), BeckOK Markenrecht, 29th ed 2022, § 8 MarkenG para 641.

[141] For example, W. Huismann (2018), Dunkle Geschäfte mit dem MSC-Siegel, available at https://www.tagesschau.de/inland/bio-fisch-guetesiegel-101.html. Accessed 14 July 2022; S. Biegger (2021), Als Bio in die Breite ging, available at https://www.tagesschau.de/wirtschaft/biosiegel-101.html. Accessed 14 July 2022.

15 Sustainability and Intellectual Property in Germany

Trademark law can thereby significantly advance the goal of a more sustainable economy. In the case of trademarks, the exclusive right does not lead to a conflict of interests such as in the case of the intellectual property rights mentioned above. The trademark is only intended to refer to a company and its products. This means that the image of a sustainable company can only be used to advertise a company that acts in the spirit of sustainability.

15.3.6 Interim Result

It becomes apparent that the various intellectual property rights can have both positive and negative effects on research and development in the field of sustainability. It turns out that not only technical gauges are affected, but also artistic and marketing-ready versions that can have a positive impact on a company's reputation and thus advance the sustainability issue within the company. However, it is also clear that while intellectual property rights have the potential to advance 'green' technologies, they have not yet been specifically designed to have such an impact. These are rather incidental sustainability effects. In the case of most intellectual property rights, there is a tension between the incentive function for the development of sustainable technologies and the inhibition function regarding the rapid dissemination of these innovations. This problem is already known and is not an explicit problem in the field of sustainability. The most important factor determining whether the positive or negative role predominates in the balance is the concrete design and scope of protection of these rights.[142]

15.4 Promoting Sustainability in the Context of Intellectual Property Rights—Outlook and Ideas

The results show that intellectual property rights certainly have the potential to promote sustainability and thus already provide good approaches in some areas. Nevertheless, the potential of intellectual property rights is far from exhausted. In the following, proposals will be examined that can contribute to further advancing sustainability through the targeted use of intellectual property rights.

15.4.1 Patent Law

As it turned out, patent law can drive sustainable innovation, but at the same time can also be an obstacle to sustainable development because of the monopolies it creates.

---

[142] A. Deren and J. Skonieczny (2022), Green Intellectual Property as a Strategic Resource in the Sustainable Development of an Organisation, available at https://www.mdpi.com/2071-1050/14/8/4758. Accessed 14 July 2022.

The goal should therefore be to reform patent law in such way that it does not become a brake on innovation but a driving force for sustainable development.

15.4.1.1 Green Impact Fund for Technology

A possible problem of patent law is that it is left to the patent holder to decide to whom the invention is granted. The only exception to this is the means of compulsory licensing. Countries in the global South are most affected by climate change, although they contribute hardly anything compared to the industrialised nations.[143] With rising population and the associated increase in emissions in these countries as well, the problem is becoming even more serious.[144]

The greatest efforts to reduce emissions are being made in the countries with the highest taxes and market prices for emissions.[145] Consequently, efforts should be made to strengthen the transfer of sustainable technologies to these regions, for example through publicly funded research and development cooperation.[146]

To break through this contradictory development, the establishment of a Green Impact Fund for Technology (GIFT) is currently being discussed. This would be a (supra-) nationally financed impact fund, in which the registered patent holders would be paid fixed amounts annually, divided according to the emissions they avoided in the previous year. In return, the patent holders would have to offer permanent royalty-free licences for the registered technology. Only the countries of the global South involved could benefit from the free licensing. Ideally, companies in these countries should be able to draw on several patented inventions relating to different components.[147] By rewarding real emission savings, a cost incentive would be set for sustainable technologies that are as effective as possible, i.e. in contrast to conventional innovation premiums, the actual feasibility would also be taken into account.[148] The basic possibility of evaluating the emission

---

[143] F. Ekardt (2010) Klimawandel und soziale Gerechtigkeit, p. 3, available at https://www.kas.de/c/document_library/get_file?uuid=5f333536-2589-7f03-c07d-1ceea90afc48&groupId=252038. Accessed 15 July 2022.

[144] A. Hollis and T. Pogge (2020), Green Impact Fund for Technology, p. 1, available at https://cpb-us-w2.wpmucdn.com/campuspress.yale.edu/dist/6/1129/files/2022/04/GIFT-White-Paper-2022-04-12.pdf. Accessed 14 July 2022.

[145] A. Hollis and T. Pogge (2020), Green Impact Fund for Technology, p. 2, available at https://cpb-us-w2.wpmucdn.com/campuspress.yale.edu/dist/6/1129/files/2022/04/GIFT-White-Paper-2022-04-12.pdf. Accessed 14 July 2022.

[146] C. Heinze, Patentrecht und Klimawandel – eine Skizze, GRUR Newsletter 2020(1), p. 7.

[147] A. Hollis and T. Pogge (2020), Green Impact Fund for Technology, p. 2, available at https://cpb-us-w2.wpmucdn.com/campuspress.yale.edu/dist/6/1129/files/2022/04/GIFT-White-Paper-2022-04-12.pdf. Accessed 14 July 2022; D. Azhgaliyeva, A. Hollis, T. Pogge, D. Rahut and Y. Yao (2022), Financing a Green Future: The Energy Transition Mechanism (ETM) and the Green Impact Fund for Technology (GIFT), available at https://www.think7.org/wp-content/uploads/2022/04/Climate_Financing-a-green-future-the-energy-transition-mechanism-ETM-and-the-green-impact-fund-for-technology-GIFT_Azhgaliyeva_Holli_Pogge_Rahut_Yao.pdf. Accessed 14 July 2022.

[148] A. Hollis and T. Pogge (2020), Green Impact Fund for Technology, pp. 2 ff., available at https://cpb-us-w2.wpmucdn.com/campuspress.yale.edu/dist/6/1129/files/2022/04/GIFT-White-Paper-

15 Sustainability and Intellectual Property in Germany

reductions attributable to the use of GIFT-registered technologies was recently confirmed, subject to a prior test phase.[149]

15.4.1.2 National Level

At national level there are various starting points for promoting sustainability in the context of patent rights too. One starting point for improvement are the patenting requirements.

15.4.1.2.1 Amendment of Patenting Requirements

The patenting of climate-friendly technologies could be stimulated, for example, by lowering the requirements for an inventive activity.[150] However, this would create the danger of 'weak patents' and thus a patent allocation that hinders innovation.[151] This is because exclusive rights would be granted for trivial modifications or obvious applications of products or processes for environmentally friendly purposes.[152] This would jeopardise legal certainty.[153] It would also hinder new market entrants and unnecessarily increase transaction costs.[154] Accordingly, a reduction of the inventive activity requirement for climate-friendly technologies is not advisable.[155] Only in the context of assessing novelty and inventive step, the sustainability-securing function of patent law in Germany could be effectively improved by considering the invention of applications of known technologies which are more climate-friendly or sustainable, as a sufficient contribution.[156]

2022-04-12.pdf. Accessed 14 July 2022; D. Azhgaliyeva, A. Hollis, T. Pogge, D. Rahut and Y. Yao (2022), Financing a Green Future: The Energy Transition Mechanism (ETM) and the Green Impact Fund for Technology (GIFT), p. 5, available at https://www.think7.org/wp-content/uploads/2022/04/Climate_Financing-a-green-future-the-energy-transition-mechanism-ETM-and-the-green-impact-fund-for-technology-GIFT_Azhgaliyeva_Holli_Pogge_Rahut_Yao.pdf. Accessed 14 July 2022.

[149] R. Block, Expert opinion of TUEV Süd (2022), available at https://cpb-us-w2.wpmucdn.com/campuspress.yale.edu/dist/6/1129/files/2022/06/TU%CC%88V.pdf. Accessed 14 July 2022.

[150] C. Heinze, Patentrecht und Klimawandel – eine Skizze, GRUR Newsletter 2020(1), p. 6.

[151] C. Heinze, Patent Law and Climate Change - Do We Need an EU Patent Law Directive on Clean Technology? GRUR Int. 2021, p. 558.

[152] G. Henry, Intellectual Property Rights and Green Technologies, p. 13, AIPPI 2010, available at https://www.aippi.fr/upload/Prix%20AIPPI/greentech-ipr-1st-academic-prize-dr.-guillaume-henry.pdf. Accessed 15 July 2022.

[153] C. Heinze, Patent Law and Climate Change - Do We Need an EU Patent Law Directive on Clean Technology? GRUR Int. 2021, p. 558.

[154] G. Henry, Intellectual Property Rights and Green Technologies, p. 13, AIPPI 2010, available at https://www.aippi.fr/upload/Prix%20AIPPI/greentech-ipr-1st-academic-prize-dr.-guillaume-henry.pdf. Accessed 15 July 2022.

[155] G. Henry, Intellectual Property Rights and Green Technologies, p. 13, AIPPI 2010, available at https://www.aippi.fr/upload/Prix%20AIPPI/greentech-ipr-1st-academic-prize-dr.-guillaume-henry.pdf. Accessed 15 July 2022.

[156] C. Heinze, Patentrecht und Klimawandel – eine Skizze, GRUR Newsletter 2020(1), p. 6.

15.4.1.2.2 Patent Exclusion

Furthermore, it is proposed to completely deny patenting of climate-damaging technologies with reference to a violation of public policy.[157] The Board of Appeal of the European Patent Office has already mentioned that the concept of public policy also includes environmental protection.[158] The benefit to mankind had to be weighed against the possible risks to the environment.[159] Inventions that are likely to seriously damage the environment can therefore be excluded from patentability.[160] In extreme cases of particularly climate-damaging inventions, the denial of patentability on the grounds of a violation of public policy certainly appears to be an appropriate remedy. However, it becomes difficult to determine when an invention is to be classified as so harmful to the environment that it is denied patentability. This would always require a case-by-case consideration and thus lead to legal uncertainty.

Furthermore, if we are not dealing with extreme cases, a balancing with other legitimate goals (e.g. energy security) is always required, so that a fine adjustment seems more appropriate than a fundamental exclusion of patentability.[161] However, this can be achieved more effectively within the framework of regulation by means of energy and environmental law (e.g. by means of emissions taxation).[162]

15.4.1.2.3 Accelerated Granting Procedure

In addition to the patenting requirements, another aspect is the patenting procedure. In some other countries, such as the United States and the United Kingdom, but also in Israel, Canada, Australia, Japan and Korea, there has been an accelerated patenting procedure for more than ten years.[163] In the United Kingdom, there is the possibility of an accelerated grant procedure for sustainable technologies (so-called 'Green Channel').[164] According to the English Patent Office, the usual time span of three to five years between application and grant of the patent was

---

[157] C. Heinze, Patentrecht und Klimawandel – eine Skizze, GRUR Newsletter 2020(1), p. 6.

[158] EPO, case T 0315/2003, *Transgenic animals v HARVARD*, ECLI:EP:BA:2004: T031503.20040706, para 10.2.

[159] EPO, case T 0315/2003, *Transgenic animals v HARVARD*, ECLI:EP:BA:2004: T031503.20040706, para 10.2.

[160] EPO, case T 0315/2003, *Transgenic animals v HARVARD*, ECLI:EP:BA:2004: T031503.20040706, para 10.2.

[161] C. Heinze, Patent Law and Climate Change - Do We Need an EU Patent Law Directive on Clean Technology? GRUR Int. 2021, p. 559.

[162] C. Heinze, Patentrecht und Klimawandel – eine Skizze, GRUR Newsletter 2020(1), p. 6.

[163] E. Lane, Building the global green patent highway: a proposal for international harmonization of green technology fast track programs, in: University of California (ed.), Berkley Technology Law Journal 2012, Vol. 27, No. 2, p. 1136.

[164] Klein, Fabian (2020), GREEN IP - A look at how sustainability influences IP and how IP can help in achieving sustainability, available at https://www.ashurst.com/en/news-and-insights/legal-updates/a-look-at-how-sustainability-influences-ip-and-how-ip-can-help-in-achieving-sustainability/. Accessed 14 June 2022.

15 Sustainability and Intellectual Property in Germany

reduced to eight to nine months in this procedure.[165] There is currently no equivalent accelerated grant procedure for sustainable technologies in Germany. While the EPO offers an accelerated granting procedure, the so-called PACE programme, this is by no means linked to sustainability requirements yet.[166] An accelerated procedure for sustainable technologies would potentially have advantages both for the applicant and for his competitors who would have legal certainty at an early stage if the application is rejected.[167] Moreover, the introduction of such a procedure would in all likelihood encourage investment in 'green' technologies.[168] Therefore, an accelerated patenting procedure for sustainable technologies should be considered in Germany.

15.4.1.2.4 Cost Savings for 'Green' Inventors

As a further consideration, one could think of relying on perks as an incentive, by reducing the costs of the application and maintenance. Similarly, as Sec. 23 (1) sentence 1 PatG does for the costs under Sec. 17 PatG in the case of a license agreement, it would be conceivable to grant reductions for patenting 'green' inventions. This would be possible, for example, by inserting a further paragraph in Sec. 23 PatG and could create a further incentive to invest in the development of sustainable inventions. Though, it should be noted that this approach does not eliminate the problems associated with the monopolisation of the 'green' invention. It would therefore be more recommendable to combine the additional reduction with the declaration of willingness to license already standardised in Sec. 23 (1) sentence 1 PatG, so that the patenting and maintenance of 'green' inventions becomes particularly attractive in terms of costs, but at the same time a dissemination of the invention is also promoted.

15.4.1.2.5 Compulsory Licence and Tacit Licence

A further approach to improve sustainability concerns licences. One way to circumvent the dissemination-inhibiting effect of patents is to introduce an obligation to license sustainability-related inventions (so-called 'compulsory licensing'). However, such scheme could significantly slow down the pace of innovation, especially if companies would have to pass on their technologies before they have recouped their

---

[165] G. Henry (2010), Intellectual Property Rights and Green Technologies, p. 15, available at https://www.aippi.fr/upload/Prix%20AIPPI/greentech-ipr-1st-academic-prize-dr.-guillaume-henry.pdf. Accessed 14 July 2022.

[166] https://www.epo.org/law-practice/legal-texts/html/guidelines/e/e_viii_4.htm. Accessed July 15 2022; C. Heinze, Patent Law and Climate Change - Do We Need an EU Patent Law Directive on Clean Technology? GRUR Int. 2021, p. 559.

[167] C. Heinze, Patent Law and Climate Change - Do We Need an EU Patent Law Directive on Clean Technology? GRUR Int. 2021, p. 559.

[168] G. Henry, Intellectual Property Rights and Green Technologies, p. 15, AIPPI 2010, available at https://www.aippi.fr/upload/Prix%20AIPPI/greentech-ipr-1st-academic-prize-dr.-guillaume-henry.pdf. Accessed 14 July 2022.

investments in research and development.[169] In addition, there is the critical circumstance that companies engaged in the development of sustainable inventions would always be exposed to a higher risk of compulsory licensing and would thus be limited in their entrepreneurial freedom of choice. As a result, interest in the development of sustainable inventions may decline, as they would always be exposed to compulsory licensing.

It therefore seems preferable to create incentives to increase the voluntary willingness to license.[170] For example, tax relief could be considered.[171] Cross-sectoral licensing seems particularly promising here.[172] A good example of cross-sector licensing is the licensing of a plant bottle technology by Coca Cola to selected companies with which Coca Cola is not in competition.[173]

Finally, the introduction of an 'implied licence' is being discussed, whereby comprehensive rights of use, e.g. for sustainable repair, are impliedly transferred with the sale of the goods in question.[174] Such legal institution has already been tried and tested in other legal systems. However, under German law there is regularly no need for the introduction of such an implied licence. After all, according to the principle of exhaustion, it is possible to repair the item or to maintain its usability as intended. In this respect, the implied licence would not change anything for the legal position of companies such as 'refurbed'.[175] On the other hand, clear guidelines on the part of case law or legislation as to how to differentiate from inadmissible complete reconstruction would be useful.[176]

---

[169] E. Eppinger, A. Jain, P. Vimalnath, A. Gurtoo, F. Tietze and R. Hernandez, Sustainability transitions in manufacturing: the role of intellectual property, COSUST 2021(49), p. 120.

[170] See E. Eppinger, A. Jain, P. Vimalnath, A. Gurtoo, F. Tietze and R. Hernandez, Sustainability transitions in manufacturing: the role of intellectual property, COSUST 2021(49), p. 120; C. Heinze, Patentrecht und Klimawandel – eine Skizze, GRUR Newsletter 2020(1), p. 7.

[171] E. Eppinger, A. Jain, P. Vimalnath, A. Gurtoo, F. Tietze and R. Hernandez, Sustainability transitions in manufacturing: the role of intellectual property, COSUST 2021(49), p. 120; C. Heinze, Patentrecht und Klimawandel – eine Skizze, GRUR Newsletter 2020(1), p. 7.

[172] E. Eppinger, A. Jain, P. Vimalnath, A. Gurtoo, F. Tietze and R. Hernandez, Sustainability transitions in manufacturing: the role of intellectual property, COSUST 2021(49), p. 120.

[173] Coca Cola Europe 2019, WHY WE'RE SHARING OUR PLANTBOTTLE TECHNOLOGY WITH THE WORLD, https://www.coca-cola.eu/news/supporting-environment/why-were-sharing-our-plantbottle-technology-with-the-world. Accessed 14 July 2022.

[174] K. Haft, G. Baumgärtel, J. Dombrowski, B. Grzimek, B. Joachim and M. Looschelders, Die Erschöpfung von Rechten des Geistigen Eigentums in Fällen des Recyclings oder der Reparatur von Waren (Q 205), GRUR Int. 2008(11), p. 952.

[175] K. Haft, G. Baumgärtel, J. Dombrowski, B. Grzimek, B. Joachim and M. Looschelders, Die Erschöpfung von Rechten des Geistigen Eigentums in Fällen des Recyclings oder der Reparatur von Waren (Q 205), GRUR Int. 2008(11), p. 952.

[176] K. Haft, G. Baumgärtel, J. Dombrowski, B. Grzimek, B. Joachim and M. Looschelders, Die Erschöpfung von Rechten des Geistigen Eigentums in Fällen des Recyclings oder der Reparatur von Waren (Q 205), GRUR Int. 2008(11), p. 952.

15.4.1.2.6 Interim Result

Patent law offers numerous opportunities to promote sustainability. Particularly, the transfer of sustainable technologies through publicly funded research and development cooperation should serve as a starting point. However, the establishment of a fast-track patenting system may also be worth considering, as examples from other countries show. Also, incentives should be set to drive the development of sustainable inventions. These can be cost savings, e.g. through tax savings or reduced application and maintenance fees.

15.4.2 Trade Secret Law

The exclusive use of data, which can often be subject to trade secret protection, poses major obstacles to sustainable development. Therefore, one of the biggest challenges in trade secret law is to focus on the use of trade secrets to promote sustainability instead of blocking it.[177] To counteract monopoly structures, properly designed access rights must be established.[178] The creation of new exclusive rights, on the other hand, could be counterproductive.[179]

15.4.2.1 Freedom of Information Rights

Existing freedom of information rights can also play a role. Both under the Freedom of Information Act (IFG),[180] but also under the Environmental Information Act (UIG),[181] which is particularly important from the point of view of sustainability, trade secrets, among other things, can be an obstacle to the disclosure of potentially ecologically relevant data (see Sec. 6 sentence 2 IFG; Sec. 9 (1) No. 3 UIG).[182] The UIG in particular does not provide for absolute protection here, but for a margin of appreciation.[183] In the case of environmental data essential for the preparation of life cycle assessments the EU Commission has proposed a balancing of the interest in the company's secrecy with the interest of the general public in the calculation of the

---

[177] P. Gailhofer and C.S. Scherf, Working Paper: Regulierung der Datenökonomie – Ansätze einer ökologischen Positionierung, Öko-Institut e.V. 2019(02), p. 21.

[178] P. Gailhofer and C.S. Scherf, Working Paper: Regulierung der Datenökonomie – Ansätze einer ökologischen Positionierung, Öko-Institut e.V. 2019(02), p. 38.

[179] P. Gailhofer and C.S. Scherf, Working Paper: Regulierung der Datenökonomie – Ansätze einer ökologischen Positionierung, Öko-Institut e.V. 2019(02), pp. 32 ff.

[180] Freedom of Information Act, (Bundesgesetzblatt 2005 I p. 2722), last amended on 9 June 2020 (Bundesgesetzblatt I p. 1328).

[181] Environmental Information Act, (Bundesgesetzblatt 2014 I p.1643), last amended on 25 February 2021 (Bundesgesetzblatt I p. 306).

[182] On this: G. Wiebe, Der Geschäftsgeheimnisschutz im Informationsfreiheitsrecht - Unter besonderer Berücksichtigung des Gesetzes zum Schutz von Geschäftsgeheimnissen, NVwZ 2019(23), p. 1705.

[183] M. Karg, in: Gersdorf and Paal (eds.), Informations- und Medienrecht, 2th ed. 2021, § 9 UIG paras 14 ff.

CO2 footprint.[184] An extension of the application scope of the freedom of information laws, for example to larger corporations that occupy a prominent position in a market with regard to their data stocks, could also be an option to make data that can be used for sustainability generally available. It would be conceivable, for example, to use the data on traffic flows available at corporations such as Alphabet to optimise public transport and to simplify the use of public transport and other resource-saving mobility solutions. This would require a corresponding right of access.[185]

15.4.2.2 Negative List

Furthermore, it is proposed to exempt certain information from protection under the law on secrecy. This could take the form of a negative list, which would exclude data that would reveal dangers to public safety and health from the protection of secrets.[186] Likewise, data that are elementary for the development of sustainable technologies, for example special environmental data, could be excluded from protection. Section 3 (2) GeschGehG already allows the use of a trade secret if this is permitted by law. Thus, there is at least the possibility of creating such exemptions for environmentally relevant data in special laws.

15.4.2.3 Economic Value

Moreover, care should be taken not to interpret the economic value of the information required in Sec. 2 No. 1a GeschGehG too broadly. Potential environmental risks that are evident from data collected by a company can be attributed an economic value insofar as public knowledge can prove damaging to a company's reputation.[187] Nevertheless, the economic value should be understood in the form of commercial exploitability through sharing of the information, since it must not be the purpose of the Trade Secret Act to support companies in environmentally damaging and thus unsustainable behaviour.[188] Ensuring a sustainable understanding of the Trade Secret Act thus remains the task of the courts.[189]

---

[184] European Commission (2020), Making the most of the EU's innovation potential - Intellectual Property Action Plan for EU recovery and resilience, p. 14, available at https://eur-lex.europa.eu/legal-content/EN/TXT/?uri=CELEX%3A52020DC0760. Accessed 14 July 2022.

[185] Cf. C. Weihe, Wertstoff Daten: Regulierung und Nachhaltigkeit, Öko-Institut e.V. 2019, available at https://www.oeko.de/e-paper/digitalisierung-konzepte-fuer-mehr-nachhaltigkeit/artikel/data-a-precious-resource. Accessed 14 July 2022.

[186] D. Levine, What can the Uniform Trade Secrets Act learn from the Bayh.-Dole Act, Hamline L. Rev. 2011, vol. 33, p. 615-647; S. Sandeen and D. Levine, in: Sarnoff (ed.), Research Handbook on Intellectual Property and Climate Change, Edward Elgar Publishing 2018, p. 357.

[187] S. Sandeen and D. Levine, in: Sarnoff (ed.), Research Handbook on Intellectual Property and Climate Change, Edward Elgar Publishing 2018, p. 358.

[188] M. Hiéramente, in: Fuhlrott and Hieramente (eds.), BeckOK-GeschGehG, 11th ed. 2022, § 2 paras 13 ff.

[189] S. Sandeen and D. Levine, in: Sarnoff (ed.), Research Handbook on Intellectual Property and Climate Change, Edward Elgar Publishing 2018, p. 358.

15.4.2.4 Interim Result

Trade secrets become an obstacle to innovation, particularly when it comes to companies who avoid making environmentally relevant data public by leaning on the Trade Secret Act. Therefore, focus should be on making such data accessible to the public by exempting it from the protection of secrets, e.g. through the creation of corresponding special laws.

15.4.3 Design Law

As previously elaborated, on the one hand, design law creates incentives for the design of innovative and environmentally friendly approaches. On the other hand, there is the objection that the use of sustainable design is reserved for the designer and is thus not available to third parties without restriction. Consequently, the interests of the right holder in his exclusive right and those of the general public in its exploitation are diametrically opposed. These must therefore be reconciled by design law in such a way that the sustainability of the design is not only promoted selectively, but also as broadly as possible.

15.4.3.1 Exceptions: Essential Interests of Environmental Protection

One approach to solving the conflicting interests is to restrict the right to the registered design as an exclusive right if essential interests of sustainability conflict with this. As long as the designer can prohibit third parties from using the design without his consent, Sec. 38 (1) DesignG, the design cannot be used on a large scale and therefore cannot effectively promote sustainability. Since this approach is completely at the expense of the designer's interests, the designer would have to be compensated financially accordingly. For this purpose, it would be conceivable to create a regulation, like Sec. 24 PatG, which allows competitors to use the sustainable design. However, it should also be noted at this point that the creation of a possibility for compulsory licensing is always accompanied by the danger of inhibiting innovation in this area, as companies are confronted with the danger of compulsory licensing. Moreover, the creation of a compulsory licence always represents a very strong intervention. Insofar as one argues for the creation of a compulsory licence, this should therefore only come into effect in exceptional cases and be linked to correspondingly strict conditions.

15.4.3.2 Destruction and Recall

Furthermore, the designer's right to choose between destroying or recalling unlawfully manufactured products or surrendering them in return for appropriate remuneration, Sec. 43 DesignG, could be reduced to the latter. On the one hand, this would result in products already manufactured not being destroyed and, on the other hand, the conflict between the designer's right to his own design and the use of the protected design by third parties would again be regulated on a financial level. Admittedly, the right to destruction is provided for under international law in Art. 46 TRIPS and can thus not be completely eliminated. Though, it is possible to achieve an appropriate result by strictly applying the principle of proportionality under Sec. 43 (4) DesignG, taking into account environmental protection. According to settled case law, already under the current legal situation, destruction is only permissible if the condition caused by infringement cannot be remedied by other means.[190] Thus, in comparison to the milder option of recall or surrender in return for appropriate remuneration, destruction is an ultima-ratio that can only be considered in exceptional cases with regard to sustainability. In this context, for example, it would be a milder remedy compared to destruction if the design-infringing product can easily be converted to a non-infringing design. Destruction is thus regularly ruled out.[191] For clarification an amendment to the law should be considered. Furthermore, this should serve as an example for international case law.

15.4.3.3 Interim Result

Design law could contribute to more sustainable economies, particularly by excluding the right to destruction.

15.4.4 Copyright

National copyright law does not play a central role in the further development of the goal of a sustainable society and economy. Thus, the coalition parties of the current German federal government also rely on European proposals for solutions to the issues of sustainability. Nevertheless, the existing provisions on limitations from Sec. 60a ff. UrhG—in the balance between individual incentive and common benefit—offer starting points for a science-friendly and thus more sustainable copyright law.[192]

Furthermore, against the background of the described balance of interests, copyright law contains the potential to focus on sustainability through lucrative remuneration situations for creative and journalistic content. The coalition agreement also provides for the creation of a legal framework in copyright law for the necessary financing issues arising directly from this.[193]

15.4.4.1 Official Works

Especially since the idealistic and commercial interests of the state, which uses taxpayers' money to design contributions to sustainability, are very small, the official works in the field of 'green' information should be expanded. This can

---

[190] See BGH, Judgment of 23 February 2006 - I ZR 27/03, GRUR 2006, 508.

[191] D. Jestaedt and H. Eichmann, in: Eichmann, Jestaedt, Fink and Meiser (eds.), DesignG und GGV, 6th ed. C.H.Beck 2019, § 43 DesignG paras 27 ff.

[192] H. Weiden, Regierungsprogramm Innovationsförderung, GRUR 2022(03), p. 153.

[193] Coalition Agreement of the Federal Government 2021, p. 98, available at https://www.spd.de/fileadmin/Dokumente/Koalitionsvertrag/Koalitionsvertrag_2021-2025.pdf. Accessed 14 July 2022.

15 Sustainability and Intellectual Property in Germany

lead to a wider dissemination of 'green' information and thus positively influence the multiplication of ideas.

15.4.4.2 Data

Data is an important part of climate research which cannot be collected by each scientist individually but must be shared with each other. Yet, copyright law does not really have any significant impact in this area. Thus—as seen above—it is rather the law on the protection of secrets that is to be consulted. However, with regard to database works under Sec. 4(2) UrhG, it would be worth considering whether the duration of protection should be reduced. It seems disproportionately long for the type of work and not very effective. Especially in the area of 'green' data. However, an international approach must be found for an appropriate exchange of data.

15.4.4.3 Interim Result

Overall, only minor changes need to be made in copyright law regarding a sustainable economy and society. However, this is because the effects are already not particularly large from the outset and thus no significant disadvantages arise.

15.4.5 Trademark Law

The role that trademark rights should play on the path to sustainability can be summarised as transparency and increasing acceptance. The purpose of using certain trademarks is already not only exhausted in a proof of origin, but also serves to communicate certain ideals and objectives of the using company as well as outstanding characteristics of the products.[194] At the same time, surveys show that consumers are willing to spend more money on products that have been produced under more sustainable conditions.[195]

These two circumstances can and should be taken advantage of: By using labels that provide reliable information about the sustainability of a company's business activities or the production conditions of a product, purchasing decisions can be motivated and acceptance for prices that have risen due to sustainability efforts can be achieved.[196] This can mitigate the 'first-mover-disadvantage' that generally hinders sustainability.[197] Special attention should be paid to reliability. Particularly, 'greenwashing' in which supposedly sustainable products are falsely labelled as

---

[194] S. Völker and G. Elskamp, Die neuen Markenfunktionen des EuGH, WRP 2010(01), p. 69.

[195] For example: EU Commission consumer survey from 2021, available at https://ec.europa.eu/commission/presscorner/detail/de/ip_21_1104. Accessed 14 July 2022.

[196] F. Klein, GREEN IP - A look at how sustainability influences IP and how IP can help in achieving sustainability, Ashurst Update 2020, available at https://www.ashurst.com/en/news-and-insights/legal-updates/a-look-at-how-sustainability-influences-ip-and-how-ip-can-help-in-achieving-sustainability/. Accessed 14 July 2022.

[197] Cf. D. Seeliger and K. Gürer, Kartellrecht und Nachhaltigkeit: Neue Regeln für Umweltschutzvereinbarungen von Wettbewerbern?, BB 2021(36), p. 2050.

such, must be avoided at all costs to prevent loss of trust and to enable 'green' brands to achieve their goals.[198] Certification marks serve as a suitable instrument for this, as already described above.[199]

15.5 Result

Intellectual property rights certainly offer potential for advancing sustainability and setting corresponding incentives.

Though, these rights must not only be thought of nationally, rather, international solutions and cooperation must be found. In this area, there is therefore a need to strengthen technology transfer, for example through publicly funded research and development cooperation, export guarantees or tax reductions.[200]

Nevertheless, to stop climate change, which is threatening the very existence of the planet, changes are needed that strengthen sustainability. To this end, it would be advisable to simplify the process of obtaining a property right for 'green' technology and thus make it more attractive. In patent law, we need to think about cost reductions, licence subsidies or accelerated granting procedures. In the interest of sustainability, the protection of secrets could be limited by more extensive freedom of information. In the case of trademark law, it is particularly important to highlight its current value and to further strengthen the appreciation of 'green' trademarks in companies. In the case of design law, the right to destruction could be restricted in favour of sustainability. Copyright, on the other hand, plays a subordinate role, allowing it to be used primarily as a supplement. Yet, it is generally true that intellectual property rights should be considered and used as a protective tool in their entirety.

On the other hand, there must be no confusion between intellectual property rights and regulatory law. Both pursue different purposes and especially intellectual property rights should not and must not primarily focus on regulation but must continue to focus on balancing the interests between the holders and society. Above all, the regulation of climate-damaging technologies can be carried out in a more targeted and effective manner within the framework of energy and environmental law.[201]

In conclusion, there is need for action by national and international legislators, but there is no need for the creation of new intellectual property rights or the complete renunciation of the existing system.

---

[198] E. Rosati, Green with... IP, Alicante News 2022, available at https://euipo.europa.eu/ohimportal/en/news?p_p_id=csnews_WAR_csnewsportlet&p_p_lifecycle=0&p_p_state=normal&p_p_mode=view&p_p_col_id=column-1&p_p_col_count=2&journalId=9135001&journalRelatedId=manual/&TSPD_101_R0=085d22110bab200025061c713b23263963b8808e44f39ecf94e5857ddb614979eec7024fcb81d4cb085173a7881430000750fb2591b956f13aeff82d3d34c0dfa96271dd65bb39bbc6da08c151d0c446aea6f7063825a349889fe389f276b1e0. Accessed 14 July 2022.

[199] R. Dissmann and S. Somboonvong, Die Unionsgewährleistungsmarke, GRUR 2016(07), p. 657.

[200] C. Heinze, Patentrecht und Klimawandel – eine Skizze, GRUR Newsletter 2020(1), p. 7.

[201] C. Heinze, Patentrecht und Klimawandel – eine Skizze, GRUR Newsletter 2020(1), p. 6.

Open Access This chapter is licensed under the terms of the Creative Commons Attribution 4.0 International License (http://creativecommons.org/licenses/by/4.0/), which permits use, sharing, adaptation, distribution and reproduction in any medium or format, as long as you give appropriate credit to the original author(s) and the source, provide a link to the Creative Commons license and indicate if changes were made.

The images or other third party material in this chapter are included in the chapter's Creative Commons license, unless indicated otherwise in a credit line to the material. If material is not included in the chapter's Creative Commons license and your intended use is not permitted by statutory regulation or exceeds the permitted use, you will need to obtain permission directly from the copyright holder.

Sustainability and Intellectual Property in Hungary

16

Bálint Halász, Ádám Liber, Dániel Arányi, Fruzsina Nagy, and Olivér Németh

16.1 Introduction

Generally, IP rights as legal instruments intend to encourage and foster innovation and creativity, as well as to safeguard values worth preserving. IP rights as intangible assets facilitate social and economic growth and are essential to fair competition. While these instruments grant exclusive rights, they are likewise subject to a number of limitations aimed at achieving the right balance between public interest and the interests of the rightsholders.

Abovementioned are the basic objectives of IP rights, however, since their foundations many economic and social challenges triggered the need for the review and adaptation of their legal framework. Achieving sustainability, as one of the most pressing problems today, inevitably desires the toolkit of IP rights, thereby entailing their reconsideration and targeted improvement along with the possibility of introducing new types of protections.

How rightsholders use the opportunities granted to them is also an important aspect of the relationship between IP rights and sustainability. It makes sense that rightsholders and users who are willing to license green and sustainable inventions, as well as copyright owners who do not fully retain their exclusive rights, make a better contribution to sustainability. Whereas, for instance, in the case of trademarks, rightsholders who are misusing 'green' and 'sustainable' labels, thereby committing greenwashing and misleading consumers, undermine sustainability efforts of committed entities.

B. Halász · D. Arányi · F. Nagy
Bird & Bird, Budapest, Hungary
e-mail: Balint.Halasz@twobirds.com

Á. Liber (✉) · O. Németh
PROVARIS Attorneys at Law, Budapest, Hungary
e-mail: Liber.Adam@provaris.hu

© The Author(s) 2024
P. Këllezi et al. (eds.), *Sustainability Objectives in Competition and Intellectual Property Law*, LIDC Contributions on Antitrust Law, Intellectual Property and Unfair Competition, https://doi.org/10.1007/978-3-031-44869-0_16

It should also be mentioned, that to achieve sustainability in an environmental, social and economic sense as well, we must reconsider our premises on the functioning of our society and the constant chase for economic growth. This brings with itself the necessity to revisit the balance that has been struck between the private interest to protect of IP rights and the public interest of access and dissemination, especially in relation to technology transfers to developing and underdeveloped countries.

In this chapter, we first look at individual IP rights, what roles they play in achieving sustainability, whether it is successful or not, and how their regulation should change in face of the current global challenges with a particular regard to Hungarian regulations. In the second part of this chapter, we outline the approach of the general Hungarian IP regulation towards sustainability, how it progresses, and in which aspects we believe it should be altered.

16.2 Sustainability Considerations of Individual IP Rights

16.2.1 Current Regulatory Standpoint

Patents are indispensable legal instruments for technological progress and socio-economic development based on innovation. In the context of sustainability, it plays an important role in all its three pillars[1]—it contributes to economic development as a highly valuable competitive tool, it helps achieving vital societal objectives by enhancing crucial solutions in the pharmaceutical and food industry, while in the form of 'green innovations', it stimulates among others climate change mitigation and resource-saving technologies thus promoting environmental protection and sustainable development.

One of the main objectives of sustainability is reducing the gap between developed and developing countries.[2] Patents both hinder and promote these efforts— patent protection indeed stimulates innovation in healthcare and pharmaceutical products, but only where it is accompanied by sufficient financial and human resources. In developing countries, however, patents eventually can cause an increase in prices.

It is undoubted that patents are crucial incentives in the field of biotechnology, which contributes to the treatment of epidemics, formerly incurable and orphan diseases, as well as to the development of new medicines and enhancing effectiveness of the ones already available. Furthermore, among other aspects, patented

---

[1] B. Purvis, Y. Mao, Y. and D. Robinson, Three pillars of sustainability: in search of conceptual origins. Sustainability Science, 14, 2019 (3), pp. 681–695, p. 682, available at https://doi.org/10.1007/s11625-018-0627-5. Accessed 9 January 2023.

[2] Transforming our world: the 2030 Agenda for Sustainable Development GA/RES/ 70/1, UN Doc A/RES/70/1 (21 October 2015, adopted 25 September 2015), p. 21.

16 Sustainability and Intellectual Property in Hungary

biotechnology plays a role in agriculture by plants becoming increasingly adaptable to the environment, thus increasing food safety.[3]

Since innovation and new technologies are relevant factors in sustainability, patents fundamentally play a positive role in achieving it. Nonetheless, fast paced economic and technological development promoted by the exclusive exploitation rights deriving from patents can likewise have an adverse effect on both society and the environment by significantly increasing the humanity's ecological footprint, thereby (already) reaching the state of over-consumption.

The law may further sustainability in patent protection by regulating the extension of protection, granting teaching and research exceptions with regard to sustainability and benefits and incentives furthering patent protection for environmentally friendly inventions.

Copyright fosters creativity by allowing authors to gain recognition and financial compensation for their work, which contributes to innovations in the field of sustainability by granting access to scientific publications through which research and essential data are made available for new discoveries, while also supporting education and teaching by providing certain exceptions and limitations under copyright protection. While reusing copyrighted works is restricted by the exclusive rights of authors, there are examples for surrendering full copyright protection, such as Creative Commons licensing[4] of works, which helps users finding and reusing works and therefore contributing to research and development.

Copyright protection of software in the field of green technology can play a crucial role in collecting and analysing relevant data, which could be used for improving existing technologies in a greener or sustainable way.

Trademarks can help consumers identify companies committed to sustainability and environmentally friendly solutions. This goal can be realised both by having a trademark referring to sustainability in a distinctive way or by acquiring one that is identified with goods and/or services of sustainable nature.

The use of collective and certification mark registered for designations related to sustainability likewise play an important role in communicating characteristics.

Trademarks, however, are neutral to sustainability, as trademark registration and maintenance are not subject to any criteria which reflect sustainability or environmentally friendly preconditions. Therefore, trademarks are useful in commercially identifying other IPs, such as green innovations on the market.

Industrial Designs can provide protection for the outer appearance of 'green' products or their parts. While an outer appearance merely determined by technical features is excluded from protection, designs serving a technical purpose, and at the same conveying a certain degree of designing freedom are not. For example, certain

---

[3]L. Tattay, A biotechnológiai találmányok jogi oltalma az Európai Közösségben. Külgazdaság Jogi Melléklet, 43, 1999 (10), pp. 125–140, p. 129.

[4]Creative Commons Licenses - About The Licenses, available at https://creativecommons.org/licenses/; Accessed 10 June 2022.

products or their parts which in their technical function have green or sustainable effects can be subject to design protection for their exterior.

Design rights are subject to a more time and cost-effective procedure than patents, but likewise can contribute to the protection of green and sustainable technologies in accordance with their legal requirements.

Geographical Indications (GI) as identifiers of products manufactured based on valuable local knowledge linked to a specific environment play a significant role in social sustainability by preserving the cultural heritage of traditional production methods developed over decades or even centuries on a given territory. GIs are linked with guaranteed quality based on the geographical place of production and strict compliance with production standards.[5]

GIs have a positive role in sustainability, as they are promoters of preserving local knowledge and identity, which is of primary importance in sustaining diversity and achieving healthy balance between the protection and utilisation of the natural environment.

Plant variety protection supports food safety objectives by the development of new variants more adaptable to the environment. As the world population is expected to grow continuously until the year 2100, while global climate change is getting worse by the day, having plant variants that provide better yields, are climate resilient and allow for the decrease in pesticides usage could prove to be an excellent instrument for humanity.

It is important to remember, though, that the development of new plant variants is often performed by large corporations in the developed world, while food security problems are (yet) more of a problem in the global South. Although more of a policy concern, striking the healthy balance between rewarding innovation with protection and profit, and disseminating the new, resilient plant varieties to countries where they are desperately needed at an affordable price is key to achieve social sustainability on a global level.[6]

Trade secrets as alternatives to IP rights are valuable business assets. Entities which develop and apply sustainable procedures, methods, practices or have such inventions that could be subjects to the acquisition of patent rights can rely on trade secret protection regarding these confidential and sensitive information. Although trade secret protection enables companies to develop best practices, solutions, and the collection of valuable data, it eventually involves withholding essential information for transitioning to more sustainable ways. In our view, trade secrets should have the role to incentivise developing sustainable best practises and collecting

---

[5]E. Vandecandelaere, L.F. Samper, A. Rey, A. Daza, P. Mejía, F. Tartanac and M. Vittori, The Geographical Indication Pathway to Sustainability: A Framework to Assess and Monitor the Contributions of Geographical Indications to Sustainability through a Participatory Process. Sustainability, 13, 2021 (14), 7535. p. 1–2. Available at https://doi.org/10.3390/su13147535. Accessed 9 January 2023.

[6]B. A. Keserű, A fenntartható fejlődés hatása a szellemitulajdon-védelem rendszerére. Dialóg Campus, Budapest, 2019, p. 218.

16 Sustainability and Intellectual Property in Hungary

valuable data, which then could serve sustainable objectives after having been licensed.

16.2.2 Necessary Changes for a Sustainable IP Regime

Patents should maximise incentives to bring out the most potential in making existing technologies greener and more sustainable, and inventing new (green and sustainable) ones. A well-developed patent system should help companies to exploit their inventions, while ensuring that they serve the economy and society at large. It is crucial that innovators and businesses (especially SMEs) are provided information on the different types of IP rights available, the nature of the protection, and the different criteria which should be fulfilled for the rights to be registered/acknowledged and thereby enforced. The Hungarian Intellectual Property Office (HIPO) should prepare and make available such informational materials and brochures and hold seminars and workshops on the proper identification and management of IP rights, as well as creating IP strategies, in the course of which practical information related to green and sustainable IP should be specifically provided.

The patent system should enable the use of patented technologies for research purposes to a justified extent based on the Bolar exemption,[7] which would facilitate more rapid development of pharmaceutical products, as well as the invention of sustainable and greener technologies by reusing knowledge from existing ones. Research institutes, universities, science and technology parks should be promoted and facilitated to carry out innovation and development related to green and sustainable technologies. In this regard, the implementation of IP policies would be necessary, as well as the provision of extensive knowledge on IP protection tools. Target areas where sustainable and green innovations are strategically needed should be centrally identified to these institutions, the realisation of which could be publicly funded. At the same time, green and sustainable R&D activities of companies should be incentivised in the form of tax deduction, and strategic IP advice should be provided, which should also involve assistance in valorisation of these assets.

Patents should serve the spread of new inventions by proper technology transfer and licensing mechanisms in place—easy licensing of sustainable technologies should be available with reasonable benefits to rightsholders. To achieve circular economy, cross-industry licencing should be promoted, where the different market players would eventually work together on more innovative solutions.[8]

For that purpose, a 'green marketplace' should be set up to facilitate both domestic and international technology transfer between innovators and businesses

---

[7] V. Munoz Tellez, Bolar Exception. In: Correa, C.M., Hilty, R.M. (eds) Access to Medicines and Vaccines. Springer, 2022, p. 136. Available at https://doi.org/10.1007/978-3-030-83114-1_5. Accessed 9 January 2023.

[8] N. Buzás, Technológiatranszfer-szervezetek és szerepük az innovációs eredmények terjedésében, SZTE GTK, JATEPRess 2002, p. 95.

seeking green and sustainable technologies. On such a platform, a searchable database of innovations and their rightsholders, as well as the potential user companies should be provided. The licensing of the technologies offered could be simplified by the provision of standardised licensing terms the parties could opt for. The royalty from licensing of green innovations could be subject to tax deduction. Moreover, apart from the licensing of already developed technologies, the platform could serve as a hub for connecting businesses having sustainable technology needs and innovators with compatible ideas of solutions, in the framework of which partnerships for the realisation of essential green technologies could be formed and supported.

Patentability conditions should involve the exclusion of inventions having a harmful effect on the environment or human, animal and plant life and health, the fulfilment of which conditions should be properly examined and enforced by IPOs. The patent system is originally technology-neutral—environmentally friendly and harmful inventions can be both subjects to patentability, however, Article 27 (2) of TRIPS mentions exclusion of inventions having an adverse environmental in the form of an exemplary list: 'Members may exclude from patentability inventions, the prevention within their territory of the commercial exploitation of which is necessary to protect public order or morality, including to protect human, animal or plant life or health or to avoid serious prejudice to the environment, provided that such exclusion is not made merely because the exploitation is prohibited by their law'.

Although currently the Hungarian Patent Act provides that inventions shall be considered unpatentable where their commercial exploitation would be contrary to public order or morality[9], there is no specific reference to the exclusion of inventions which could pose a serious harm to life and the environment itself. It is not excluded that environmental effects should be part of the assessment, as under the existing provisions patent examiners can assess these aspects, but—contrary to the provisions of TRIPS—currently these are not mentioned in the exemplary list of non-patentable inventions.

Building from international examples, in 2015 the HIPO started a 'green patent' qualification programme, the aim of which is to incentivise applications that have the potential to develop solutions to achieve sustainability. When submitting a patent application, the HIPO checks whether the invention belongs to a pre-determined category of 'green' technologies, in which case it puts the application on a fast track and rewards the inventor with a relatively quick application procedure.[10] Unfortunately, the programme did not live up to its expectations, neither in Hungary, nor in any other country: only a fraction of all patent applications gets into the green patent fast track programme.

---

[9] Section 6 (2) of the Hungarian Act XXXIII of 1995 on the Patent Protection of Inventions ('Hungarian Patent Act').

[10] Jedlik Plan National Strategy for the Protection of Intellectual Property (2013) Budapest; p. 191–192. Available https://www.mtmt.hu/system/files/jedlik-terv.pdf. Accessed 31 August 2022.

16 Sustainability and Intellectual Property in Hungary

Copyright regulatory framework should be able to create a reasonable balance between the interests of the rightsholders and the public, who would benefit from the access to copyright protected works in education, research and in the acquisition of professional knowledge.

Copyright should be a tool in achieving social sustainability by assisting in social inclusion and granting equal opportunities. In this regard, education is of fundamental importance, to which copyright could contribute by providing access to information one would need for their personal development and well-being. Therefore, copyright should have a role in maximising creativity by granting exclusive right to creators, as well as including exceptions and limitations to education, research, and personal use.

In addition to tackling social inequalities, fair access to copyrighted materials is likewise important to have skilled professionals who could have the knowledge and potential of creating future innovations. Increase in the average minimum level of education and having qualified professionals is the enabler of development— professionals should have the most up-to-date knowledge possible, which would be hindered by copyright regulation of restrictive nature. If access is costly or even impossible, little knowledge is available from primary sources.

Social sustainability goals such as education for all, up to date knowledge for professionals, as well as research to support scientific works and development could be greatly supported by broad statutory free use exception of education and research, and means to rightsholders to share and thereby authorise use of their copyright protected works contributing to social, environmental, medical, etc. studies as granting free access and reuse of related scientific research data and findings, which could be enabled by public funding and sponsorships.

Copyright should likewise be a flexible tool to contents made available online. Digital materials are essential to online learning and research, as well as the possibility of sharing and accessing digital contents of libraries and archives. In case of the latter, digitalisation helps preserving works for future generations.

Copyright regarding green innovations should contribute by solid regulations on software protection. First, green software supports sustainability in minimising the environmental impact of technologies and is aiming at energy efficiency.[11] Moreover, software as part of technologies can contribute by collecting and analysing data, which could help improvement of efficiency, emission, resource allocation, etc. inter alia in the field of manufacturing, agriculture, construction. Sui generis database protection should likewise have a role in the use and diffusion of data collected and systematised to identify the need for further green and sustainable improvements and to help their realisation by further innovations. Database protection incentivises obtaining, verifying and presenting data which could be beneficial to industries in transitioning to greener technologies.

---

[11] S. Podder, A. Burden, S. K. Singh and R. Maruca, How Green Is Your Software? – Sustainable Business Practices, Harvard Business Review, 2020. Available at https://hbr.org/2020/09/how-green-is-your-software (Accessed 11 June 2022).

The development, use and licensing of green software should be promoted by incentives, such as tax deduction from the royalty, and providing subsidies and tenders to companies for the acquisition of green software. Similarly to the green patent marketplace, there could be a database for connecting developers and potential users of software supporting green technologies and objectives, e.g. analytics of emission, energy efficiency. In this regard, there should be openness to discuss whether copyright framework should be adapted in the view of new technologies, as artificial intelligence, Internet of Things, which could likewise play an important role in businesses sustainability goals and successful green transition.

One could also make an argument that if we aimed at achieving social sustainability and decreasing the wealth gap in society, revisiting the current term of copyright protection could be a useful instrument, as in the last decades private interests have dominated the development of copyright. While it is important to maintain the incentive for creativity, the present regulations that offer protection decades postmortem auctoris do nothing to incentivise creative authors, rather they provide a powerful tool for rightsholders, especially large corporations keep hold of their accumulated intangible wealth. Reducing the term of protection to a fix period so that works may become public domain and differentiating between the term given to individuals and legal persons could be a way to go.

Trademarks should contribute to sustainability goals by raising consumer awareness and helping consumers identify entities which are committed to green and sustainable ways. Certification and collective marks should also be assets to distinguish companies, products and services which serve sustainable objectives.

The registration of collective and certification marks referring to sustainability should be subjects to a discounted registration fee. Moreover, the crucial role they could play in sustainability should be communicated by HIPO, especially in the view of certification marks of sustainable fabrics and materials which could contribute to achieving a more sustainable fashion industry.

However, trademarks that falsely refer to 'green', 'eco', 'sustainable', etc. goods and services, aka greenwashing, should be subject to cancellation by a third party on the grounds of deceiving consumers. It is not feasible to assess such intentions during examination, however, there should be a mechanism to identify greenwashing trademarks after they are registered. Consumer protection authorities could regularly observe—either ex officio or by notice—trademarks of rightsholders whose activities do not reflect the sustainable and green characteristics or efforts their trademark display. Consumer protection authorities should also facilitate to file a cancellation action against a trademark the use of which would thus deceive consumers.

Industrial Designs as 'design patents' should be assets to sustainability by protecting the design of green and sustainable technologies fulfilling registrability conditions, in a more time- and cost-effective way.

In respect of design rights, information on the scope of protection should be made available to businesses having an interest in the registrability of green and sustainable product appearance. Green designs are often in connection with the texture and/or materials of products, as they feature recycled, natural, environmentally

16 Sustainability and Intellectual Property in Hungary

friendly materials. Displaying such materials in the application could be challenging, as examiners should be able to determine the subject of the application, as well as novelty and individual character being preconditions of protection. To enable the design protection of green and sustainable texture and material, HIPO should support businesses by providing practical manuals on the display of designs acceptable and information on the relevant criteria taken into consideration regarding the registration of green and sustainable designs. Design application system should adopt to the easy display and facilitation of green and sustainable designs, which could be defined by new tendencies, e.g. 3D printed objects.

Easy and cost-effective acquisition of design rights would play a significant role in achieving a sustainable fashion industry—it incentivises the use of sustainable fabric and materials, and by the enforcement of exclusive rights it contributes to the fight against counterfeit products and fast fashion businesses, which contribute greatly to increasing humanity's ecological footprint.

It should also promote circular economy by protecting the design of easy-assembly or disassembly goods, as well as their spare parts which could enhance lifespan of products as enabling their reparation. Likewise, goods and their packaging which are reusable and are made of recycled, eco-friendly or natural materials should be subjects to design protection.

In fashion industry, designs should play a crucial role in tackling fast fashion copycats, as well as counterfeit products. Those are often linked with socially and environmentally harmful manufacturing practices and serious overconsumption, and the seizure and destruction of the latter involves environmentally detrimental procedures hindering sustainability objectives.[12]

Geographical Indications (GI) should preserve local production/manufacturing methods and practices well developed over time. The different entities whose product is subject to GI protection are serving social sustainability objectives by supplying the domestic and international community with high-quality products and creating jobs for local people.

GIs also assist in preserving the natural environment which is crucial to environmental sustainability—maintaining the resources and productive capacity of the land is important for farmers and producers, which in turn likewise serves the interests of future generations.

To better support social sustainability goals, however, besides food, agricultural products, wines and beverages, emphasis should also be put on the protection of non-agricultural products of local manufacturers. These products represent high quality and durability and they usually are subjects to sustainable production methods. HIPO and other IPOs as well should raise wider awareness of the possibility of the GI protection of non-agricultural products—from cutlery to clothes and

---

[12] J. Soentgen, Disposing of counterfeit goods: unseen challenges. WIPO Magazine, 2012 (6), pp. 25–28. Available at https://www.wipo.int/wipo_magazine/en/2012/06/article_0007.html. Accessed 31 August 2022.

footwear, local skills and know-how on their production represents a great value and their recognition and preservation should be promoted.

Plant variety protection should have the role of enhancing food security by promoting the development and dissemination of plants more adopted to the environmental adversity of climate change, as well as reducing the environmental impact of agriculture.[13]

To achieve seed diversity adapted to the adverse effects of climate change, there should be economic incentives for the development of new and improved climate-proof plant varieties. HIPO and other government institutions could offer valuable aid to Hungarian researchers and enterprises in the agricultural sector for the protection and dissemination of new varieties, primarily via consultations and helping them find partners whom they can cooperate with internationally. World-wide initiatives, primarily involving national governments, regional cooperations and UN institutions, such as the FAO, should be drafted and executed to disseminate the newly developed plant varieties to countries where they are actually needed. We should also pay attention to provide the new varieties to farmers at a price they can afford and not to prohibit them to reuse the seeds from the previous year's crop.

16.2.3 Further Rights To Be Adopted

The protection of traditional knowledge and traditional cultural expressions of local communities should be considered both on a national and regional level. As time goes by, there are fewer and fewer people actively using traditional knowledge for agricultural, economical, or even medical purposes. Preserving the know-how, skills and practices of local people would contribute to biodiversity, the protection, restoration, and sustainable use of nature, which could improve the quality of life of local communities and of society itself. Protection of traditional cultural knowledge, such as music, ceremonies, dance, handicraft, and other expressions of folklore contribute to cultural diversity and preserving local identity, which is an essential factor of achieving social sustainability.[14]

While some of these values could be protected by traditional IP rights, such as GI, there are many of them that are worth preserving but would otherwise fall out of the scope of the existing IP rights. Exclusive rights should serve the community itself by incentivising and assisting them to maintain their valuable traditions, as well as these particular know-hows, methods and practises should be collected in searchable databases to be well-preserved for future generations.

---

[13] S. Mariani, Law-Driven Innovation in Cereal Varieties: The Role of Plant Variety Protection and Seed Marketing Legislation in the European Union. Sustainability, 13, 2021 (14), 8049, p. 1–2. Available at https://doi.org/10.3390/su13148049. Accessed 9 January 2023.

[14] M. Shafi, Geographical Indications and Sustainable Development of Handicraft communities in Developing Countries. The Journal of World Intellectual Property, 25, 2022 (1), pp. 122–142, p. 123. Available at https://doi.org/10.1111/jwip.12211. Accessed 9 January 2023.

Genetic resources, especially those originating from developing and underdeveloped countries, and indigenous communities should also be given protection to prevent exploitation from economically more powerful actors, and to avoid that those who have discovered them and been using them for generations would be locked out of utilisation. This new type of protection should be a negative one: genetic resources are primarily heritage of the local community and of humanity at large, hence no one should receive exclusive rights for them.[15]

Artificial intelligence is still a source of many headaches for legal professionals and legislators alike, and the field of IP is no exception to that. It is still unclear whether the algorithms themselves should be covered by copyrights or patents, while the outputs of the AI can only be categorised as trade secrets at the moment.[16] There are a number of problems with the current solution, not least the fact that AI algorithms often make decisions about our daily lives, and they are going to have even more influence on us in the future, yet we do not know the process of they operate and come up with their decisions. A patent application process, for example, would aid us a lot in understanding AI and its implications on our lives.

From the point of sustainability, however, an even more pressing matter is the fact that since AI systems and their solutions are more often than not categorised as trade secrets, they are nearly impossible to be utilised throughout the world, as it would constitute a grave financial misstep to give up on a trade secret from a company's point of view. The AI could be a rather powerful instrument for humanity to combat issues from climate change to social inequality, as they can provide us with possibilities on a previously unimaginable scale. To achieve this, it is necessary to incentivise dissemination of AI systems by offering protection via IP rights. This could be done by creating a sui generis right for AI, that would encompass the algorithm itself and its outputs, or by widening the scope of patents, to include both categories. It is also important to strike a balance between the extension of the protection and the need not to strangle new developments by abusing the newfound rights.

While modifications to the current IP regime could start on a national-regional level regarding AI-related IP rights, the foreseeable impact and benefit of the AI in connection with sustainability may prove desirable enough so that national legislators commence the work in an international environment in the first place.

---

[15] B. A. Keserű, A fenntartható fejlődés hatása a szellemitulajdon-védelem rendszerére. Dialóg Campus, Budapest, 2019, p. 212–225.

[16] K. Foss-Solbrekk, Three routes to protecting AI systems and their algorithms under IP law: The good, the bad and the ugly. Journal of Intellectual Property Law & Practice, 16, 2021 (3), pp. 247–258. Available at https://doi.org/10.1093/jiplp/jpab033. Accessed 9 January 2023.

16.3 IP Framework in Hungary

As a preliminary remark (and perhaps as an early conclusion) to this section, it can be stated that the Hungarian legislation, and by extent the Hungarian Intellectual Property Office ('Szellemi Tulajdon Nemzeti Hivatala', 'HIPO'), still has some room for improvement concerning the sustainability-related considerations of IP rights and their regulations. If nothing else, the fact that while HIPO issues an annual report on the current state of IP rights in Hungary, it does not contain any reference to sustainability aspects of these rights.[17]

As previously presented, following international examples, there exists a fast-track 'green patent' procedure in HIPO, but on one hand, it is barely utilised by market actors, and on the other, HIPO does not advertise the possibility at all. The option itself is of good intention, but in its current form it does not serve its purpose, incentivising and aiding the development of 'green' solutions, very well. It would be worth reconfiguring the programme by catching the attention of innovators via media campaigns and by offering financial incentives: tax deduction, reduced procedural fee or direct funding, especially for SMEs could prove fruitful. However, in the current form, the procedure is of very little meaning.

As far as the incentivisation of research, development and innovation is concerned, there are a number of tenders available for SMEs, universities, and businesses overall from EU and domestic funds—such as the National Research, Development and Innovation Fund (NRDIF). Some of these tenders were/are especially dedicated to sustainable innovations, e.g. the funding schemes of Horizon Europa programme, however, the majority are promoting general R&D activities. NRDIF likewise offers funds to natural persons and businesses for the registration, maintenance and valuation of IP rights.

In this regard, HIPO provides an R&D qualification procedure, in the framework of which it assesses whether specific projects of businesses qualify as research and development, the positive outcome of which would constitute as the basis of eligibility for tax incentives and fundings.[18] Moreover, HIPO offers free consultation to small and medium-sized enterprises with tailored guidance on how they can consciously manage their intellectual property.

Apart from consultation, raising awareness of SMEs on the importance of developing and protecting their intellectual property, is a possibility that can be performed with limited financial and operative investment, but can yield great results. Understanding of IP, its benefits and procedures is not really widespread in Hungary, even in the entrepreneurial sphere, while on the other hand, many sustainable developments come from start-ups and SMEs. Programmes for these actors are already up and running, which is excellent, but more emphasis should be put on

---

[17] K. Botos-Penyigey, Tények & Adatok 2021, In: S. Tulajdon and N. Hivatala (ed) https://www.sztnh.gov.hu/sites/default/files/hipo_annualreport_2021_web.pdf. Accessed 31 August 2022.

[18] Kutatás fejlesztési tevékenység minősítése - Hungarian Intellectual Property Office. Available at https://www.sztnh.gov.hu/sites/default/files/kftm_hun_web_final.pdf. Accessed 31 August 2022.

16 Sustainability and Intellectual Property in Hungary

improving the IP-consciousness of those who are just starting out in setting up a business.

Innovation—be it scientific or more business orientated—is promoted through both project-based and more general or strategic funds, which could contribute to sustainability. As these innovations qualify as intellectual property and most likely are/will become subjects to IP rights, it can be concluded that these incentives contribute to IP's role in sustainability. Nevertheless, as there are a negligible number of incentivisation schemes offered directly to the acquisition, maintenance, valuation of green IP rights, this role should be further improved.

Another important consideration of IP regulations is enforcement of the rights that rightsholders already possess. Having enforceable exclusive rights deriving from a patented technology is the driving force of innovation. Nevertheless, for the diffusion of green technology, pharmaceutical and biotechnology-based products, so for the good of the industrial, societal, economic and environmental transition to sustainability, licensing is an essential instrument and should be highly promoted. It should therefore be facilitated that the balance between the interest of rightsholders and the interest of public is achieved. Voluntary licensing should be promoted and governed by non-discrimination. Sustainable and green technology can serve public objectives if it is transferred in and across industries which could implement these technologies and develop new and advanced ones.

Proper enforcement of copyright, trademarks and designs against counterfeit products would contribute to all three aspects of sustainability. It has an adverse impact on market sales and employment, as well as it has a negative effect on the environment, let alone posing health and safety issues to consumers. Public enforcement authorities should be well-equipped—and should be in communication with rightsholders—to tackle the import and offering for sale of such goods.

IP enforcement efforts against counterfeit and infringing products, which could pose serious health risks to consumers and are harmful to all three pillars of sustainability, are supported by the work of the National Board for Counterfeit in Hungary ('Hamisítás Elleni Nemzeti Testület', 'HENT'). HENT is responsible for drafting and enforcing the national strategy and action plans against counterfeiting in coordination with law enforcement authorities, in the course of which they are organising awareness-raising and information campaigns and training, as well as preparing amendments regarding IP enforcement legislation.

While the board's objectives do support IP rights in their role of promoting sustainability, the reports of HENT on the current situation in the fight against counterfeiting show that there is still room for improvement, e.g. regarding the citizens' awareness on the infringing nature of these products and the risks they might pose.[19] Moreover, cooperation and data sharing between rightsholders and

---

[19] E. Hamisítás, N. Testület, S. Tulajdon and N. Hivatala, Hamis termékek kereslete Magyarországon, 2019: A hamis termékekkel és az illegális forrásból származó szerzői jog által védett tartalmakkal kapcsolatos lakossági attitűdök alakulása 2009–2019 között, 2019 p. 27–46. Available at https://www.hamisitasellen.hu/wp-content/uploads/2019/11/HENT_TARKI_2019_% C3%89VES_omnibusz.pdf. Accessed 31 August 2022.

law enforcement authorities could still be enhanced, and the introduction of new technologies that can help detect infringements or assist authorities in other ways should be likewise promoted.

It should also be noted that destruction of counterfeit products seized by law enforcement authorities could be highly pollutant, as it sometimes involves incineration of the goods. It should therefore be examined how developing more sustainable practices in this context could be realised. For example, rightsholders should be incentivised to compromise and, once the product has been de-infringed, to offer the products for charitable purposes. Furthermore, recycling should also be considered—it should be properly executed in the framework of a dedicated process, by certified entities which guarantee that infringing products do not re-enter the market.

Apart from other IP rights, GIs are never subjects to licensing, as they are connected to a particular territorial origin and a product specification—producers who do not fulfil these conditions should not be able to use the GI. Enforcement of GIs against infringing third-party use is highly important and should be well-facilitated by authorities and courts, as—in addition to a well-functioning control system—it contributes to assuring the proper qualities and characteristics the registered GI should own. Given that GIs can have many rightsholders at the same time, it is difficult for them to act in a timely and effective manner in case of an infringement, let alone in many cases they are not aware of their rights regarding enforcement. Therefore, rightsholders of a GI should be properly informed and facilitated—even by public funds and incentives—to act not only to obtain a GI but also to monitor a market and take enforcement actions.

16.4 How We Would Amend the Hungarian IP Framework

All IP rights currently available in Hungary support the role of sustainability in a way that they promote innovation, development, preservation, and diversity which are the core factors of sustainability. However, there are certain aspects which should still be improved in order for these rights to better contribute to sustainability goals.

First, awareness should be raised on how different IP rights can contribute to sustainability objectives and why it is crucial that these are properly identified, registered (if applicable) and maintained. In this regard, HIPO should publish informational materials, studies and reports to demonstrate how intellectual property is currently affecting sustainability—e.g. the volume of green patents or designs registered—and in what way this role could be enhanced. Together with other actors, such as the Chamber of Commerce or the competent ministry, HIPO should also focus on educating entrepreneurs from SMEs and the start-up sphere, especially those who are just starting out, to raise their awareness on IP rights and sustainable solutions.

Second, acquisition of IP rights which are identified as having sustainable or green characteristics and effects, should be made easier by reduction of official fees and/or enhancing the already existing fast track procedures. The latter could be essential in case of pharmaceutical products and biotechnology to facilitate a quicker market entry. Regarding green designs and trademarks, especially certification and collective marks, maintenance should be subject to reduced maintenance fees.

Incentivisation of creating intellectual property which has a role in achieving sustainability goals should be further improved via funding schemes. These programs—in addition to boosting inventiveness—would raise awareness of and promote the necessity of creating more sustainable solutions.

Although it would go beyond national borders, authorities and legislators should commence working on the development of AI-related IP rights to provide the opportunity for AI systems and sustainable solutions developed by the AI to disseminate around the globe as soon as possible.

As mentioned in Sect. 16.2.2, there should be a searchable database of green intellectual property rights and a platform operated for the purpose of connecting inventors/rightsholders and businesses which intend to license these rights. Moreover, it could serve as a platform to seek partnerships for the realisation of green and sustainable projects.

Furthermore, for sustainability reasons, the law should make it easier to expand the public domain with copyrighted works. Accordingly, the copyright owner should be permitted to voluntarily abandon his/her copyright prior to the expiration of the work's copyright term and thereby the work could become part of the public domain. This opportunity is not given yet under Hungarian laws.

Overall, sustainability objectives require international cooperation, as they are a globally pressing issue. With regard to the improvement of the IP system to better support sustainability goals, a unified EU-level approach and coordination would be ideal, although most of the advancements mentioned earlier could be achieved at a national level.

Sustainable solutions require simple, economical procedures and a territorially far-reaching protection, which is why the European Unitary Patent is a great step towards achieving sustainability. As of completing this chapter, Hungary has not yet ratified the UPC Agreement. We can only hope that Hungary follows other EU Member States' example and ratifies the agreement as soon as possible.

We consider that the role of IP rights in sustainability is better promoted through incentives and benefits, with a more practical approach, as well as lower-level legislation, rather than through international treaties. There is, however, need for new *sui generis* forms of IP rights when discussing sustainable development: cultural heritage, traditional knowledge and genetical resources cannot be protected well enough under the current IP regime, and these subject matters should have their place in the relevant international IP framework. Moreover, it would usually be in the interest of developing or underdeveloped countries to create new forms of protection for them, thus it would be preferred to get it done in a multilateral arrangement.

16.5 Conclusion

As we established in this chapter, sustainability considerations play a significant role regarding IP rights, as they are legal tools for promoting the creation of and preserving valuable assets which could be put at the service of achieving sustainability. On the other hand, while IP rights are an excellent instrument, that incentivise development, it could also very well hinder it, if the legislation we enact degrade them to a simple tool for private interests. Should we truly wish to achieve a sustainable future, the current IP regime has to be revisited and we must strike a balance between private and public interests once again.

Open Access This chapter is licensed under the terms of the Creative Commons Attribution 4.0 International License (http://creativecommons.org/licenses/by/4.0/), which permits use, sharing, adaptation, distribution and reproduction in any medium or format, as long as you give appropriate credit to the original author(s) and the source, provide a link to the Creative Commons license and indicate if changes were made.

The images or other third party material in this chapter are included in the chapter's Creative Commons license, unless indicated otherwise in a credit line to the material. If material is not included in the chapter's Creative Commons license and your intended use is not permitted by statutory regulation or exceeds the permitted use, you will need to obtain permission directly from the copyright holder.

Sustainability and Intellectual Property in Italy

17

Marina Cristofori

17.1 Introduction

The future of the Earth depends on protecting biodiversity. The term Biodiversity is commonly used to "describe the number and variety of living organisms on Earth" and can be considered synonymous with "life on Earth." Biodiversity is a fundamental resource for our survival, and an economic and social asset.

Italy has one of the most significant biodiversity assets in Europe, both in terms of the total number of animal and plant species and the high rate of endemism. Thanks to its geological, biogeographical, and socio-cultural history, as well as its central position in the Mediterranean basin, Italy is home to about half of the plant species and about a third of all the animal species currently found in Europe.

Despite this richness, biodiversity in our country is rapidly declining as a direct or indirect consequence of human activities. According to the Red List of ecosystems of Italy[1] which follows the criteria established by the IUCN, the World Union for the Conservation of Nature, Italy is indeed a country with many ecosystems at risk.

In the light of the findings of the Red Lists, as well as increasing pressures on biodiversity, there is an urgent need to define more incisive and effective actions to reverse the trend over the next decade and sustainability certainly plays a pivotal role.

Underlying these changes is the digital revolution—the so-called "Industry 4.0"—which affects both the production and distribution sectors. At the same time, consumers have also changed their approach towards more sustainable, more personalized, more "social-oriented" needs and demands.

---

[1] Natural Capital Committee (2021), Fourth Report on the State of Natural Capital in Italy.

M. Cristofori (✉)
Studio Legale Jacobacci & Associati, Milan, Italy
e-mail: mcristofori@jacobacci-law.com

© The Author(s) 2024

P. Këllezi et al. (eds.), *Sustainability Objectives in Competition and Intellectual Property Law*, LIDC Contributions on Antitrust Law, Intellectual Property and Unfair Competition, https://doi.org/10.1007/978-3-031-44869-0_17

Therefore, companies are forced to modify and adapt their business models, and traditional "linear" economic system is losing its appeal in favor of new "circular" solutions.

This is leading to a progressive shift from a linear economy to a circular economy, which is "a model of production and consumption, which involves sharing, leasing, reusing, repairing, refurbishing and recycling existing materials and products as long as possible."[2]

In this regard, the concept of sustainability is fundamental, meaning the adoption of business models to support projects capable of generating positive results in many ways, such as on the environment, on social governance, and on people's well-being and health—leading to greater respect for natural balances, with the right approach that is essential and decisive for the very survival of the planet and those who live on it, as the current pandemic has shown.

17.2 Whether IP Has a Role in Sustainability

IP certainly plays a very important role in sustainability.

Since 2015, the United Nations has defined an ambitious set of 17 sustainable development goals in the so-called "2030 Agenda."[3] In this regard, at European level, the Commission has adopted a specific Action Plan on Industrial and Intellectual Property (IP)[4] aimed at strengthening the resilience and supporting the economic recovery of the EU, by making the most of the potential of the IP sector.

In this process of implementing the circular economy and sustainability, among the many tools needed to meet the challenge and achieve the desired results, a key role is played by the IP system which must be valued and protected for its positive effects on strategies oriented towards a green and blue transition based on innovation, competitiveness, value generation, and sustainability.

It is, therefore, completely wrong to claim that, today, IP rights represent a barrier and a sort of brake on the use of innovation, also because without this instrument there would be no "innovation" at all.

This is why, to enable and accelerate the transition towards true sustainable development, the driving force provided by the IP system and the related protections is necessary, with the possible corrective measures provided for in the exceptions to the exclusive right of the owner on his own patent (a cue offered by the recent lively debate on vaccines highlighted by the pandemic crisis) when states can grant and thus make available to third parties compulsory licenses on patents when they appear

---

[2]"Circular economy: definition, importance and benefits | News | European Parliament." *www.europarl.europa.eu.* December 2, 2015.

[3]"UN 2030 Agenda" signed on September 25, 2015 available at https://sdgs.un.org/2030agenda.

[4]"Action Plan on Industrial and Intellectual Property" available at https://ec.europa.eu/commission/presscorner/detail/en/ip_20_2187.

as "essential" to combat emergencies and dependent on innovation and "critical" technologies.

Politics must offer companies tools and incentives for industrial research, development, and technological innovation, in order to be and remain competitive, seeking to create a closer relationship with consumers, a collaborative and participatory economy, which implies the necessary support of digital technologies, such as the Internet of Things (IoT), big data, blockchain, and artificial intelligence.

The circular economy will thus provide consumers with high quality, functional, safe, efficient, and affordable products, with a longer lifetime and designed to be reused, repaired, or recycled. They will also benefit from new sustainable services, such as product-as-service models to improve their quality of life and increase their knowledge and skills.

In this route IP protection is fundamental because companies, to maintain their competitive advantage, must be able to maximize the exploitation of intangible assets (such as: trademarks, designs, software, know-how, inventions, artistic and cultural creations, R&D, processes, and company data) and, at the same time, protect them from being appropriated by current or future competitors.

The IP system should therefore be reviewed in the light of the new changes dictated by the circular economy, considering the development of new technologies and enabling companies to share knowledge.

The most suitable measures to achieve the above targets include the licensing system, the so-called licensing of essential patents referring to the fundamental elements of the digital transformation or of new eco-sustainable materials or of the process for obtaining them, improving their transparency and predictability in a so-called *frand fair, reasonable and non-discriminatory* approach, and also through the *pooling* measures that make it possible to combat abusive monopolies or anti-competitive conduct.

17.3 The Role That IP Should Play in Sustainability

IP has a very important role in sustainability, which is different depending on the specific IP right concerned.

17.3.1 Trademarks

Distinctive signs, and specifically trademarks, constitute a real communication tool, as they convey a message to those who interact with them. With regard to innovations, trademarks particularly indicate the commercial origin of the product or service, allowing the public to associate that innovative or sustainable quantum with one company rather than another, allowing it to become an added value in the marketplace, directing the choices of buyers who want that value and encouraging companies to compete in providing that added value, thus also environmental sustainability. The company adopting a trademark that convey sustainability, gains a competitive advantage, visibility, and an economic return on its investments, leading it to produce in an increasingly sustainable manner. The balance between the protection of competition and exclusivity is ensured by the rules that penalize business communications liable to mislead the public, going so far as to impose the forfeiture of a trademark that has deceptive meanings.[5]

Moreover, there are eco-brands owned by qualified entities that carry out checks on the production of those authorized to use them, through co-branding agreements. Once again, the market and intellectual property are at the service of sustainability, bringing benefits to both the consumer and the company.

17.3.2 Designs

Design plays a fundamental role in the development of goods whose production is guided by and reflects the principles of the circular economy.

In fact, through appropriate design—so-called eco-design—innovation in a circular and sustainable sense starts already at the stage of the "design conception" and "development" of a product or service, to reduce its environmental impacts.

Indeed, the Italian Association of Industrial Design (ADI), which established the ADI Compasso d'Oro Award, the oldest but above all the most prestigious design award in the world, recently established a new award open to international companies and designers that have applied a design culture to production methods that are both advanced and culturally aware of the material and immaterial qualities of the products involved, and that is responsible towards the individual, society, and the environment and which proposes new ethically sustainable forms of behavior that are advantageous for the communities, populations, and economies involved.[6]

17.3.3 Patents

The protection of technical innovation, and thus the incentive to realize new technical solutions that are compatible with the protection of the environment and the future of the world, is one of the most important tools provided by intellectual property to sustainable development. In this sense, exclusivity represents a fundamental incentive to innovation, since the patent system guarantees, on the one hand, that only the patent holder can enjoy and dispose of the right to implement his invention (giving him an exclusive right, which, moreover, in many cases, is convenient for him to exploit through generalized licensing policies, as typically happens with patents that become standards for which licenses are granted to anyone who requests them under FRAND conditions, i.e., Fair, Reasonable and Not

---

[5]C. Galli, Diritti di proprietà intellettuale per la crescita sostenibile, Studi Cattolici, 2020.

[6]Full program of ADI International Compasso d'Oro available at https://www.adi-design.org/compasso-d-oro-internazionale.html.

17 Sustainability and Intellectual Property in Italy

Discriminatory) and, on the other hand, that the conceptual content of the patented invention becomes in the "public domain," from the first day of publication of the relevant application, fueling competition and ensuring a fruitful circulation of information for competitors and researchers.

17.4 How Should This Role Be Pursued

A robust and widespread IP system, particularly a patent system, stimulates innovation and development of technologies that effectively address climate change and support the emergence of the green economy, which is increasingly at the center of global policy debates.

"Innovate for a green future": this is the theme chosen by WIPO in 2020 to celebrate World Intellectual Property Day.[7] Working for a more environmentally sustainable future is, in fact, an imperative that technological innovation and intellectual property shall pursue. The development and deployment of new technologies, whether revolutionary or capable of de-carbonizing existing ones and providing solutions for the more sustainable management of resources, is indeed of paramount importance in addressing climate change while responding to the growing demand for energy and natural resources on a global scale.

And so, the mission is to know how to make the most of Italy's enormous potential: those related to quality productions, increasingly green, inseparable from the changes towards decarbonization and circularity of production, distribution, and consumption models; those in which Italy has achieved levels of excellence, such as waste recycling, a pillar of the circular economy, energy efficiency and renewable energy sources strategic in the energy transition towards a climate-neutral economy.

On February 10, 2022[8] the Italian Ministry of Economic Development issued a Decree containing the terms, conditions, and procedures for granting contributions to support programs and initiatives targeted by the Sustainable Growth Fund, aimed at the ecological and circular transition in the areas of the "Italian Green New Deal." These are non-repayable contributions and subsidized financing. The overall budget of the measure is EUR 677,875,519.57.

Expenses strictly functional to the realization of the investment programs, relating to the purchase of new tangible and intangible fixed assets, are eligible for aid if they concern industrial machinery and equipment; computer programs and licenses relating to the use of the tangible assets; acquisition of environmental certifications. For investment projects aimed at improving the company's energy sustainability, expenses relating to consulting services directed toward the definition of energy diagnosis are also eligible.

---

[7] Available at https://www.wipo.int/ip-outreach/en/ipday/2020/green_future.html.

[8] Italian Ministerial Decree dated February 10, 2022 published on the Official Journal on April 2, 2022, also available at https://www.gazzettaufficiale.it/eli/id/2022/04/02/22A02042/sg.

Sustainable innovation programs involving industrial research, experimental development, and/or, limited to SMEs, the industrialization of research and development results, are eligible for the support of facilitative interventions with particular regard to the objectives of: reduction of plastic use and replacement of plastics with alternative materials; urban regeneration; sustainable tourism; decarbonization of the economy; adaptation and mitigation of risks on the ground from climate change.

Industrialization activities must have high innovation and sustainability content, and be aimed at diversifying the production of an establishment through additional new products or radically transforming the overall production process of an existing establishment; furthermore, they must include investment in tangible assets and may be eligible separately or together with an industrial research and experimental development project as part of an integrated program submitted for facilities under this measure.

To improve sustainability, it might not be necessary to create new IP rights but that existing ones can also be used. For example, trademarks perform an important guarantee function ensuring that consumers are confident in what they buy, or that the product that they are buying complies with certain standards. It is, therefore, increasingly common to see logos indicating that a product is certified by a particular organization, such as the trademark Ecolabel, which is a European trademark of ecological quality for consumer goods and services. The birth of the trademark dates back to the establishment of European Regulation No. 880 in 1992, recently updated by the new Regulation No. 1980/2000 of 21 September 2000. Products with the "daisy trademark" are quality products and services that respect the environment.

Our system, such as most legal systems, provides specific types of trademarks for this purpose, known as certification marks and collective marks.

A similar result can be achieved through the use of the marks by environmental associations which, having become aware of an innovation in environmental matters transposed into a product or service for which a given mark is used, make use of that mark, within the limits of a real descriptive need, to report on the results of its evaluations, positive or negative, on its ecological compatibility, and also on the truthfulness of the messages that the owner company disseminates.

In this way, companies will be encouraged to create products that are as environmentally friendly as possible, also to avoid negative publicity resulting from the judgement of the environmental associations.

Regarding the enforcement of IP rights, I believe that if there were no patents and therefore no patent enforcement there would be no technical progress and therefore no innovation.

Of course, the most suitable measures to achieve sustainability include the licensing system, the so-called licensing of essential patents referring to the fundamental elements of the digital transformation or of new eco-sustainable materials or of the process for obtaining them, and through the pooling measures that make it possible to combat abusive monopolies or anti-competitive conduct.

17 Sustainability and Intellectual Property in Italy

17.5 The Success of IP in Its Role for Sustainability

I believe that the current IP rights provided by the Italian regulation can also be tooled to support sustainability, although there should be more tax reduction from the government and more incentive to use licenses and pooling systems.

For example, to identify Italian patent applications concerning green technologies, the IPTO (Italian Patent and Trademark Office) adopted the WIPO methodology, which is based on the IPC Green Inventory, identifying as eco-innovations those patents classified with at least one IPC (International Patent Classification) code belonging to the Green Inventory.[9] The choice was dictated by the exclusive use of the IPC classification system for Italian patent applications.

Although the eco-innovation definition is very broad and, in any case, not exhaustive, the IPTO decided to use all the fields identified by the WIPO Inventory to reduce the risk of excluding patents pertaining to potentially sustainable technologies.

From 2009 to 2018, eco-sustainable inventions averaged 9.6% of total patents filed in Italy. In the last decade, confirming a rather sustained activity of Italian companies in the search for technological innovations attentive to environmental sustainability, the percentages, globally, attribute to eco-inventions between 5–10% of total patent filings.[10] As in the rest of Europe, companies in Italy have pursued strategic investments and initiatives to develop technologies that improve environmental sustainability.

If a detailed analysis of Italian patents is made by placing them in the seven technological fields as catalogued in the IPC Green Inventory, it emerges that Italian patent applications are mainly concentrated in three areas: Alternative Energy Production; Waste Management and Energy Conservation. These fields cover more than 60% of the entire dataset. Patent activity therefore confirms that our country is taking an active part in the development of the green economy with interesting performances in technologies involving the reduction of pollutants and the reuse of secondary materials, as well as in technologies aimed at improving energy efficiency and the use of renewable sources.

Moreover, in 2018 IPTO launched a new database dedicated to biotechnological inventions.[11] Biotechnology and Life Sciences are technologies that have refined and evolved to play a major role in the modernization of European industry. They include techniques capable of providing breakthrough solutions and new applications in various industries such as healthcare and pharmaceuticals, animal health, textiles, chemicals, plastics, paper, fuel, food, and feed processing.

---

[9]"Data on the number of green patents in Italy. A UIBM analysis of patents filed in the field of eco-sustainable technologies" available at https://www3.wipo.int/wipogreen/en/docs/green_patents_in_italy.pdf.

[10]Ibid.

[11]Available at https://uibm.mise.gov.it/index.php/it/banca-dati-nazionale-della-invenzioni-bioteconologiche.

According to the estimates of the Organization for Economic Co-operation and Development ("OECD"), their exploitation will have a huge impact on the growth of the entire world economy in the future, contributing to sustainable development, improved public health, and environmental protection.

The main applications of biotechnology in the EU economy are in the health and pharmaceuticals sector (with the discovery and development of advanced drugs, therapies, diagnostics, and vaccines), the agriculture, animal husbandry, veterinary products and aquaculture sector (by helping to improve animal nutrition producing vaccines for livestock and enabling advances in diagnostics for the detection of various diseases), and the industrial and manufacturing processes sector (where biotechnology has led to the use of enzymes in the production of detergents, pulp and paper, textiles, and biomass by using fermentation and biocatalyst of enzymes instead of traditional chemical synthesis, it has been possible to achieve greater process efficiency, reducing energy and water consumption, and decreasing the production of toxic waste).

The national database of biotechnological inventions developed within the Directorate General for the Protection of Industrial Property - Italian Patent and Trademark Office contains information on Italian patent applications classified in the technical field of biotechnology as defined by the OECD. The data comes from the IPTO computerized register and patent applications are identified on the basis of the first IPC assigned by the examiner during the prior art search and indicated in the relevant report.

The database provides a complete picture of Italian biotech innovation by means of statistical tables illustrating various aspects of patenting in this sector, from the number of filings to geographical distribution and the most innovative companies. The system is built with an open perspective; in fact, the user, in addition to searching by application number, title, or classification, can download the set of Biotech patent applications present in the BD in csv or Excel format.

The recent measures taken by the Italian government are of course a great incentive for companies to invest in sustainability. These measures should be renewed annually and coupled with incentives for companies that demonstrate that they file IP rights that meet sustainability criteria.

For example, fees reduction could be granted when registering patents and designs that implement sustainability and meet certain requirements. Moreover, the government could arrange for tax reductions of companies that are able to prove that they have invested in sustainability in the tax year.

Concerning the enforcement of IP rights regarding sustainability, the Courts should encourage settlement agreements to grant licenses to those who guarantee that they meet the necessary requirements to use the patents/trademarks/designs.

The application of patent law to the field of biotechnological inventions faces many issues not seen in other technological fields, therefore the activity directed to the preparation of contracts for the economic exploitation of patents in this area is delicate.

Moreover, the data regarding the Italian current situation of technology transfer processes of universities is rather critical but, in the light of the upcoming reforms, it

17 Sustainability and Intellectual Property in Italy

is expected that that will be evaluated and new solutions prepared aimed at fostering intellectual protection, technology transfer, and economic exploitation of patents.

17.6 Improving the Success of IP's Role in Sustainability

Improving the industrial property protection system is a necessary change to strengthen the competitiveness of the production system, an objective outlined by the strategic lines of intervention on industrial property, for the three-year period 2021–2023, provided by the 23 June 2021 decree, signed by the Italian Minister of Economic Development.[12] Another strategic objective is the implementation of the "European Patent with Unitary Effect"[13] which will aim to strengthen Italy's role in European and international industrial property contexts.

In the strategic lines of action, attention is drawn to the need to adopt a series of interventions aimed at promoting the culture of innovation and enhancing and protecting intellectual property, which is seen as a key element in mastering the digitization process that is sweeping our society and, specifically, our production system.

The importance of industrial property is also explicitly confirmed by the National Recovery and Resilience Plan (NRRP),[14] which announces the reform of the industrial property system, which is seen as crucial for the protection of ideas, work development, and what is generated by innovation, and for ensuring a competitive advantage for those who have been active in this area.

The NRRP provides a package of investments and reforms divided into six missions. The Plan promotes an ambitious reform agenda, particularly those covering public administration, justice simplification, and competitiveness.

The Plan is fully consistent with the six pillars of the Next Generation EU with regard to the planned investment shares for green (37%) and digital (20%) projects.

The resources allocated in the Plan amount to 191.5 billion euros, divided into six missions:

– Digitalization, innovation, competitiveness, and culture—40.32 billion
– Green revolution and ecological transition—59.47 billion
– Infrastructure for sustainable mobility—25.40 billion
– Education and research—30.88 billion

---

[12] Ministerial Decree issued on June 23, 2021 by the Italian Ministry of Economic Development.

[13] The European Patent with Unitary Effect is regulated by the EU Regulation No 1257/2012 of the European Parliament and of the Council of 17 December 2012 implementing enhanced cooperation in the area of the creation of unitary patent protection and by EU Council Regulation No 1260/2012 of 17 December 2012 implementing enhanced cooperation in the area of the creation of unitary patent protection with regard to the applicable translation arrangements.

[14] The National Recovery and Resilience Plan (NRRP) was transmitted by the government to the European Commission on 30 April 2021. On 13 July 2021, Italy's NRRP was finally approved by a Council Implementing Decision.

- Inclusion and cohesion—19.81 billion
- Health—15.63 billion

To finance additional interventions, the Italian government has approved a Supplementary Fund with resources of 30.6 billion euros. In total, investments under the NRRP and the Supplementary Fund amount to EUR 222 billion.

Relative to the Green Revolution and Ecological Transition, the projects included in the mission aim to foster the country's green transition by focusing on energy produced from renewable sources, increasing resilience to climate change, supporting investment in research and innovation, and incentivizing sustainable public transport.

For these lines of intervention, 1.25 billion is allocated to strengthen investment in the main sectors of the green transition, including by fostering industrial reconversion processes and new entrepreneurship.

One billion investments for renewables and batteries aims to develop industrial supply chains in the photovoltaic, wind, and battery sectors through three main lines of action, namely, the establishment of a giant plant for the construction of innovative high-efficiency photovoltaic panels, the construction of an industrial plant for the production of flexible panels for wind power, and the construction of another giant "ultra-modern" (4.0) plant in the battery sector.

Returning to the subject of industrial property, Italy is adopting a number of strategies to strengthen its promotion and enforcement, such as the upcoming reform of the Industrial Property Code, which is considered essential for the adaptation of the regulatory framework to the developments being foreshadowed by the above-mentioned lines of action.

Another novelty, which has been recently approved by the Italian Parliament, concerns the protection of innovation in universities and at public research institutions:[15] ownership of inventions will no longer be granted to the researchers who developed them, but to the respective structures in which they operate and, only in the event of inaction by the structure in question, will it be attributed to the researcher.

This is an important and long-awaited reform, which was promoted, among others, also by Federchimica Assobiotec, the national association for the development of biotechnologies. The latter, together with the reform of the so-called "Professor's Privilege," also proposed the creation of a Tech Transfer Competence Center (TTCC) for the Life Science field which should take care of organizing and integrating the competences developed at the level of the individual Italian regions, guaranteeing the research centers and other entities (such as universities, hospitals, etc.) the possibility of being autonomous, equipping themselves with the necessary competences to provide concrete support also in the area of intellectual property management. The centers should also make it possible to achieve a minimum

---

[15] Italian Law no. 102/2023.

amount of material, not duplicating competences, and making use of the resources that the regions already have.

Lastly, efforts are being made to create a new telematics support infrastructure for the management of all industrial property applications and titles, using the latest technology, and to be complemented by the implementation of the Unitary Patent System, thus optimizing the "Supplementary Protection Certificates" system.

In conclusion, with regard to the provision of improved aspects within the European system, certainly the implementation of the "European Patent with unitary effect" is of fundamental support. It will be granted by the European Patent Office (EPO) and will enable its holder, through the payment at the EPO of a single renewal fee, to simultaneously obtain patent protection in the 25 participating EU countries.

This patent will not replace the national patent, but will be pre-eminent in case of cumulation of protections with the national patent. The European Patent with unitary effect will be operational only after the International Agreement on the Unified Patent Court (UPT) enters into force.

17.7 Conclusions

Investing in innovation is very important as there is a direct positive relationship between R&D investment and a country's growth. Innovation is a complex and articulated process that starts with an idea and evolves through research, development, production, and commercialization.

Moreover, it can be the starting point for new ideas and new innovations resulting from the superior contribution that other individuals or companies can make.

This creates a true virtuous circle whereby innovation generates new innovation, with benefits not only for those who produce it, but also for the direct or indirect stakeholders involved and for the entire planet.

Open Access This chapter is licensed under the terms of the Creative Commons Attribution 4.0 International License (http://creativecommons.org/licenses/by/4.0/), which permits use, sharing, adaptation, distribution and reproduction in any medium or format, as long as you give appropriate credit to the original author(s) and the source, provide a link to the Creative Commons license and indicate if changes were made.

The images or other third party material in this chapter are included in the chapter's Creative Commons license, unless indicated otherwise in a credit line to the material. If material is not included in the chapter's Creative Commons license and your intended use is not permitted by statutory regulation or exceeds the permitted use, you will need to obtain permission directly from the copyright holder.

Sustainability and Intellectual Property in Malta

18

Philip Mifsud and Sasha Muscat

18.1 Intellectual Property and Sustainability in the Maltese Islands

18.1.1 A Brief Overview of Existing Legislation

In Malta, intellectual property (hereinafter referred to as 'IP') is regulated by various laws both at EU level—either through transposition or else by virtue of their direct applicability—as well as locally. For the purposes of responding to the questionnaire, we have focussed on the local legislative framework and local landscape (save where we have felt it necessary to expand our response to include the EU). The '*main*' laws regulating IP in Malta are the Copyright Act (Chapter 415 of the Laws of Malta), the Trademarks Act (Chapter 416 of the Laws of Malta), the Patents and Designs Act (Chapter 417 of the Laws of Malta), the Intellectual Property Rights (cross-border measures) Act (Chapter 414 of the Laws of Malta), the Enforcement of Intellectual Property Rights (Regulation) Act (Chapter 488 of the Laws of Malta), the Trade Secrets Act (Chapter 589 of the Laws of Malta) and the Commercial Code (Chapter 13 of the Laws of Malta). Additionally, Malta is also a party to various international treaties, conventions and agreements including the World Trade Organisation's Agreement on Trade-Related Aspects of Intellectual Property Rights (TRIPS), the Berne Convention, the Patent Cooperation Treaty, the European Patent Convention and the Unified Patent Court and Regulation.

Although there is clearly an abundance of legislation which serves to regulate IP and related rights in Malta, the Maltese legislator has yet to enact any regulations, acts, bills, legal notices, by-laws or other legal instruments which specifically regulate the emerging relationship between IP and sustainability. While Malta

---

P. Mifsud (✉) · S. Muscat
Ganado Advocates, Valletta, Malta
e-mail: pmifsud@ganado.com; smuscat@ganado.com

© The Author(s) 2024

P. Këllezi et al. (eds.), *Sustainability Objectives in Competition and Intellectual Property Law*, LIDC Contributions on Antitrust Law, Intellectual Property and Unfair Competition, https://doi.org/10.1007/978-3-031-44869-0_18

does have its own Sustainable Development Act (Chapter 521 of the Laws of Malta), the Act makes no reference to IP. Nonetheless, over recent years there have been schemes and incentives which have been established, as well as broad guidance documents which have been published, at a local level, which indicate that Malta is growing increasingly conscious of the vital role that IP can play when it comes to promoting and increasing sustainable solutions for the future.

18.1.2 Existing Local Initiatives, Programmes and Strategies: A General Perspective

One initiative which has been taken at local level is the annual *'Malta Intellectual Property Awards'*, which was first launched in 2009. The scope of the awards, which are organised by the Commerce Department of the Maltese Ministry for the Economy, Investment and Small Business, is to recognise and encourage the development of innovative ideas and products which have a degree of potential that is both sustainable and unique. Through the initiative, the Maltese Ministry has sought to promote the creation of IP in Malta, including patents, trademarks and designs which can benefit society in innovative and sustainable ways.[1] As of 2022, the amount of available funding for the awards is 60,000 euros and Maltese nationals, groups of Maltese individuals, as well as entities established in Malta may participate and are eligible for funding. Furthermore, the awards are divided into four wide-ranging categories, which are: Creative Innovation, Scientific Innovation, Technological Innovation and Emerging Innovation, and which allow for a broad range of innovative, sustainable ideas to be considered.[2] Examples of novel IP which has managed to secure funding over the years through these awards are, inter alia, (i) the creation of an integrated offshore energy store (in 2017), (ii) the creation of a mechanism to teach sustainable development through gaming (also in 2017), (iii) the development of an idea for a multitrophic polyculture food production system for Malta (in 2020), and (iv) an evaporative cooling method for lithium-ion batteries (in 2021). The initiative shows that Malta has been, for a number of years, placing heightened focus on the protection and promotion of ideas and creations which can serve to enhance sustainability, and these efforts are clearly bearing fruit. Moreover, the initiative plays a crucial role in providing persons who might not have the financial ability or the necessary know-how to protect their IP with an avenue to develop and simultaneously protect their ideas.

The Maltese government, through its dedicated ministries and departments, has also on a number of occasions supported the development of innovative, sustainable projects in Malta through financial grants for research, testing, and the creation of infrastructure. One example is the Malta Council for Science and Technology's

---

[1] Malta Intellectual Property Awards, 2022. In: Commerce.gov.mt. https://commerce.gov.mt/en/Awards/Pages/Malta-Innovation-Awards.aspx. Accessed 13 May 2022.

[2] Ibid.

'FUSION' programme, which is a Maltese funding programme that supports local research and innovation—particularly for ideas which aim to improve the quality of life, and provides the necessary support for researchers and technologists to turn their innovative ideas into market-ready realities. [3] In previous years, the funding has been granted for the development of a sustainable water treatment plan for local hotels which was developed by a Maltese engineer, and for a research project carried out by a group of Maltese researchers to develop efficient storage systems for energy generated on structures at sea. Although the scheme does not specifically target the protection of sustainable IP, it can certainly be said that most of the projects which fall to be considered under the programme have sustainable characteristics, and are also likely to be patentable due to their inventive aspects.

Notably, a consultation document published in 2016 by the Maltese Ministry for Sustainable Development, the Environment and Climate Change entitled *'Greening our Economy – Achieving a Sustainable Future'*,[4] emphasises the need for local policy action to be taken to reduce the costs associated with acquiring IP rights, particularly for small and start-up firms. While the consultation document encourages *'a greener economy'* and *'enhance[d] sustainability'*, it also explains that these aims are unlikely to be achieved if barriers to entry, particularly in relation to IP, are not removed. Furthermore, the consultation document also cites, as an area for consideration, the need to provide adequate training to persons working in emerging fields. Specifically, the protection of IP in the field of new energy technologies through the provision of part-time courses and internal company training to improve the competencies of the work force in IP institutions.

Clearly the Maltese government and local government authorities already recognise that IP plays a key role in enhancing sustainability, whether the sustainable component or aim of the IP in question relates to environmental, social, economic, technological or even to other forms of sustainable innovation. While the strategies, schemes, and initiatives which have been and which continue to be pursued at a local level might not appear to be on a grand scale in comparison with other larger and wealthier jurisdictions, in our view Malta's efforts are commensurate with its size, amount of resources, and available funds which can be allocated towards this scope.

---

[3] FUSION - MCST. In: MCST. https://mcst.gov.mt/ri-programmes/fusion/. Accessed 7 November 2022.

[4] MSDEC, Greening our economy – achieving a sustainable future 2022. Available at https://meae. gov.mt/en/public_consultations/msdec/documents/green%20economy/consultation%20document %20-%20green%20economy.pdf. Accessed 7 November 2022.

18.2 The Role of Intellectual Property in the Journey Towards Sustainability

18.2.1 Introduction

In our view, IP should continue to become more relevant in the achievement of sustainable solutions. Although, as we have elaborated upon in Sect. 18.1, the importance of sustainability has already started to intersect with IP rights and related notions in Malta, we perceive that the relationship between the two will continue to grow in the years to come. On a global level, it is already evident that sustainability has moved to the fore both on a social level as well as at policy level. Laws are being enacted to better ensure sustainable practices are adopted while at the same time, governments and supranational authorities are committing towards initiatives which have sustainability, climate protection, and the protection of the environment at the heart of such initiatives. It is therefore reasonable to expect that innovators and creators will develop ideas that better serve these central themes. Accordingly, IP will become all the more critical in ensuring that such innovative solutions and ideas are protected.

18.2.2 The Role of Intellectual Property in Malta, and How It Should Be Pursued

From the foregoing, it is evident that IP should play a primary role in not only enhancing and promoting sustainability but also in incentivising people to create such solutions. As we have elaborated, Malta has been making efforts in recent years to support sustainable innovation on a local level and it is expected that Malta will seek to continue to incentivise people to pursue these objectives in the future. In fact, in a press release published by the Maltese Ministry for the Economy, Investment and Small Businesses, it was emphasised that the Maltese government has plans to encourage local innovation which serves the common good, and wants to provide increased assistance to individuals and entities that are in the process of developing innovative or technological ideas that can contribute further to the economy and society in general. [5]

To pursue these ends, we would hope to see Malta supporting sustainable innovation not just through financial incentives—as it has already done, but also through other forms of incentives. While it is recognised that financial incentives are key to the innovative process, there are also other aspects which will help bolster sustainable innovation. Financial aid on its own will only ensure progress to some extent but coupled with other elements—such as education, technical, strategic and

---

[5] Press Release by the Ministry for the Economy, Investment and Small Businesses, 2020. Available at https://www.gov.mt/en/Government/DOI/Press%20Releases/Pages/2020/April/25/pr200751en.aspx. Accessed 20 May 2022.

legal support, it could be ensured that there is much further development. The aim should be to create a one-stop shop for innovators to have access to funding, as well as information and advice on the IP registration processes, the bounds of such protection, etc. Such additional aspects could potentially facilitate innovative development in the field of sustainability and remove any unnecessary barriers for persons or entities seeking to develop sustainable IP, such as through the reduction of costs or waiting time associated with obtaining the IP rights in question.

18.2.3 Creating Change Through New Policy and Legislative Frameworks

It is pertinent to note that in 2015, Malta, along with all other United Nations Member States, adopted the 2030 Agenda for Sustainable Development (hereinafter referred to as the '2030 Agenda') which aims to *'achieve a better and more sustainable future for all'*.[6] The 2030 Agenda is based on 17 sustainable development goals (hereinafter referred to as 'SDGs') which call for urgent action by all countries to address issues regarding poverty, health, education, inequalities, economic growth, climate change and the preservation of oceans and land ecosystems. In our view, IP is of relevance and can play a critical role in achieving the SDGs, despite the fact that the 2030 Agenda does not itself directly refer to the notion of IP. The only direct reference to intellectual property in the 2030 Agenda framework is in Goal 3, which mentions the relevance of the Agreement on Trade Related Aspects of Intellectual Property Rights (hereinafter referred to as the 'TRIPS Agreement') in the context of the public health aim to provide medication and vaccines to all. In a 2021 Report entitled *'WIPO and the Sustainable Development Goals: Innovation Driving Human Progress'*,[7] WIPO explains the role that IP and innovation can have, particularly in achieving SDG 3 (dealing with good health and well-being), SDG 6 (dealing with clean water and sanitation), SDG 7 (dealing with affordable and clean energy), SDG 8 (dealing with decent work and economic growth), SDG 9 (dealing with industry, innovation and infrastructure) and SDG 13 (dealing with climate action), among others.[8] In our view, the role pursued by IP to achieve these goals should be one which, inter alia, advances the development of green technologies and eco-friendly products, supports a healthy environment that improves the quality of life, promotes sustainable agricultural and economic development, and creates new jobs in the process. Having strong IP rights and effective IP legislative frameworks in place are necessary to achieve this, namely because

---

[6] 2030 Agenda for Sustainable Development. Available at https://www.itu.int/en/ITU-D/Statistics/Pages/intlcoop/sdgs/default.aspx. Accessed 7 November 2022.

[7] (2021) World Intellectual Property Organization (WIPO) https://www.wipo.int/edocs/pubdocs/en/wipo_pub_1061.pdf. Accessed 20 May 2022.

[8] WIPO Illustrates How Innovation, Intellectual Property Can Support, 2018, available at Sdg.iisd.org. https://sdg.iisd.org/news/wipo-illustrates-how-innovation-intellectual-property-can-support-sdgs/. Accessed 7 November 2022.

creators and innovators will reasonably expect to be granted proper protection for their hard-earned efforts and a return on their investments.

Currently, Malta is revising its national sustainable development strategies and has also launched inter-ministerial consultations with the aim of incorporating the 2030 Agenda and the SDGs within its local policy,[9] and is developing a comprehensive action plan on a local level.[10] As Malta pursues different courses of action it should also give consideration to whether its intellectual property frameworks could benefit from being improved or amended (as we will elaborate upon further in our responses to the questions below), as in our view, having an effective and modern local intellectual property system will serve not only as a driver of innovation, but can also have the effect of fostering sustainable economic, environmental, social and cultural growth both in Malta and abroad.

18.2.4 Creating Change by Incentivising Sustainable Innovation

In our view, the role of intellectual property in the journey towards sustainability should be directed mainly towards supporting sustainable innovation. The provision of incentives and the creation of legislative frameworks which encourage sustainable innovation can, however, have the indirect effect of discouraging innovation which lacks a sustainable impact. However, creating a distinction between the type of protection that 'unsustainable innovation' attracts versus 'sustainable innovation' is not something that should be handled on a legislative framework as IP laws are there to protect based on a set of objective criteria. Additionally, it is entirely possible that a 'sustainable solution' may later be deemed 'unsustainable' which creates potential for future issues on how such protection may have been applied. Thus, incentives, and benefits, must necessarily only extend to those sustainable solutions such that there is not a situation where 'unsustainable innovation' is rewarded — to the extent possible.

Within the IP industry, Malta currently has in place a number of financial and tax-related incentives, as well as other IP-related schemes; however, unfortunately, many of them fail to specifically apply to or target IP which has some form of sustainable element or which pursues a sustainable end. Section 18.1 further elaborates upon the incentives which are currently available in Malta that relate to IP and sustainability concurrently. However, in our view, the necessity for Malta to develop new ways to encourage this type of IP cannot be overstated, particularly when we consider the fact that the vast range of technology, medicines, and

---

[9] European Environment Agency. Malta country profile - SDGs and the environment, 2020 https://www.eea.europa.eu/themes/sustainability-transitions/sustainable-development-goals-and-the/country-profiles/malta-country-profile-sdgs-and. Accessed 21 May 2022.

[10] Malta Voluntary National Review on the Implementation of the 2030 Agenda, 2018 https://sustainabledevelopment.un.org/content/documents/20203Malta_VNR_Final.pdf.

renewable energy inventions which are available on the market today would not be possible without the existence of robust IP protection.

One recommendation could be to establish a new dedicated local branch within the Malta Industrial Property Registrations Directorate (the 'MIPRD'), which would be entrusted with handling IP filings, renewals and other requests solely in relation to those types of IPR, i.e. IPR having a proven sustainable goal or element which could potentially be determined based on a list of objective criteria. Through the creation of these dedicated branches, the relevant procedures would become more streamlined, and members of staff could be provided with specialised training to be equipped to handle specific requests more efficiently. As a result, IP right holders can expect to have their requests expedited since they will not go through the standard local channels. We have already seem streamlining of certain functions by other registries and authorities in Malta with great success and it would be expected that this would also work at the MIPRD . Additionally, while registration fees for IP in Malta are not too costly, reduced fees for sustainable solutions and business operating in a sustainable manner could also be applied to further incentivise this type of development.

In fact, this role should be a twofold approach where the focus is both on sustainable solutions which are developed as well as businesses which operate in a sustainable manner. Targeting the latter through, e.g., lower tax brackets, government funding opportunities, etc. will also ensure that it is not only the innovative sector of a country's population which is targeted as all businesses will be able to participate by ensuring they develop, practice and (indirectly) promote sustainable methods.

Moreover, we do not think it necessary to alter the existing conditions required under the law to acquire IP rights, but rather we think it would be fruitful to provide favourable conditions to those persons or entities that, in filing their patent applications, or other IPR applications, satisfy certain sustainability requirements.

18.3 Malta's Current Intellectual Property Framework

18.3.1 The Current Landscape

A variety of IP rights exist in Malta, namely trademarks, copyrights, design rights, patents, trade secrets, and protected geographical indications. From a sustainability perspective however, the IP rights available in Malta cannot be said to be playing any specific role in driving sustainability, nor are they pursuing any specifically sustainable goals.

From a purely legal perspective, the IP rights available in Malta fulfil their traditional roles of granting legal protection to the endeavours of inventors and creators, while also restraining third parties from doing any acts which might prejudice or violate those rights. However, from a sustainability perspective, in our view the IP rights available in Malta cannot really be said to be playing a specific role in driving sustainability, nor are they pursuing any specifically sustainable goals.

The main reason for this view is that as yet, Malta has not implemented any system or framework which requires that an applicant fulfils any environmentally, economically or socially sustainable criteria to be granted an IP right, and the success or otherwise of an IP filing is mainly determined on a 'first-to-file' basis.

18.3.2 Creating Further Rights to Support Sustainability

In our view, an IP right which could be created in Malta is the utility model, which is a patent-like IP right typically used to protect inventions that improve or adapt existing products, from being commercially exploited without the consent of the right holders. The term 'utility model' only appears once in Maltese law, in Article 78 of the Patents and Designs Act (Chapter 417 of the Laws of Malta), however as of today Malta has no legal regime for the utility model.

The main reason why this new IP right could be beneficial for Malta is that the conditions that must be fulfilled to obtain utility model protection are much less rigorous than those required to obtain patent protection. For an invention to qualify for patent protection, it must be new, involve an inventive step, be capable of industrial application and not specifically excluded from protection. In our view, what makes utility models attractive is that they cover inventions that may fail to meet the formalities required for patient protection, but which nonetheless warrant some form of protection.[11] Therefore, the individuals or entities behind sustainable inventions which could potentially fall short of qualifying for patent protection, will not risk having their sustainable ideas overlooked and will still be in a position to obtain legal protection through the utility model. In addition, Malta does not see a lot of patentable work, particularly since it is relatively small in comparison with neighbouring states. Therefore, having an effective utility model could also supplement the low patent work in Malta, while encouraging those who may be deterred from developing their ideas, due to non-patentability, to obtain legal protection for their ideas.[12]

18.3.3 Enforcement of IP Rights in Malta

Generally, the rigorous enforcement of IP rights would have an indirect benefit, as prospective innovators and licensees would have more peace of mind in knowing that any infringements would be actively and effectively pursued. That being said, enforcement mechanisms should not be the primary driver towards promoting sustainability but rather should be a by-product of Malta having an environment which actively promotes sustainable practices.

---

[11] N. Sultana, Utility Models: A Comparative Analysis with a View to Determining the Feasibility of Introducing this Intellectual Property Regime in Malta, 2016, pp. 37-46.
[12] Ibid.

18 Sustainability and Intellectual Property in Malta

In our view, a positive recommendation could be the introduction of a form of fast-tracked 'green IP' system in Malta. Green IP would be protection granted for environmentally sustainable inventions and technology which are designed to stop or prevent environmental degradation and promote environmental sustainability. Examples include technology concerning waste, wind power, geothermal energy, solar energy, and tidal energy, to name a few. The creation of a local system that processes green IP applications in priority over other applications, within a considerably shorter period of time and through specially trained personnel, could serve as an incentive for this type of innovation and could generate both environmental and economic benefits for Malta. Research and technical studies have shown that Malta, particularly due to its geographical location and climate, has a strong potential to produce wind energy[13] and solar energy[14] for example, so the creation of such a system is likely to maximise Malta's renewable energy potential, while achieving an environmentally and economically sustainable goal.

Although the MIPRD does provide information on the different types of IP Rights through its website and there is information on other government websites providing information regarding sustainability, the two are not always directly linked. Therefore, carrying out such an assessment is dependent on the individual who is gathering such information and it would be beneficial if these two were to be linked more closely.

18.3.4 The Role of Intellectual Property Authorities in the Journey Towards Sustainability

When it comes to the local procedures in place for the registration of IP rights, Malta like many EU Member States, has digitised most of its processes allowing applicants to submit applications to register their trademark, patent, or design and other requests through an online form at https://ips.gov.mt/. This has made the application procedure easier and more efficient—not to mention that the public also has access to view and conduct searches of local trademarks and patents online via the Malta Trademark Register and the Malta Patent Register respectively. Notwithstanding this, the time periods involved between the date of filing and the date of registration of an IPR may be lengthy; with the average time for registering a Maltese trademark being three to five months, Maltese patents being around eighteen months, and Maltese designs being around two months. Additionally, IPR related fees imposed by the Maltese Commerce Department are not excessive and not a barrier to entry for a prospective applicant.

---

[13] I. Camilleri, Wind energy good prospect for Malta, In: Times of Malta 2009. https://timesofmalta.com/articles/view/wind-energy-good-prospect-for-malta.261375. Accessed 3 June 2022.

[14] C. Pille, Renewable Energies: Malta's location, a little-used energy advantage - Friends of the Earth Malta, 2020. In: Friends Of the Earth Malta. https://foemalta.org/blog/renewable-energies-maltas-location-a-little-used-energy-advantage/. Accessed 3 June 2022.

18.3.5 The Way Forward

Naturally, there is always room for improvement, both with regards to the duration for registration of IP Rights (where possible) as well as the duration vis-à-vis proceedings before courts. Although the Maltese courts ensure the issuance of injunctions very swiftly to ensure the parties to a case are not prejudiced, greater expediency for court proceedings to be concluded would always be welcomed. This is naturally the case irrespective of whether that IP right is pursuing a sustainable goal or otherwise. It is pertinent to note that the European Commission (the 'Commission') identified these as obstacles which are being faced in Member States generally and in 2020, it published an Action Plan on Intellectual Property to implement specific IP-related policies which provide persons with wider access to IP rights, and with the aim of removing challenges and barriers to entry that undermine creative and innovative efforts—particularly, as the Commission explains, in times of serious health and economic crises. In a Q&A concerning the Action Plan, the Commission identified IPR industries as major contributors to the EU's economy and to the generation of employment within the EU, particularly in renewable energy and low-carbon energy-intensive industries, and therefore sought to also encourage green transitions.[15] The Action Plan has encouraged Member States, including Malta, to take action through participation in the Unitary Patent System; which aims at simplifying patent protection in the EU, reducing related costs and strengthening legal certainty, as well as participation in financing programmes; targeted at helping SMEs to better manage their IP assets while granting them access to strategic IP advice, to name a few.[16]

18.4 Promoting National and International Cooperation

To reiterate what we have discussed in Sect. 18.3, in our view, IP rights and IP authorities in Malta could be improved through inter alia the establishment of dedicated branches within our local IP departments, the creation of a local green-IP scheme, the introduction of the utility model as a new local IPR, as well as through the promotion of further schemes, financial or otherwise, which serve to encourage various types of sustainable innovation such as 'FUSION' and the 'Malta Intellectual Property Awards'.

To implement these changes, amendments to existing local legislation will need to be made, for example to introduce the concept of a utility models and to provide

---

[15]European Commission, Action Plan on Intellectual Property - Questions and Answers, 25 November 2020. https://ec.europa.eu/commission/presscorner/detail/en/qanda_20_2188. Accessed 4 June 2022.

[16]E. Hoss, European Commission's Action Plan on Intellectual Property, 2021, available at https://www.elvingerhoss.lu/publications/european-commissions-action-plan-intellectual-property#:~:text=On%2025%20November%202020%2C%20the,field%20and%20helping%20small%20and. Accessed 4 June 2022.

for its regulation under the Patents and Designs Act (Chapter 417, Laws of Malta). The introduction of new legislation might also be required to provide for the regulation of the green IP, potentially through a dedicated Legislative Act or subsidiary legislation. Furthermore, for the aforementioned dedicated branches within the local IP institutions to function efficiently and effectively, the provision of specialised training to members of staff will be of utmost importance, so that staff can acquire specialised knowledge and be equipped to handle specified IP rights requests.

Most of the ideas that we elaborated upon in our responses to the previous questions, would require efforts that are taken mainly on a local level in Malta. Having said this, a concerted effort on an international scale, from the various state parties of the international treaties and conventions pertaining to IP but also from non-state parties around the world is required to raise awareness and effectively promote sustainable IP globally. Such a holistic approach at a global level will ensure homogenous development of sustainable initiatives which tie in with IP. In our view, this can only be achieved once more states understand that having effective intellectual property rights and sound intellectual property regimes and policies are key components for sustainable development and innovation.

18.4.1 The Role of International Treaties and Agreements

Although today there are numerous international treaties which regulate IP rights— many of which Malta is also a party to, it can decisively be said that the TRIPS Agreement is to date the most comprehensive and extensive multilateral agreement on intellectual property, with 164 states from all over the world being parties to it.

In spite of its large number of signatories, including those from less developed and developing countries, the TRIPS Agreement has been criticised as being incompatible with the achievement of socially sustainable goals by developing countries, such as the reduction of poverty or the increase of public health, to name a few. While in theory the TRIPS Agreement does meet these aims, its implementation in practice has at times come under fire from critics who have argued that in practice the manner in which the TRIPS Agreement is implemented can have a detrimental impact certain countries which may still be developing certain IP regimes.[17]

For example, it has been argued that the TRIPS Agreement imposes excessively strict protection of intellectual property rights, which in turn has the effect of placing access to essential and life-saving medication further out of reach for those in developing countries, by decreasing access or increasing the costs to obtain those medications.[18] In fact, initiatives similar to those observed in the competition law

---

[17] M. Banda (2019) https://www.wto.org/english/tratop_e/trips_e/colloquium_papers_e/2019/chapter_9_2019_e.pdf.

[18] J. Subhan, Scrutinized: The TRIPS Agreement and Public Health, available at https://www.ncbi.nlm.nih.gov/pmc/articles/PMC2323529/. Accessed 7 November 2022.

sphere (predominantly matters relating to FRAND licensing) would be a step in the right direction to ensure that IP, international treaties regulating IP and sustainable initiatives reach the desired result both in theory as well as in practice.

In our view, the TRIPS Agreement could benefit from being amended to create specific criteria for managing compulsory licensing in relation to IP (similar to standard setting with certain essential patents) which is of great sustainable benefit on a global or national level. This type of increased flexibility will ensure that creators are sufficiently protected and renumerated while also ensuring that member states benefit from such sustainable innovations.[19]

18.5 Conclusion

In conclusion, it should be highlighted that the sustainability goals that we seek to achieve through IP cannot be achieved through national efforts alone, irrespective of how effective, modern or environmentally and socially conscious a nation's IP frameworks might be. Although it is crucial for each state to participate and contribute its own efforts, the best outcomes are not achieved through national efforts in isolation from other states—particularly when we speak of the efforts of smaller nation states such as Malta. Global cooperation is necessary. Although in recent years, the concepts of sustainable development and the effective protection of IP rights have rightly become more of a priority for many states, we must keep in mind that most environmental, economic and social impacts are not confined to a state's territory, and more states—including Malta—must continue to adopt modern IP policies that facilitate the transition to a more sustainable world.

Open Access This chapter is licensed under the terms of the Creative Commons Attribution 4.0 International License (http://creativecommons.org/licenses/by/4.0/), which permits use, sharing, adaptation, distribution and reproduction in any medium or format, as long as you give appropriate credit to the original author(s) and the source, provide a link to the Creative Commons license and indicate if changes were made.

The images or other third party material in this chapter are included in the chapter's Creative Commons license, unless indicated otherwise in a credit line to the material. If material is not included in the chapter's Creative Commons license and your intended use is not permitted by statutory regulation or exceeds the permitted use, you will need to obtain permission directly from the copyright holder.

---

[19] R. Cardwell, The Effects of the TRIPS Agreement on International Protection of Intellectual Property
Rights, 2012.

Sustainability and Intellectual Property in Sweden

19

Martin Zeitlin

19.1 Background

Intellectual property (IP) is considered to be of major importance in the modern economy and is, as a rule, a prerequisite for the dynamic development of culture, technology, and economic prosperity.[1] It is not clear what role IP has in the pressing global issue of sustainability and the green transition and if it, for that matter, even should have a certain role. This chapter briefly addresses this matter from a Swedish IP perspective.

19.1.1 European Conformity

Sweden is since 1995 a full member of the European Union (EU). IP has naturally been the subject of comprehensive harmonisation efforts with the pursued aim of market integration in the EU (and the integrated European Economic Area thereto). The EU as of today has explicit competence to legislate on issues concerning IP law pursuant to Article 118(1) of TFEU.[2] The legislative discretion awarded to Sweden is accordingly limited in the field of IP law. Swedish courts and public authorities are bound to follow the case law of the Court of Justice of the European Union (CJEU) to guarantee compliance with EU law in accordance with the fundamental principle

---

[1] S. Wolk, Immaterialrätten då, nu och i framtiden, Svensk Juristtidning 2016, p. 130.

[2] The Treaty on the Functioning of the European Union of 26 October 2012, OJ C 326/47.

M. Zeitlin (✉)
Advokatfirman MarLaw, Stockholm, Sweden
e-mail: martin.zeitlin@marlaw.se

© The Author(s) 2024
P. Këllezi et al. (eds.), *Sustainability Objectives in Competition and Intellectual Property Law*, LIDC Contributions on Antitrust Law, Intellectual Property and Unfair Competition, https://doi.org/10.1007/978-3-031-44869-0_19

of the effectiveness of EU law[3] and the principle of EU-consistent interpretation.[4] In other words, Swedish law effectively mirrors EU law in the field of IP and the issues raised herein cannot be treated as a national matter by Sweden. It should already at this point be mentioned that the European Commission will propose a new strategy to ensure that IP remains a key enabling factor for the circular economy and the emergence of new business models.[5] As such, the issue of sustainability in IP is essentially a European question outside the scope of Swedish public policy. Nonetheless, the existence of a Swedish perspective is naturally not entirely precluded due to EU law.

19.1.2 Defining Sustainability

Sustainability cannot easily be defined. As one easily might associate the term with superior environmental performance, the term itself may imply a wide array of aspects outside the scope of the environmental context. For instance, the Sustainable Development Goals of the 2030 Agenda for Sustainable Development as adopted by the United Nations member states in 2015[6] may serve as a basis for defining sustainability. The Sustainable Development Goals include a vast set of different goals including the mitigation of corruption, promotion of gender equality, to name a few. Within the context of IP rights, these goals could altogether be highly relevant. However, to delimit the topic of this chapter, the topic of sustainability will be treated within the context of the environment only.

Even if the scope of sustainability is limited to only cover environmental aspects, the scope is nevertheless wide. An overall purpose of environmental sustainability can be summarised as the mitigation or complete removal of harmful effects on the environment. Such harmful effects are mainly the emission of greenhouse gases, destruction of ecosystems, impairments to wildlife and ecological diversity, and similar effects. Reducing a single harmful effect on the environment may give rise to an increase of harmful effects elsewhere. If one country decreases the emissions while upholding the same level of overall consumption, the emissions of another (exporting) country may very well increase. A highlighting example from the Nordics of the difficulties with sustainable production is the forest industry. The forest industry has the capacity to deliver products that can substitute fossil intensive alternatives and thereby reduce the carbon footprints of certain products. At the same time, the forest itself is a complex ecosystem. Massive plantation of certain

---

[3] The Treaty on the European Union of 9 May 2008, OJ C 115/13, Article 4(3).

[4] M. Bobek, The effects of EU law in the national legal systems. In Barnard and Peers (eds), European Union Law, Oxford University Press 2014, p. 153.

[5] European Commission, Communication from the European Commission to the European Parliament, the Council, the European Economic and Social Committee and the Committee of the Regions of 11 March 2020, COM(2020) 98 final, para 6.3.

[6] UN General Assembly, Transforming our world: the 2030 Agenda for Sustainable Development, 21 October 2015, A/RES/70/1.

fast-growing trees cause detriment to ecological diversity and the cutting of trees stops the uptake of carbon in the atmosphere—even if the end users can access a product based on renewable resources with a smaller carbon footprint than the substituted more fossil intensive product. The forests' role in mitigation of climate change is an act of balancing where the benefits of substitution stand in conflict with the interest of preservation. A fundamental problem of this act of balancing is that the discourse has turned polemical and becomes increasingly difficult to overview.[7]

The intricate ecological aspects of sustainability can be elaborated in infinity. Perfect sustainability is nearly unfathomable. Any consumption is contingent on production and production is united with impacts due to the use of resources. The impacts are to varying extents damaging to some, perhaps even conflicting, aspects of sustainable metrics. We could effectively create the most beneficial impact by ceasing to consume and ultimately to live. Now, this is not a manifesto against contemporary and affluent humanity. Notwithstanding, the fact of the matter is that perfect and constant sustainability is nearly unfathomable as long as there is extensive consumption. To put it differently, sustainability is not a final destination but rather an ongoing journey. It is indeed challenging to address the subject of IP and sustainability when sustainability itself arguably is an act of balancing and contingent on an ever-evolving discourse. Having said that, the role of IP and the role of law in general should, as a starting point, be viewed in light of the said inherent challenges of sustainability.

19.2   The Current Role of IP in Sustainability

The international rapporteur has rightfully so emphasised that IP rights are intrinsically neutral on the issue of sustainability. Sustainability characterised as the public policy aim of today emerged after the creation of the global IP system. It is already clear from this point that IP was not created for a specific role in the matter of sustainability. Even so, its creation does not preclude an emerged role in sustainability as of today.

The fundamental purpose of IP can be summarised to stimulate and protect creative efforts including innovations by enabling persons to protect and individualise their associations, products, and investments.[8] In the Swedish legal tradition, the interests of authors in terms of copyright have in these matters been given a particular emphasis. The interests of authors lie within the making of 'a contribution to the cultural development to the extent his talent warrants'.[9] The same reasoning could be applied to anyone making a contribution to sustainable innovation and

---

[7]M. Rummukainen, Skogens klimatnyttor – en balansakt i prioritering, CEC Syntes Nr 06 2021, p. 28.

[8]U. Bernitz et al, Immaterialrätt och otillbörlig konkurrens, Jure 2017, p. 1.

[9]Government inquiry, SOU 1956:25. Upphovsmannarätt till litterära och konstnärliga verk, p. 85.

improvements. This gives rise to the question of under what circumstances IP can be attributed to a specific role in sustainability work as of today.

19.2.1 Swedish Public Policy on Sustainability

Climate change is consistently prevalent on the political agenda in Sweden. For instance, in 2017, the Swedish Climate Act (SFS 2017:720) was enacted. Among other things, the law stipulates in its fifth section that the government must present political climate action plans every four years. The role of IP rights in relation to combatting climate change is neither mentioned in the preparatory works nor in the act itself.[10] When the Swedish government finalised its first political climate action plan in late 2019, IP rights were also not mentioned.[11] Nevertheless, the action plan sheds some light on the issue of IP and sustainability in terms of public policy: Goods and services affect the climate throughout the whole life cycle—during manufacture, use, re-use, recycle of material, disposal of waste, and transportation. It is important that consumers in Sweden have the possibility to make informed and conscious decisions when buying goods and services. To decrease the emissions related to private consumption, it is important to facilitate for single consumers to consume more sustainably and make it economically viable to make the right choices. The change in consumer behaviour sends important signals to the market for the transitioning to fossil-free production.[12]

The scope of the aim quoted above could encompass the use and direction of IP rights. It can be concluded that the government's emphasis on consumer power and the consumers' access to information for informed decisions is ostensible. Nonetheless, it must be noted that IP rights are clearly omitted.

As mentioned above, the 2030 Agenda may serve as a concrete expression of global sustainability undertakings. When the Swedish Delegation for the 2030 Agenda presented its proposals and assessments for Sweden's implementation of the Sustainable Development Goals in 2019, IP rights were also omitted.[13] The Swedish government did not mention IP rights either when it presented its proposal regarding the implementation to the parliament.[14] The Swedish Intellectual Property Office (Patent- och registreringsverket) made a submission, as is customary in the Swedish legislative process, regarding the proposed implementation.[15] The Swedish

---

[10] Preperatory work, Prop. 2016/17:146. Ett klimatpolitiskt ramverk för Sverige.

[11] Preperatory work, Prop. 2019/20:65. En samlad politik för klimatet – klimatpolitisk handlingsplan.

[12] Ibid., p. 62 (our free translation).

[13] Government inquiry, SOU 2019:13. Agenda 2030 och Sverige: Världens utmaning – världens möjlighet.

[14] Preperatory work, Prop. 2019/20:188. Sveriges genomförande av Agenda 2030.

[15] SEIPO, submission in the Government Office's dossier no. M2019/00661/S of 14 June 2019 (dossier no. AD-411-2019/), p. 2.

19 Sustainability and Intellectual Property in Sweden

Intellectual Property Office (SEIPO) did not criticise the fact that IP rights were omitted in the proposal although the authority observed that IP rights play an important part for the undertakings pursuant to the 2030 Agenda. SEIPO quoted its strategic goals encompassing its pursued sustainability aim (freely translated below) in its submission: By promoting the insight and capacity to make use of intellectual property, SEIPO contributes to a strongly innovative and sustainable society.[16]

SEIPO's position is clearly linking innovation to contributions to a sustainable society. Conclusively, innovation is considered to facilitate sustainability.

19.2.1.1 The Position of the Intellectual Property Office

Part of SEIPO's stated mission is to ensure that 'IP that may be protected should also lead to a sustainable society'. The said part of its mission is, in SEIPO's own words, 'challenging' since the exercise of public authority by the office must remain legally 'neutral'. Swedish public authorities are constitutionally obligated to adhere to a principle of objectivity in their exercise of power pursuant to Chapter 1 Section 9 of the Swedish Instrument of Government (SFS 1974:152). An example of an action contrary to said principle is if an authority makes immaterial considerations in its decision-making.[17] Accordingly, any material changes to the law would diminish the risk of a public authority making immaterial considerations. Although not explicitly stated by SEIPO, a legislative change might be appropriate in this regard to allow differentiated treatment of IP based on sustainability performance. Additionally, SEIPO is legally bound by the government's instruction concerning, e.g., application fees for IP rights. All applicants are, pursuant to the government's current instructions, treated equally and there is no differentiation of fees based on sustainability or size of the business.[18] The lack of public incentives in terms of IP application fees does not impede public financial incentives entirely. Sweden has an innovation agency named Vinnova with the adapted vision to make Sweden 'an innovative force in a sustainable world'. The public authority invests approximately three billion SEK annually in research and innovation.[19] Its issued instructions include to contribute to sustainable growth and the strengthening of Sweden's power to compete by facilitating the use of research and promotion of innovation.[20] Therefore, even if SEIPO cannot incentivise sustainable business models in Sweden, there might be public funding available for such businesses.

SEIPO has itself defined some of its sustainability undertakings. Notably, the Swedish authority considers patent information to be a valuable tool for the

---

[16] Id. p. 2.

[17] T. Bull, Objektivitetsprincipen. In: Marcusson (ed), Offentligrättsliga principer, Iustus 2012, p. 113.

[18] Government Regulation SFS 1967:838.

[19] Vinnova website, Our mission [2022], available at https://www.vinnova.se/en/about-us/vart-uppdrag/. Accessed on 7 July 2022.

[20] Government Regulation SFS 2009:1101, Section 1 para 2.

monitoring of the Sustainable Development Goals. It can for instance, in SEIPO's words, be used as basis for prioritisation and follow-up on progress (at a macro level).[21] One project by SEIPO where IP meets sustainability is a particular collaboration with the Swedish Chemicals Agency. The project involves the processing of patent data with the help of artificial intelligence to, for instance, detect potential hazardous exposure and to monitor trends in different technical fields. If successful, the Swedish Chemicals Agency's aspiration is to ensure more accurate forecasting and to detect potentially new chemical hazards earlier than before. Both SEIPO and the Swedish Chemical Agency can benefit from the implementation of the assessed artificial intelligence methods in their operations according to the description of the joint project.[22]

The international contributions of Sweden in terms of foreign aid to countries and organisations are explicitly mentioned as a sustainability undertaking by SEIPO. The global contribution by SEIPO mainly evolves around the initiative 'WIPO Green'.[23] The World Intellectual Property Organization (WIPO) considers technology solutions for greener growth to already exist. In WIPO's view, that technology must be scaled up and investments must be channelled into the most promising inventions. According to WIPO, technology must also be transferred to those who can use it on the ground and WIPO is altogether working for the promotion of an efficient global market for environmentally friendly technologies. WIPO has under these conditions launched the initiative 'WIPO Green'—a marketplace and 'networking forum'. It holds a database of green technologies and network of partners and experts to help parties looking to commercialise, license, access, or distribute green technology.[24] WIPO Green is an example of IP being used for sustainability according to SEIPO. Moreover, SEIPO, similar to WIPO, argues that patents can facilitate technology transfer since the patents themselves provide a defined scope of transfer.[25] On a final note, it can be mentioned that SEIPO regularly promotes sustainable businesses in its communication as examples of successful utilisation of IP rights.[26]

---

[21] SEIPO, submission in the Government Office's dossier no. M2019/00661/S of 14 June 2019 (dossier no. AD-411-2019/1012), p. 4.

[22] The Swedish Chemicals Agency, PM 1/22. Dataanalys av patentinformation med hjälp av artificiell intelligens.

[23] SEIPO, submission in the Government Office's dossier no. M2019/00661/S of 14 June 2019 (dossier no. AD-411-2019/1012).

[24] WIPO, WIPO and the Sustainable Development Goals, reference no. 1061E/2021.

[25] SEIPO, submission in the Government Office's dossier no. M2019/00661/S of 14 June 2019 (dossier no. AD-411-2019/1012), p. 4.

[26] See, e.g., descriptions of sustainable training wear (SEIPO [2022] website, available at https://www.prv.se/sv/foretagare/foretagare-berattar/casall/. Accessed on 1 July 2022), digital platforms for training and management of sustainable business strategies (SEIPO [2022] website, available at https://www.prv.se/sv/foretagare/foretagare-berattar/pure-act/. Accessed on 1 July 2022), and renewable energy solutions (SEIPO [2022] website, available at https://www.prv.se/sv/foretagare/foretagare-berattar/absolicon%2D%2Dfran-sol-till-fjarrvarme/. Accessed on 1 July 2022).

19.2.2 European Public Policy on Sustainability

The European Commission has made similar considerations as the Swedish government in its action plan for the Circular Economy. The Commissions states that 'the choices made by millions of consumers can support or hamper the circular economy'. In addition, the Commission considers price to be a key factor affecting purchasing decisions and thereby encourages Member States to incentivise, using economic instruments such as taxation, to ensure that product prices reflect environmental costs.[27] Importantly, the Commission's action plan of 2015 does not mention IP at all. According to the Commission's later action plan of 2020, the regime for intellectual property needs to be fit for the digital age and the green transition and support EU businesses' competitiveness. The Commission will propose an Intellectual Property Strategy to ensure that intellectual property remains a key enabling factor for the circular economy and the emergence of new business models.[28]

It can be noted that, although the Commissions emphasises the role of IP in the green transition, there is no evaluation whether IP is fit for that purpose. The Intellectual Property Strategy, as mentioned in the introduction above, is yet to be proposed at the time of writing and its content is not publicly known.

19.2.3 International IP Policy on Sustainability

The basis for the modern IP system can be found in two international agreements; the Paris Convention for the Protection of Industrial Property of 1883, last amended on 28 September 1979 (the Paris Convention) and the Agreement on Trade-Related Aspects of Intellectual Property Rights, adopted in Marrakesh on 15 April 1994 (the TRIPS Agreement). Article 1 of the Paris Convention defines 'Industrial Property' (i.e. IP). From the wording of said article, IP should be understood in the broadest sense and encompasses patents, utility models, industrial designs, trademarks, service marks, trade names, indications of source or appellations of origin, and the repression of unfair competition. All of the mentioned rights except for repression of unfair competition are fundamentally proprietary rights. Neither the preamble nor any article of the Convention contains any reference to sustainability. It can easily be concluded that the creation of the foundation of the IP system did not describe any form of IP as having a role in sustainability.

The TRIPS Agreement does, unlike the Paris Convention, contain one reference to the environment. Under Article 27(2) of the TRIPS Agreement, it is possible to exclude inventions from patentability if the exclusion is necessary to, inter alia,

---

[27] European Commission, Communication from the European Commission to the European Parliament, the Council, the European Economic and Social Committee and the Committee of the Regions of 2 December 20215, COM(2015) 614 final, p. 6.

[28] European Commission, Communication from the European Commission to the European Parliament, the Council, the European Economic and Social Committee and the Committee of the Regions of 11 March 2020, COM(2020) 98 final, para 6.3.

avoid serious prejudice to the environment. The notion of 'prejudice to the environment' in this regard can be seen as a clarification that such acts constitute an infringement of *ordre public*.[29] The exclusion of inventions contrary to *ordre public* and public moral from patentability is also stipulated in Chapter 1 Section 1 c of the Swedish Patent Act (SFS 1967:837). The environment is not mentioned as an example of *ordre public* although the list is considered non-exhaustive.[30] The general exclusion from patentability under Swedish law does not entail that SEIPO shall examine the practical use of a certain patent. Instead, the exception can only be applied to inventions that per se cannot be used in any other way than contrary to law and public morality. A widely cited example of this exception from patentability under Swedish law is equipment for conducting torture.[31] There is no example in Swedish case law of an invention being deemed non-patentable due to its severe environmental impact. Nevertheless, it is clear that the TRIPS Agreement indeed does enable such considerations by patent offices.

19.2.3.1 Technology Transfers

The signatories to the TRIPS Agreement consist of countries across all continents with profound differences in welfare. It is already from the preamble of the TRIPS Agreement apparent that IP right are 'private rights' and that the agreement must recognise the 'special needs' of least-developed member countries. Under Article 66(2) of the TRIPS Agreement: Developed country Members shall provide incentives to enterprises and institutions in their territories for the purpose of promoting and encouraging technology transfer to least-developed country Members in order to enable them to create a sound and viable technological base.

Technology transfers are, in other words, encouraged to less developed countries subject to the TRIPS Agreement. The issue of the TRIPS Agreement and transfer of technology specifically for the benefit of the environment was first raised by Ecuador to the Council for Trade-Related Aspects of Intellectual Property Rights:[32] For the assertion that 'intellectual property rights' guarantee the promotion of innovation and promote the timely and widespread dissemination of the industrial applications of such innovations is questionable:[24] [sic!] What is clear, however, is that for many countries, especially the most vulnerable countries in which ESTs (environmentally-sound technology) are most needed for adaptation and/or mitigation of climate change, it may plausibly be asserted that the patent system as currently designed can restrict the dissemination of such technologies through monopolisation or abuse

---

[29]Å. Hellstadius, A Quest for Clarity – Reconstructing Standards for the Patent Law Morality Exclusion, Stockholm University 2015, p. 163.

[30]Preperatory work, Prop. 2003/04:55. Gränser för genpatent m.m. - genomförande av EG-direktivet om rättsligt skydd för biotekniska uppfinningar, p. 143.

[31]M. Levin and Å. Hellstadius, Lärobok i immaterialrätt, Norsteds juridik, 2019, p. 287.

[32]Communication of Ecuador to the Council for Trade-Related Aspects of Intellectual Property Rights of 27 February 2013, IP/C/W/585.

of exploitation rights by right holders or excessive additional costs resulting from payment of royalties for voluntary licensing of ESTs.[33]

Since Ecuador raised the issue before the TRIPS Council, it has been discussed on several occasions. The WTO emphasises in this regard that the TRIPS Agreement aims to contribute to promoting technological innovation and its transfer and dissemination.[34] Nonetheless, the criticism by Ecuador concerning IP being used for monopolisation is highly relevant and a recurring topic of debate. Similar criticism of the monopolisation of inventions particularly due to patents have been expressed by countries in relation to equipment necessary to combat the COVID-19 pandemic.[35] One role of IP accordingly appears to be the facilitation of transfer of vital technology—especially between countries with different economic capabilities.

19.2.4 The Current Role of IP in Sustainability

In light of the above, the discussions on IP in relation to sustainability appears to be nearly non-existent in Sweden, a bit more prevalent in the European context, and mostly discussed in the international context with reference to technology transfer. One way of assessing the matter of IP's role is to conclude that the regulators do not appear to treat IP as having a specified role in the green transition other than as a facilitation of innovation. As such, technology transfers are clearly emphasised as being of pertinence in relation to sustainability. Technology transfers pertain to the exercise of IP rights rather than the holding of IP rights. This distinction is vital in the way that the holding itself is neutral although the application may not be. IP rights (except for unfair competition) are ultimately, as clearly stated in the Paris Convention, proprietary rights of exclusive nature. It cannot be assumed that the holding of proprietary rights per definition entails an enforcement of such rights in a way that precludes others from exploiting such rights. As such, it is rather the exercise of IP rather than the holding of IP that affects the facilitation of technology transfer. Derived from the perspective that IP rights are proprietary rights (i.e. awarded to a holder), IP rights are neither created nor construed as having any certain role in sustainability. This is further supported by the fact that if an invention is detrimental to the environment, it may nonetheless be patented and when the patent protection ultimately expires, the invention will remain detrimental to the environment. In other words, the existence of IP is in this regard completely immaterial—it neither enhances nor decreases the sustainability of the invention. It is also immaterial whether the holder of the IP is the original owner or controller (e.g. exclusive licensor).

---

[33] Ibid, para 19.

[34] World Trade Organization (WTO) website (2022), Climate change and TRIPS, available at https://www.wto.org/english/tratop_e/trips_e/cchange_e.htm. Accessed on 10 July 2022.

[35] Communication of India and South Africa to the Council for Trade-Related Aspects of Intellectual Property Rights of 2 October 2020, IP/C/W/669.

On a final note, the issue raised in relation to the TRIPS Agreement regarding the monopolisation of innovation due to IP could be further addressed. When the monopolising aspect of IP generally is questioned, it is not the proprietary nature itself that is criticised but rather the abuse or exploitation of IP in the way that it limits access to innovation. This inherent conflict and the need for balancing competing interests thereof is indeed relevant. However, within the Swedish perspective, access to IP is overall well-functioning. This can likely be attributed to the Sweden's relative affluence. From a national level, the monopolising aspect of IP is not a significant issue in terms of sustainability in Sweden. Even if the monopolisation can be criticised for possibly inhibiting the transfer of technology, it should be emphasised that IP itself consolidates information that enables effective transfers of technology. Also, the way IP consolidates information enables for data insights and monitoring of environmental performance as is the case with the collaboration between SEIPO and the Swedish Chemicals Agency mentioned above. Without the existence of patents and the patent registry thereof, such extensive collaboration would not be possible.

19.3 IP's Contribution to Unsustainability

Since IP rights, as described above, are considered to contribute to innovation and thereby to sustainability, it can be questioned if IP also could pose as a barrier to innovation. This could, for instance, be the case if the characteristics of certain IP per se hinders transfer of technology in some regard.

19.3.1 Barriers to Innovation

The effective transfer of technology and innovation is a pillar of the modern IP system and a possible and inherent obstacle to technology transfer is trade secrets. Trade secrets are protected under the Trade Secrets Act (SFS 2018:558) in Sweden. Swedish law does not stipulate any date of expiry for trade secrets.[36] The act transposes the European Directive (EU) 2016/943 which also does not stipulate any date of expiry.[37] Before the directive was transposed, Swedish law stipulated that trade secrets that have been subject to any form of litigation could only be kept secret for a period of 20 years. Under the current law, there is no such limitation at all.[38] Trade secrets are not mentioned per se in the Paris Convention although the

[36] Swedish Supreme Court, judgment of 28 June 1999, reference NJA 1999 s. 469, Nordbanken v NS.

[37] Directive (EU) 2016/943 of the European Parliament and of the Council of 8 June 2016 on the protection of undisclosed know-how and business information (trade secrets) against their unlawful acquisition, use and disclosure, OJ L 157/1.

[38] Preperatory work, Prop. 2017/18:200. En ny lag om företagshemligheter, p. 104.

19 Sustainability and Intellectual Property in Sweden

provisions on unfair competition (Article 10bis) include trade secrets in a broad sense.[39] The characteristics of trade secrets deviate from traditional IP rights in the way that it is not an exclusive proprietary right since several companies may hold the same trade secrets independently. Furthermore, the information constituting trade secrets is not exclusive although the holding of trade secrets entails a protection from unlawful acquisition, use, and disclosure.[40] Trade secrets are nevertheless largely treated as a proprietary right—at least in the sense that trade secrets are licensed and accounted for as valuable IP. Trade secrets are in principle valid in perpetuity and can effectively hinder technology transfer—which is a prerequisite for spreading innovation in many cases. If a patent expires, the transfer of the protected invention to third parties may not occur in the absence of technology transfer (as widely seen in the pharmaceutical sector). In other cases, the publication of patents can limit the possible scope of trade secrets thus warranting the transfer of technology once the patent expires. There is nevertheless a risk that a perpetual protection of trade secrets is far-reaching and can constitute a barrier to technology transfers.

Although technology and innovation underpin the modern IP framework, there is one apparent exception to that form of reasoning. Protected geographical indications are basically, especially in the European context, treated as IP.[41] The current European Regulation on protected geographical indications for agricultural products and foodstuff as of today effectively protects the commercial value of traditional means of production.[42] These types of rights protect local traditions rather than innovation. Protected geographical indications can in that regard be viewed as the antithesis of otherwise innovation-friendly IP. From the perspective of the environment, it can be emphasised that there are no environmental requirements imposed for eligibility for protection as geographical indication.[43] If a producer would deviate from any specified traditional method of production in favour of a more environmentally sound alternative, the protected geographical indication cannot be used.

19.3.2 Corrective Measures in Cases of Infringement

Aside from the holding of IP, the enforcement of IP may in a limited regard cause detriment to the environment. In the event of infringement, IP rights can pursuant to EU law result in the destruction of fully functional goods if it is ordered as a

---

[39] Cf. TRIPS Agreement, Article 39(1).

[40] C. Wainikka, Företagshemligheter – en introduktion, Studentlitteratur 2010, p. 13.

[41] CJEU, case C-159/20, Commission v Denmark (AOP Feta), ECLI:EU:C:2022:561, paras 51–52.

[42] Regulation (EU) No. 1151/2012 of European Parliament and of the Council of 21 November 2012 on quality schemes for agricultural products and foodstuff, OJ L 343/1, Recitals 1-2. Also see the definition of 'Traditional' in Article 3(3) that necessitates that usage must be proven for at least 30 years.

[43] Ibid., Article 5.

corrective measure due to the infringement.[44] Destruction as a corrective measure is also warranted by the TRIPS Agreement.[45] When the EU directive was transposed in Sweden, no environmental considerations were made in relation to destruction of infringing goods.[46] From an environmental point-of-view, it can easily be argued that the destruction of fully functional goods is a major waste of resources and such destruction would have been impossible without IP law. As the law is implemented in Sweden, there is no clear legal foundation for any court to make any environmental considerations while determining whether to order the destruction of infringing goods. It could possibly be argued that such considerations can be made within the framework of proportionality. Furthermore, whenever destruction as a corrective measure is ordered by a Swedish court, the form of destruction is not decided by the court. Instead, the court has to decide whether the infringing party shall bear the costs for the destruction.[47] Accordingly, there are no rules in place ensuring that infringing goods are recycled or otherwise destroyed in a way that is most gentle on the environment.

19.3.3 The Neutrality of IP vis-à-vis Sustainability

As already stated above, the holding of IP itself is immaterial to environmental considerations. Some aspects of the exercise of IP may per se be at risk hampering innovation (which, as already described, appears to be premier role of IP in sustainability). Trade secrets can restrict the transfer of technology and are not limited in time. At the same time, trade secrets are not exclusive proprietary rights and do not preclude reverse engineering thus enabling competitors to utilise the same protected know-how. Moreover, the existence of patents entails that some know-how inevitably becomes excluded from the scope of trade secrets. Know-how that is not patented and constitutes trade secret can be protected in perpetuity which may be questioned in terms of facilitating transfer of technology. In that sense, there is a risk that the lack of date of expiry for trade secrets may impair transfer of technology and accordingly being negative for innovation and sustainability. Aside from this point, IP must be considered welcoming towards innovation in the way that progress and improvements are never foreign to protection. Innovation is not per se positive for sustainability. However, the introduction of cross-industry environmental improvements while preserving industrial production necessitates innovation in some form. The one exception where IP intrinsically is not welcoming towards

---

[44] Directive 2004/48/EC of the European Parliament and of the Council of 29 April 2004 on the enforcement of intellectual property rights, the destruction of infringing goods can be ordered as a corrective measure, OJ L 195/16, Article 10(1)(c).

[45] TRIPS Agreement, Article 46.

[46] Preparatory work, Prop. 2008/09:67. Civilrättsliga sanktioner på immaterialrättens område - genomförande av direktiv 2004/48/EG, p. 256.

[47] See, e.g., Swedish Patent and Market Court of Appeal, judgment of 22 March 2019, case no. PMT 5885-18, Daniel Wellington AB v Ur & Penn AB.

19 Sustainability and Intellectual Property in Sweden

innovation is protected geographical indications. If innovation in the conservative sector of agricultural products and foodstuff should be incentivised, the legislative framework should accommodate environmental improvements in production. As of today, it cannot be ruled out that protected geographical innovations impair the possibility of improvements in production methods hence becomes a vivid example of how IP can be negative for sustainability. Another example of IP being negative for sustainability is the destruction of infringing goods. Such negative environmental effects associated with the destruction of infringing goods could easily be mitigated if the courts would be obligated to, at least to some extent, adhere to the European waste hierarchy.[48]

19.4 Empowerment of Consumer Choices

Even if the holding of IP itself is neutral on the topic of sustainability, the exercise of IP may nonetheless be subject to legal consequences at least indirectly promoting sustainability. Based on the considerations of both the Swedish government and the European commission, the choices of consumers have an incredible potential to contribute to the green transition. In this regard, the practice of greenwashing can be detrimental to the consumers' abilities to make informed decisions and to the consumers' role in the green transition entirely. Greenwashing raises the issue of unfair competition which is not a regular IP right of proprietary nature. Notwithstanding, it is indeed often treated as IP and is also mentioned in both the Paris Convention and TRIPS Agreement.

19.4.1 Sustainability in Unfair Competition

Greenwashing has become a highly relevant topic lately. The European Commission has in a recent proposal decided to define greenwashing as the making of a misleading *environmental claim*.[49] In Article 1 of the Commission's proposed directive, the following definition is provided: '(o) "environmental claim" means any message or representation, which is not mandatory under Union law or national law, including text, pictorial, graphic or symbolic representation, in any form, including labels, brand names, company names or product names, in the context of a commercial communication, which states or implies that a product or trader has a positive or no impact on the environment or is less damaging to the environment than other products or traders, respectively, or has improved their impact over time'.

---

[48] Directive 2008/98/EC of the European Parliament and of the Council of 19 November 2008, OJ L 312/3, Article 4.

[49] European Commission, Proposal for a Directive of the European Parliament and of the Council amending Directives 2005/29/EC and 2011/83/EU as regards empowering consumers for the green transition through better protection against unfair practices and better information of 30 March 2022, COM(2022) 143 final, p. 17.

A few observations should be made from the mentioned proposal. First, it is evident that the EU regulator views greenwashing as an issue necessitating legislative attention. Second, it should be noted that 'environmental claim' is a term encompassing the use of elements that also may constitute proprietary IP rights (such as labels, brand names, etc.). In other words, the exercise of IP rights may be restricted if the exercise is considered to misleadingly communicate an environmental benefit. In Sweden, this is not only the case with reference to the proposed EU legislation. For instance, the Swedish Patent and Market Court held in 2021 that the use of the prefix 'ECO' in a registered trademark used in marketing is considered an environmental claim. The court found the defendant's explanation that the prefix 'eco' referred to economic benefits to be immaterial in relation to the average consumer's perception of the marketing as implying environmental benefits.[50] The European Commission has also stressed that trademarks cannot be used to circumvent the requirements when communicating environmental claims.[51] This is of particular significance since the amount of trademarks containing environmental references is clearly increasing.[52]

With reference to the above, the utilisation of trademark rights may in practice be inhibited when used in marketing based within the scope of unfair commercial practices (also known as unfair competition law or marketing law). The Unfair Commercial Practices Directive[53] is transposed in the Swedish Marketing Act (SFS 2008:486). Contrary to the Unfair Commercial Practices Directive, the Marketing Act does not only apply to commercial practices unfair towards consumers only, as stipulated in Article 3(1) of the Directive, but also to practices unfair towards other businesses, i.e. as a rule competitors.[54] In Sweden, any trader affected by the unfair business practice has *locus standi* and competitors may accordingly bring proceedings against each other due to unfair business practice.[55]

19.4.1.1 Environmental Benchmarking and Labels

The environmental footprints of goods and services are highly difficult to assess in the absence of impartial benchmarking to facilitate comparisons. One way of benchmarking is through the use of independent third-party audits and authorisations that use a standardised benchmarking framework to assess environmental impacts of goods and services. Any misappropriation of such authorisations may constitute

---

[50] Swedish Patent and Market Court, judgment of 22 June 2021, case no. PMT 2976-20, Konsumentombudsmannen v Ecoswe AB.

[51] European Commission, Guidance on the interpretation and application of Directive 2005/29/EC of the European Parliament and of the Council concerning unfair business-to-consumer commercial practices in the internal market, OJ C 526/1, para 4.1.1.3.

[52] European Union Intellectual Property Office (EUIPO), Green EU trade marks – Analysis of goods and services specifications 1996-2020, 2021, doi: 10.2814/900650, p. 19.

[53] Directive 2005/29/EC of the European Parliament and of the Council of 11 May 2005, OJ L 149/22.

[54] Swedish Marketing Act (SFS 2008:486), Section 1.

[55] Ibid., Section 47.

infringement of the third party's trademark rights. SEIPO has explicitly mentioned such use of trademark rights as a way that IP may be used to support the commercialisation of sustainable goods and services.[56] Aside from trademark law, such misappropriation can easily be remedied as an unfair commercial practice. Under Section 10 para 2 of the Swedish Marketing Act, a trader may not, among other things, mislead consumers with reference to its own or any other trader's qualifications. This includes the use of independent authorisations. For instance, if an authorisation is revoked and is nevertheless subsequently used by a trader, such marketing is misleading.[57] Annex 1 to the Unfair Commercial Practices Directive contains commercial practices which in all circumstances are considered unfair. Section 4 of the Swedish Marketing Act stipulates that said Annex 1 is directly applicable under Swedish law. Articles 3 and 4 of the Annex clearly prohibit false use of approvals, endorsements, or authorisations. An overall purpose of the directive in this regard is to prevent traders from unduly exploiting the trust which consumers may have in self-regulatory codes.[58] In other words, environmental authorisations are duly protected under unfair competition law. Misleading use of environmental claims are considered particularly serious violations of unfair competition law.[59]

19.4.1.2 Burden of Proof in Marketing Communications

It is well established under Swedish law that a trader must be able to substantiate each communicated factual claim used in advertisements (also known as the *reversed burden of proof*). The principle is not explicitly laid out in Swedish law although it is by all means established in case law.[60] The reversed burden of proof can also be derived from Article 12(a) of the Unfair Commercial Practices Directive. When communicating environmental claims, the facts used by the trader must prevail under a particularly high degree of scrutiny to satisfy the reversed burden of proof.[61] A trader's communication of a general or vague environmental claim entails a burden of proof covering all interpretations of the marketing in the eyes of the average consumer.[62] Documentation to substantiate an environmental claim

---

[56] SEIPO, submission in the Government Office's dossier no. M2019/00661/S of 14 June 2019 (dossier no. AD-411-2019/1012), p. 4.

[57] Swedish Market Court, judgment of 20 April 2006, reference MD 2006:10, Centrala Antennföreningen CANT Aktiebolag v D. R. / Skaraborgs antenn o satellit.

[58] European Commission, Guidance on the interpretation and application of Directive 2005/29/EC of the European Parliament and of the Council concerning unfair business-to-consumer commercial practices in the internal market, C 526/1, para 2.8.4.

[59] Swedish Patent and Market Court of Appeal, judgment of 25 March 2022, case no. 13193-20, Hunton Fiber AB & Hunton Fiber AS v Föreningen Swedisol.

[60] Preperatory work, Prop. 1994/95:123. Ny marknadsföringslag, p. 153.

[61] Swedish Market Court, judgment of 7 May 2004, reference MD 2004:12, Konsumentombudsmannen v Ford Motor Company Aktiebolag.

[62] Swedish Market Court, judgment of 1 June 2011, reference MD 2011:13, KTF Organisation Aktiebolag v NPT Sweden AB.

should be available as soon as the claim is used in marketing.[63] Certain know-how underpinning a trader's environmental performance can be subject to secrecy in court proceedings and records.[64] Under the basic principles of procedural law, the secrecy cannot limit the litigants' access to information in an ongoing trial.[65] It could be feared that the transparency hereto can enable abuse of unfair competition law in the way that competitors may attempt to access each other's trade secrets due to the reversed burden of proof. Even if information disclosed in litigation can be kept confidential by the courts, it is inevitable that the claimant will access the confidential information in the course of the proceedings and the adversarial process thereof. This could potentially incentivise abuse of litigation due to the claimant's granted access to the defendant's otherwise confidential information warranted by the reversed burden of proof. Such fears nevertheless appear to be unfounded in the Swedish context. Instead, the reversed burden of proof serves as a solid obstacle to greenwashing.

19.4.2 Facilitation of the Circular Economy

As mentioned above, consumption is inherently associated with the use of resources. The only way to circumvent the use of resources while maintaining consumption would be to reuse as much goods as possible instead of producing new ones. Restrictions on resale of used goods could limit consumers' access to used goods in a way that could impair the circular economy. Luckily, the exclusive right to sale of goods subject to IP rights is exhausted after the goods have been put on the European market (with the consent of the rightholder). The exhaustion doctrine can be categorised as a demarcation of IP protection in favour of the free movement of goods and services. The underlying principle of Swedish law is, in accordance with EU law, that products or services that have been exhausted in relation to IP may also be sold and advertised in the European market.[66] The underlying reason for exhaustion is, rather self-explanatory, to diminish the risk of complete monopolisation with regards to production, sales, and repairs through the holding of IP.[67] Different licenses regarding use of IP rights in general may be in conflict with competition law if IP rights are exercised to an extent surpassing what is legally permissible.[68] Furthermore, the overall consumption by consumers may also be reduced if goods have a long lifespan. Inevitably, most goods break at some point. There is no explicit

---

[63] European Commission, Guidance on the interpretation and application of Directive 2005/29/EC of the European Parliament and of the Council concerning unfair business-to-consumer commercial practices in the internal market, OJ 2005 C 526/1, para 4.1.1.5.

[64] Swedish Public Access to Information and Secrecy Act (SFS 2009:400), Chapter 36 Section 2.

[65] Similar considerations can also be found in Article 43(1) of the TRIPS Agreement.

[66] S. Arnerstål, Varumärkesanvändning, Norstedts Juridik 2018, p. 170.

[67] L. Pehrson, Varumärken från konsumentsynpunkt, Liber 1981, p. 338.

[68] Preperatory work, Prop. 1992/93:56. Ny konkurrenslagstiftning, p. 70.

right to repair in Sweden and the issue has been acknowledged, inter alia, in a governmental inquiry on the advancement of the circular economy. The inquiry stated that it is common in the IT-sector to limit access to codes, instructions and the same, and to only permit access to authorised parties.[69] However, the inquiry did not make any suggestions for amendments to the current legislation in this regard. In addition to securing a right to repair, it would indeed be beneficial for the circular economy to facilitate a well-functioning market for spare parts for goods. Technological innovation such as 3D-printing enables cost efficient production of spare parts that arguably could prolong the lifespan of sold goods. The production of spare parts without the consent of the rightholder will in many cases constitute infringement of IP rights. This gives rise to the question whether restrictions on the enforcement of IP rights should be introduced as further discussed below.

19.4.3 Consumer Choices in a Successful Role for Sustainability

The green transition naturally necessitates investments and significant amount of the costs are incurred on businesses. Ultimately, the business of all businesses is business. If green investments are profitable, the green transmission is ultimately doable. However, if there is a general narrative dominated by greenwashing, the sincere green investments may go unnoticed among the consumers. Unfair competition can in this regard serve as a mechanism that hinders competitors from misappropriating the use of sustainability without the necessary investments being made. The fact that competitors may litigate unfair commercial practices can further safeguard the markets from greenwashing. An observation that can be made from Swedish case law is that litigation regarding misleading environmental claims to a large extent has been initiated by competitors. Unfair competition, if viewed as a form of IP, can ensure that unsustainable businesses do not misrepresent their own environmental performance hence incentivising businesses to address sustainability properly. This necessitates that unfair competition is duly enforced. The efficient enforcement of unfair competition is a clear example of how IP can contribute to sustainability. Moreover, the limitation of the exclusive rights associated with IP due to exhaustion ensures that goods can be re-used for the benefit of the circular economy. An explicit right to repair could benefit the circular economy further.

---

[69]Government inquiry, SOU 2017:22. Från värdekedja till värdecykel – så får Sverige en mer cirkulär ekonomi, p. 181.

19.5 Discussion on Possible Improvements in IP for Encouraging Sustainability

This chapter has up to this point mainly discussed if IP has a role in sustainability and the proprietary nature of IP's effects on sustainability. A closely related question hereto is whether IP should have any specific role at all in sustainability. Given that IP's close relationship with innovation is established, the question regarding IP's role in sustainability implies that sustainability in some regard deviates from the general concept of innovation. The existence of deviations in this regard is likely contingent on the nature of each industry sector. Different industries simply have different inherent challenges in relation to sustainability and the relationship between sustainability and IP is arguably equally affected by the nature of such inherent challenges. Additionally, the fact that innovation is closely associated with economic growth can itself serve to distinguish an inherent clash between IP and sustainability. This in the sense of a harsh reality where the modern economy basically necessitates growth and growth itself necessitates the consumption of natural resources and gives rise to emissions. Nonetheless, within the aspect of holding IP, there is no such clear reason to distinguish sustainability from innovation in general. If it were to be distinguished, it could be argued that a *sui generis* type of IP should be introduced specifically for the promotion of sustainability. However, since there is no need for such distinction, there is accordingly no vacuum that needs to be with the introduction of an additional proprietary right for the protection of sustainable innovation. As such, there is no clear need to introduce a sui generis IP right for sustainability.

IP rights may have contributed to environmental decay in the sense that the most damaging innovation at some points have been patented. When the patents have expired, the products created thereof continue to be damaging. In that sense, it could be argued that the exclusivity of the patents may have restricted the spread of innovation for some time thus slowing down its damaging effects. When there is a substitute to the invention, such improvements may again be patented. The patent for the substitute could restrict its distribution thus slowing down the introduction of the improved substitute. One might then argue that all IP rights in relation to substitutions and improvements having positive environmental impacts should be confined to allow a swifter green transmission. In this regard, the need for protection of IP rights as a driving force for innovation should be properly balanced with the growing need for transfer of sustainable innovation.

19.5.1 How IP Rights Can Better Contribute to Sustainability

The introduction of sustainability in IP entails new complexities and challenges to an already dynamic and complex area of the law. IP being neutral in relation to the subject-matter of innovation could herewith also be one of its virtues. The IP system is old and well-established and cannot be swiftly changed materially. Unfair competition does play a major role for IP in relation to sustainability and further enforcement serves to benefit the environment. The major inherent obstacles to sustainability identified in this chapter are protected geographical indications and the destruction of infringing goods without consideration of the waste hierarchy. Aside from this, any material changes to the IP system would entail changes to international law and require significant considerations. As described in the background above, sustainability cannot be easily defined and different aspects of sustainability may be in conflict in relation to each other. It is in this regard not appropriate to affix certain aspects of sustainability to IP when such aspects may prove to be less efficient or relevant in the future. Therefore, any direction of IP to support sustainable or to refrain from supporting unsustainable innovation should be considered with utmost concern. Such direction in terms of public policy is easily achieved with financial incentivisation which more effortlessly can be evaluated and tweaked than IP. Swedish public policy is falling behind in this regard due to lack of financial incentivisation. Moreover, the incentivisation of technology transfer and promotion of other instruments than IP such as open source could be introduced. If such progress is to be incentivised, the Swedish regulators would need, as a starting point, to promote and investigate the potential of IP for sustainable development. That is not the case as of today.

19.5.1.1 An Introduction of Conditions for Acquiring and Subsisting of IP

If any IP would need to be evaluated on sustainability before it is acquired, such requirement would in principle necessitate that IP offices assess a particular IP right *a priori,* i.e. during the examination procedure. It would be a heavily intricate task to assess whether specific goods or services in part or in whole can be considered sustainable. The workload of the offices would easily become extensive and likely give rise to a significant volume of litigation thereto. Therefore, it could be much more viable to examine IP *a posteriori,* i.e. once it has been created and commercially utilised. The effectiveness of imposing environmental conditions for subsisting IP rights can nevertheless be questioned. If, for instance, a patent is invalidated due to its environmental detriment, it does not effectively stop the use of the invention. It would be more efficient to prohibit the use of the invention itself rather than the IP underpinning it. Any invalidation or revocation could also backlash in the way that third parties could start utilising the invention without the risk of infringement hence contributing to a more widespread use of the environmentally detrimental invention. Accordingly, imposing conditions pertaining to sustainability on patents in order to subsist is likely not appropriate.

One IP right that appears to be more suitable for invalidation or revocation due to unsustainability is trademarks. As mentioned above, the use of green terms in trademarks is increasing. A feasible way of incentivising the proper use of environmental symbols and claims (thus mitigating greenwashing) could be to stipulate a requirement similar to the doctrine of genuine used in trademark law. Under Chapter 3 Section 2 of the Swedish Trademarks Act (SFS 2010:1877), a five year or older trademark may be invalidated if it has not been used for the goods and services it was registered for. The regulation transposes Articles 16 and 19 of the

Trademarks Directive.[70] It is the trademark proprietor that bears the burden of proof concerning genuine use of the trademark.[71] A similar requirement for the accurateness of environmental claims used in trademarks would not bestow any heavy administrative burden on the examining authority. It would also entail a grace period to properly implement and assess the environmental impact of the goods and services of the trademark. Such requirement would essentially be in line with the general burden of proof under Swedish marketing law concerning the use of environmental claims. Additionally, such requirement would be in line with the general aim of helping consumers make informed and conscious decisions. At the same time, it might be disproportional to add such ground for invalidity since the use of the trademark nevertheless would, in principle, be subject restrictions on environmental claims under unfair competition law. Adding an additional ground for invalidation or revocation can also be in conflict with the TRIPS Agreement and Paris Convention.[72] Regardless, since unfair competition law already entails a burden of proof on the trader when communicating environmental claims, a case regarding unfair competition could be tried in conjunction with the use of environmental trademarks. Even if enforcement of unfair competition law has great potential in terms of facilitating sustainable development, the risk of losing a trademark may serve as an additional incentive for compliance and improve the consumers' overall trust in environmental labels. In Sweden, cases of unfair competition and trademark law are both litigated before the patent and market courts.[73] Accordingly, there are well-founded reasons to introduce a new ground for invalidation or revocation specifically for trademarks misrepresenting environmental performance and to try such cases in conjunction with unfair competition law.

19.5.2 Possible Restrictions on Enforcement of IP in Favour of Sustainability

If an environmentally detrimental product is protected by IP, any improvements of the product could theoretically constitute infringement—particularly with regard to patents. Whenever an improvement of a pre-existing patented invention in terms environmental performance is assessed in terms of infringement, the assessment could incorporate a favourable perspective for the incentivisation of improvements. In other words, some forms of infringements could be justified based on the environmental improvement emanating thereof. This could incentivise research into improvements of innovation by mitigating the risk that such improvements

---

[70] Directive (EU) 2015/2436 of the European Parliament and of the Council of 16 December 2015, OJ L 336/1.

[71] CJEU, case C-183/21, Maxxus Group GmbH & Co. KG v Globus Holding GmbH & Co. KG, ECLI:EU:C:2022:174, para 36.

[72] See, in particular, Article 15(2) of the TRIPS Agreement.

[73] Swedish Act on Patent and Market Courts (2016:188), Section 4.

would constitute infringement. This line of reason could perhaps also be applied to spare parts and substitutions of unsustainable parts of inventions. As already stated, the need for protection of IP rights as a driving force for innovation should be properly balanced with the growing need for transfer of sustainable innovation. Non-waivable provisions governing IP licensing in favour of the environment could be imposed. This could, for instance, be applied in relation to spare parts for the benefit of the circular economy. From a Swedish perspective, such legislative additions would need to be addressed within the context of the EU and also in the international arena to enable transfer of technology to less affluent countries. The extensive considerations necessary to draft a conclusive proposal with the proper balancing of interests thereto cannot be done to any satisfactory extent in this chapter. Nevertheless, these preliminary suggestions may be further assessed and serve as basis for further considerations.

19.6 Concluding Remarks

In this chapter, it is concluded that the standing of IP in relation to IP is a complicated issue as it is. Consequently, suggestions for improvements are equally—if not even more—complicated. The assessed room for improvement in this chapter and the suggestions thereto should be viewed as talking points for further discussions rather than anything of conclusive nature. The challenges of global warming and the green transition are truly global and pressing. To overcome these immense challenges, all of society should partake. The role of IP is not given and IP practitioners should seize the opportunity to construct a solid narrative concerning the role of IP. In all circumstances, the changes necessary for the green transition will affect IP and the absence of IP practitioners in the discourse risks resulting in unfavourable outcomes. The invitation to discuss this issue in this comparative context based on different jurisdictions is accordingly highly welcomed and needed.

Open Access This chapter is licensed under the terms of the Creative Commons Attribution 4.0 International License (http://creativecommons.org/licenses/by/4.0/), which permits use, sharing, adaptation, distribution and reproduction in any medium or format, as long as you give appropriate credit to the original author(s) and the source, provide a link to the Creative Commons license and indicate if changes were made.

The images or other third party material in this chapter are included in the chapter's Creative Commons license, unless indicated otherwise in a credit line to the material. If material is not included in the chapter's Creative Commons license and your intended use is not permitted by statutory regulation or exceeds the permitted use, you will need to obtain permission directly from the copyright holder.

Sustainability and Intellectual Property in Switzerland

20

Eugenia Huguenin-Elie

20.1 Introduction

Threatened by climate change and suffering from inequalities, our planet and the humanity need sustainable development. '*A world in which poverty and inequity are endemic will always be prone to ecological and other crises. Sustainable development requires meeting the basic needs of all and extending to all the opportunity to satisfy their aspirations for a better life*'.[1] Sustainable development, as a meta-project, aims at developing solutions while securing actual development without causing harmful consequences to the environment and the humankind to offer, ultimately, the ability of future generations to meet their own needs.[2]

Adopted by all 193 UN member states on 25 September 2015 and entered into force in 2016, the 2030 Agenda for Sustainable Development ('**UN 2030 Agenda**') and its 17 Sustainable Development Goals ('**SDGs**') constitute a major milestone for sustainable development: they set forth the new global and universally applicable frame of reference for sustainable development that shall include social, economic and environmental dimensions.[3]

---

[1] UN-Secretary General/World Commission on Environment and Development (1987) Report of the World Commission on Environment and Development, Chapter 2, §4, available at: https://digitallibrary.un.org/record/139811?ln=fr. Accessed 22 November 2022.

[2] UN-Secretary General/World Commission on Environment and Development (1987) Report of the World Commission on Environment and Development, Chapter 2, §1, available at: https://digitallibrary.un.org/record/139811?ln=fr. Accessed 22 November 2022.

[3] UN General Assembly (2015) Resolution 'Transforming our world: the 2030 Agenda for Sustainable Development' (70/1), available at: https://undocs.org/en/A/RES/70/1. Accessed 22 November 2022.

E. Huguenin-Elie (✉)
Lenz & Staehelin, Geneva, Switzerland
e-mail: eugenia.huguenin-elie@lenzstaehelin.com; eugeniahugueninelie@protonmail.com

© The Author(s) 2024
P. Këllezi et al. (eds.), *Sustainability Objectives in Competition and Intellectual Property Law*, LIDC Contributions on Antitrust Law, Intellectual Property and Unfair Competition, https://doi.org/10.1007/978-3-031-44869-0_20

Intellectual Property ('**IP**') aims at protecting creations of the mind, such as inventions; literary and artistic works; designs; and symbols, names and images used in commerce. IP has an important role to play in helping companies and people protect creative and innovative products and services that are the result of major investments. IP offers also visibility, attractiveness and value of products and services on the market, but also access to technical and business information and knowledge.

With respect to sustainable development, IP can represent either a critical incentive or a hindrance to innovation and creativity. The purpose of this report is to go beyond the positions of principle and to underline the current and prospective IP protection and uses that would create positive impact to achieve sustainability. Convinced by IP's role for sustainability, the World Intellectual Property Organisation ('**WIPO**')'s statement should be kept in mind: '[o]*nly through human ingenuity will it be possible to develop new solutions that: eradicate poverty; boost agricultural sustainability and ensure food security; fight disease; improve education; protect the environment and accelerate the transition to a low-carbon economy; increase productivity and boost business competitiveness*'.[4] The role of IP in sustainable development shall therefore be a global effort to tackle challenges in relation to biodiversity, traditional knowledge, climate change, transfer of clean technologies, agrobiotechnology and food security.[5]

Through the analysis of the various IP rights enshrined in Swiss IP laws, this contribution aims at discussing the role of these legal instruments and/or other measures taken or proposed in Switzerland to encourage sustainable enterprises and innovations.

After describing the different IP rights protected under Swiss law and their respective/current role in sustainability (*see* Sect. 20.2 below), we will provide an overview of the measures and initiatives proposed in Switzerland (*see* Sect. 20.3 below). Further, a few thoughts on the success of IP in its role for sustainable development and the potential improvement to fulfil its role will be shared before concluding this report (*see* Sect. 20.4 below).

---

[4] WIPO, The Impact of Innovation, WIPO and the Sustainable Development Goals, at: https://www.wipo.int/sdgs/en/story.html. Accessed 22 November 2022.

[5] IPI, Intellectual property and sustainable development, at: https://www.ige.ch/en/intellectual-property/ip-and-society/sustainable-developmentenvironmentagriculture.   Accessed 22 November 2022.

20 Sustainability and Intellectual Property in Switzerland

20.2 Swiss IP Legal Framework Current Role in Sustainability

20.2.1 Preliminary Remarks

By essence, IP plays an essential role in sustainable development as it aims to provide creators with legal protection for their ideas. Offering protection to ideas and entitling companies and people to exploit them shall incentivise them to engage means for innovation. Promoted under Goal 9 of the SDGs, innovation and environmentally sound technologies are necessary to increase resource-use efficiency and to adopt 'clean' industrial processes to make infrastructure and industries sustainable.[6] Innovation is therefore an important tool for achieving environmental, economic and social sustainability and has also an impact on many other SDGs, such as Goals 2 (Zero Hunger), 3 (Good Life and Well-Being), 6 (Clean Water and Sanitation), 7 (Affordable and Clean Energy), 8 (Decent Work and Economic Growth), 11 (Sustainable Cities and Communities) and 13 (Climate Action).[7]

In Switzerland, unlike other areas (e.g. territory planning or equality), only a few provisions of the laws protecting IP rights have yet been amended to specifically further incentivise sustainable development. However, the Swiss IP legal framework offers means that can be useful for innovation to become (rapidly) economically and socially profitable. These provisions and means will be further described in the following sections.

20.2.2 Swiss IP Legal Framework

Under Swiss law, IP consist of:

- Copyright, which is protected by the Federal Act on Copyright and Related Rights (Copyright Act,[8] '**CopA**');
- Design, which is protected by the Federal Act on the Protection of Designs (Designs Act,[9] '**DesA**');

---

[6]UN General Assembly (2015) Resolution 'Transforming our world: the 2030 Agenda for Sustainable Development' (70/1), p. 14, 20 and 21 (Goal 9: Build resilient infrastructure, promote inclusive and sustainable industrialization and foster innovation), available at: https://undocs.org/en/A/RES/70/1. Accessed 22 November 2022.

[7]Edson Kuzma, Luccas Santin Padilha, Simone Sehnem, Dulcimar José Julkovski, Darlan José Roman, The relationship between innovation and sustainability: A meta-analytic study, Journal of Cleaner Production, Volume 259, 2020; WIPO, The Impact of Innovation, WIPO and the Sustainable Development Goals, available at: https://www.wipo.int/sdgs/en/story.html. Accessed 22 November 2022.

[8]RS 231.1, available at: https://www.fedlex.admin.ch/eli/cc/1993/1798_1798_1798/en. Accessed 22 November 2022.

[9]RS 232.12, available at: https://www.fedlex.admin.ch/eli/cc/2002/226/en. Accessed 22 November 2022.

- Patent, which is protected by the Federal Act on Patents for Inventions (Patents Act,[10] '**PatA**');
- Trademark, which is protected by the Federal Act on the Protection of Trademarks and Indications of Source (Trademark Protection Act,[11] '**TPA**');
- Plant varieties, which are protected by the Federal Law on the Protection of New Plant Varieties;
- Topographies and semi-conductor products, which are protected by the Federal Law on the Protection of the Topographies of Semiconductor Products.

Although not granting an absolute right to its owner (as the above IP rights), IP shall also include *trade secrets* (protected by several provisions, i.e. Articles 321a et 418d of the Swiss Code of Obligations, Articles 162 and 273 of the Swiss Criminal Code as well as Article 4 (c) of the Federal Law against Unfair Competition).

The following subsections will focus on few amendments entered into force recently in the PatA and PatO (*see* Sect. 20.2.3 below), in the TPA (*see* Sect. 20.2.4 below) and in the CopA (*see* Sect. 20.2.5 below) to implement sustainable development objectives.

20.2.3 PatA and PatO

Where innovation can be a solution, patent protection should be considered. Patents are indeed directly relevant to technology with a bearing on sustainability and such knowledge shall benefit to everyone to effectively engage in a sustainable environment. In Switzerland, the PatA and its Ordinance of application [the Ordinance on Patents[12] ('**PatO**')] set forth the conditions to protect inventions and contain useful tools to implement SDGs.

20.2.3.1 Compulsory Licences

Articles 40 to 40*d* of the PatA provide for the granting of compulsory licences particularly (*i*) for inventions that present a public interest (Art. 40 PatA), (*ii*) for inventions in the field of semiconductor technology (Art. 40*a* PatA), (*iii*) for inventions that are intended to be used as research tools (Art. 40*b* PatA), (*iv*) for inventions concerning a diagnostic product or procedure for humans (Art. 40*c* PatA) and (*iv*) for the export of pharmaceutical products (Art. 40*d* PatA).

Compulsory licences limit the rights of the patent holder and must therefore be seen as an exception regulation which must be applied with restraint. They are

---

[10]RS 232.14, available at: https://www.fedlex.admin.ch/eli/cc/1955/871_893_899/en. Accessed 22 November 2022.

[11]RS 232.11, available at: https://www.fedlex.admin.ch/eli/cc/1993/274_274_274/en. Accessed 22 November 2022.

[12]RS 232.141, available (in French) at: https://www.fedlex.admin.ch/eli/cc/1977/2027_2027_202 7/fr. Accessed 22 November 2022.

20 Sustainability and Intellectual Property in Switzerland

however particularly appropriate when there is a latent risk of abuse of a dominant position and can guarantee access to and development of research results in problematic cases.[13] In this scenario, their effectiveness lies less in the frequency with which they are imposed than in the pressure they can exert in favour of an amicable solution.[14] As regards the compulsory licences that present a public interest (Art. 40 PatA) or for the export of pharmaceutical products (Art. 40*d* PatA), the aim is however to enable private as well as public stakeholders to have access to technology at an affordable price.

Switzerland has adopted Art. 40*d* PatA to implement the World Trade Organization's ('**WTO**') decision to allow WTO Member States with sufficient pharmaceutical production capacity to provide for a compulsory licence for the manufacture and export of patented pharmaceuticals under clearly defined conditions.[15] This measure aimed at enabling countries (particularly developing countries) with insufficient or no in-house production capacity to have access to patented pharmaceuticals at reasonable price, when they need them to fight serious public health problems, such as HIV/AIDS or malaria.[16]

Article 40*d* para 1 PatA provides that:

> [a]ny person may bring an action before the court to be granted a non-exclusive licence for the manufacture of patent-protected pharmaceutical products and for their export to a country that has insufficient or no production capacity of its own in the pharmaceutical sector and which requires these products to combat public health problems, in particular those related to HIV/AIDS, tuberculosis, malaria and other epidemics (beneficiary country).

As already stated by William O. Hennessey in 1996, '[l]*icensing can also foster the diffusion of environmental technologies to other geographical markets which desperately need them but which remain beyond the reach of the technology*

---

[13] Swiss Federal Council's Message of 23 November 2005 concerning the amendment of the Patent Act and the Federal Decree approving the Patent Law Treaty on the Law of Patents and the Implementing Regulations, BBl 2006 1, p. 73, available (in French) at: https://www.fedlex.admin.ch/filestore/fedlex.data.admin.ch/eli/fga/2006/1/fr/pdf-a/fedlex-data-admin-ch-eli-fga-2006-1-fr-pdf-a.pdf. Accessed 22 November 2022.

[14] Swiss Federal Council's Message of 23 November 2005 concerning the amendment of the Patent Act and the Federal Decree approving the Patent Law Treaty on the Law of Patents and the Implementing Regulations, BBl 2006 1, p. 133, available (in French) at: https://www.fedlex.admin.ch/filestore/fedlex.data.admin.ch/eli/fga/2006/1/fr/pdf-a/fedlex-data-admin-ch-eli-fga-2006-1-fr-pdf-a.pdf. Accessed 22 November 2022.

[15] Swiss Federal Council's Message of 23 November 2005 concerning the amendment of the Patent Act and the Federal Decree approving the Patent Law Treaty on the Law of Patents and the Implementing Regulations, BBl 2006 1, p. 106 ff., available (in French) at: https://www.fedlex.admin.ch/filestore/fedlex.data.admin.ch/eli/fga/2006/1/fr/pdf-a/fedlex-data-admin-ch-eli-fga-2006-1-fr-pdf-a.pdf. Accessed 22 November 2022.

[16] WTO, Decision of 30 August 2003, WT/L/540, Implementation of Paragraph 6 of the Doha Declaration on the TRIPS Agreement and Public Health, at: https://docs.wto.org/dol2fe/Pages/FE_Search/FE_S_S009-DP.aspx?CatalogueIdList=51809,2548,53071,70701&CurrentCatalogueIdIndex=1. Accessed 22 November 2022.

developer.[17] Thus, this type of measure is directly in line with a logic of sustainability. However, it is regrettable that these compulsory licences are very little used and their contours very little (or not yet) tested in court.

20.2.3.2 Disclosure of Source

In implementing the SDGs, the role of innovation and technology is essential; scientific knowledge is central to solve economic, social and environmental issues. However, innovation is not the only form of knowledge, and closer links between science and other systems of knowledge are made to address sustainable development problems at the local level such as natural resources management and biodiversity preservation. Traditional societies have nurtured and refined systems of knowledge in very diverse fields (i.e. agriculture, botany, ecology, geology, health, meteorology, and psychology) that represent an enormous wealth.[18] However, traditional knowledge is often only transmitted orally and is therefore not documented. In the past, several patents were granted for inventions that were based on and/or used traditional knowledge, and thus should have been considered not sufficiently inventive. In most of the cases, such 'not inventive' patents were granted because patent authorities had no means to access such knowledge and to determine whether there was prior art as regards this knowledge.

It was therefore acknowledged that the sharing of commercial and other benefits related to the use of genetic resources and the related traditional knowledge can cause numerous issues. To address them, several international instruments have been concluded, including in particular the Convention on Biological Diversity ('**CBD**'), the Bonn Guidelines, and the International Treaty of the Food and Agriculture Organization ('**FAO**'). In the context of the CBD, it was decided to set an International Regime on Access and Benefit Sharing. Among the measures taken, particular requirements with respect to patent application have been included to disclose certain information in patent applications so as to increase transparency and prevent 'not inventive' patents. The purpose was to ensure that the sharing of the benefits arising from the use of genetic resources and the related traditional knowledge allow the providers of such knowledge (particularly developing countries and indigenous and local communities) to benefit from the patent system.[19]

In this context, Switzerland submitted several proposals to WIPO bodies and working groups to introduce a 'disclosure of source' of genetic resources and

---

[17] William O. Hennessey, Sustainable Development: A "Win-Win" for Licensing and for the Environment, at: https://law.unh.edu/sites/default/files/media/2018/09/sustainable-development-win-win-licensing.pdf. Accessed 22 November 2022.

[18] International Council for Science, Series on Science for Sustainable Development No. 4, p. 3, at: https://unesdoc.unesco.org/ark:/48223/pf0000150501. Accessed 22 November 2022.

[19] WIPO, Intergovernmental Committee on Intellectual Property and Genetic Resources, Traditional Knowledge and Folklore, 11th Session Geneva, July 3 to 12, 2007, Document submitted by Switzerland, WIPO/GRTKF/IC/11/10, § 9 ff., at: https://www.wipo.int/edocs/mdocs/tk/en/wipo_grtkf_ic_11/wipo_grtkf_ic_11_10.pdf. Accessed 22 November 2022.

traditional knowledge in patent applications.[20] The aim was ultimately to achieve four policy objectives: *transparency*, *traceability*, *technical prior art* and *mutual trust* (also known as the 'four T's').[21]

The proposals, which Switzerland worked for on an international level, have also been implemented on a national level: Article 49*a* PatA (as well as Article 138 let. b PatA for international applications) obliges the patent applicant to provide information regarding the source of genetic resources and related traditional knowledge in their patent application. Under Article 81*a* PatA, anyone who intentionally violates such obligation is liable to a fine of up to CHF 100,000 and the court may order the publication of the judgment.

20.2.3.3 Fast Track

Like many other countries (and more recently Singapore[22]), Article 63 PatO offers patent applicants the possibility to request an accelerated procedure. It provides that: '[t]*he applicant may request that the substantive examination be undertaken under an accelerated procedure. Until the expiry of 18 months from the date of filing or priority, such a request may only be made if the formal requirements set out in Articles 46 to 52 are fulfilled'*. An additional fee of CHF 200 is required to request this procedure (Annex 3 of the Swiss IP Office Ordinance on fees).

In Switzerland, patent applications usually undergo a substantive examination which is carried out by the Swiss IP Office and usually begins about three years after filing.

By speeding up a patent grant, the accelerated procedure may help companies investing to find innovative solutions to address pressing issues, such as climate change, food security and public health, to give visibility to their project as well as protecting it. It might however be adapted to further encourage the development of sustainable inventions (by exempting applicants from the surcharge).

20.2.4 TPA

The more consumers are environmentally conscious and wish to buy products and/or services that are sustainable, the more certification marks have an important role to

---

[20] IPI, Biodiversity and sustainable development, Disclosure of source, at: https://www.ige.ch/en/law-and-policy/international-ip-law/ip-organisations/wipo/biodiversity-and-sustainable-development/disclosure-of-source. Accessed 22 November 2022.

[21] WIPO, Intergovernmental Committee on Intellectual Property and Genetic Resources, Traditional Knowledge and Folklore, 11th Session Geneva, July 3 to 12, 2007, Document submitted by Switzerland, WIPO/GRTKF/IC/11/10, § 12, at: https://www.wipo.int/edocs/mdocs/tk/en/wipo_grtkf_ic_11/wipo_grtkf_ic_11_10.pdf. Accessed 22 November 2022.

[22] Intellectual Property Office of Singapore, Registry of Patents, Circular No. 2/2020: Launch of the SG Patent Fast Track Programme on 4 May 2020, at: https://www.ipos.gov.sg/docs/default-source/resources-library/patents/circulars/(2020)-circular-no-2-launch-of-sg-patent-fast-track-programme-on-4-may-2020-(final).pdf. Accessed 22 November 2022.

play. They both contribute to ensure that products and services comply with certain standards and allow the consumer to distinguish green or environmentally friendly products and services. Therefore, it is increasingly common to witness the use of marks indicating that a product is certified by a specific organisation or has a specific geographical origin and possess qualities, characteristics and/or a reputation that are due to that origin. Such marks are provided by most legal systems around the world and known as certification marks, collective marks or guarantee marks.[23] In Switzerland, both guarantee marks (Art. 21 TPA) and collective marks (Art. 22 TPA) exist.

As regards appellations of origin ('**AOs**') and geographical indications ('**GIs**'), those collective rights are protected under Articles 47–50*f* TPA and shall help producers obtain a fair compensation on their efforts in building the reputation of traditional origin-based products. In this specific regard, the Swiss IP Office participates in cooperation projects with other countries to help them setting a legal framework that would support businesses and therefore improve the socioeconomic situation of the project countries (*see* Sect. 20.3.2.2 below).

20.2.4.1 Certification Marks

Article 21 para 1 TPA provides that: '[a] [certification] *mark is a sign that is used by several undertakings under the supervision of the proprietor of the mark and which serves to guarantee the quality, geographical origin, the method of manufacture or other characteristics common to goods or services of such undertakings*'. The proprietor of the certification mark must allow its use for goods or services that possess the common characteristics set forth under the regulations governing the use of the mark in return for equitable remuneration (Art. 22 para 3 TPA).

Certification marks are used by several companies under the control of its proprietor, with the aim of guaranteeing that goods and services have specific characteristics (i.e. quality, geographical origin, manufacturing methods, etc.).[24] Certification marks are also intended to distinguish goods or services of a group of companies (those subject to the trademark regulations) from those originating from companies outside that group.[25]

It is worth mentioning that the proprietor of such mark (or of an undertaking with which he has close economic ties) may not use it for the goods or services he offers on the market (Art. 21 para 2 TPA): as the proprietor has to monitor the use of the label, it is considered that he cannot monitor himself and shall avoid any conflict of interest.[26] The certification marks conditions of use must be determined in regulations that must be approved by the Swiss IP Office (Art. 23 para 1 TPA).

---

[23] WIPO, World Intellectual Property Day 2020 – Innovate for a Green Future, at: https://www.wipo.int/ip-outreach/en/ipday/2020/Articles/sustainable_trademark.html. Accessed 22 November 2022.

[24] Commentaire romand de la Propriété intellectuelle, Claudia Maradan, *ad* Art. 21 TPA No. 9.

[25] Commentaire romand de la Propriété intellectuelle, Claudia Maradan, *ad* Art. 21 TPA No. 10.

[26] Commentaire romand de la Propriété intellectuelle, Claudia Maradan, *ad* Art. 21 TPA No. 38.

They set out the common characteristics of the goods or services subject to guarantee and also provide for an effective control mechanism and adequate sanctions (Art. 23 para 2 TPA).

As of now, less than 250 certification marks are registered (or subject to IPI review for registration) in Switzerland,[27] they can represent an important tool to support environmental awareness and implement SDGs.[28] Indeed, when certification marks are used to distinguish green or sustainable products and services, companies and holders of such certification mark draw consumers to their products and/or services. The use of such products and/or services shall in turn contribute to generating positive environmental impact as they comply with standards and requirements which have been tested.

20.2.4.2 Collective Marks

Article 22 TPA provides that: '[a] *collective mark is a sign of an association of manufacturing, trading or service undertakings which serves to distinguish the goods or services of the members of the association from those of other undertakings*'. Whereas guarantee marks ensure the presence of specific common properties of the goods and services they designate, collective marks serve to attest the membership to 'an association of manufacturing, trading or service undertakings'. It may have a guarantee function, but this is not an essential function of the collective mark.

At international and national levels, collective marks are also used as a tool for promoting sustainable development goals.[29] However, from a consumer perspective, collective marks offer less guarantees compared to the guarantee marks: not only do they only serve to prove membership of a designated grouping, but holders of such mark are not subject to control procedures neither to sanction should they not comply with the specific requirements deriving from the use of such mark.

---

[27] In contrast, 268,367 individual trademarks are registered (or filed for registration), at: https://database.ipi.ch/database-client/search/query/trademarks. Accessed 22 November 2022.

[28] Among others, here is a selection of certification marks registered in Switzerland with this purpose: GO CARBON FREE (fig.; registration No. CH 773790) RAINFOREST ALLIANCE PEOPLE & NATURE (fig.; application No. 04887/2021); V (fig.; from the V-Label, registration No. 730408, 730408 and 730724), at: https://database.ipi.ch/database-client/register/detail/trademark/1206918516, https://database.ipi.ch/database-client/register/detail/trademark/1206863474 and https://database.ipi.ch/database-client/register/detail/trademark/1205804258. Accessed 12 January 2023.

[29] Among others, here is a selection of collective marks registered in Switzerland with this purpose: CLIMATE-NEUTRAL certified (fig.; registration No. 1614741); biokreis (fig.; registration No. 1416324); green (fig.; registration No. CH 729446), at: https://www3.wipo.int/madrid/monitor/en/showData.jsp?ID=ROM.1614741, https://www3.wipo.int/madrid/monitor/en/showData.jsp?ID=ROM.1416324 and https://database.ipi.ch/database-client/register/detail/trademark/1205895576. Accessed 12 January 2023.

20.2.5 CopA

Copyrighted works have also their role to play in protecting green technologies, especially as regards scientific research. Those creative works are necessary to the development of sustainable solutions and shall be disseminated to support innovation but also education (in this particular regard, *see* also Sect. 20.3.2.4 below).

In this respect, a new restriction—Article 24*d* CopA—has been adopted and entered into force on 1 April 2020 with the aim of facilitating access to scientific research and to incentivise it.

Article 24*d* CopA provides that: '[f]*or the purposes of scientific research, it is permissible to reproduce a work if the copying is due to the use of a technical process and if the works to be copied can be lawfully accessed*' (para 1); '[o]*n conclusion of the scientific research, the copies made in accordance with this Article may be retained for archiving and backup purposes*' (para 2) and an exception for computer programs '[t]*his Article does not apply to the copying of computer programs*' (para 3).

Under certain conditions, the reproduction of a work for science research and its saving for archiving and backup purposes are now allowed. In the research field, the technique of 'text & data mining' is often used.[30] It consists of an automatic technical process treating big volumes of documents and data with the objective of analysing texts and compiling information so to establish interconnections and correlations among them, including copyrighted works (such as texts, audios, images and other kind of data).[31] Reproducing, copying and retaining copyrighted works would normally require the authors' consent in the form of individual licenses. This restriction authorises now under certain conditions the reproduction of works for science purposes. It allows also to save the works used for archiving and backup purposes, as long as it is necessary to verify the research results and procedure.[32]

---

[30] Swiss Federal Council's Message of 22 November 2017 concerning the amendment of the Copyright Act, the approval of two World Intellectual Property Organisation treaties of Intellectual Property and their implementation, BBl 2018 559, p. 594, available (in French) at: https://www.fedlex.admin.ch/filestore/fedlex.data.admin.ch/eli/fga/2018/184/fr/pdf-a/fedlex-data-admin-ch-eli-fga-2018-184-fr-pdf-a.pdf. Accessed 22 November 2022.

[31] Swiss Federal Council's Message of 22 November 2017 concerning the amendment of the Copyright Act, the approval of two World Intellectual Property Organisation treaties of Intellectual Property and their implementation, BBl 2018 559, p. 594 f., available (in French) at: https://www.fedlex.admin.ch/filestore/fedlex.data.admin.ch/eli/fga/2018/184/fr/pdf-a/fedlex-data-admin-ch-eli-fga-2018-184-fr-pdf-a.pdf. Accessed 22 November 2022.

[32] Swiss Federal Council's Message of 22 November 2017 concerning the amendment of the Copyright Act, the approval of two World Intellectual Property Organisation treaties of Intellectual Property and their implementation, BBl 2018 559, p. 594 f., available (in French) at: https://www.fedlex.admin.ch/filestore/fedlex.data.admin.ch/eli/fga/2018/184/fr/pdf-a/fedlex-data-admin-ch-eli-fga-2018-184-fr-pdf-a.pdf. Accessed 22 November 2022.

20 Sustainability and Intellectual Property in Switzerland

20.3 Specific and Current Measures for Sustainability in Switzerland

20.3.1 Mandate of the Swiss IP Office re Swiss 2030 Sustainable Development Strategy

If Swiss IP laws offer a few means to promote sustainable initiatives, as of now, the Swiss government has essentially set out high level measures strategies in this prospect. In its 2030 Sustainable Development Strategy ('**2030 SDS**') adopted on 23 June 2021, the Swiss Federal Council outlines the priorities it intends to follow to implement the 2030 Agenda over the next ten years.[33] The 2030 SDS sets out the guidelines for the Swiss government's sustainability policy and positions sustainable development as an important requirement for all federal policy areas.[34] Under these guidelines, all federal units are called upon to participate in the implementation of 2030 SDS within the scope of their responsibilities. The Federal Council has therefore adopted the 2021–2023 Action Plan for the 2030 Sustainable Development Strategy ('**2021–2023 Action Plan**'), which gives concrete form to the strategy with a selection of 22 new measures at the federal level. Each of these measures includes a description, the main objectives, the responsible federal units and those involved in their implementation.[35] Section 19 of the 2021–2023 Action Plan provides that the autonomous units shall take proper measures to strengthen sustainable development in their strategic objectives.

In May 2022, the Swiss Federal Council has therefore issued specific and strategic goals for the Swiss IP Office, as an autonomous unit of the Federal Administration.[36] According Section 13 of this roadmap, the Swiss IP Office is expected to contribute to the implementation of the UN Agenda 2030 in the area of sustainable development and to pursue a sustainable strategy based on ethical principles. The Swiss IP Office shall identify areas of sustainable development where it has a significant influence and set targets based on the 17 SDGs.[37] Based on this mandate, the Swiss IP Office has undertaken a number of specific measures to

---

[33] Swiss Federal Council 2030 Sustainable Development Strategy, 23 June 2021, available at: https://www.are.admin.ch/are/en/home/sustainable-development/strategy/sds.html. Accessed 22 November 2022.

[34] Swiss Federal Council 2030 Sustainable Development Strategy, 23 June 2021, available at: https://www.are.admin.ch/are/en/home/sustainable-development/strategy/sds.html. Accessed 22 November 2022.

[35] Swiss Federal Council 2021-2023 Action Plan for the 2030 Sustainable Development Strategy, 23 June 2021, available (in French) at: https://www.are.admin.ch/are/fr/home/developpement-durable/strategie/sdd.html. Accessed 22 November 2022.

[36] Swiss Federal Council Strategic goals 2022-2026 for the Swiss IP Office, 18 May 2022, available (in French) at: https://www.fedlex.admin.ch/eli/fga/2022/1332/fr. Accessed 22 November 2022.

[37] Swiss Federal Council Strategic goals 2022-2026 for the Swiss IP Office (§13), 18 May 2022, available (in French) at: https://www.fedlex.admin.ch/eli/fga/2022/1332/fr. Accessed 22 November 2022.

contribute to and implement the SDGs. Not only the Swiss IP Office is part to international cooperation programs (*see* Sects. 20.3.2.1 and 20.3.2.2) but has also taken more local measures (*see* Sects. 20.3.2.4, 20.3.2.5 and 20.3.2.6).

20.3.2 Measures

20.3.2.1 WIPO GREEN Partnership[38]

Since the end of 2019, the Swiss IP Office is an official partner of WIPO GREEN. Launched in 2013 by WIPO, WIPO GREEN is a marketplace for sustainable technology and offers an online platform which contributes to the diffusion of green technologies.[39] This platform is a free database which compiles green technologies, a network of interested parties, as well as specific WIPO GREEN projects. It aims at supporting global efforts by connecting providers with seekers of such technologies.[40] The Swiss IP Office actively contributes to the diffusion of green technologies (for example with assisted patent searches)[41] and is part of the Core Committee, which consists of a selection of WIPO GREEN partners from different backgrounds reunited to make recommendations on the strategy, priority areas and topics, as well as partnerships to implement the goals of WIPO GREEN.[42] The partnership with WIPO GREEN is also part of the Swiss IP Office's contribution to the achievement of the UN's sustainable development goals.

20.3.2.2 Cooperation Projects[43]

Sustainable development needs global and widespread transfer of clean technologies and knowledge.[44] To mitigate the effects of climate change, there is yet an urge to

---

[38] IPI, WIPO GREEN – a marketplace for green technologies, at: https://www.ige.ch/en/law-and-policy/international-ip-law/ip-organisations/wipo/biodiversity-and-sustainable-development/wipo-green. Accessed 22 November 2022.

[39] WIPO GREEN, The Marketplace for Sustainable Technology, at: https://www3.wipo.int/wipogreen/en/. Accessed 22 November 2022.

[40] WIPO GREEN, WIPO GREEN Database of Innovative Technologies and Needs, at: https://wipogreen.wipo.int/wipogreen-database/database. Accessed 22 November 2022.

[41] IPI, WIPO GREEN – a marketplace for green technologies, at: https://www.ige.ch/en/law-and-policy/international-ip-law/ip-organisations/wipo/biodiversity-and-sustainable-development/wipo-green, Accessed 22 November 2022.

[42] WIPO GREEN, Core Committee, at: https://www3.wipo.int/wipogreen/en/partners/core_committee.html. Accessed 22 November 2022.

[43] IPI, Development cooperation, at: https://www.ige.ch/en/law-and-policy/development-cooperation. Accessed 22 November 2022.

[44] Guillaume Henry, Joël Ruet, Matthieu Wemaëre (INPI), Sustainable development & INTELLECTUAL PROPERTY Access to technologies in developing countries, p. 11, at: https://www.inpi.fr/sites/default/files/web_sustainable-development_and_ip_inpi_20151125.pdf. Accessed 22 November 2022.

20 Sustainability and Intellectual Property in Switzerland

deploy environmentally sound technologies on a large-scale and rapidly in all sectors and all countries, whether developed or developing.[45]

The Swiss IP Office has engaged in technical cooperation with developing countries in the field of IP. Such technical cooperation projects include providing assistance to the competent authorities in their operations, providing training to the employees and drafting legislation.[46] These cooperation projects aim at creating an environment that is more conducive to business and therefore improving the socio-economic situation of the project countries. This involves, in particular, (*i*) to develop technology and knowledge transfer, (*ii*) to prevent piracy and trade of counterfeit goods (e.g. medicines, auto parts and batteries) and (*iii*) to protect IP in an appropriate, effective, calculable and easily applicable manner. This involves weighing up the economic benefits and the interests of all those affected by IP rights.[47]

In the performance of these technical cooperation projects, the Swiss IP Office is committed to international standards set out in the Paris Declaration on Aid Effectiveness.[48] Among others, the Swiss IP Office shall contribute to '*enhancing the individual responsibility and independence of project countries and partner institutions (**Ownership**)*' and make the projects '*perennial and* [are] *geared towards the national development strategies of project countries and the needs of partner institutions (**Alignment**)*'. Moreover, '[t]*he results achieved continue to have an impact after the projects have ended (**Sustainability**)*'.[49]

As of today, the Swiss IP Office is supporting more than ten projects across the world under the program named '**Global Program for Intellectual Property Rights**' funded by the Swiss State Secretariat for Economic Affairs ('**SECO**').[50] Since its launch in June 2018, this program has supported the following countries: Albania, Colombia, Myanmar, Peru, Serbia and South Africa. Additional activities

---

[45] Guillaume Henry, Joël Ruet, Matthieu Wemaëre (INPI), Sustainable development & INTELLECTUAL PROPERTY Access to technologies in developing countries, p. 11, at: https://www.inpi.fr/sites/default/files/web_sustainable-development_and_ip_inpi_20151125.pdf.   Accessed 22 November 2022.

[46] IPI, Development cooperation, at: https://www.ige.ch/en/law-and-policy/development-cooperation. Accessed 22 November 2022.

[47] IPI, Development cooperation, Objectives and working principles, at: https://www.ige.ch/en/law-and-policy/development-cooperation/objectives-and-principles. Accessed 22 November 2022.

[48] OECD, at: https://www.oecd.org/dac/effectiveness/parisdeclarationandaccraagendaforaction.htm. Accessed 22 November 2022.

[49] IPI, Development cooperation, Objectives and working principles, at: https://www.ige.ch/en/law-and-policy/development-cooperation/objectives-and-principles. Accessed 22 November 2022.

[50] IPI, Development cooperation, Current Projects, Global Program for Intellectual Property Rights, at: https://www.ige.ch/en/law-and-policy/development-cooperation/current-projects/global-program-for-intellectual-property-rights. Accessed 22 November 2022.

and projects are being planned in Albania, Benin, Iran, Colombia, Myanmar, Peru, Serbia, South Africa and Tunisia.[51]

For example, since January 2020, the IPI has entered into a cooperation agreement with Iran [Iranian-Swiss Intellectual Property Project ('**IRSIP**')]. According to the IPI, *'the overall objective of the IRSIP is to strengthen the Iranian IP system to improve business competitiveness and commercialization of intellectual property assets, and to make a positive impact on Iran's economic development'*.[52] In this context, one of the main objectives of the IRSIP was to enable Iran to capitalise on its internal resources, for example in the field of GIs, and to provide assistance to the national GIs system.[53]

20.3.2.3 Technology Transfer Agreements Templates[54]

When accessible, technical data can contribute to technological innovation and is of major economic interest. However, these data are often held by private entities either voluntarily or due to a lack of sharing solutions. Concerned by researchers, businesses and civil society's interests in setting the conditions for the legal, secure and fair sharing of technical data, the Swiss Federal Council commissioned the IPI to issue a report on access to non-personal data in the private sector.[55]

Therefore, the IPI has entrusted various specialised stakeholders to carry out different analysis. Among them, one had the objective of proposing standardised agreements to facilitate exchanges of technical data held by private sector actors to other organisations. After identifying in a report the major stakeholders' needs and concerns with respect to this question (i.e. loss of control over shared data, protection of interests when sharing, complexity of the legal framework, absence of good practices, etc.),[56] three model of agreements were proposed covering three types of cases:[57]

---

[51] IPI, Development cooperation, Current Projects, Global Program for Intellectual Property Rights, at: https://www.ige.ch/en/law-and-policy/development-cooperation/current-projects/global-program-for-intellectual-property-rights. Accessed 22 November 2022.

[52] IPI, Iranian-Swiss Intellectual Property Project – IRSIP, at: https://www.ige.ch/fileadmin/user_upload/recht/entwicklungszusammenarbeit/factsheet_iran.pdf. Accessed 22 November 2022.

[53] IPI, Iranian-Swiss Intellectual Property Project – IRSIP, at: https://www.ige.ch/fileadmin/user_upload/recht/entwicklungszusammenarbeit/factsheet_iran.pdf. Accessed 22 November 2022.

[54] IPI, IP and society, Data processing and data security, Access to non-personal data in the private sector, at: https://www.ige.ch/en/intellectual-property/ip-and-society/data-processing-and-data-security. Accessed 22 November 2022.

[55] IPI, IP and society, Data processing and data security, Access to non-personal data in the private sector, at: https://www.ige.ch/en/intellectual-property/ip-and-society/data-processing-and-data-security. Accessed 22 November 2022.

[56] Juliette Ancelle/Michel Jaccard (id est), Explanatory Report – Model Agreements for Sharing Technical Data, p. 2 f., at: https://www.ige.ch/fileadmin/user_upload/recht/gesellschaft/e/IPI_Explanatory_Report_id_est_avocats_August_2020.pdf. Accessed 22 November 2022.

[57] Juliette Ancelle/Michel Jaccard (id est), Explanatory Report – Model Agreements for Sharing Technical Data, p. 4, at: https://www.ige.ch/fileadmin/user_upload/recht/gesellschaft/e/IPI_

- The one-time provision of data covered by an Agreement for the Transfer of Technical Data;
- Access to a data feed or the regular provision of long-term data covered by a Subscription Agreement for Access to Technical Data;
- The exchange of data between parties covered by an Agreement for the Exchange of Technical Data.

Such initiative (although not directly related to IP) contributes to transfer of technologies and therefore implementing SDGs.

20.3.2.4 National Open Access Strategy[58]

Out of the 17 SDGs, ten need constant scientific input. Considering that these goals have to be achieved globally, it is essential to remove restrictions to disseminate research outputs widely. Governments as well as international organisations such as UNESCO have acknowledged this connection and officially recognise open access ('**OA**') has a major role to support and achieve sustainable social, political, and economic development.[59] UNESCO *'believes that OA has a fundamental role to support the SDGs and is committed to making OA one of the central supporting agendas to achieve the SDGs'.*[60]

In 2016, on behalf of the State Secretariat for Education, Research and Innovation and in collaboration with the Swiss National Science Foundation, swissuniversities (the Swiss Conference of Rectors of Higher Education Institutions) developed a national strategy for Open Access. On 31 January 2017, the Swiss National Strategy on Open Access has been approved by the plenary assembly of swissuniversities and has been subsequently endorsed by the governing board of the Swiss University Conference. It sets forth the following objective: '[...] *by 2024, all scholarly publication activity in Switzerland should be OA, all scholarly publications funded by public money must be freely accessible on the internet'.*[61]

A significant step taken to achieve this goal was the signature of the four-year contract OA agreement with John Wiley & Sons in 2021. It shall offer and *'provide students and researchers from over 40 Swiss higher education and research*

Explanatory_Report_id_est_avocats_August_2020.pdf. Accessed 22 November 2022. Model Agreements, accessible here: https://www.ige.ch/en/intellectual-property/ip-and-society/data-processing-and-data-security. Accessed 22 November 2022.

[58] State Secretariat for Education, Research and Innovation (SERI), Open Science, at: https://www.sbfi.admin.ch/sbfi/en/home/services/publications/data-base-publications/s-n-2020-3/s-n-2020-3g.html. Accessed 22 November 2022.

[59] UNESCO, Open Access to Scientific Information, at: https://en.unesco.org/themes/building-knowledge-societies/open-access-to-scientific-information. Accessed 22 November 2022.

[60] UNESCO, Open Access to Scientific Information, at: https://en.unesco.org/themes/building-knowledge-societies/open-access-to-scientific-information. Accessed 22 November 2022.

[61] swissuniversities, Swiss National Strategy on Open Access, §4, at: https://www.swissuniversities.ch/fileadmin/swissuniversities/Dokumente/Hochschulpolitik/Open_Access/Open_Access_strategy_final_e.pdf. Accessed 22 November 2022.

institutions access to the Wiley Online Library with over 1,450 journals. At the same time, this agreement enables authors affiliated to these institutions to make their scientific Articles published in the hybrid journals openly accessible immediately upon publication'.[62]

20.3.2.5 Innosuisse: Swiss Innovation Agency[63]

As the Innovation Agency for Switzerland, Innosuisse's role is *'to promote science-based innovation in the interest of the economy and society in Switzerland'.*[64]

Innosuisse action aims at accelerating the transfer of knowledge from academia to industry and to help innovations and start-ups to strengthen the competitiveness of Swiss SMEs and start-ups on the market, therefore contributing to the sustainable development and prosperity of Switzerland. More concretely, Innosuisse's support is threefold: (*i*) to promote networking and expanding knowledge, (*ii*) to help implementing innovation projects and (*iii*) to help found and establish start-ups.[65] Should Innosuisse have approved an application for funding and when providing support for the implementation of the project, Innosuisse offers—in collaboration with the Swiss IP Office—two free services: the Assisted Patent Search and the Assisted Patent Landscape Analysis enabling applicants to find out whether their innovation can be patented and gain valuable insights into their innovation.[66]

20.3.2.6 Swiss Cleantech Report Contribution[67]

Published for the third time, the purpose of the Swiss Cleantech Report is to highlight the efforts being invested by the Swiss industry to overcome the challenges presented by climate change. This publication aims at promoting and broadcasting these technological innovations and concrete solutions brought to the market. To better illustrate Switzerland position on an international level, the Swiss IP Office shares its observations and statistics on the Swiss cleantech patent landscape.[68]

---

[62] Swissuniversities, swissuniversities and Wiley sign transformative Open Access agreement, 29.04.2021, at: https://www.swissuniversities.ch/en/news/press-releases/swissuniversities-and-wiley-sign-transformative-open-access-agreement. Accessed 22 November 2022.

[63] Innosuisse, at: https://www.innosuisse.ch/inno/en/home.html. Accessed 22 November 2022.

[64] Innosuisse, Mission, at: https://www.innosuisse.ch/inno/en/home/about-us/mission.html. Accessed 22 November 2022.

[65] Innosuisse, Get started, p. 3, at: https://www.innosuisse.ch/dam/inno/en/dokumente/ueberuns/Angebotsflyer/innosuisse-angebotsuebersicht.pdf.download.pdf/Innosuisse_Angebotsuebersicht_EN.pdf. Accessed 22 November 2022.

[66] Innosuisse, Assisted patent searches, at: https://www.innosuisse.ch/patent. Accessed 22 November 2022.

[67] Swiss Cleantech Report, at: https://swisscleantechreport.ch. Accessed 22 November 2022.

[68] Swiss Cleantech Report, 3rd ed., p. 18-19, at: https://swisscleantechreport.ch/order-your-copy-3rd-edition/. Accessed 22 November 2022.

20.3.3 Prospective Measures

According to the latest information available to us, the Swiss IP Office has set up a new working group whose aim is to deal with issues related to sustainable development and to evaluate new measures to support green innovation/technology.

Among these measures that the Swiss IP Office is currently further evaluating are the following ones:

– be even more active in the field of international cooperation and specifically in green technologies (*see* also Sect. 20.3.2.2 above);
– set up more workshops on green innovation issues and the role of intellectual property in this context;
– facilitate access to patent searches (e.g. support provided by Innosuisse, *see* Sect. 20.3.2.5 above);
– set up platforms for exchange between the various players by offering training, events and prizes/grants to promote green technologies;
– set up a 'recycling' and 'upcycling' programme for certain counterfeit goods that have been seized, rather than destroying them.

20.4 Conclusion

Overall, despite global efforts to support the implementation of sustainable policies and practices aligned with the SDGs, many initiatives are being overwhelmed by the reality of the market. Notably, the WTO decision on the waiver of IP rights under the exceptional circumstances driven by the COVID-19 pandemic illustrates the outstanding challenges to the inclusion of sustainability considerations in IP laws.[69]

That being said, Switzerland and the Swiss IP Office (as well as other countries, including for instance Singapore, Japan and the US) have undertaken multiple steps to integrate and contribute to the realisation of the SDGs. This is highlighted by numerous measures leading to cooperation, transfer of technology or knowledge (*see* Sect. 20.3.2 above). Some of the existing IP regulations have been progressively adapted to integrate specific incentives, but most importantly to support access to science, or even vulnerable individuals. There is, however, room for improvement.

Switzerland should follow the implementation of recent amendments to IP laws facilitating the access to patent protection or creating green technology classification

---

[69] World Trade Organization, Decision of 17 June 2022 (Document WT/MIN(22)/W/15/Rev.2), at: https://docs.wto.org/dol2fe/Pages/SS/directdoc.aspx?filename=q:/WT/MIN22/W15R1.pdf& Open=True. Accessed 22 November 2022.

schemes for patent, as well as providing assistance to entrepreneurs and raising awareness on IP's role for sustainability. In this context, the efforts of WIPO GREEN and its stakeholders should be commended, and we further hope that the policies and programs launched will be the right means to accelerate the green transition.

Open Access This chapter is licensed under the terms of the Creative Commons Attribution 4.0 International License (http://creativecommons.org/licenses/by/4.0/), which permits use, sharing, adaptation, distribution and reproduction in any medium or format, as long as you give appropriate credit to the original author(s) and the source, provide a link to the Creative Commons license and indicate if changes were made.

The images or other third party material in this chapter are included in the chapter's Creative Commons license, unless indicated otherwise in a credit line to the material. If material is not included in the chapter's Creative Commons license and your intended use is not permitted by statutory regulation or exceeds the permitted use, you will need to obtain permission directly from the copyright holder.

Sustainability and Intellectual Property in United Kingdom

21

Ben Hitchens, Caitlin Heard, and Joel Vertes

21.1 The Role of Intellectual Property in Sustainability

21.1.1 Introduction

Intellectual property ('IP') currently plays a strong role in sustainability, although its relationship with, and impact on, environmental, social and economic factors could be strengthened further. For example, at least six of the United Nation's Sustainable Development Goals,[1] namely zero hunger, clean water and sanitation, affordable and clean energy, industry, innovation and infrastructure, sustainable cities and communities and responsible consumption and production, overlap directly with most 'definitions' of sustainability and will require a range of new technologies to implement. Technological innovation geared towards environmental or sustainable factors (which could also loosely be referred to as 'green' technology) will therefore be pivotal in achieving the UN's goals.

IP (whether falling within the category of patents, trademarks, copyright, designs or some other 'neighbouring' right) plays an important role in the commercialisation

---

The authors of this report wish to thank Alexander Parkin, a trainee at CMS Cameron McKenna Nabarro Olswang LLP, London, for his invaluable contribution.

[1] UN Department of Economic and Social Affairs Sustainable Development Goals, available at https://sdgs.un.org/goals. Accessed 19 November 2022.

B. Hitchens (✉) · C. Heard · J. Vertes
CMS Cameron McKenna Nabarro Olswang LLP, London, UK
e-mail: Ben.Hitchens@cms-cmno.com; Caitlin.Heard@cms-cmno.com;
Joel.Vertes@cms-cmno.com

© The Author(s) 2024
P. Këllezi et al. (eds.), *Sustainability Objectives in Competition and Intellectual Property Law*, LIDC Contributions on Antitrust Law, Intellectual Property and Unfair Competition, https://doi.org/10.1007/978-3-031-44869-0_21

and protection of such technology. Given that IP is ultimately an asset which can be commercialised, it has the potential to generate income which in turn can be used to further innovate and develop the technologies required for the pursuit of sustainable goals. IP rights bestow exclusive rights in connection with specific products, services, processes, information, which—in the absence of some form of reward—would not otherwise be economically viable to pursue.

However, there are various issues concerning IP that can also pose limitations or obstacles to progress around Environmental, Social and Corporate (ESG) ESG goals, which are examined later in this chapter.

Fundamentally, the role of IP in sustainability differs across patents, trade secrets, trademarks, copyright and design rights, as broadly set out below.

21.1.2 Patents

Patents are the seminal example of how IP can contribute to sustainable practices. As a powerful monopoly right, patents allow a greentech business to license products/processes that enhance environmental sustainability (i.e. new technology), but create revenue streams in the meantime.

The process of seeking patent protection requires applicants to disclose their innovation in detail and this can allow others to benefit from, and build on, that innovation. If a patent is granted, the patent owner can prevent unauthorised use of the innovation while the patent remains in force, but after the patent expires the technology is free for anyone to use. In the case of environmentally beneficial technology this is clearly advantageous.

Patents can be used to protect (and, consequently, disclose) important climate-change mitigation technologies for energy, transport and construction, as well as environmental management and water-related adaptation technologies. Mitigation technologies aim to reduce greenhouse gas emission, increase energy efficiency, improve resource use, minimise waste and improve reuse and recycling. Referred to as low-carbon technologies, they generate relatively lower $CO2$ emissions than fossil-fuel energy. In transport, an example would be electric vehicles. In energy production, examples include solar photovoltaic (PV) energy, wind turbines and coal-powered power plants fitted with carbon-capture storage facilities. Carbon dioxide removal and capture technologies, for example power plant storage facilities, reduce $CO2$ emission by capturing and storing gases either in reservoirs (geological, terrestrial or in the ocean) or in products such as wood.

The World Intellectual Property Report 2022[2] observes that patent filings can in themselves be a valuable data source, particularly in identifying trends in sustainable research and development, which may then inform the technological direction

---

[2] WIPO Report 2022 – The Direction of Innovation, available at https://www.wipo.int/edocs/pubdocs/en/wipo-pub-944-2022-en-world-intellectual-property-report-2022.pdf. Accessed 19 November 2022.

pursued by other entities/organisations. For example, p. 68 of WIPO's report illustrates the sharp increase in patents filed after 2000 in the sector of so-called clean technologies. Further analysis shows that they are associated with renewable energy sources, such as solar, wind and fuel cells, which are like batteries that do not run down or need recharging. Trends in patent filings can therefore be observed and assessed, giving an insight into the future direction of sustainable technology.

Examples of the use of patents within the field of sustainable innovation include Nike applying for patent protection for methods for forming footwear using recycled plastics,[3] and Allbirds, a footwear and apparel company, filing a patent application for a shoe including a 'continuous knit textile comprising eucalyptus fiber'.[4]

While patents have a normal lifespan of 20 years, they can create a legacy that extends far beyond their expiry. In 1871 Charles Goodyear, Jr. was awarded a patent (US111197A[5]) for a machine for stitching boots and shoes which allowed the creation of the 'Goodyear Welt'. This manufacturing technique is still used today, albeit with refinements over the years.

Goodyear welted shoes are considered desirable as they are relatively waterproof, but the construction also allows for relatively easy resoling and repair. This means that shoes using a Goodyear welt are more likely to be repaired and used for longer, thereby reducing wastage and minimising the environmental and financial impact of disposal.

While it is unlikely that environmental considerations were at the forefront of Charles Goodyear's mind when he devised his machine, the resulting patent is a good example of an early sustainable innovation that was disclosed and became freely available for general benefit. Many of the patents filed in the field of green/sustainable technology will eventually become of general utility to the world at large, being periodically improved/refined, and contributing to the pursuit of sustainable practices.

On the other hand, there are those who argue the opposite. Some contend that patent rights *inhibit* widespread adoption of green/sustainable technology. Patents create a monopoly right, and if that monopoly is not made available to other parties for exploitation, for example, via licensing, development and adoption of sustainable practices could be unduly delayed pending expiry of the patent in question. We address these arguments in greater detail in this report.

21.1.3 Trade Secrets

Over recent years, trade secrets have been the focus of IP reform in the EU (including in the United Kingdom). Following the adoption of the Trade Secret Directive 2016/943 in 2016, many EU member states have upgraded their national protection

---

[3] https://patents.google.com/patent/WO2019232024A1. Accessed 19 November 2022.

[4] https://patents.google.com/patent/WO2020223382A1. Accessed 19 November 2022.

[5] https://patents.google.com/patent/US111197A. Accessed 19 November 2022.

regime, including the UK via the Trade Secrets (Enforcement, etc.) Regulations 2018 (SI 2018/597). In the UK, trade secrets legislation supplements the equitable cause of action of breach of confidence.

Trade secret protection is available for any information that is not generally known, commercially applicable and of value due to its secrecy. Green innovations are thus eligible, but commercial or financial information can also be protected. There is no registration process, but the trade secret owner must have applied reasonable protection measures to keep the information secret.

The Regulations confirm that protection (in the form of infringement proceedings) is only available against unlawful acquisition, use or disclosure of the information—not for the 'content' itself. Since only one of these three acts need occur for infringement, it is possible for infringement to take place in circumstances where a Trade Secret was lawfully acquired but subsequently unlawfully used or disclosed.

Under common law/equity, threatened or actual unauthorised use of confidential information can constitute a breach of confidence. Further, in circumstances where a recipient of information comes across that material 'lawfully', i.e. through the discovery of apparently discarded papers in a public place, subsequent disclosure of that information can still constitute a breach of confidence, if the recipient has actual or constructive knowledge of its confidentiality.

Trade secret protection is therefore less 'complete' (and harder to enforce) compared to patent protection. Nevertheless, the trade secrets regime/breach of confidence can be especially helpful in stages prior to obtaining patent protection, or as 'supplemental' protection, e.g. for aspects that do not need to be disclosed in a patent application.

Confidential information and trade secrets can be among the most valuable assets available to a business. A competitive edge in the marketplace may rely on a business having certain information which its competitors do not, and being able to maintain the secrecy of that information.

Treating a technological development in the field of sustainability as confidential information may be a preferable route in circumstances where an invention may not be patentable or enforceability may be difficult, such as where it would be challenging to prove a competitor is using an invention. There are also considerable licensing opportunities for the technology transfer of valuable know-how in combination with other IP rights, including patents. This can enable developers of green technology to commercialise their know-how through a third party or via collaborative arrangements, including as envisaged by WIPO GREEN. WIPO GREEN is an online platform for technology exchange operated by the World Intellectual Property Office. It supports global efforts to address climate change by connecting providers and seekers of environmentally friendly technologies. Through its database, network and acceleration projects, it helps to promote the development and sharing of green technology innovation.

Examples of valuable and proprietary confidential information or trade secrets in this context include source code and algorithms for applications which track food source provenance or provide access to clean water reserves.

AI creations and the data/datasets that inform those outputs are likely to be critical to the world's ambitions to deliver a sustainable future. Raw data in particular will enable technological innovation to be driven by statistical analysis. While everyone agrees data will be of tremendous value, and while *collections* of data can be protected through database rights, there is currently no 'traditional IP' protection regime to cover the data itself adequately, which is consistent with the notion of the idea/expression dichotomy. The idea/expression dichotomy is a well-established principle that underpins IP law, namely that while underlying ideas are not protectable, the expression of those ideas in writing can be protected. The trade secrets and confidential information regimes are therefore critical to the protection and commercialisation of IP in the field of sustainable innovation.

21.1.4 Trademarks

Trademark registrations are granted to signs that are capable of designating the commercial origin of specific goods and services. Such signs may comprise words or logos, but also three-dimensional shapes, colours or patterns, or even sounds and surface structures. Trademark registrations are obtained via a simple online application process operated by the United Kingdom Intellectual Property Office.[6] In addition, the United Kingdom bestows national unregistered rights to signs that benefit from goodwill/reputation under the law of passing-off.

Trademarks already play a balanced role in sustainability. They ensure that words that are descriptive of sustainable practices will remain free for use by all traders, so that no one is precluded from using such words to market their own initiatives in the field, which would otherwise have a chilling effect on innovation. Further, the trademark protection regime enables businesses to carve out specific words/logos as distinctive badges of origin that will enable them to foster consumer recognition and loyalty, thereby stimulating growth and enabling further revenue to be invested in the development of sustainable practices.

Trademarks often perform functions other than just indicating commercial origin. They convey attitudes, meanings or expectations—both of the product and of the customer. Trademarks are therefore indispensable to communicating that a product is 'green', sustainable or otherwise environmentally friendly. Although trademarks may not be purely descriptive, they can in theory incorporate easily-understandable terms such as 'green', 'eco', or 're' (for recycling) to indicate the broad nature of the goods/services provided (although the extent to which the inclusion of such terms affords protection is limited).

As in the case of patents (discussed in Sect. 21.1.2 above), trademark filing data can itself also be a useful indication of innovative development, which can

---

[6]IPO – trademark application portal, available at https://trademarks.ipo.gov.uk. Accessed 19 November 2022.

subsequently signpost the future direction of specific fields of technology.[7] By way of example, the European Union Intellectual Property Office's (EUIPO) report on Green EU trademarks details the percentage of so-called 'green' EUTMs by product group, finding that over 40% related to energy conservation, 17.7% to pollution control and 9.7% apiece to energy production and transportation.[8]

The trademarks regime also has an important role in combatting so-called 'greenwashing'. In essence, 'greenwashing' describes an attempt to make consumers believe that a company/organisation is doing more to protect the environment than it really is.

In its recent report on Green EU trademarks, the EUIPO examined the frequency with which terms contained in trademark specifications reflect environmental and sustainable goods/services. According to the EUIPO's report,[9] of the approximately 46,700 EUTM applications received by EUIPO in 1996, the first year of operation, 1,588 were green trademarks. Since then, the increase in green trademarks has been continuous, except for 2001 and between 2011 to 2014. In 2020, the number of green EUTMs filed approached 16,000.

Recently, the European Commission conducted a website analysis, an exercise carried out each year to identify breaches of EU consumer law in online markets. In its report, the European Commission revealed that over half of the green claims identified in its screening exercise lacked evidence. Further, in over 40% of instances, it is believed that the claims made may be regarded as false or deceptive and may therefore amount to unfair commercial practice under EU law.[10] It is reasonable to assume that a similar proportion of the so-called 'green' trademark filings listed above may also constitute specious representations concerning the applicant's environmental/sustainable credentials.

Insofar as greenwashing and trademark protection are concerned, UK trademark legislation already prohibits the registration of signs which are of such a nature as to deceive the public; for instance, as to the nature and quality of the goods and services.[11] Moreover, misleading *use* of a trademark can result in the revocation of a trademark registration under Section 46(1)(d) Trade Marks Act 1994.

Certification marks—which exist as a specific type of trademark—can also fulfil a significant role in the pursuit of sustainable goals. Certification marks guarantee specific characteristics of certain goods and services, namely by indicating that all such goods/services comply with given standards. The owner of the certification mark is responsible for defining such standards/characteristics; all products/services

---

[7] See S. Mendonça, T. Pereira and M. Godinho, Trademarks as an Indicator of Innovation and Industrial Change, Research Policy Volume 33, Issue 9, November 2004, pp. 1385–1404.

[8] EUIPO – Green EU trademarks, available at https://euipo.europa.eu/tunnel-web/secure/webdav/guest/document_library/observatory/documents/reports/2021_Green_EU_trade_marks/2021_Green_EU_trade_marks_FullR_en.pdf, p. 22. Accessed 19 November 2022.

[9] Ibid.

[10] EC – Screening of websites for 'greenwashing': half of green claims lack evidence, available at https://ec.europa.eu/commission/presscorner/detail/en/ip_21_269. Accessed 19 November 2022.

[11] Section 3(3)(b) Trade Marks Act 1994.

that feature the certification mark must therefore meet those standards to benefit from continued 'membership' of the group. Consequently, certification marks can reassure consumers that the goods/services they purchase are aligned with sustainable practices, or are manufactured according to environmentally friendly principles.

Trademarks therefore currently play an important role in the field of sustainability. Indeed, brand owners already use trademarks to communicate specific 'green' values to consumers about the nature of the goods/services being provided.

21.1.5 Copyright

Like other IP rights, copyright is inherently agnostic as to whether the works in question are sustainable or not. However, certainly copyright still has a role to play. Given that we are living in the digital age, a great deal of technological innovation is computer-driven. Copyright can therefore protect software/algorithms that analyse and help to develop more sustainable industrial processes/systems, for example, computer-driven solutions to create more energy-efficient methods and tools.

As a licensable right, copyright-protected software can be commercialised through a licensing model. The ability to derive revenue from software will help incentivise creators to develop further solutions and/or improve existing products. Equally though, the idea/expression dichotomy and the general preclusion from protection of ideas and functionality within the field of computer software, will ensure that rightsholders are unable to unfairly leverage their rights to prevent competition/investment by third parties in similar/identical fields.

The present toolkit of IP rights in the UK is well equipped to provide protection for software-driven technologies, both in the green sector and elsewhere. But it seems likely that sustainable innovation in the coming decades will largely be driven by two major trends: Artificial intelligence and big data. Can results stemming from these be protected as well?

21.1.5.1 Artificial Intelligence

For the protection of works in connection with artificial intelligence, two separate scenarios need to be distinguished. First, inventions and creations made *in the field* of artificial intelligence, which can enjoy the 'regular' protection, mostly via patents, copyrights, or trade secrets. Much more discussion is, however, necessary in connection with the scenario whereby an AI system *itself* makes a creation. Examples are manifold — from the creation of church windows by random colour patterns[12] to improved drinking mugs[13] created by 'inventing machines'.

---

[12] Wikipedia – Cologne Cathedral Window, available at https://en.wikipedia.org/wiki/Cologne_Cathedral_Window. Accessed 19 November 2022.

[13] Artificial Inventor – Patent Applications, available at http://artificialinventor.com/patent-applications. Accessed 19 November 2022.

As of now, such AI creations or computer-generated works (CGW) are afforded radically different treatment in different territories. There are broadly three approaches:

(i) CGW are afforded copyright protection even where there is no human intervention
(ii) Attribution of authorship where there has been human involvement to some degree (e.g. the press of button to commence an operation)
(iii) No copyright protection for CGW if there is no human intervention

Under UK copyright law [Section 9(3) of the Copyright, Designs and Patents Act (CDPA)], the legislation as it stands provides that: 'In the case of a literary, dramatic, musical or artistic work which is computer-generated, the author shall be taken to be the person by whom the arrangements necessary for the creation of the work are undertaken'.

By contrast, the continental European approach has been to say that copyright only applies to original works, and that originality must reflect the 'author's own intellectual creation' (ECJ, case C-5/08 *Infopaq International A/S v Danske Dagbaldes Forening*).

There are certainly arguments that greater—and more internationally consistent—protection for CGW would help incentivise the creation and deployment of AI-driven technologies more broadly.

21.1.5.2 Data and Databases

The capture and analysis of data is the very life force behind innovation—humans and computers alike need data from which to learn. It is apparent that the underlying software technology supporting industrial and technological development has led to unprecedented volumes of data being generated, measured and recorded. Especially in the field of sustainability, those datapoints are critical—from measuring energy efficiency to carbon emissions, from biological impact data to wind power data—our ability to learn is growing exponentially.

By contrast, the world of IP has never dealt especially well with generic, raw data. Data is not (at the time of writing) an IP right, per se. It is not necessarily a copyright work. It is also not necessarily even confidential information or a trade secret. So how do companies ensure they are deriving value from their datasets—whether internally through learning and improvement, or externally through licensing that data to third parties?

Both under UK and EU law, the data may be protected if has been systematically selected and arranged in a database, and if someone has invested financially in the collation, verification or presentation of that data. In that instance, the rights-owner may own database rights which protects the data from being extracted and re-utilised without consent.

In the UK, separately to the database right (which protects the data), rights-owners can also benefit from the database copyright regime which protects the *structure* of a database (but not the individual datapoints within it). To benefit from this, the selection and arrangement of the data points must be the author's own intellectual creation.

There is no current proposal to amend these two complementary rights (database right and database copyright). For now, anyone seeking to protect data arising from their sustainable innovation must either:

– Structure the data in a database with the aim that it is protected by database right or database copyright;
– Seek to wrap the data in confidentiality—such that one can prevent third parties accessing or disclosing the data to third parties without consent.

21.1.6 Designs

The United Kingdom's designs regime is complex, consisting currently of four distinct sets of rights, namely registered designs, unregistered design rights, continuing unregistered designs and supplementary unregistered designs. Continuing unregistered designs and supplementary unregistered designs arose post-Brexit to combat a perceived imbalance of rights in the United Kingdom.

UK-registered designs protect the appearance of the whole or part of a product resulting from the lines or contours, colours, shape, texture or material of the product itself, or the product's ornamentation. A 'product' is defined as any industrial or handicraft item, including parts intended to be assembled into a complex product (provided they remain visible during normal use), packaging and graphic symbols, and typographic typefaces.

The design must be new; in other words, the design must be different to known designs or differ in more than only immaterial details, and it must have individual character (i.e. it creates a different overall impression on the informed user to known designs).

Unregistered design rights cover the shape or configuration, whether internal or external, of the whole or part of an article. Surface decoration is excluded. The design must be original and not commonplace in qualifying countries (the United Kingdom and countries with reciprocal arrangements with the United Kingdom).

Design rights play a current role in the field of sustainability, as they can be used to secure protection for 'green' products or their parts. Although parts that consist of purely technical features are excluded from protection, parts that merely serve a technical purpose but are also open to a degree of design freedom are not. For example, the specific designs of parts of cars or aeroplanes that reduce air resistance and thus lead to better fuel efficiency may be eligible for design protection, but the design right protects the form and not the function of those parts.

Design rights also play an important role with regard to spare parts, and thus have a large influence on sustainability of products. Where products can be repaired—something that especially the EU will require manufacturers to ensure to an ever-increasing degree—design right protection of spare parts could secure the spare part markets for original equipment manufacturers. However, especially where spare parts must have the same design as the original ('must-match' parts), design right protection is often excluded. This is true, e.g. for EU design rights, and similar initiatives are planned in e.g. Germany and France.

In January 2022, the United Kingdom Intellectual Property Office ran a 'Call for Views' consultation on the reform of the UK designs framework.[14] It invited submissions from design industry stakeholders and IP professionals on possible improvements to the current system. As part of that consultation, the UKIPO proposed to extend the spare parts exemption to supplementary unregistered designs, which is directed at supporting the right to repair agenda and increasing the lifespan of products. In support of its proposal, the UKIPO referred specifically to the UK government's commitment to achieving net zero by 2050.[15]

Unregistered design rights also currently play, at least indirectly, an important role in the restriction of unsustainable practices. Unregistered design rights are utilised frequently in the fashion industry, as they do not entail any registration requirement and are therefore well-suited to fast-moving items made available during a limited sales window. Unregistered designs will assist particularly as a deterrent to preclude unlawful reproductions of clothing articles, mass-produced in low-cost jurisdictions, which would otherwise expend significant amounts of water and natural resources. In the United Kingdom, the deterrent has been strengthened further by the decision in *Original Beauty & others v G4K Fashion & others*,[16] in which the court made a substantial award totalling over £450,000, including an additional damages uplift of 200%. The decision sounds a stark warning to any business contemplating the direct copying of a competitor's products on the basis of a 'calculated risk' approach that assumes any award of damages would be relatively modest. While the judge in that case may not have taken into account the desire to promote sustainable practices, it is hoped that the financial penalties that may await such infringers could help to dissuade unsustainable manufacturing in the fashion industry.

Design rights are often overlooked by businesses seeking to protect their intellectual investment in new products or new technologies. However, design rights can offer particularly attractive benefits over other forms of IP right. This includes, for example, speed and simplicity of registration at a relatively low cost. Businesses should take into account that design rights may discourage copying, even more so as the increased availability of 3D printing means even complex products can be easily reproduced.

As demonstrated, IP undeniably currently plays a role in sustainability. It plays a role in sustainable innovation. It plays a role through IP related legal instruments. It

---

[14] IPO – reviewing the designs framework: calls for views, available at https://ipoconsultations. citizenspace.com/ipo/reviewing-the-designs-framework-call-for-views/. Accessed 19 November 2022.

[15] GOV.UK – Net Zero Strategy: Build Back Greener, available at https://www.gov.uk/government/ publications/net-zero-strategy. Accessed 19 November 2022.

[16] https://www.bailii.org/ew/cases/EWHC/Ch/2021/3439.html. Accessed 19 November 2022.

21 Sustainability and Intellectual Property in United Kingdom

plays a role in branding. It spills into various sectors from Real Estate, where energy efficient buildings are increasingly on the agenda to Automotive, where the likes of Tesla and Toyota have released their patents as part of the effort to tackle climate change.

21.2 Impact of IP on Sustainability

As with most things, there are both positive and negative sides to IP's relationship with sustainability. Without the constraints of IP rights, companies, governments, institutions, etc. could more easily adopt technology that benefits the environment and/or has some other sustainable benefit. However, we consider that in a market economy with free competition, the use of IP as an incentive to develop/create is critical to the UK's outlook on sustainability and serves as a force for good.

21.2.1 Positives

The harsh reality is that the lowest cost method of manufacturing, retailing, transporting, etc. is rarely the most environmentally friendly solution. Without the benefit of a (quasi) monopoly, it would be economically restrictive for businesses and institutions to develop brand new technologies solely targeting sustainable practices. Sustainability in an environmental context cannot be (currently) separated from economic benefit. To put it simply, unless publicly-funded or charitable, many organisations would be unable to maintain significant expenditure in the field of sustainability without some form of economic return. IP rights enhance that process by creating the conditions for the creation of revenue streams to maintain research/ development in the field of sustainability.

Positives range from the sharing of IP to advance sustainable developments, to shifting consumer demands having an impact on sustainable innovation. For example, the exchange of registered IP and unregistered IP (e.g. trade secrets, know-how, etc.) helps to diffuse sustainable technology.

21.2.1.1 Accessibility and Transparency of IP Registries

A positive can also be found in the accessibility of IP registries and the transparency of the patents regime. UK registered IP applications and registrations are openly published and accessible for all without fees/minimal fees. Publication is particularly helpful in relation to patents, which disclose details of the underlying inventions. In filing a patent, the applicant agrees to disclose their invention to the world, in exchange for a 20-year monopoly right. While the temporarily monopolistic nature of the process could be construed as a negative (see below) the transparent nature of the register and the very fact of having to disclose in the patent claims how the invention works means that in due course (either under licence or on expiry of the patent) others will be able to use and develop such IP.

21.2.1.2 The Green Channel

A further positive, specifically with regard to patents, is the UKIPO Green Channel. In 2009, an accelerated processing of patent applications was introduced. The channel is open to all patent applications that can make a reasonable assertion of having an environmental benefit, with explanations as to how the applications are environmentally friendly and which actions they wish to accelerate in the application process (examination, publication, etc.). Around 2873 requests have been made since its introduction, for inventions such as 'Method for producing fuel using renewable methane' and 'Paver with solar panel'. The hope is that the Green Channel can continue to stimulate the discovery and production of sustainable inventions.

21.2.1.3 Specific Positives of Designs Regime

The flexibility and openness of design rights in particular, which allow for protection of the aesthetic configuration of a product, encourages innovation and is an advantage for designers of environmentally friendly products, as design rights may be available for a broad range of innovations, from new types of furniture to innovative packaging.

A particular positive on design rights in the UK is that spare parts are excluded from protection, complementing the extension of the so-called 'right to repair' measures introduced under the Ecodesign for Energy-Related Products and Energy Information Regulations 2021.[17] The rationale for the regulations is to minimise the disposal of electrical goods by enhancing access to spare parts. The spare parts exemption ensures that creators of a design cannot maintain a monopoly over certain 'fixable' parts of their design. Consequently, others can create 'spare parts' for a design and thus keep a product in its life-cycle for longer, rather than creating such a monopoly right over a design which would allow the designer to strong-arm consumers into buying an entirely new version of the design, should it undergo wear and tear. The environmental benefits of the spare parts exemption are also clear.

21.2.1.4 Certification Marks

In the field of trademarks, certification marks—i.e. those that guarantee specific characteristics, such as the Fairtrade logo, or the Red Tractor Assurance logo for British food and drink—play a positive role in the pursuit of sustainability. Such marks can instantly communicate to consumers the sustainable credentials of particular goods/services, better informing consumer choice and leading to more sustainable purchasing decisions.

---

[17] https://www.legislation.gov.uk/uksi/2021/745/contents/made. Accessed 19 November 2022.

21.2.2 Negatives

Negative aspects of IP's relationship with sustainability concern typically the perceived financial gain arising from competition to protect ideas, and the absence of competition itself, which might otherwise stimulate further (and faster) growth in the sector in question.

It could be argued that the patent system as a whole is flawed. Patent systems typically support resource-strong companies more than SMEs or new entrants/start-ups (e.g. purchasing power over valuable patents), meaning that the patent system can operate as a barrier to market entry through structural discrimination and stifle the development of green tech and IP. As mentioned above, without an effective licensing regime in place, or appropriate incentives to encourage parties to make patented technology widely available, the widespread adoption of green tech could be hindered pending expiry of relevant patents.

A further negative aspect of patents are cross-licence agreements. Such agreements raise entry barriers for new players by creating corridors of knowledge for only certain companies in any given field. Such issues may also give rise to competition law concerns.

21.3 Impact of Differing Approaches by Owners/Controllers of IP Rights

The role played by IP will inevitably differ depending on the attitude/approach of an owner/controller of an IP right. Businesses that leverage IP rights for maximum profitability will weaken the role that IP plays in sustainability, by driving up price and increasing the risk associated with innovation in the field. If the IP rightsholder does not then reinvest the proceeds of the litigation/negotiation into the development of future-facing sustainable technology/practices, the initial investment in the IP is undermined from the perspective of sustainability.

By contrast, the licensing of innovative IP in the field of sustainability on fair and reasonable terms will enable licensees to benefit from the proprietor's investment and insight in an economically viable way that will benefit their business and the environment (i.e. by enabling them to maintain an economically sustainable business), while also generating revenue for the proprietor that can be reinvested in the creation of new technologies.

On an even more extreme level, the examples of Tesla and Toyota are worthy of highlight. Elon Musk, the CEO of Tesla stated in 2014 that in an effort to fight climate change, Tesla 'will not initiate patent lawsuits against anyone who, in good faith, wants to use our technology...Tesla Motors was created to accelerate the advent of sustainable transport...If we clear a path to the creation of compelling electric vehicles, but then lay intellectual property landmines behind us to inhibit others, we are acting in a manner contrary to that goal'.[18] Similarly, in 2019 Toyota opened up a number of its patents in the field of electrification and fuel-cell technology on a royalty-free basis, albeit with a charge for 'technical support'.[19]

21.4 Normative Role of IP in Sustainability

21.4.1 Introduction

The role that IP should play in sustainability depends on one's attitude to market economics and free competition. In an ideal world, technology/products/services in the field of sustainability could be free for all to use, to enhance the wellbeing of the planet/society. However, in the absence of significant state intervention/investment, it is difficult to conceive of a situation where an entity/company/organisation would be sufficiently incentivised to develop such technology for the benefit of the environment (for example). In itself, the practice would be unsustainable, as it would entail significant investment without return. Therefore, whatever one's position on the merits of a democratised economy, in the absence of a centralised government-funded sustainability initiative, it does not appear to be a viable model within which to promote sustainability as a concept.

Being able to maintain an enterprise while avoiding long-term depletion of natural resources likely requires proactive innovation. Because such proactive innovation requires the invention of ideas, and because IP is a field concerning the protection and distribution of ideas, this paper asserts that IP should have a role in sustainability. That is, IP should have a role in protecting and distributing ideas which facilitate the maintaining and continuing of processes while avoiding long-term depletion of natural resources.

21.4.2 Incentivisation and Co-operation

As outlined above, IP can be a powerful force for good in the pursuit of sustainability in the fields of environment, economy and society. At its best, it can balance rights to incentivise innovation while safeguarding one's ability/right to use existing technology to implement/enhance environmental sustainability. By incorporating the concepts of novelty, obviousness, originality, commonplace, descriptiveness/distinctiveness, IP rights carve out existing inventions, designs, works, words and ensure that they can be used to pursue unrestricted technological development

---

[18] Tesla – All Our Patent Are Belong To You, available at https://www.tesla.com/en_GB/blog/all-our-patent-are-belong-you. Accessed 19 November 2022.

[19] Toyota Promotes Global Vehicle Electrification by Providing Nearly 24,000 Licenses Royalty-Free, available at https://global.toyota/en/newsroom/corporate/27512455.html. Accessed 19 November 2022.

without the risk of infraction or financial burden in the form of licensing. So as much as they are in place to protect such inventions, designs, etc., the law of intellectual property is equally critical in establishing what is not protectable, thereby liberating such inventions/designs/works, etc. for public consumption.

Against that background, the role of IP to incentivise the creation of ground-breaking and novel technology with the promise of financial reward that can be reinvested in the development of technology/works with a sustainable aim, or concessionary fees, is a realistic and viable role.

In addition, established firms (incumbents) need to move beyond legacy, less sustainable tech and IP. Incumbents typically operate less sustainable 'legacy' technologies and IP which are vulnerable to lock-in periods and large switching costs. This, in turn, can lead to substantial costs and changes required to become more sustainable. Co-operation between these incumbents and new entrants would help develop innovation. This could manifest through new green IP developed by such entrants being licensed for use or sold to the incumbent, or made in collaboration, or through acquisitions. The resources enjoyed by incumbents give them greater opportunities to develop green IP manufacturing processes.[20] The IP framework should facilitate and encourage such collaboration.

21.4.3 The Sustainability Dichotomy

In our view, while the role of IP should be directed to supporting sustainable innovation only, it should not seek to penalise the protection of apparently unsustainable technology. Unless premised by governmental action in the form of regulation/amendments to existing statutes, the subsistence of IP should not be excluded specifically from innovation that may currently be considered unsustainable. Exceptions to IP protection already exist and are written into statute (i.e. not descriptive, generic, business models, immoral marks, etc.). Further, not only can viewpoints shift (blue hydrogen, for example, is now considered an environmentally unfriendly energy source, which may in fact hinder the government's ambitions to achieve net zero), but in the absence of government intervention, IP should remain neutral, in the sense that it should not exclude from protection innovation that may not be obviously sustainable.

Moreover, it is submitted that the question of whether a practice/innovation is currently sustainable will need careful consideration, as the burden of classifying a specific work/invention as sustainable should not fall on the shoulders of an IP examiner, who may otherwise be overburdened or underqualified to reach a determination. If the burden is to fall on IP offices, Government will need to produce detailed guidance to enable examiners to accurately assess the sustainability of a specific work/invention, etc. Alternatively, the rightsholder could submit a formal

---

[20] E. Eppinger et al., Sustainability transitions in manufacturing: the role of intellectual property, Current Opinion in Environmental Sustainability Vol. 49, April 2021, pp. 118–126.

declaration supported by a statement of truth attesting to the sustainability of the subject matter of the IP right in question.

21.4.4 Pursuing This Role

When examining IP's role in the support of sustainable innovation, there is some overlap between the influence played by the acquisition of IP and the conditions necessary for IP rights to subsist.

21.4.4.1 Incentives and the Registration of IP Rights

From the perspective of acquisition, further incentives can be introduced to support sustainable innovation, although care must be taken to ensure (i) that the relevant criteria are as objective and fair as possible and (ii) that the task of determining whether the criteria have been met does not place an undue burden on the UKIPO, which is already stretched.

The UKIPO has already adopted a future-facing approach with the introduction and successful maintenance of the 'Green Channel' for patent applications.[21]

The Green Channel for patent applications was introduced on 12 May 2009. The service allows applicants to request accelerated processing of their patent application if the invention has an environmental benefit. The applicant must make a request in writing, indicating: (i) how their application is environmentally friendly and (ii) which actions they wish to accelerate: search, examination, combined search and examination, and/or publication.

The service is available to patent applicants who make a reasonable assertion that the invention has some environmental benefit. If, for example, the invention is a solar panel or a wind turbine then a simple statement is likely to be enough. However, a more efficient manufacturing process which uses less energy is likely to need more explanation. The UKIPO will not conduct any detailed investigation into these assertions, but will refuse requests if they are clearly unfounded. Applications will only be accelerated when requested by the applicant; there is no automatic Green Channel entry for particular areas of technology.

The Green Channel service may be requested in relation to applications filed prior to 12 May 2009 as well as in relation to those filed after this date. The written request can be made electronically as a covering letter using the Office's online patent filing services. The request can be made at the same time as filing the application, or can be made on a later date, quoting the application number.

In 2012, the United Kingdom government introduced the Patent Box regime, which is a beneficial tax regime to encourage development and exploitation of patents in the UK. Profits under the Patent Box regime are taxed at 10% instead of

---

[21] GOV.UK – Green Channel, available at https://www.gov.uk/guidance/patents-accelerated-processing#green-channel. Accessed 19 November 2022.

the main corporation tax rate of 19%.[22] Pure or embedded patent income, including profits from the sale of products deriving their value from patents are included in the regime. However, the regime does not cover profits from the sale of products made by a patented process which are not themselves protected by a patent, or profits from services based on patented innovation. In addition, and of significance to the field of sustainability, is the exclusion of income arising from oil extraction activities or oil rights. It does not operate as a direct penalty to development in the field of oil extraction; however, by excluding traditional, polluting, sources of energy, the Patent Box regime is a relatively early example of a government incentive scheme that seeks to encourage technological development in green technologies.

In 2021, the UKIPO announced the imminent launch of an 'IP Access' fund, directed at helping businesses grow and rebound from the Covid outbreak. The scheme is intended to support businesses to manage and commercialise their IP so that they can, in turn, unlock the inherent value subsisting in those assets to further develop their business. The scheme is offered in conjunction with the UK government's existing 'IP Audit Plus' programme. Under that programme, the UK government contributes £2500 towards the cost of an IP audit undertaken by an approved audit partner, up to a total cost of £3000. The IP Access fund offers grants of up to £5000 to small and medium-sized enterprises who have been awarded part-funding for an IP audit between April 2020 and March 2022. The grant funding can be used by the successful applicant for a broad selection of measures, including the management and commercialisation of its IP, or for engaging professional IP services in both the UK and further afield.

Within the field of copyright, at present there are no incentivisation schemes directed specifically at ESG issues — in the sense that copyright arises automatically, and is therefore sector-agnostic as to protection. There also does not appear to be any particular prospect of (or benefit to) a change in this regard.

The one area where additional thought could perhaps be considered is with regard to the *exceptions* to copyright protection. Unlike in the US, the UK has a narrow 'fair dealing' regime. The defences to copyright are therefore very specific and include:

(i) Non-commercial research and private study
(ii) Text and data mining for non-commercial research
(iii) Criticism, review and reporting current events
(iv) Teaching or certain other educational purposes
(v) Helping disabled people
(vi) Time-shifting
(vii) Parody, caricature and pastiche

Of the list above, those exceptions designed to help disabled people already achieve an ESG-related purpose—namely improving social fairness and access to materials. For example, the following are not infringements:

---

[22] Corporation tax rate correct as of July 2022.

(i) making braille, audio or large-print copies of books, newspapers or magazines for visually-impaired people
(ii) adding audio-description to films or broadcasts for visually-impaired people
(iii) making sub-titled films or broadcasts for deaf or hard of hearing people
(iv) making accessible copies of books, newspapers or magazines for dyslexic people

The defences around text and data mining mean that, provided you already have lawful access to the original work, it is not an infringement to use automated analytical techniques to analyse text and data for patterns, trends and other useful information. Currently, the exception applies only to circumstances where such techniques are for the purposes of non-commercial research. However, the UKIPO recently issued a 'Call for Views' in connection with legislative changes required to accommodate and enhance the growth of artificial intelligence in the UK. In its response to the consultation, the UKIPO has confirmed that it intends to introduce a new defence for text and data mining ('TDM'), expanding the existing exception to cover TDM 'for any purpose' (albeit still with a requirement for the data in question to be accessed lawfully). As the exception has the potential to damage the business models of organisations that aggregate and licence data for commercial purposes, the precise wording and implementation of the defence will no doubt be subject to further scrutiny.

With regard to environmental issues, one could see an argument for adding ESG-related uses to the existing list of 'fair use' defences. However, such an approach would be difficult to implement. First, we have discussed elsewhere the difficulty in defining the thresholds for sustainability. While some innovations are plainly greentech-related, others are less so—for instance, a less harmful or more efficient method of burning fossil fuels. Comparatively such technology would be preferable to the less efficient/harmful methods that exist today. But others would argue the overall detrimental effects of burning fossil fuels means that they should not benefit from ESG-related incentives, whether financial or legal.

There is also the issue of measuring the significance of the improvement. Generally, innovation happens in small increments more than vast leaps. Would a tiny incremental improvement in a technology (e.g. a 1% increase in efficiency) justify what would otherwise be infringements of copyright?

It is not clear how ESG-related changes could easily be added to the exceptions to copyright protection. The existing exceptions for non-commercial research provide a fair balance between the interests of copyright and data owners and the public interest in using ESG data for non-commercial research to achieve sustainable goals.

21.4.4.2 The Conditions for IP Rights to Subsist

We do not believe that the conditions for IP subsistence should be amended to facilitate the protection of apparently sustainable inventions, designs, works, etc. The conditions for subsistence are—correctly—agnostic concerning the specific underlying industry or technical sector. If conditions were revised to favour inventions/works that are presented as 'sustainable', an unfair advantage would be afforded to innovation merely because it could be used for the pursuit of sustainable ambitions. Moreover, there is no guarantee that the proprietor of the resultant right will—having secured protection through a facilitated process—use its invention for the pursuit of sustainable goals. For example, many inventions and designs have dual use capability, including outside of the field of environmental sustainability. The proprietor could also choose to leverage its rights for purely commercial purposes, i.e. to pursue licensing arrangements. The lowering of barriers to entry could even have unintended consequences, namely increased litigation, which may consequently impact adversely on the pursuit of sustainability.

By incorporating the concepts of novelty, obviousness, originality, common-place, descriptiveness/distinctiveness, IP rights carve out existing inventions, designs, works, words and ensure that they can be used to pursue unrestricted technological development without the risk of infraction or financial burden in the form of licensing. So as much as they are in place to protect such inventions, designs, etc., the law of intellectual property is equally critical in establishing what is not protectable, thereby liberating such inventions/designs/works, etc. for public consumption. It is our view that the conditions for IP rights to subsist should not, therefore, be revised to facilitate protection of 'sustainable' inventions, works, designs, marks, etc. This position is distinguishable from our views in connection with the potential revision of the underlying procedures/processes that regulate IP protection, or the desire for additional incentives to encourage green IP applications.

Rather than revising the conditions for IP subsistence, the solution for demonstrating sustainable origin/provenance is better addressed by regulatory labelling and information requirements, forcing companies to declare where and how the components, ingredients, parts of their products are sourced, so that consumers can reach a transparent purchasing position based on the organisation's green credentials.

21.4.4.3 Creation of New IP Rights to Support Sustainability

It is difficult to conceive of what types of additional IP rights might be created to achieve sustainable objectives. The selection of registered and unregistered rights which exist are designed to cover the full range of possible innovations—from software to mechanical devices, and from chemical formulae to plant varieties. The meshing of these rights has been developed over years of jurisprudence and legislative development, and therefore seeking to add a brand-new right into the mix will not be straightforward.

To the extent that there are any lacunae in the current IP protection regime, theoretically one might consider a *'patent-light'*. As it stands, the cost and timeframe involved in protecting patents can be extremely burdensome, especially on start-ups and SMEs. With ballpark costs of £10,000 just to get a patent on file—before having any meaningful sense of its likely path to registration—that can pose challenges for access to justice. Similarly, it typically takes 18 months before a patent application finishes its initial examination and is published.

One idea may be to create a slimmed down version of a patent, possibly borrowing from the German Utility Model. It could be cheaper and more rapid, perhaps in return for having a narrowed scope of protection. This might amount to something less than the traditional patent monopoly right. Or, it could simply be a shorter period of protection—e.g. 5–10 years instead of 20 years. If one were to limit the newly-conceived right to ESG-related inventions, the result may be to incentivise innovation while making access to IP protection more feasible. As with other issues above, one of the challenges here would be defining what types of innovation would qualify. Moreover, while a *'patent-light'* may serve to secure protection more affordably, it would not reduce the costs of infringement proceedings themselves, as the validity of the IP right would no doubt be challenged by way of a counter-claim. Consequently, any cost savings secured obtaining protection may quickly be exhausted in any ensuing litigation. *'Patent-light'* protection may also increase the volume of disputes, encouraging the proprietors of such rights to leverage their registrations to secure licensing revenues. As a result, the merits of introducing any such right appear to be limited.

An alternative option would be the development of some new right protecting raw (or processed) data. However, the implications of such a change would impact every sector, not only ESG-related ones. It is unfeasible, in our view, to seek to make new IPR rights sector-specific—simply because (as we have seen with music and mobile phones) sectors converge in unexpected and unpredictable ways as technology rapidly develops.

21.4.4.4 The Enforcement of IP Rights Against Third Parties

The idea of a party issuing infringement proceedings against a company/entity in relation to its unlicensed use of IP for a sustainable purpose is likely to instantly arouse strong emotions. Some may question the morality of contesting unlicensed third-party use of an innovation/invention with genuine environmental benefits, for example. In those circumstances, it is easy to take the view that enforcement may impact on sustainability by limiting competition and thereby restricting innovation in a specific or even adjacent field that may have environmental benefits. However, while there are specific exceptions to the rule (most notably patent trolls, whose business model is to extract licensing revenue/financial settlement, rather than to develop the underlying IP), in our view a rightsholder's ability to enforce its IP is in fact critical to the pursuit of a sustainable future.

As discussed earlier in this report, in the absence of significant government intervention, it would be financially unviable for the vast majority of companies/ organisations to invest in the development of sustainable technology/innovation, without some opportunity for reward. The ability of a rightsholder to prevent others from piggybacking onto their investment (often made at an initial loss), which would otherwise enhance competition and weaken the rightsholder's opportunity to create first-mover advantage and carve out a market for the product/service in question, is therefore a critical tool and one which in fact supports/strengthens sustainable innovation.

Of course, a small number of extremely large tech corporations are in the position of being able to take on new projects that are focused on promoting sustainable development, despite the significant start-up costs. In those circumstances, the scale of projects undertaken are of such a scale and expense that even if third party intent to appropriate the relevant technology existed, the practical ability to do so would be extremely limited.[23] Therefore, in some (limited) instances the need for enforcement will become less pressing.

The relationship of enforcement and sustainability is also contingent on the IP right at issue. By their very nature, trademarks are unlikely to enable proprietors to prevent third party use of technology, materials and ideas.

Trademarks are premised on the principle of origin—namely an ability to designate the goods/services of a particular undertaking. Consequently, while logos and, in some instances, product shapes are registrable as trademarks, the functionality and innovation they embody cannot be protected under the trademarks regime. As a result, it is highly unlikely that enforcement based on earlier trademark rights would impact on competition, innovation and sustainability. In the event that a third party were to acquire trademark rights in a word or logo that is somehow descriptive of the characteristics of a sustainable principle or feature, the mechanism exists under Section 47(1) Trade Marks Act 1994 to invalidate the registration of that mark for descriptiveness/lack of distinctiveness.

Similarly, the law of copyright cannot be used to prevent the common acquisition of ideas and functionality. The idea/expression dichotomy reinforces that copyright subsists in the expression of a work in some physical form, i.e. in the words on a page. However, the scope of protection afforded to those words (to continue the previous example) is limited to the appropriation of *the words* themselves, either in whole or a substantial part thereof. As a result, copyright protection does not extend to the ideas/concepts underlying those words. Likewise, while copyright will subsist in computer software in the form of source code and object code as a literary work under the CDPA 88, infringement of that right will arise only where the source code/object code itself has been deliberately or sub-consciously copied, again either in whole or substantial part. Consequently, the mere appropriation of the functionality or idea on which an application or software platform is based will not represent an infringement of the copyright subsisting in the source code/object code. The enforcement of copyright is therefore less likely to impact on competition and thus restrain innovation in the field of sustainability.

21.5 Success (or Otherwise) of IP's Role in Sustainability

21.5.1 IP Supporting Sustainability

As outlined above, we consider that the United Kingdom's IP rights, in which we include trade secrets and breach of confidence, are broadly sufficient to support sustainable goals. A significant degree of balance is already ingrained in the relevant

---

[23] See, for example: https://www.apple.com/uk/newsroom/2020/09/apple-expands-renewable-energy-footprint-in-europe.

legislation, which will safeguard against the monopolisation of ideas, works, descriptive words/terms, etc. Consumers should ultimately be given greater visibility in respect of a company's green credentials, but this should not be a burden of the IP regime, rather an impetus for government regulation.

21.5.1.1 Technological Development and the Relationship with IP

The pace of technological change threatens to complicate the legal landscape, particularly in relation to computer-generated designs/works, which are—in theory—protectable under the law of designs and copyright. Section 2(4) of the Registered Designs Act 1949 (RDA) states that the author of a design shall, 'In the case of a design generated by computer in circumstances such that there is no human author, [be] the person by whom the arrangements necessary for the creation of the design are made'. A similar provision exists in relation to design rights at Section 214(2) of the Copyright Designs and Patents Act 1988 (CDPA). That section provides that 'In the case of a computer-generated design the person by whom the arrangements necessary for the creation of the design are undertaken shall be taken to be the designer'. It is noted that the provisions for computer generated works in the field of designs effectively mirror the position in connection with the authorship of computer-generated works under the law of copyright.

Although the legislature should be commended for attempting to anticipate future technological developments, the provisions as drafted are not sufficient to address the complex processes that underlie the relevant technology, particularly within the context of works 'created' by tools/applications powered by artificial intelligence. The generation of such works by artificial intelligence in practice entails numerous inputs/processes, including training datasets, training algorithms, model architecture, neurons, weights and thousands of layers. Each of these inputs/processes is likely to be overseen/managed by numerous human operators. However, the formulation in the RDA and CDPA distils the author/designer to a single 'person by whom the arrangements necessary for the creation of the design are undertaken'. Quite aside from the practical challenge of determining who has made a relevant arrangement (when there may be a large team of human operators responsible for the AI tool), the provisions as drafted are evidently insufficient to address the technological complexities of computer-generated works, which may in fact have numerous contributors.

In the context of unregistered designs and copyright, the grant of protection for computer-generated rights is inconsistent with the operative provisions regulating 'originality' and the subsistence of such rights more generally. Irrespective of whether 'originality' is construed according to the criteria of 'skill, labour and judgment' of the author, or through the European lens of the author's own intellectual creation, both concepts involve the exercise of input that is tied directly to an author's contribution. In the context of computer-generated works, the author will be the person who made the arrangements necessary for the creation of the design. In practical terms, this may be the programmer who writes the algorithm, or the person who trains the relevant model. However, neither of those individuals could be said to have exercised skill, labour or judgment in connection with the creation of the design itself, as opposed to their specific inputs into the AI program that 'created' the work or design: their contributions will be too remote. Yet the originality criterion is tied specifically to the design, rather than a process, the execution of which leads to the design. Thus, there is a disconnect between the authorship requirement regulating computer generated works in the context of unregistered design rights and copyright and the subsistence of an original design/work itself. To enhance certainty (and the risk that many such works/designs may not be entitled to protection), greater clarity is required, whether in the form of legislative amendment or judicial precedent.

There are also grounds to argue that protection for computer generated designs/works should be excluded for policy reasons. An AI-enabled tool that is used to create designs/works could have the capacity to generate a significant number of works in a very short period of time. Subject to satisfying the requisite novelty/originality threshold, the author of those works (whoever that might be) is then placed in a powerful position, as they will have at their disposal a constantly expanding portfolio of unregistered design rights and/or copyrights, which in the UK will give rise to at least a 10-year monopoly period in relation to designs and a 50-year monopoly period in relation to copyright. It is therefore possible that the extension of such protection for computer generated designs/works will lead to the rise of design/copyright trolls, thereby inhibiting innovation. Based on the increasing importance of artificial intelligence within the field of sustainability and climate change, there is a significant risk that unwittingly the current legal framework applying to computer-generated designs/works may in fact have a chilling effect on innovation within the field of sustainability.

In March 2021, the UKIPO issued a 'call for views' in relation to copyright and its relationship with artificial intelligence. Particularly, the UKIPO asked whether copyright protection for computer-generated works without a human author should be retained. The UKIPO reported the findings of its inquiry in July 2022. The UKIPO concluded that changes to the law may have unintended consequences and therefore confirmed that it would maintain the status quo. Given the rapid speed of advance in AI technology, it is submitted that the law should be kept under review so that any challenges posed by the issues identified above can be mitigated promptly, if required.

21.5.1.2 Information and Data from IP Registering Authorities

As discussed above, the UKIPO offers a 'Green Channel' for patent applications, which allows applicants to request accelerated processing of their patent application if the invention has an environmental benefit. The UKIPO also operates a database that enables a search for published applications and granted patents which have been accelerated under the Green Channel, thereby enabling third parties to assess the role of IP in the field of sustainable innovation.[24]

---

[24] IPO – Published Green Channel Patent Applications, available at https://www.ipo.gov.uk/p-gcp.htm. Accessed 19 November 2022.

While it is possible to search for UK trademarks that may feature words that are relevant to the field of sustainability, there is no specific search function operated by the UKIPO that would allow a user to identify 'green' terms within an application/registration's specification of goods and services. Similarly, it is not possible to search for certification marks according to the underlying characteristics that they guarantee. The current search function for United Kingdom registered designs is also limited, and allows users to search only according to owner name or registration number.

The EUIPO takes a proactive position in publicising the role of IP in the field of sustainability and climate change. For example, in September 2021, the EUIPO published a report on green EU trademarks, providing data on the number of applications incorporating green terms, along with historical trends.[25] While it is acknowledged that the resources (financial and human) available to the EUIPO far exceed those of the national IP offices, it is submitted that enhanced public engagement by the UK regulatory authorities would be beneficial in identifying the benefit of IP rights in supporting sustainability. The UK government has specifically highlighted the integral role that IP will play in its pursuit of net zero.[26] It should therefore ensure that further information on green IP is made available publicly, thereby encouraging scrutiny in the system and promoting the benefits of protection to prospective applicants. As stated by the UK government in its Innovation Strategy, IP will play a central role in creating the right environment to meet the challenges posed in pursuit of net zero. An effective IP system gives confidence to businesses, creators and investors that their ideas will be protected and that a financial return can be secured.[27]

21.5.1.3 Procedures and Fees of the IP Right Granting Authorities

While the current designs regime in place in the United Kingdom may be beneficial for certain actors, it also undermines access to justice for others. Currently, the UKIPO does not carry out substantive examination of design applications to verify whether they are 'new and have individual character'—the onus is on the applicant to consider these matters before filing. Therefore, the grant of a registration does not guarantee its validity, and the novelty and individual character of a registered design must be challenged at the enforcement stage, or in separate cancellation proceedings.

While a registered design can be obtained cost-effectively and quickly, it is often difficult for the applicant and third parties to understand the validity and scope of the resulting registration. Consequently, in the context of a dispute, significant costs can be expended on the identification and assessment of prior art, thereby reducing the

---

[25] Ibid. 7.

[26] CIPA – Minister Solloway says IP 'key' to tackling climate change, available at https://www.cipa.org.uk/news/mp-solloway-says-ip-key-to-tackling-climate-change/. Accessed 19 November 2022.

[27] Department for BEIS – UK Innovation Strategy, available at https://assets.publishing.service.gov.uk/government/uploads/system/uploads/attachment_data/file/1009577/uk-innovation-strategy.pdf. Accessed 19 November 2022.

value of the design registration itself. Taken within the broader ambition for a sustainable future, the grant of designs without examination for novelty/individual character could be seen to facilitate the grant of anti-competitive (or 'bad faith') registrations that may be used to unfairly extract financial compensation or prevent access to certain technology. In each instance, innovation will suffer. However, any changes to the examination procedure should be balanced carefully against the evident current benefits of cost and time.

It is submitted, therefore, that the UKIPO could consider maintaining current design examination practice, but with the introduction of specific grounds for objection (whether opposition or cancellation) based on bad faith. In those circumstances, a proprietor of a design that could be employed to perpetuate sustainable practices, but who chooses instead to leverage the right merely for financial purposes, could be challenged. The balance of justice would be preserved, however, as the proprietor would have ample opportunity to demonstrate with evidence that its actions are nonetheless consistent with honest commercial practices. It would fall to the Court/tribunal to determine whether the active hindrance of sustainable pursuits through the enforcement of IP rights would amount to bad faith, noting that public perception of honest commercial practices will shift over time and according to local and global economic and environmental conditions.

The UKIPO ceased to conduct substantive examination of designs in 2006 to harmonise practice with the EUIPO. As rights granted by the EUIPO do not, following the expiry of the Brexit transition period, apply to the UK, there is no longer a need for harmonisation in design practice. The suggestions outlined above could therefore be implemented on a national level, without seeking international comity.

While, within the field of trademarks, the concept of bad faith was regulated historically by decisions of the Court of Justice of the European Union (CJEU), following Brexit the English courts have freedom to develop the jurisprudence as they see fit. While it is not anticipated that significant departures will be made from existing 'retained' EU case law, the possibility exists that the concept of bad faith could expand to incorporate questionable practices that undermine sustainable goals. Again, any such development in the law can be undertaken on a national level.

It is proposed that the UKIPO consider augmenting its fee regimes by introducing a concession for certification marks that seek to guarantee the environmental/sustainable credentials or provenance of specific goods/services. The supporting evidence that is required to supplement a certification mark, namely the specific regulations governing use, can often be time-consuming and costly to produce, particularly once the cost of legal review is incorporated. As certification marks have particular utility in guaranteeing the ecological and/or sustainable characteristics of specific products/services, which in turn will encourage consumers to base their purchasing choices on sustainable concerns, it is submitted that the current fee regime could be qualified to facilitate certification marks with green credentials.

21.5.1.4 IP Incentivisation Schemes

The Green Channel has achieved a reasonable measure of success. From the publicly-available data released by the UKIPO, it is clear that the number of Green Channel applications has generally been increasing each year. In 2010 (the first full year of the Green Channel), the number of Green Channel applications recorded was 260. In 2020 (the last year for which data is currently available), the number stood at 402. However, more should be done to publicise the Green Channel: as an overall proportion of total patent applications, Green Channel applications amounted to only 1.19% in 2010 and 1.95% in 2020. The Green Channel not only provides an expedited route to protection, but also no doubt acts as a powerful marketing tool for the purposes of acquiring customers or funders. Both sets of parties will be more willing to invest in the product knowing that it has been ratified by the UKIPO as a 'green' technology.

Given that SMEs typically identify costs as the principal barrier to securing registered IP protection, the government's financial interventions are helpful, particularly the IP Access fund and IP Audit Plus scheme. However, anecdotally at least, it appears that many of the prospective beneficiaries of such schemes are simply unaware of their existence. More investment in the promotion of the government's IP incentive schemes would therefore be welcome.

Large corporations already have access to significant funds for the purposes of protecting and enforcing their IP rights. Further, studies on who is carrying out the majority of innovation in low-carbon emission technologies find that most of the disruptive technologies—those that make existing technologies obsolete, as happened in telecommunications with mobile phones—originate from small firms rather than incumbent large ones. By way of example, Climeworks, a start-up spun-off from the Swiss Federal Institute in Zurich, built the world's largest 'directly from the air' carbon capture and storage plant in Iceland. Completed in the summer of 2021, the Orca plant is expected to collect 4000 tonnes of $CO2$ per year, which it will store underground. Like many disruptive technologies in this field, Orca is expensive to operate and may not return a profit for some time. The example demonstrates why start-ups and SMEs may be reluctant to invest in developing ground-breaking innovation of this nature, or expending valuable (and limited) financial resources on IP protection.

The government's incentive schemes are therefore rightly directed at start-ups and SMEs. However, even with the funding made available, the costs of securing patent protection in particular, as well as enforcement more broadly, are often prohibitive for such businesses. Formal IP protection is often sought consequently at a later stage of development, typically once seed funding is obtained. By that stage, the IP may be invalid for lack of novelty if it has been disclosed to the public, or third-party appropriation may have already occurred. Both outcomes are potentially damaging to a company's future ability to commercialise its idea/invention and, thus, the government's pursuit of net zero based on innovation in the field of green tech.

In August 2021, as part of the UK Hydrogen strategy, the Department for Business, Energy & Industrial Strategy launched its consultation on the Net Zero

21 Sustainability and Intellectual Property in United Kingdom

Hydrogen Fund—a proposed £240 million fund to be delivered between 2022 and 2025. The fund is designed to support businesses that are building new electrolytic hydrogen production facilities. Although not marketed specifically, the aims of the fund would cover any associated IP prosecution/enforcement in that field. It is submitted, therefore, that where government funding is allocated on a sector-specific basis (i.e. directed at innovation in green energy sources), the relevant department should ensure that the protection/enforcement of IP is highlighted as a potential beneficiary of such funds.

21.5.1.5 IP Enforcement Systems/Authorities

To the extent that improvements could be made, we consider that these can be confined to the procedure that underlies the enforcement of IP rights.

In our view, the current system is still broadly effective. However, that is not to say that the system is perfect. Based on recent experience, it is clear that start-ups and SMEs are disproportionately disadvantaged when making use of the IP enforcement framework, particularly on the issue of costs. Indeed, it is well-established that a party with a clear financial advantage can often leverage its position to secure settlement on favourable terms that do not reflect legal reality. Even with the application of discounts and innovative fee structures, the cost of retaining counsel, drafting particulars and prosecuting the case can ultimately dissuade an aggrieved party from pursuing a case, or at least influence it to take a less bullish approach.

In our experience, in cases where there is a clear mismatch in financial resource, the party with the fiscal advantage can often fail to approach the proceedings constructively with the overriding objective in mind, i.e. by hedging its position on the assumption that the counterparty will be unable to sustain the proceedings from a financial perspective. While the cost caps applying to actions in the United Kingdom's Intellectual Property Enterprise Court ('IPEC') were broadly welcomed on the grounds that they force parties to secure better and more realistic merits assessments at the outset of a matter, the counter-position is that such costs caps may force less well-off claimants to settle claims more quickly to avoid inevitable irrecoverable costs. The UK is currently consulting on increasing the costs cap at the IPEC from £50,000 to £60,000, to account for inflationary pressures.

In May 2021, the Competition and Markets Authority (CMA) published new draft guidance to support businesses that wish to make green claims while complying with consumer protection legislation. As part of that guidance, the CMA found that 40% of green claims made online could be misleading to consumers.[28] Both the CMA and Advertising Standards Agency (ASA) have reiterated that green claims will become a key focus for enforcement efforts in the future. Care should therefore be taken to ensure that any green claim is carefully written and can be fully supported by evidence.

---

[28]GOV.UK – Global sweep finds 40% of firms' green claims could be misleading, available at https://www.gov.uk/government/news/global-sweep-finds-40-of-firms-green-claims-could-be-misleading. Accessed 19 November 2022.

The role of the CMA and ASA are critical to ensuring the accuracy of such claims, and the consequent impact on consumer behaviour. As outlined in the CMA's report, half of UK consumers take environmental considerations into account when buying products. The potential for a misleading claim to positively influence consumer behaviour is therefore material. Competition based on green credentials can only be fair to the extent that such claims are based on evidence. The enforcement regimes policed by the CMA and ASA are therefore critical to ensuring that organisations that offer genuinely 'green' goods/services are not unfairly prejudiced and are given the opportunity to thrive based on their sustainable agendas. Only by ensuring a level playing field will such organisations continue to feel incentivised to pursue sustainable/green innovation.

21.6 Conclusion

The role of IP in the field of sustainability is complex, but certainly positive. As outlined above, the existing IP framework in the UK is broadly sufficient to cater for sustainable innovation. The various IP rights (both traditional and non-traditional) are flexible enough to encompass the majority of innovation that could be classified as 'sustainable', ranging from data and software to formulae and industrial machinery. In our view, the conditions for IP subsistence should remain sector and technology agnostic, and indeed already incorporate sufficient checks and balances to ensure that sustainable initiatives can prosper, while safeguarding fair competition. In the same vein, we discuss—but ultimately dismiss—the addition of new IP rights in the fields of patents and data. While there is certainly a need to further support SMEs and start-ups in particular to commercialise their innovations at lower cost, which will enhance protection and incentivise development in green technology, the law as currently framed is broadly effective.

Indeed, rather than revising the legal conditions for subsistence (or introducing new rights), greater attention should be paid to the systems and procedures that underpin the examination and enforcement regimes. Access to justice is a significant concern, particularly in the context of IP enforcement and notwithstanding the introduction of the UK's Intellectual Property Enterprise Court. While the UK has introduced various incentive programmes directed at encouraging innovation, it is our view that more needs to be done to publicise the availability of these schemes to the very companies they were established to benefit. By way of example, additional fee concessions could be introduced into the trademarks regime, specifically to balance the cost of securing protection for certification marks, which can play a particularly important role in the pursuit of a sustainable future.

21 Sustainability and Intellectual Property in United Kingdom

Open Access This chapter is licensed under the terms of the Creative Commons Attribution 4.0 International License (http://creativecommons.org/licenses/by/4.0/), which permits use, sharing, adaptation, distribution and reproduction in any medium or format, as long as you give appropriate credit to the original author(s) and the source, provide a link to the Creative Commons license and indicate if changes were made.

The images or other third party material in this chapter are included in the chapter's Creative Commons license, unless indicated otherwise in a credit line to the material. If material is not included in the chapter's Creative Commons license and your intended use is not permitted by statutory regulation or exceeds the permitted use, you will need to obtain permission directly from the copyright holder.

Printed in the United States
by Baker & Taylor Publisher Services